S0-AIO-283

Methods in Enzymology

Volume 170
NUCLEOSOMES

METHODS IN ENZYMOLOGY

EDITORS-IN-CHIEF

John N. Abelson   Melvin I. Simon

DIVISION OF BIOLOGY
CALIFORNIA INSTITUTE OF TECHNOLOGY
PASADENA, CALIFORNIA

FOUNDING EDITORS

Sidney P. Colowick and Nathan O. Kaplan

QP601
M49
V.170

Methods in Enzymology

Volume 170

Nucleosomes

EDITED BY

Paul M. Wassarman

DEPARTMENT OF CELL AND DEVELOPMENTAL BIOLOGY
ROCHE INSTITUTE OF MOLECULAR BIOLOGY
ROCHE RESEARCH CENTER
NUTLEY, NEW JERSEY

Roger D. Kornberg

DEPARTMENT OF CELL BIOLOGY
STANFORD UNIVERSITY SCHOOL OF MEDICINE
STANFORD, CALIFORNIA

20225254

ACADEMIC PRESS, INC.
Harcourt Brace Jovanovich, Publishers

San Diego   New York   Berkeley   Boston
London   Sydney   Tokyo   Toronto

NO LONGER THE PROPERTY
OF THE
UNIVERSITY OF R.I. LIBRARY

COPYRIGHT © 1989 BY ACADEMIC PRESS, INC.
All Rights Reserved.
No part of this publication may be reproduced or transmitted in any form or
by any means, electronic or mechanical, including photocopy, recording, or
any information storage and retrieval system, without permission in writing
from the publisher.

ACADEMIC PRESS, INC.
San Diego, California 92101

United Kingdom Edition published by
ACADEMIC PRESS LIMITED
24-28 Oval Road, London NW1 7DX

LIBRARY OF CONGRESS CATALOG CARD NUMBER: 54-9110

ISBN   0-12-182071-8   (alk. paper)

PRINTED IN THE UNITED STATES OF AMERICA
89 90 91 92   9 8 7 6 5 4 3 2 1

Table of Contents

Section I. Chromatin and Nucleosomes: Preparative Techniques

Section II. Chromatin and Nucleosomes: Analytical Techniques

Section III. Histones: Preparation and Analysis

Contributors to Volume 170

Article numbers are in parentheses following the names of contributors.
Affiliations listed are current.

S. G. BAVYKIN (20), *Englehardt Institute of Molecular Biology, USSR Academy of Sciences, Moscow 117984, USSR*

MARIA BELLARD (16), *Institut de Chimie Biologique, Faculté de Médecine, 67085 Strasbourg, France*

A. V. BELYAVSKY (20), *Englehardt Institute of Molecular Biology, USSR Academy of Sciences, Moscow 117984, USSR*

W. F. BRANDT (23), *UCT-FRD Research Centre for Molecular Biology in the Department of Biochemistry, University of Cape Town, Cape Town, South Africa*

R. S. BROWN (22), *Department of Microbiology and Molecular Genetics, Harvard Medical School and Howard Hughes Medical Institute, Cambridge, Massachusetts 02138*

RICHARD R. BURGESS (30), *Department of Biochemistry, University of Wisconsin, Madison, Wisconsin 53706*

MICHAEL BUSTIN (11), *Laboratory of Molecular Carcinogenesis, National Cancer Institute, National Institutes of Health, Bethesda, Maryland 20892*

CHARLES R. CANTOR (10), *Department of Genetics and Development, College of Physicians and Surgeons, Columbia University, New York, New York 10032*

IAIN L. CARTWRIGHT (18), *Department of Molecular Genetics, Biochemistry & Microbiology, University of Cincinnati College of Medicine, Cincinnati, Ohio 45267*

ROGER CHALKLEY (30), *Department of Molecular Physiology and Biophysics, Vanderbilt University, Nashville, Tennessee 37212*

LEONARD H. COHEN (25), *Institute for Cancer Research, Philadelphia, Pennsylvania 19111*

R. DAVID COLE (24), *Department of Biochemistry, University of California, Berkeley, California 94720*

PIERRE COLIN (8), *ENSPS, 67000 Strasbourg Cedex, France*

MATT COTTEN (30), *Institute for Molecular Pathology, Himmelpfortgasse 1/3, A-1010 Vienna, Austria*

MICHAEL E. CUSICK (15), *Department of Medical Biochemistry, Texas A & M College of Medicine, College Station, Texas 77843*

JOAN-RAMON DABAN (10), *Departament de Bioquímica i Biologia Molecular, Facultat de Ciències, Universitat Autònoma de Barcelona, 08193-Bellaterra (Barcelona), Spain*

MELVIN L. DEPAMPHILIS (15), *Department of Cell and Developmental Biology, Roche Institute of Molecular Biology, Roche Research Center, Nutley, New Jersey 07110*

ANN DEAN (3), *Laboratory of Cellular and Developmental Biology, National Institute of Diabetes and Digestive and Kidney Diseases, National Institutes of Health, Bethesda, Maryland 20892*

GUY DRETZEN (16), *Institut de Chimie Biologique, Faculté de Médecine, 67085 Strasbourg, France*

K. K. EBRALIDSE (20), *Englehardt Institute of Molecular Biology, USSR Academy of Sciences, Moscow 117984, USSR*

SARAH C. R. ELGIN (18), *Department of Biology, Washington University, St. Louis, Missouri 63130*

WILLIAM T. GARRARD (6), *Department of Biochemistry, The University of Texas Southwestern Medical Center at Dallas, Dallas, Texas 75235*

G. P. GEORGIEV (21), *Department of Nucleic Acids Biosynthesis, Engelhardt Institute of Molecular Biology, USSR Academy of Sciences, Moscow 117984, USSR*

ANGELA GIANGRANDE (16), *Istituto di Genetica, Università di Bari, Bari, Italy*

JOEL M. GOTTESFELD (17), *Department of Molecular Biology, Research Institute of Scripps Clinic, La Jolla, California 92037*

H. J. GREYLING (23), *UCT-FRD Research Centre for Molecular Biology in the Department of Biochemistry, University of Cape Town, Cape Town, South Africa*

TIMOTHY M. HERMAN (4), *Department of Biochemistry, Medical College of Wisconsin, Milwaukee, Wisconsin 53226*

JEAN-CLAUDE HOMO (8), *U184 de l'INSERM, LGME du CNRS, 67085 Strasbourg Cedex, France*

WOLFRAM HÖRZ (31), *Institut für Physiologische Chemie, Physikalische Biochemie und Zellbiologie der Universität München, 8000 München 2, Federal Republic of Germany*

SUE-YING HUANG (6), *Department of Biochemistry, The University of Texas Southwestern Medical Center at Dallas, Dallas, Texas 75235*

BRET JESSEE (29), *Department of Biochemistry, University of Rochester, Rochester, New York 14642*

V. L. KARPOV (20), *Englehardt Institute of Molecular Biology, USSR Academy of Sciences, Moscow 117984, USSR*

A. KLUG (22), *Laboratory of Molecular Biology, MRC, Cambridge CB2 2QH, England*

ROGER D. KORNBERG (1), *Department of Cell Biology, Stanford University School of Medicine, Stanford, California 94305*

JAN W. LAPOINTE (1), *Department of Cell Biology, Stanford University School of Medicine, Stanford, California 94305*

RONALD A. LASKEY (27), *Cancer Research Campaign Molecular Embryology Group, Department of Zoology, University of Cambridge, Cambridge CB2 3EJ, England*

RONALD W. LENNOX (25), *Institute for Cancer Research, Philadelphia, Pennsylvania 19111*

G. G. LINDSEY (23), *UCT-FRD Research Centre for Molecular Biology in the Department of Biochemistry, University of Cape Town, Cape Town, South Africa*

YAHLI LORCH (1), *Department of Cell Biology, Stanford University School of Medicine, Stanford, California 94305*

LEONARD C. LUTTER (13), *Molecular Biology Research, Henry Ford Hospital, Detroit, Michigan 48202*

A. D. MIRZABEKOV (20), *Englehardt Institute of Molecular Biology, USSR Academy of Sciences, Moscow 117984, USSR*

SYLVIANE MULLER (12), *CNRS, Institut de Biologie Moléculaire et Cellulaire, 67084 Strasbourg, France*

S. A. NEDOSPASOV (21), *Unit of Eukaryotic Gene Expression, Englehardt Institute of Molecular Biology, USSR Academy of Sciences, Moscow 117984, USSR*

BERND NEUBAUER (31), *Institut für Physiologische Chemie, Physikalische Biochemie und Zellbiologie der Universität München, 8000 München 2, Federal Republic of Germany*

HANS NOLL (5), *Department of Genetics, University of Hawaii School of Medicine, Honolulu, Hawaii 96815*

MARKUS NOLL (5), *University of Zurich, Institute for Molecular Biology II, 8093 Zurich, Switzerland*

PIERRE OUDET (2, 8), *Institut de Chimie Biologique, Faculté de Médecine, 67085 Strasbourg Cedex, France*

DAVID S. PEDERSON (3), *Department of Microbiology, University of Vermont School of Medicine, Burlington, Vermont 05405*

O. V. PREOBRAZHENSKAYA (20), *Englehardt Institute of Molecular Biology, USSR Academy of Sciences, Moscow 117984, USSR*

PHILIPPE RAMAIN (16), *Institut de Chimie Biologique, Faculté de Médecine, 67085 Strasbourg, France*

EVELYNE REGNIER (2), *U184 de l'INSERM, LGME du CNRS, 67085 Strasbourg Cedex, France*

J. D. RETIEF (23), *UCT-FRD Research Centre for Molecular Biology in the Department of Biochemistry, University of Cape Town, Cape Town, South Africa*

D. RHODES (22, 27), *Laboratory of Molecular Biology, MRC, Cambridge CB2 2QH, England*

J. DE A. RODRIGUES (23), *UCT-FRD Research Centre for Molecular Biology in the Department of Biochemistry, University of Cape Town, Cape Town, South Africa*

PATRICK SCHULTZ (8), *U184 de l'INSERM, LGME du CNRS, 67085 Strasbourg Cedex, France*

S. SCHWAGER (23), *UCT-FRD Research Centre for Molecular Biology in the Department of Biochemistry, University of Cape Town, Cape Town, South Africa*

LINDA SEALY (30), *Department of Molecular Physiology and Biophysics, Vanderbilt University, Nashville, Tennessee 37212*

B. T. SEWELL (23), *UCT-FRD Research Centre for Molecular Biology in the Department of Biochemistry, University of Cape Town, Cape Town, South Africa*

A. N. SHAKHOV (21), *Unit of Eukaryotic Gene Expression, Englehardt Institute of Molecular Biology, USSR Academy of Sciences, Moscow 117984, USSR*

V. V. SHICK (20), *Englehardt Institute of Molecular Biology, USSR Academy of Sciences, Moscow 117984, USSR*

AKIKO SHIMAMURA (29), *Department of Biology, University of Rochester, Rochester, New York 14627*

ROBERT T. SIMPSON (3), *Laboratory of Cellular and Developmental Biology, National Institute of Diabetes and Digestive and Kidney Diseases, National Institutes of Health, Bethesda, Maryland 20892*

JOSÉ M. SOGO (7), *Institut für Zellbiologie, Eidgenössische Technische Hochschule-Hönggerberg, CH-8093 Zürich, Switzerland*

ARNOLD STEIN (28), *Department of Biological Sciences, Purdue University, West Lafayette, Indiana 47907*

FRITZ THOMA (7), *Institut für Zellbiologie, Eidgenössische Technische Hochschule-Hönggerberg, CH-8093 Zürich, Switzerland*

JEAN O. THOMAS (19, 26), *Department of Biochemistry, University of Cambridge, Cambridge CB2 1QW, England*

MARC H. V. VAN REGENMORTEL (12), *CNRS, Institut de Biologie Moléculaire et Cellulaire, 67084 Strasbourg, France*

C. VON HOLT (23), *UCT-FRD Research Centre for Molecular Biology in the Department of Biochemistry, University of Cape Town, Cape Town, South Africa*

PAUL M. WASSARMAN (15), *Department of Cell and Developmental Biology, Roche Institute of Molecular Biology, Roche Research Center, Nutley, New Jersey 07110*

ETIENNE WEISS (2), *U184 de l'INSERM, LGME du CNRS, 67085 Strasbourg Cedex, France*

CHRISTOPHER L. F. WOODCOCK (9), *Department of Zoology and Program in Molecular and Cellular Biology, University of Massachusetts, Amherst, Massachusetts 01003*

ABRAHAM WORCEL (29), *Department of Biology, University of Rochester, Rochester, New York 14627*

CARL WU (14), *Laboratory of Biochemistry, National Cancer Institute, National Institutes of Health, Bethesda, Maryland 20892*

Preface

Presentation of a collection of methods used in the study of nucleosomes is timely for a number of reasons. Experimental approaches have been introduced or refined and new principles discovered in more than a decade of research on nucleosomes. These approaches and principles are applicable to unsolved problems such as the basis of sequence-specific protein–DNA interaction and the mechanism of transcription control. There is also the hope of progress in studies of chromosome structure beyond the level of the nucleosome based on methods and principles from previous work.

Our objective in assembling this volume has been to facilitate the further application of basic methods and principles. Contributors to the volume have tried, wherever possible, to describe procedures in a simple stepwise, or even generic, fashion that should prove useful at the laboratory bench. We encouraged a degree of overlap between chapters so as to present multiple versions of commonly used procedures.

Perhaps the most widely applicable approaches used in studies of nucleosomes relate to digestion with nucleases. Artifacts arising from sequence specificities of nucleases may be avoided by cleavage with nonenzymatic reagents. Products of digestion are conveniently resolved as intact nucleoprotein complexes in gels or in sucrose gradients. Among the many methods of analyzing nucleosomes, protein–protein and protein–DNA cross-linking stand out for their potential importance in future studies. Although most of these approaches were devised before the discovery of the nucleosome, they were refined and their utility better appreciated as a result of their application to nucleosomes.

The most general principles revealed in studies of nucleosomes relate to the structure and bending of DNA. Digestion with DNase I discloses the periodicity of the double helix, while positions of nucleosomes mapped along a DNA sequence depend on the intrinsic flexibility of the sequence. The wrapping of DNA around the histone octamer in the nucleosome serves as a paradigm for other surface-wrapped structures and sets a kind of baseline of sequence-independent protein–DNA interaction for comparison with cases of sequence specificity.

Further studies of chromatin will entail functional analyses at the level of transcription and replication and structural investigations of isolated minichromosomes. Methods of assembling chromatin and the use of immunological and fluorescent probes, in addition to electron microscopy, will doubtless prove important as well. In these areas especially we may expect improvements, with current methods leading to the development of more powerful procedures.

We extend our thanks to the authors for their contributions as well as for their cooperation and patience during the preparation of this volume. We also extend special thanks to Mrs. Alice O'Connor for her tireless efforts in assembling and editing this book. Without her assistance and advice, our job would have been considerably more difficult and the final product less useful.

The impetus for a volume on nucleosomes was provided by the late Dr. Nathan O. Kaplan. We hope he would have approved of the final product.

PAUL M. WASSARMAN
ROGER D. KORNBERG

METHODS IN ENZYMOLOGY

Section I

Chromatin and Nucleosomes: Preparative Techniques

[1] Preparation of Nucleosomes and Chromatin

By ROGER D. KORNBERG, JAN W. LAPOINTE, and YAHLI LORCH

Chromatin is usually isolated as fragments of chromosomal material, except for small plasmids and viral genomes, which may be isolated as intact minichromosomes. The main considerations in the isolation procedure are the nature of the starting material, the solution conditions, the methods of fragmentation and purification, and the criteria for preservation of the native state. All these considerations have been affected by the advent of the nucleosome idea[1] and by the attendant use of nucleases to reveal and release nucleosome material.[2-6] At the same time, current isolation procedures are primarily limited by a lack of structural information and lack of functional assays beyond the level of nucleosome. In this chapter, we focus on principles of isolation that seem likely to apply even when more demanding structural and functional tests of the product become available. We begin with a general discussion that emphasizes these principles and then present some examples of detailed procedures.

General Considerations

Starting Material

The procedures described here are applicable to all eukaryotic cell sources that have been tested. The main concern in breaking the cells is to avoid damage to chromatin caused by shearing (see below). This may be accomplished by keeping the cell nuclei intact. Cells in suspension, such as cultured cells, blood cells, and yeast spheroplasts, are lysed by addition of a nonionic detergent or by suspension in a hypotonic medium and homogenization with a hand-held or motor-driven grinder. Solid tissues are simply homogenized with a motor-driven grinder. The lysate or homogenate may be used directly in the preparation of chromatin, or nuclei may first be purified by sedimentation to remove soluble ("cytosolic") material and large contaminants such as unbroken cells.

[1] R. D. Kornberg, *Science* **184**, 868 (1974).
[2] R. J. Clark and G. Felsenfeld, *Nature (London), New Biol.* **229**, 101 (1971).
[3] C. G. Sahasrabuddhe and K. E. Van Holde, *J. Biol. Chem.* **249**, 152 (1974).
[4] D. R. Hewish and L. A. Burgoyne, *Biochem. Biophys. Res. Commun.* **52**, 504 (1973).
[5] M. Noll, *Nature (London)* **251**, 249 (1974).
[6] M. Noll and R. D. Kornberg, *J. Mol. Biol.* **109**, 393 (1977).

Copyright © 1989 by Academic Press, Inc.
All rights of reproduction in any form reserved.

The chief concern in the isolation of nuclei is with the loss of chromatin components. There is ample precedent for the leakage of nuclear proteins, for example, nucleoplasmin, which is localized to the nucleus *in vivo,* according to indirect immunofluorescence, but which is readily lost upon isolation.[7] No doubt there are weakly or transiently associated chromatin components that behave in a similar manner. The importance of such components and the conditions needed for their retention during the isolation of nuclei will only be apparent when new structural and functional assays are available.

Solution Conditions

During cell breakage and nuclear isolation, at least five components may be required in the solution. First, a multivalent cation, such as Mg^{2+}, Ca^{2+}, spermine, or spermidine, is needed to prevent the swelling of chromatin and consequent rupture of nuclei, especially in solutions of low ionic strength. At ionic strengths above about 0.1, where chromatin containing H1 is generally insoluble, there may be less need for a multivalent cation, but this has not been investigated. The ionic strength is usually maintained with NaCl or KCl, although acetate salts may be preferable for some purposes.[8]

A second component of the solution is often EDTA or both EDTA and EGTA. The purpose of such chelating agents is to inhibit nucleases, for example, the endogenous Ca^{2+}- and Mg^{2+}-dependent nuclease of rat liver.[4] Where EDTA and EGTA have been used, spermine and spermidine were added to satisfy the multivalent ion requirement.[4] It remains to be seen whether polyamines alone are sufficient or whether there is an absolute requirement for Mg^{2+}, or another metal ion, for some structural or functional purpose.

A third important solution component is one or more protease inhibitors.[9] Sodium bisulfite (50 mM) and phenylmethylsulfonyl fluoride (PMSF, 1 mM) are most commonly used to prevent proteolysis of histones. Other protease inhibitors that have been used in the isolation of nuclei and chromatin include diisopropyl fluorophosphate (DFP, 1 mM), pepstatin (2 μM), leupeptin (0.6 μM), chymostatin (2 μg/ml) and benzamidine (2 μM). Proteolysis and also nuclease action may be minimized by the maintenance of a somewhat elevated pH, perhaps because many hydrolytic enzymes are of lysosomal origin.

[7] G. Krohne and W. W. Franke, *Cell Res.* **129,** 167 (1980).

[8] N. Lue and R. D. Kornberg, *Proc. Natl. Acad. Sci. U.S.A.* **84,** 8839 (1987).

[9] S. Panyim, R. H. Jensen, and R. Chalkley, *Biochim. Biophys. Acta* **160,** 255 (1968).

A fourth solution component that is sometimes included is 0.15–0.35 M sucrose, which may serve two purposes. When added to cells swollen in hypotonic solution prior to homogenization or when included in media of low ionic strength, sometimes used for homogenization of solid tissues, the sucrose may restore tonicity and thus reduce the swelling and rupture of lysosomes and consequent release of hydrolytic enzymes. When included in media used for homogenization, sucrose may reduce the shear rate and help prevent the rupture of nuclei.

Finally, an inert polymer is an important solution component in the preparation of yeast nuclei. While no rationale was given when this was first introduced, the polymer may help protect the yeast nuclei, which seem particularly fragile, against rupture. An additional, possibly related reason for the use of a polymer was recently put forward[8]: the polymer reduces the leakage of nuclear components by an excluded volume effect.[10] While Ficoll is customarily used in the preparation of yeast nuclei, other polymers, such as polyvinyl alcohol and polyethylene glycol, may also provide the desired crowding effect. Polymers have been employed to date only in the preparation of yeast nuclei, but they may be of value for preserving the structure and contents of nuclei from other sources as well.

Following the preparation of nuclei, the chromosomal material is cleaved with nucleases, and the resulting chromatin fragments are purified as described below. The solution conditions required for nuclease digestion generally include 1–5 mM Ca^{2+} or Mg^{2+}, an ionic strength of 0.01–0.2, and a pH near neutrality. As already mentioned, nuclei are stable under these conditions, and minor adjustments can probably be made in the future as structural and functional considerations arise. The conditions for release of chromatin fragments from nuclei and subsequent manipulations are more problematic. Large fragments are best released by disrupting the nuclei, which can be accomplished without shearing by the removal of salts, especially multivalent cations. Chromatin containing H1 is soluble under these conditions, and it swells, bursting the nuclei. The chromatin must be maintained under these conditions during purification, since the restoration of physiologic salts causes precipitation. The requirement for exposure to low salt conditions is a major deficiency of current procedures for chromatin isolation. Swelling and solubilization of chromatin are presumably associated with a loss of higher order structure. Functional attributes may be lost as well, owing to the dissociation or subtle alteration of chromatin components. The extent of such damage and whether it is reversible remain to be seen.

---

[10] C. Tanford, in "Physical Chemistry of Macromolecules." Wiley, New York, 1961.

Fragmentation and Purification

Chromatin must be fragmented during preparation to enhance its solubility, to avoid the deleterious effects of shearing (see below), and to allow the isolation of chromosomal regions of interest. The fragments should probably be less than about 150 nucleosomes, although the critical lengths for solubility and for escape from damage by pipetting, mixing, and so forth have not been determined. (The critical lengths will probably depend on the nature and degree of preservation of higher order structures.)

Fragmentation must be brought about by cleavage with enzymatic or chemical reagents (rather than by shearing). Enzymes such as micrococcal nuclease and restriction endonucleases are preferred, since they cut only linkers between nucleosomes and tend to cut across both DNA strands. (Other enzymes, such as DNase I, cut within core regions of nucleosomes and introduce single-strand breaks.) Linkers in transcribed regions are cut somewhat more rapidly than the average,[11] but those in the most condensed states of chromosomes are not cut much more slowly.[12] Intact nuclei and metaphase chromosomes are readily digested. Virtually any region of a chromosome can be excised, with the sole limitation that the products of digestion will generally be heterogeneous in length. In the case of micrococcal nuclease digestion, the linkers in a given region are cleaved more or less at random, and fragment length heterogeneity is unavoidable. In the case of restriction enzyme digestion, the heterogeneity depends on the locations of nucleosomes along the DNA sequence in question. If these locations are variable, as is often the case, then a particular restriction site will be exposed in a linker in some genomes but protected in a core in others. Only if steps are taken to avoid such variability will the same fragments be derived from every genome.

The methods available for purifying chromatin fragments are very limited. The most useful are hydrodynamic methods, such as sucrose gradient sedimentation and gel filtration. Fragments are resolved on the basis of length and are freed of contamination by lower and higher molecular weight material. Sucrose gradients are employed for resolving nucleosome dimers, trimers, and higher oligomers (Noll and Noll, this volume [5]), while gel filtration is most convenient for large-scale preparations.[13] Fractionations on the basis of solubility have been described for separating H1-containing fragments[14] and for enriching transcriptionally active mate-

[11] M. Bellard, F. Gannon, and P. Chambon, *Cold Spring Harbor Symp. Quant. Biol.* **42**, 779 (1977).
[12] J. L. Compton, R. Hancock, P. Oudet, and P. Chambon, *Eur. J. Biochem.* **70**, 555 (1976).
[13] L. C. Lutter, *J. Mol. Biol.* **124**, 391 (1978).
[14] A. Prunell and R. D. Kornberg, *J. Mol. Biol.* **154**, 515 (1982).

rial,[15] and such steps may prove useful as parts of more extended purification schemes. Fractionation on the basis of buoyant density in media such as Nycodenz[16] and various affinity chromatographic methods have been reported (Dean, Pederson, and Simpson, this volume [3]), but experience with these approaches is still limited. A major concern, where isolations of specific chromosomal regions are attempted, is with the quantity of starting material. The sequence of interest must be present in many copies, either highly reiterated in the genome or carried on a multicopy episomal DNA element. The possibility, mentioned above, of the loss or perturbation of chromosomal components (as well as the adventitious association of nonchromosomal material) is a concern through any purification procedure and can only be addressed with suitable assays and tests of significance.

Criteria for Preservation of the Native State

Only the most rudimentary tests of structure are available, and functional assays are altogether lacking. The primary structures of histones and DNA are checked by gel electrophoresis. H1 is usually the first histone to be degraded,[9] followed by H3.[17] Ratios of these histones to H4 (with quantitation by densitometry or by counting of radioactive material) probably provide the most reliable measure of degradation, since H4 is well resolved in gels and is comparatively stable. The DNA in chromatin fragments may be checked for nicks by denaturation and analysis in 7 M urea- or formamide-containing gels.[18]

The structure of chromatin at the level of nucleosomes is best checked by brief micrococcal nuclease digestion and gel electrophoresis of the resulting DNA fragments. The pattern of bands obtained from isolated chromatin should appear identical to that from the starting nuclei. Damage caused by shearing is revealed by an increase in the background beneath the band pattern.[17] In the extreme, there are no bands but only a continuous smear. The effect of shearing is attributed to the stretching of chromatin fibers and greater exposure of linker regions. The stretching may even cause some unraveling of DNA from the periphery of nucleosomes and expansion of linker regions. The loss or removal of H1 has a similar effect

[15] J. M. Gottesfeld, W. T. Garrard, G. Bagi, R. F. Wilson, and J. Bonner, *Proc. Natl. Acad. Sci. U.S.A.* **71**, 2193 (1974).
[16] D. S. Pederson, M. Venkatesan, F. Thoma, and R. T. Simpson, *Proc. Natl. Acad. Sci. U.S.A.* **83**, 7206 (1986).
[17] M. Noll, J. O. Thomas, and R. D. Kornberg, *Science* **187**, 1203 (1975).
[18] T. Maniatis, E. F. Fritsch, and J. Sambrook, *in* "Molecular Cloning: A Laboratory Manual." Cold Spring Harbor Lab., Cold Spring Harbor, New York, 1982.

on the nuclease digestion pattern, presumably for a similar reason, of extending linker regions. The depletion of H1 may also be revealed by the appearance of new bands in the digestion pattern, corresponding to multiples of 146 base pairs (bp), rather than the usual nucleosome repeat.[19] The new bands are attributed to sliding and interaction of histone octamers along the DNA, with the elimination of some linker regions. This effect is enhanced by incubation at elevated ionic strength and temperature (see below).

The structure of small chromatin fragments can also be assessed by analysis directly in gels, without prior extraction of the DNA.[20] The mobility of the nucleoprotein complexes is dependent on the presence of H1 and may also be sensitive to modifications, conformational changes, association of additional components, and the like. The degree of such sensitivity remains to be determined. Both agarose and acrylamide gels have been used with buffers of low or moderate ionic strength [Tris–EDTA (TE) or half-strength Tris–borate–EDTA (TBE) buffer]. Inclusion of 30% glycerol and electrophoresis in the cold have been reported to preserve the association of H1 and may stabilize interactions of other proteins as well (Huang and Garrard, this volume [6]). Nucleoprotein complexes may be detected in gels by staining, by Western blotting, and by DNA blotting and hybridization. Western and DNA blotting allow comparison of specific chromosomal regions between isolated chromatin and intact nuclei, and, if no damage has occurred during isolation, the results should be the same.

Low-angle X-ray diffraction and electron microscopy have provided evidence for coiling of chains of nucleosomes into fibers 300 Å thick.[21] Results of analytical ultracentrifugation have been interpreted along similar lines.[22] These physical methods are more specialized than the biochemical ones just mentioned. Moreover, the criterion of 300 Å coiling may not be applicable to all chromosomal regions. But further analysis and findings of this nature are what is needed to guide the isolation of fully native chromatin.

For lack of discriminating structural and functional tests, it cannot be said how best to store chromatin preparations for preservation of the native state. We freeze all preparations in liquid nitrogen and maintain them at −70°. Others prefer storage at 4° in the presence of 0.02% sodium azide.[13]

[19] C. Spadafora, P. Oudet, and P. Chambon, *Eur. J. Biochem.* **100,** 225 (1979).
[20] A. Varshavsky, V. V. Bakayev, and G. P. Georgiev, *Nucleic Acids Res.* **3,** 477 (1976).
[21] G. Felsenfeld and J. McGhee, *Cell* **44,** 375 (1986).
[22] D. L. Bates, J. G. Butler, and J. O. Thomas, *Eur. J. Biochem.* **119,** 469 (1981).

Procedures

Long Chromatin Fragments

The length of chromatin fragment that can be prepared is limited by solubility. We generally determine the extent of nuclease digestion required to solubilize each preparation. Such a determination is illustrated in the procedure given here for rat nuclei, with digestion by micrococcal nuclease.*

Nuclei are prepared by a version of published methods.[4] Rats of any age and either sex are sacrificed by suffocation with carbon dioxide (for example, in a chamber containing dry ice). Livers are removed, rinsed in cold buffer A, minced with a scissors, and homogenized in virtually no buffer (at most, a few milliliters of buffer A) in a motor-driven, Teflon–glass tissue grinder in the cold. Four or five up-and-down strokes with the motor at top speed are sufficient to pulverize all the livers. The homogenate is filtered through several (4–8) layers of cheesecloth, to remove any remaining bits of liver and connective tissue. (It is helpful to wring the cheesecloth to recover all the homogenate.) The filtrate is layered over 10 ml of a 1:1 mixture of buffers A and B in 50-ml plastic tubes (four tubes for 60 g liver) and centrifuged for 15 min at 10,000 rpm in a Beckman JA-21 rotor at 4°. The supernatants are decanted, and the pellets are suspended in buffer B (120 ml for 60 g of liver). The suspension is layered over 9 ml of buffer B in Beckman SW27 tubes (four tubes for 60 g liver) and centrifuged for 90 min at 27,000 rpm at 4°. Semisolid material on the surface is removed with a spatula, and the supernatants are decanted. The pellets are suspended in buffer C (4–5 ml per tube) and centrifuged for 5 min at 5000 rpm in a Beckman JA-21 rotor at 4°. The pellet is suspended in 0.1 ml of buffer C per gram of starting liver. The absorbance at 260 nm of 5 μl of suspension diluted in 2 ml of 1 N NaOH is typically about 0.5.

The minimum level of digestion with micrococcal nuclease for the conversion of chromatin to a soluble form is determined in a set of trials. An example of a protocol and results are tabulated below.

---

* Buffer A: 15 mM Tris-Cl, pH 7.5, 60 mM KCl, 15 mM NaCl, 0.15 mM 2-mercaptoethanol, 0.15 mM spermine, 0.5 mM spermidine, 2 mM EDTA, 0.5 mM EGTA, 0.34 M sucrose. Buffer B: same as buffer A except 0.1 mM EDTA, 0.1 mM EGTA, 2.1 M sucrose. Buffer C: same as buffer A except with 10 mM NaHSO$_3$ and without EDTA and EGTA. Buffer D: 10 mM NaHSO$_3$, pH 7.5, 1 mM EDTA. Micrococcal nuclease: 500 μM units (Sigma, St. Louis, MO) are dissolved in 0.85 ml 5 mM Tris-Cl, pH 7.5, 2.5×10^{-5} M CaCl$_2$, to give 50 A_{260} units/μl, and stored in aliquots at $-20°$.

	Trial			
Step	1	2	3	4
Nuclei (μl), 228 A_{260} units/ml	100	100	100	100
0.1 M CaCl$_2$	1	1	1	1
Incubate 2 min at 37°				
Micrococcal nuclease (μl), [50 units/μl]	0.5	0.5	0.5	1
Incubate (sec) at 37°	15	30	60	60
0.25 M EDTA (μl)	4	4	4	4
Chill, microfuge 2 min				
Suspend pellet in buffer D (ml)	0.2	0.2	0.2	0.2
Microfuge 2 min				
A_{260} of supernatant ($\times 10^{-2}$)	0.02	0.48	0.54	0.56

The transition to solubility evidently occurred between 15 and 30 sec of digestion. The transition point could have been determined more precisely, for the purpose of obtaining the longest possible soluble chromatin fragments, by digestion with less enzyme for longer periods. In the case shown here, sample 3 was scaled up 30-fold (with centrifugation for 5 min at 5000 rpm in a Beckman JA-21 rotor and suspension of the pellet in 1.5 ml of buffer D) to give 216 A_{260} units of chromatin fragments with a weight-average length of about 15 nucleosomes. Lengths of soluble chromatin up to 150 nucleosomes have been reported, and the size range of the fragments has been narrowed by fractionation in a sucrose gradient.[22]. Lengths are determined by extraction of the DNA from chromatin samples with phenol–chloroform and analysis in agarose gels, as described.[22]

Nucleosome Oligomers

Nucleosome monomers, dimers, and higher oligomers may be prepared by micrococcal nuclease digestion and sedimentation in sucrose gradients. Trials are performed as described above, except with higher levels of digestion, to obtain as much as possible of the desired oligomer in the mixture before fractionation. For example, for the preparation of monomers and dimers, three levels of digestion were tried, 1 μl of enzyme for 1 and 2 min, and 2 μl for 2 min. On the basis of the band patterns in an agarose gel, the middle level was chosen and was scaled up directly. Material from 0.5 ml of nuclei, suspended after digestion in the same volume of buffer D, was sedimented in a 5–28.8% isokinetic sucrose gradient (Noll and Noll, this volume [5]; a linear gradient gives essentially the same results) in buffer D for 30 hr at 24,000 rpm in a Beckman SW27 rotor at 4°. There is invariably a degree of overlap of oligomers between fractions, especially for higher oligomers (Fig. 1). Homogeneous preparations of monomers, dimers, trimers, and tetramers have been obtained by sedimenting pooled fractions in a second sucrose gradient.

Core Particles

The removal of histone H1 from chromatin before micrococcal nuclease digestion facilitates the conversion to core particles.[6] The method was refined by Lutter,[13] and a version of the procedure is given here. H1 is removed by gel filtration in 0.45 M NaCl (except in the case of chicken erythrocyte chromatin, for which 0.6 M NaCl is required to remove both H1 and H5). A solution of long chromatin fragments (3 ml, 432 A_{260} units), prepared as described above, is made 0.45 M in NaCl by the addition of a high-salt mixture (0.31 ml 5 M NaCl, 75 μl 1 M Tris-Cl, pH 7.5, 30 μl 1 M NaHSO$_3$, 12 μl 0.25 M EDTA) dropwise with stirring in ice. The solution becomes turbid and then clears as the salt concentration is raised. The mixture is filtered through a column of Sepharose 4B (2.5 ×

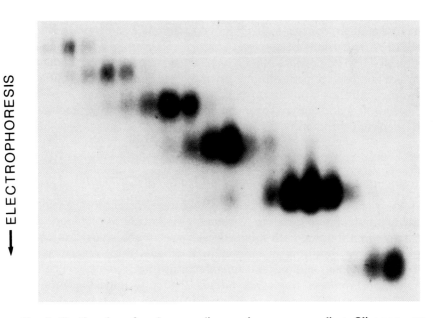

FIG. 1. Fractionation of nucleosome oligomers in a sucrose gradient. Oligomers were prepared, sedimented, and analyzed in an agarose gel, as described in the text. Nucleosome monomers, dimers, and so forth are displayed from right to left.

45 cm) in 0.45 M NaCl, 25 mM Tris-Cl, pH 7.5, 10 mM NaHSO$_3$, and 1 mM EDTA at 15 ml/hr at 4°. Fractions (4 ml) absorbing at 260 nm (typically fractions 25–35) are pooled, dialyzed for 4 hr against 1 liter 10 mM Tris-Cl, pH 7.5, 10 mM NaHSO$_3$, and 0.1 mM EDTA with one change (reducing the conductivity 10-fold), dialyzed against solid sucrose (concentrating the material about 5-fold), and dialyzed overnight against the same Tris-Cl–NaHSO$_3$–EDTA buffer as before, giving 334 A_{260} units of H1-depleted chromatin in 12.5 ml.

The amount of micrococcal nuclease needed for the conversion to core particles is determined in trial digestions with 50 μl H1-depleted chromatin, 0.5 μl 1 M CaCl$_2$, and 25 units micrococcal nuclease for 0.25, 0.5, 1, and 2 min at 37°. Digestion is stopped by the addition of 2 μl 0.25 M EDTA, followed by extraction of the DNA with phenol–chloroform, ethanol precipitation, and electrophoresis in a 5% polyacrylamide gel in TBE buffer.[18] The digestion time that leaves a trace of dimers (lane 30 in Fig. 2A) but leads to no material of less than 146 bp (lane 75 in Fig. 2A and lane 150 in Fig. 2B) is used in the full-scale preparation. In the example described here, 1.5 ml of H1-depleted chromatin, supplemented with 15 μl 1 M CaCl$_2$, is preincubated for 2 min and digested with 750 units micrococcal nuclease for 0.5 min at 37°. Digestion is stopped with 75 μl 0.25 M EDTA, and core particles are purified by sedimentation in a 5–30% linear sucrose gradient containing 0.55 M NaCl, 10 mM Tris-Cl, pH 7.5, 10 mM NaHSO$_3$, and 1 mM EDTA for 16 hr at 27,500 rpm in a Beckman SW28 rotor at 4°. (The purpose of a high salt concentration in the gradient is to ensure the displacement of nuclease and other contaminants, but a lower salt concentration would probably do as well.) Peak fractions are dialyzed, concentrated, and dialyzed again, as described above for the preparation of H1-depleted chromatin.

Restriction Endonuclease Cutting and Ligation of Chromatin

Restriction enzymes can be used to excise chromosomal regions of interest and also to produce chromatin fragments for ligation. In the example that follows, both micrococcal nuclease and restriction enzymes are employed, resulting in chromatin fragments with, on average, one 3′-phosphoryl end and one end suitable for ligation.

Long chromatin fragments (40 μl, 3640 A_{260} units, weight-average length ~15 nucleosomes) are digested in 0.2 ml of 0.1 M NaCl, 6 mM Tris-Cl, pH 7.5, 6 mM MgCl$_2$, 6 mM 2-mercaptoethanol, 0.1 mg/ml BSA, 10 mM NaHSO$_3$, and 1 mM PMSF with 50 units each of *Alu*I, *Hae*III, and *Rsa*I for 3 hr at 37°. The digest is sedimented in a 5–30% linear sucrose gradient, containing 50 mM Tris-Cl, pH 7.5, 0.1 mg/ml BSA, 10 mM

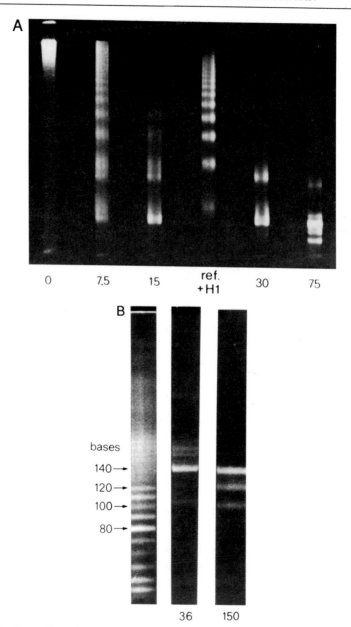

FIG. 2. Conversion of H1-depleted chromatin to core particles. The data shown are illustrative and do not correspond with the experiment described in the text. Digestion was for 30 sec with units/ml of micrococcal nuclease indicated beneath the lanes. DNA fragments were analyzed in polyacrylamide gels without (A) or with (B) 99% formamide. (Reproduced from Ref. 6, with permission.)

$NaHSO_3$, and 1 mM PMSF, for 19 hr at 35,000 rpm in a Beckman SW41 rotor at 4°. Fractions (0.5 ml) are collected from the bottom, and aliquots (20 μl) of fractions 1–15 are extracted and analyzed in a 1.5% agarose gel. Peaks of nucleosome monomers, dimers, and trimers are generally in a single fraction (typically fractions 10, 6, and 3, respectively) and are dialyzed and concentrated as described above for H1-depleted chromatin.

We have used blunt-ended nucleosome monomers, prepared in this way, for ligation to a 465-bp XhoI–SmaI fragment containing an SP6 RNA polymerase promoter. Monomers (100 μl, 0.06 A_{260} units) and the XhoI–SmaI fragment (2.5 μg, filled in at the XhoI end with [α-^{32}P]dTTP to give 2.4 $\times$ 10^6 cpm/μg) were combined in 0.2 ml containing 20 mM Tris-Cl, pH 7.5, 0.5 mM ATP, 4 mM MgCl$_2$, 10 mM dithiothreitol, 0.2 mg/ml BSA, 20 mM NaHSO$_3$, 0.5 mM EDTA, and 1 mM PMSF and treated with 2000 units T4 DNA ligase (New England BioLabs, Beverly, MA) for 40 hr at 4°. The mixture was made 5 mM in EDTA and fractionated in a sucrose gradient, as described above for the preparation of blunt-ended monomers. Aliquots (20 μl) of the gradient fractions were extracted and analyzed by electrophoresis, in a 4% polyacrylamide gel in TBE buffer, and autoradiography, to identify the 650-bp product of ligating the nucleosome monomer to the SP6 promoter fragment. This product was largely in a single fraction (typically fraction 10 or 11), with little contamination by the promoter fragment joined to itself or the free fragment.

Acknowledgments

This research was supported by grants from the National Institutes of Health to R.D.K.

[2] Preparation of Simian Virus 40 Minichromosomes

By PIERRE OUDET, ETIENNE WEISS, and EVELYNE REGNIER

The characterization of the chromatin organization of genes requires the availability of a purified homogeneous population of structures in a soluble form in order to pursue biochemical or structural analysis. For a number of years, the nuclear infection intermediates of the SV40 virus have offered a unique opportunity to obtain clear and reliable information on the packaging, replication, and transcription processes in eukaryotic cells.

The SV40 genome offers a compact, complete system containing an origin of replication and two sets of genes with their extensively studied

Copyright © 1989 by Academic Press, Inc.
All rights of reproduction in any form reserved.

control elements.[1,2] At an intermediate stage during the course of infection, the SV40 genome is packed into a chromatin-type structure called the SV40 minichromosome. In this structure, the 5243-base pair (bp), double-stranded, circular DNA[3] is associated with cellular histones in 20–24 nucleosomes. During the process of infection, the SV40 genome is present in 200,000 copies per cell.

The minichromosomes can be extracted from infected cells and purified under various conditions, including a physiological ionic environment.

One of the disadvantages of the SV40 system, however, is that during the viral cycle, the minichromosomes are rapidly transformed into previrions and virions.[4] The preparation of purified SV40 minichromosomes involves the infection of cells in culture, the preparation of nuclei, extraction and purification of the minichromosomes, as well as controls to verify the integrity of the major associated proteins and of the superhelical DNA.

Infection of CV1 Cells

Cells of the permissive monkey cell line CV1 are maintained and replated regularly in Dulbecco modified Eagle medium (DMEM, Gibco, Grand Island, NY) supplemented with 5% fetal calf serum (Gibco) and antibiotics (Gibco) (penicillin 500 IU/ml, streptomycin 0.1 mg/ml, and gentamicin 0.4 mg/ml). The cells are cultivated at 37° under a 5% CO_2 atmosphere. Every 2 months, the cells are regenerated from a new tube of frozen cells (stored in liquid nitrogen).

Virus stocks (twice-plaque-purified strain 776[3]) are prepared from infected cells, stopped by rapid freezing at −20° when they start to lyse. The cells and medium are then melted and frozen 3 times, and the lysate is collected and stored at −20°. This solution contains about 10^9 PFU (plaque-forming units) per milliliter.

CV1 cells are replated 72 hr before infection at a density of 350,000 cells per petri dish of 9 cm in diameter (Fig. 1). This dilution produces a homogeneous layer of cells at 30% confluence. Immediately before infection, 10 ml of culture medium is removed by aspiration. The virus stock dilution (0.5 ml, corresponding to 10 PFU per cell) is added to each petri

[1] M. L. DePamphilis and P. M. Wassarman, in "Organization and Replication of Viral DNA" A. S. Kaplan, ed., p. 37. CRC Press, New York, 1982.

[2] S. McKnight and R. Tjian, Cell 46, 795 (1986).

[3] W. Fiers, R. Conteras, G. Haegman, R. Rogiers, A. van Heuverswyn, J. van Herreweghe, G. Volckaert, and M. Ysebaert, Nature (London) 273, 113 (1978).

[4] J. Tooze, "DNA Tumor Viruses." Cold Spring Harbor Lab., Cold Spring Harbor, New York, 1982.

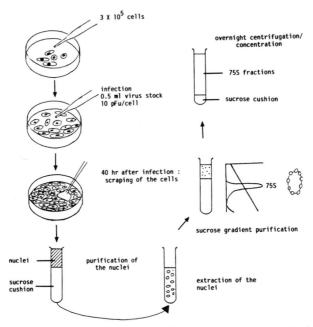

FIG. 1. General scheme for infection, labeling, extraction, and purification of SV40 minichromosomes.

dish, and the dishes are placed in an incubator, agitating gently by hand from time to time. After 1 hr, 9.5 ml of fresh medium is added to each petri dish.

In Vivo Labeling Procedures

DNA Labeling

To follow the purification of the minichromosomes, they are labeled by the incorporation of radioactive nucleotides. The medium is removed 16 hr postinfection, and 10 ml of fresh medium, containing either 0.5 μCi of [^{3}H]thymidine or 0.25 μCi of [^{14}C]thymidine (25 Ci/mmol or 53 mCi/mmol, respectively) is added and left until the cells are harvested.

RNA Pulse-Labeling

The transcribing minichromosomes can be detected by labeling the growing RNA chain(s) attached to them. The cells are washed rapidly 38 hr postinfection with 10 ml of preheated TBS buffer (137 mM NaCl, 5.1 mM

KCl, 0.7 mM Na$_2$HPO$_4$, 50 mM Tris-HCl, pH 7.5), then pulse-labeled for 5 min to 1 hr by 2 mCi of [^{3}H]uridine (20 Ci/mmol) in 1 ml of medium per plate.

Protein Labeling

The proteins associated with the SV40 minichromosomes can be identified after *in vivo* labeling. For short incubations (30 min – 4 hr), the incubation medium is removed, the cells are washed twice with DMEM − Met (DMEM containing no methionine) and incubated with 5 ml of DMEM − Met, supplemented with 2% dialyzed fetal calf serum. After 30 min, 0.2 mCi of [^{35}S]methionine (1100 Ci/mmol) per plate is added and left for 30 min to 4 hr. At the end of the labeling period, the medium is removed and the cells collected. For long labeling periods (> 5 hr) the depleted medium is complemented by 1/20 its volume (0.25 ml) with complete DMEM.

Stock Solutions

0.2 M ethylenediaminetetraacetic acid (EDTA) (Merck, Darmstadt, FRG), pH 7.0
0.1 M triethanolamine-HCl (TEA-HCl) (Merck) pH 6.8, 7.4, and 7.8
10% Triton X-100 (Sigma, St. Louis, MO)
0.1 M dithiothreitol (DDT) (Sigma)
0.1 M phenylmethylsulfonyl fluoride (PMSF) (Sigma)
500 μg/ml chymostatin (Sigma)
5 M NaCl (Merck)
0.1 M MgCl$_2$ (Merck)
2.5 M KCl (Merck)
0.2 M HEPES (N-2-hydroxyethylpiperazine-N'-2-ethanesulfonic acid) (Sigma), pH 7.8
2 M Tris-HCl (Sigma), pH 7.5
Phosphate-buffered saline (PBS) (Biochrom, KG)

Preparation of SV40 Minichromosomes in the Presence of 130 mM NaCl and EDTA

Forty hours after infection, approximately 10 μg of purified minichromosomes can be obtained per 9-cm-diameter plate in the presence of 130 mM NaCl and EDTA[5,6]

[5] A. J. Varshavsky, O. H. Sundin, and M. J. Bohn, *Nucleic Acids Res.* **5**, 3469 (1978).
[6] R. Fernandez-Munoz, M. Coca-Prados, and M. T. Hsu, *J. Virol.* **29**, 612 (1979).

Solutions

Lysis buffer:
   10 mM EDTA
   0.25% Triton X-100
   2 mM DTT
   0.2 mM PMSF
   10 mM TEA-HCl, adjusted to pH 6.8 (check before use)
Extraction buffer:
   10 mM EDTA
   0.13 M NaCl
   0.25% Triton X-100
   2 mM DTT
   0.2 mM PMSF
   1 μg/ml chymostatin
   10 mM TEA, adjusted to pH 7.8 before use
Purification buffer:
   0.13 M NaCl
   0.1 mM EDTA
   2 mM DTT
   0.2 mM PMSF
   10 mM TEA-HCl, adjusted to pH 7.4
Storage buffer:
   50 mM NaCl
   2 mM DTT
   0.2 mM PMSF
   10 mM TEA, pH 7.4

Methods

Forty hours after infection, the cells are collected and the nuclei pre-pared. The incubation medium is completely removed by aspiration. The plates are rinsed twice with 5 ml of PBS (Biochrom KG) supplemented with 5 mM EDTA. All subsequent operations are conducted at 4°. Add 3 ml of lysis buffer (pH 6.8) on each plate. Incubate for 10 min in a cold cabinet (4°). Increase the ionic strength to 0.13 M NaCl by the addition of 0.45 ml of 1 M NaCl.

Following this step, the material is kept in 0.13 M NaCl. Ther material (essentially broken cells) is scraped with a rubber policeman, collected, and centrifuged immediately for 5 min at 1000 g. The pellet is resuspended in a volume corresponding to 2 ml for 15 plates of lysis buffer containing 0.13 M NaCl. The nuclei are purified by centrifugation through a 5-ml sucrose cushion of 0.3 M sucrose made in lysis buffer containing 0.13 M

NaCl. The tubes are centrifuged 10 min at 2000 g. The pellet is resuspended in 1.8 ml of extraction buffer (pH 7.8) for 15 plates, incubated 2 hr on ice, and agitated gently by hand from time to time. The tubes are centrifuged 10 min at 16,000 g. The supernatant containing the minichromosomes is applied on a linear 5–30% sucrose gradient made in the purification buffer. The tubes are centrifuged in a Beckman SW41 rotor for 90 min at 40,000 rpm and 0.5-ml fractions are collected from the top or the bottom (to avoid possible contamination by pelleted virions); then 0.025-ml aliquots are counted in order to localize the SV40 minichromosome peak.

The DNA peak fractions are pooled and pelleted through a 30% sucrose cushion (0.2 ml) made in the purification buffer by 15 hr of centrifugation at 25,000 rpm in a Beckmann SW60 rotor. This final pellet is carefully resuspended in 20–50 μl of storage buffer. If the minichromosomes from 15 plates were concentrated per tube, an SV40 DNA concentration of more than 1 mg/ml is normally obtained at this step.

We have found that the change in pH between the lysis and the extraction buffers is important, at least in our hands, to obtain a high and reproducible recovery of minichromosomes. An increase in pH or ionic strength induces a precocious lysis of the nuclei and formation of a chromatin aggregate from which the minichromosomes can no longer be extracted.

Minichromosomes Prepared in 130 mM NaCl and MgCl$_2$

The method is essentially the same as described above except that the EDTA is omitted in all buffers. The cells are washed twice in PBS. All subsequent procedures are performed in the presence of 0.1 mM MgCl$_2$. This is, in fact, the highest concentration providing intact SV40 minichromosomes with no linear forms visible on a DNA gel stained with ethidium bromide.[6]

Minichromosomes Prepared in Hypotonic Conditions

Solutions

Purification buffer:
   5 or 50 mM KCl
   0.5 mM EDTA or 0.1 mM MgCl$_2$
   1 mM DTT
   0.2 mM PMSF
   10 mM HEPES, adjusted to pH 7.8

Storage buffer:
  50 mM KCl
  2 mM DTT
  0.2 mM PMSF
  10 mM HEPES, pH 7.5

Methods

All operations are carried out on ice. Remove the medium by aspiration. Rinse the cells twice with the buffer. Remove carefully as much as possible of the second wash solution. Normally 0.1 ml of buffer is left per Petri dish. Scrape and collect the cells. Lysis is achieved by 5 strokes of a Dounce homogenizer with a tight-fitting B pestle. The lysate is centrifuged 5 min at 1000 g. The pellet obtained from 15 plates is resuspended in 1.8 ml of the purification buffer complemented with 1 μg/ml chymostatin. The extraction occurs during a 2-hr incubation with gentle agitation from time to time. The solution is centrifuged 10 min at 16,000 g. The supernatant is applied on a linear 5–30% sucrose gradient made in the same buffer. Fractions of 0.5 ml are collected, and 0.025-ml aliquots are counted. The SV40 DNA peak fractions are pooled and concentrated on a 0.2 ml cushion of 30% sucrose made in the purification buffer. The final pellet is resuspended in 25 or 50 μl of the storage buffer.

Characterization of the Purified Minichromosomes

In most of our experiments, cells were collected 40 hr postinfection. This time was chosen as a compromise in order to obtain a high level of SV40 DNA, a low level of infectious viruses, and a high amount of extractable minichromosomes sedimenting around 75 S (Fig. 2).[4] The SV40 DNA is essentially superhelical as verified by DNA gel electrophoresis,[7] and the proteins are controlled by gel electrophoresis in the presence of sodium dodecyl sulfate (SDS).[8] When observed by electron microscopy, around 25% of the minichromosomes present a nucleosome-free region[7] and are sensitive to single cut restriction enzyme (Fig. 3) regardless of the extraction used.

The efficiency of the different methods is shown in Table I. It is clear that the highest amount is obtained at physiological ionic strength and in the presence of EDTA. The presence of a significant DNA peak sedimenting at 75 S is generally attributed to dissociated previrion structures formed

[7] E. Weiss, D. Ghose, P. Schultz, and P. Oudet, *Chromosoma* **92,** 391 (1985).
[8] U. K. Laemmli, *Nature (London)* **227,** 680 (1970).

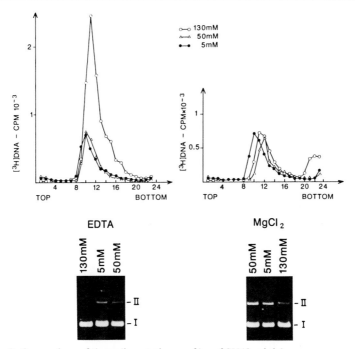

FIG. 2. Comparison of the sedimentation profiles of SV40 minichromosomes on sucrose gradients. The position and shape of the different peaks are identical and correspond to the presence of SV40 DNA (extractions realized in the presence of 130 mM NaCl, 50 or 5 mM KCl). In all conditions described, we reproducibly observed the shoulder on the heavy side of the 75 S peak. Collection of the fractions starting from the bottom of the gradient gives an identical result. The extractions in the presence of a low concentration of divalent cations (0.1 mM MgCl$_2$) produce more nicked forms, FII, and some linear molecules, FIII. The low ionic strength conditions (50 and 5 mM KCl) in the presence of magnesium increase the accessibility to the endonucleases present in the nuclear lysate or alternatively releases more DNases from the purified nuclei.

in the presence of chelating agents and DTT.[9,10] The fact that, even in presence of divalent cations, a second significant peak in the bottom of the gradient was not reproducibly observed and that the increase in minichromosome extraction in the presence of EDTA and DTT correlates to an increased heterogeneity in the mass of particles (owing to the higher amount of bound VP1 proteins; see ref. 11 and our unpublished results)

[9] J. N. Brady, C. Lavialle, and N. P. Salzman, *J. Virol.* **35**, 371 (1980).
[10] E. Fanning and I. Baumgartner, *Virology* **102**, 1 (1980).
[11] P. Schultz, E. Weiss, P. Colin, E. Régnier, and P. Oudet, *Chromosoma* **94**, 189 (1986).

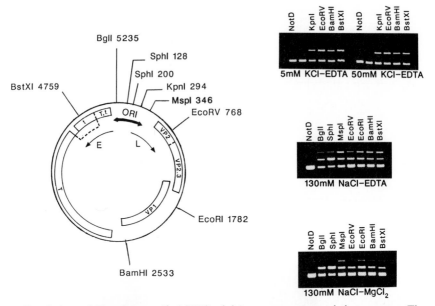

FIG. 3. Accessibility of the purified SV40 minichromosomes to restriction enzymes. The purified minichromosomes were digested in storage buffer complemented with 5 mM MgCl$_2$. A 5-fold excess of enzyme is added (5 times more enzyme than necessary to digest the same amount of naked DNA), and the digestion is realized during a 30-min incubation at 37°. In all the different conditions of extraction used, roughly 30% of the DNA molecules are linearized, suggesting the presence of a subpopulation showing a similar sensitivity to the restriction enzymes on the early or late region of the SV40 genome. A slightly higher sensitivity of the ORI region is observed reproducibly.

TABLE I

EFFECT OF IONIC CONDITIONS ON EFFICIENCY OF
SV40 MINICHROMOSOME EXTRACTION

Condition	Extraction efficiency (%)[a]	
	0.1 mM EDTA	0.1 mM MgCl$_2$
130 mM NaCl	30	12
50 mM KCl	8	8
5 mM KCl	8	8

[a] The efficiency of the recovery of SV40 mini-chromosomes is expressed as the percentage of SV40 DNA found in the 75 S peak when compared to the Hirt's extract, corresponding to the SV40 DNA present in infected cells 40 hr postinfection.

would suggest that under these conditions additional particles are extracted corresponding to a different previral population.

In all cases, the minichromosomes sediment as a 75 S peak with a shoulder on the heavy side (Fig. 2). As checked by protein or DNA gels, we did not observe significant differences between the various regions of the peak: the FII (nicked or relaxed forms) to FI (superhelical form) ratio and the histone to late protein ratio were identical in the light, center, and heavy fractions, as probed by ethidium bromide and Coomassie blue staining, respectively.

The various methods of extraction produce different protein patterns, as evidenced by the different ratios of VP1 to core histones (Fig. 4). That minichromosomes are purified and not virions is evidenced by a ratio of

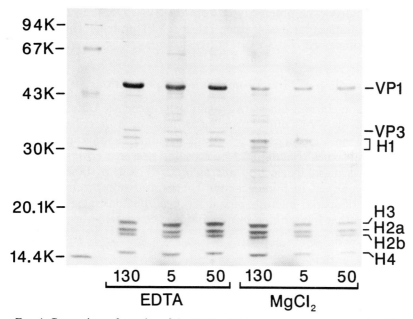

FIG. 4. Comparison of proteins of the SV40 minichromosomes prepared under different ionic conditions. Aliquots of the proteins copurified with the SV40 DNA were applied to a SDS gel. Comparisons of the intensities of the VP1, H1, and H3–H2a–H2b–H4 bands shows that, as compared to the four core histones, the viral late protein VP1 is present in a clearly lower amount on the minichromosomes extracted in presence of magnesium. The intensities of these bands in the presence of 130 mM NaCl, 50 mM KCl, or 5 mM KCl under EDTA or $MgCl_2$ isolation conditions were determined. Taking into account the different molecular weights, 0.75 and 1.0 H1 histones are present per nucleosome in EDTA and $MgCl_2$ minichromosomes, respectively. The presence of magnesium in the extraction decreases the amount of VP1 from 1.4 to 0.5 molecules per nucleosome. Varying the ionic strength between 130 and 5 mM does not seem to change the protein patterns significantly.

histone H1 to the other four histones which is similar to the one observed for cellular chromatin. There are essentially 0.8 to 1 H1 per octamer.[12]

Depending on the extraction conditions, different SV40 late protein (in which VP1 is the major protein) to histone ratios are reproducibly observed (see legend to Fig. 4). The heterogeneity of the SV40 minichromosome population as previously described,[11] arises presumably as a result of the extraction at different stages of the virion assembly process. Extraction in the presence of divalent cations producing a lower amount of minichromosomes with a lower VP1 content can simply be explained by the preferential extraction of the corresponding subpopulation since we did not observe exchange of VP1 under the conditions used.

Higher ionic strength extractions (0.2–0.4 M NaCl) can also be used to provide a higher percentage of extracted material.[13] These conditions cause proteins to dissociate progressively from the SV40 DNA, as visualized by the rapid disappearance of histone H1 and VP1 proteins when the ionic strength is increased.

Characterization of Minichromosome Subpopulations

SV40 minichromosomes are actually heterogeneous. Different subpopulations of the SV40 minichromosomes have been described including replicative intermediates,[14–16,17] transcriptional complexes,[18–21] and minichromosomes containing a nucleosome-free area covering the ORI region as visualized by electron microscopy.[22,23] None of these subpopulations was purified to homogeneity. The gapped minichromosomes are molecules of special interest since it has been demonstrated by different approaches that transcribing minichromosomes contain a gap or nucleosome-free ORI region.[24,25]

[12] R. D. Kornberg and J. O. Thomas, *Science* **184**, 86511 (1974).
[13] B. Hirt, *J. Mol. Biol.* **26**, 365 (1967).
[14] R. T. Su and M. L. DePamphilis, *J. Virol.* **28**, 53 (1978).
[15] M. M. Seidman and N. P. Salzman, *J. Virol.* **30**, 600 (1979).
[16] H. J. Edenberg, M. A. Waquar, and J. A. Huberman, *Nucleic Acids Res.* **4**, 3083 (1977).
[17] E. A. Garber, M. M. Seidman, and A. J. Levine, *Virology* **90**, 305 (1978).
[18] M. H. Green and T. L. Brooks, *Virology* **72**, 110 (1976).
[19] P. Gariglio, R. Llopis, P. Oudet, and P. Chambon, *J. Mol. Biol.* **131**, 75 (1979).
[20] H. J. Edenberg, *Nucleic Acids Res.* **8**, 573 (1980).
[21] R. Llopis and G. R. Stark, *J. Virol.* **44**, 864 (1982).
[22] S. Saragosti, G. Moyne, and M. Yaniv, *Cell* **20**, 65 (1980).
[23] E. B. Jakobovits, S. Bratosin, and Y. Aloni, *Nature (London)* **285**, 263 (1980).
[24] M. Choder, S. Bratosin, and Y. Aloni, *EMBO J.* **3**, 1243 (1984).
[25] E. Weiss, C. Ruhlmann, and P. Oudet, *Nucleic Acids Res.* **14**, 2045 (1986).

It is possible to obtain fractions containing up to 80% minichromosomes containing a nucleosome-free region or gap using the preferential sensitivity of these molecules to restriction enzymes. Purified minichromosomes are digested with an excess of restriction enzymes cutting inside the ORI region like *Bg*lI, *Kpn*I, and *Msp*I or outside like *Eco*RI and *Bam*HI. The product of the digestion is then fractionated on a 5–30% sucrose gradient identical to the one previously used for the purification. The linearized gapped molecules are found in the light part of the gradient.[25]

Potentially powerful procedures to isolate specific fractions have been developed using the specificity of antibodies.[26-29] Affinity chromatography has also been used to purify chromatin fragments accessible to restriction enzymes.[30] Transcribing minichromosomes or replicative forms have not yet been isolated because of the low percentage of such complexes (< 1% of the total population) and the significant amount of contaminants that are copurified. The use of powerful purification procedures for the isolation of SV40 minichromosome subpopulations will provide significant quantities of homogeneous material and allow the characterization of structural modifications involved in chromatin replication and gene regulation. These templates may also be used in *in vitro* assays for transcription and replication.

Acknowledgments

We thank J. C. Homo, C. Werlé, and B. Boulay for the illustrations used in this chapter. We are grateful to P. Schultz and G. Richards for critical readings of the manuscript. This work was supported by grants from the INSERM, CNRS, and the Ministère de la Recherche et de l'Enseignement Supérieur (Décision No. 86T0371).

[26] J. Reiser, J. Renart, L. V. Crawford, and G. R. Stark, *J. Virol.* **33,** 78 (1980).
[27] M. Segawa, S. Sugana, and N. Yamaguchi, *J. Virol.* **35,** 320 (1980).
[28] L. C. Tack and M. L. DePamphilis, *J. Virol.* **48,** 281 (1983).
[29] L. C. Tack, J. H. Whright, and E. G. Gurney, *J. Virol.* **58,** 635 (1986).
[30] J. L. Workman and J. P. Langmore, *Biochemistry* **24,** 7486 (1985).

[3] Isolation of Yeast Plasmid Chromatin

By Ann Dean, David S. Pederson, and Robert T. Simpson

Introduction

Recombinant DNA technology has allowed insight into the mechanisms of control of gene expression in eukaryotic cells (see Ref. 1 for review). Cloned genes have provided the assay material necessary to identify and, in some cases, isolate *trans*-acting factors thought to be important in gene regulation. Hybridization methods, together with nuclease or chemical digestion, have shown differences in the chromatin organization of specific genes related to the functional state. Still lacking is a complete description of the composition and structure of a regulated gene in the transcribed and repressed states, one that includes histones and other structural proteins in addition to components of the transcription machinery. We describe here methods for isolating episomal yeast genes as chromatin, a biological system that has the potential to accomplish this goal.

Stinchcomb *et al.*[2] discovered that certain yeast sequences allowed DNA to be maintained as episomal, amplified elements. The *propositus* DNA, the TRP1ARS1 circle, is also called YARp1.[3] It is 1453 base pairs (bp) in length and contains the coding sequence for *N*-phosphoribosylanthranilate isomerase (TRP1) and an autonomously replicating segment (ARS1) sequence (Fig. 1A). The plasmid is amplified; about 100 copies are present per cell on average. The plasmid DNA is packaged as chromatin. Nuclease digestion and indirect end-label analyses have been used by Thoma *et al.*[4] to establish a structure of the plasmid chromatin. Our goal is to insert regulated yeast genes into the TRP1ARS1 plasmid, isolate plasmid chromatin when the genes are transcribed or repressed, and compare their composition and structure. We also hope to insert *cis*-acting DNA sequences into the plasmid DNA and use these constructions as rescue vehicles for isolation of *trans*-acting regulatory factors.

A number of derivatives of the TRP1ARS1 circle have been constructed to serve as cloning vectors. A partial restriction map of the one we currently often employ is shown in Fig. 1B. It consists of pBR322 with the

---

[1] D. S. Pederson, F. Thoma, and R. T. Simpson, *Annu. Rev. Cell Biol.* **2,** 117 (1986).

[2] D. T. Stinchomb, K. Struhl, and R. W. Davis, *Nature (London)* **282,** 39 (1979).

[3] V. A. Zakian and J. F. Scott, *Mol. Cell. Biol.* **2,** 221 (1982).

[4] F. Thoma, L. W. Bergman, and R. T. Simpson, *J. Mol. Biol.* **177,** 715 (1984).

*Bam*HI and *Eco*RI sites eliminated by cutting, filling overhanging ends, and recircularizing the plasmid DNA. The TRP1ARS1 circle was linearized and cloned into the *Hin*dIII site of the vector. The *Nae*I site at 1067 map units of the yeast DNA has been replaced by a *Bam*HI site to use in adding other yeast genes. Finally, a segment of pUC19 containing the *Escherichia coli lac* operator has been inserted in the *Eco*RI site in the 5′-flanking region of the TRP1 gene. After inserting a gene or DNA

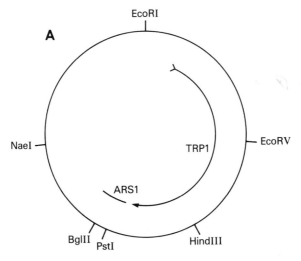

Fig. 1. Maps of typical plasmids. (A) The TRP1ARS1 circle or YARp1.[3] (B, p. 28) pBRAT9, a pBR322 derivative that lacks *Eco*RI and *Bam*HI sites in the vector DNA and contains a modified TRP1ARS1 segment cloned at the *Hin*dIII site. (C, p. 28) AT9, the modified TRP1ARS1 circle. In all maps the TRP1 gene is indicated as an arrow, and the position of the autonomously replicating segment near the 3′ end of the gene is shown. In (B), bacterial plasmid sequences are the thin line while the yeast plasmid segment is shown as a heavy line. Construction: A cloning vector lacking *Nae*I sites was constructed by cutting pBR322 with *Bam*HI and *Ava*I, Klenow filling, and blunt-end circularization. TRP1ARS1 sequences were excised from a pBR322 clone with *Eco*RI and subcloned into the *Nae*I-less plasmid. This was cut at the unique TRP1ARS1 *Nae*I site at 1067 map units and 10 bp *Bam*HI linker inserted. The yeast sequences of this chimeric plasmid were excised with *Eco*RI and purified by gel electrophoresis. This DNA was treated with calf intestinal alkaline phosphatase and ligated in the presence of an excess of a gel-purified fragment containing the *lac* operator from pUC19 (*Hin*dIII – *Hae*II fragment polished with Klenow polymerase) with *Eco*RI linkers added. After ligation, the mixture was cut with *Hin*dIII, and linear molecules about 1.5 kilobase pairs in length were isolated by agarose gel electrophoresis. These were then cloned into the *Hin*dIII site of pBR322; *Bam*HI and *Eco*RI sites were previously removed from the vector by cutting, Klenow filling, and recircularization. These manipulations yielded the plasmid shown in (B). Digestion of this plasmid with *Hin*dIII, followed by circularization at low DNA concentrations and transformation of yeast, yields the yeast plasmid shown in (C).

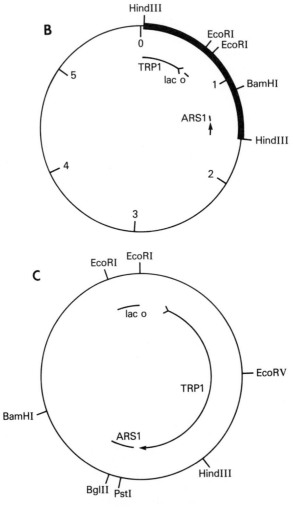

FIG. 1. *(continued)*

sequence of interest into this chimeric plasmid and amplifying it in *E. coli,* the plasmid DNA, prepared in any standard fashion, is cut to completion with *Hind*III, circularized with DNA ligase (polydeoxyribonucleotide synthase) at low DNA concentrations, and used to transform *Saccharomyces cerevisiae* by standard methods. Selection is based on the use of a *trp1* host and screening by plating on tryptophan-deficient media. A chimeric shuttle plasmid can also be passed directly to yeast without removing the bacterial plasmid sequences. However, we prefer to use DNA circles that are as small as possible to enhance the signal to noise ratio for plasmid-associated proteins and facilitate chromatin structure mapping; therefore, we routinely use constructions that contain only yeast sequences and the *lac* operator.

While the plasmid DNA is amplified and constitutes about 1% of total yeast DNA, DNA makes up only about 2% of the total nucleic acid in *S. cerevisiae.* This necessitates a 5000-fold enrichment of plasmid DNA relative to total nucleic acid in order to achieve biochemical purity of the plasmid chromatin. We have devised two purification schemes for isolating yeast plasmid chromatin. The first uses conventional biochemical methods and the second, protein–nucleic acid affinity and immunologic reagents. It is important to note that, so far as we have examined, chromatin produced by the two methods appears to be indistinguishable. Criteria include histone content, topological linking number for the plasmid DNA, and micrococcal nuclease digestion patterns. In this chapter, we present the methods and then discuss the advantages and disadvantages of each.

Biochemical Purification Method

In isolating plasmid chromatin by conventional fractionation methods,[5] the initial purification of nuclei is an important feature, serving to remove much of the RNA and mitochondria and to concentrate plasmid chromatin. Plasmid chromatin is eluted from nuclei and the eluate filtered on a Sephacryl S-300 column with the goal of removing contaminating nucleases and proteases. The excluded fraction from the sizing column is then subjected to two rounds of isopycnic centrifugation in Nycodenz, leading to a substantial purification of the chromatin. Another sizing column removes Nycodenz, and a final sucrose gradient yields pure plasmid chromatin. Because the isopycnic centrifugation steps also concentrate plasmid chromatin and precede the sucrose gradient step, it is possible to maintain or exceed a plasmid concentration of 1 nM throughout the isolation and reduce the possibility of nucleosome dissociation. The buffers used often reflect compromises demanded by the isolation. For instance, the nucleus isolation buffer minimizes leakage of plasmid chromatin whereas the elution buffer permits leakage but avoids nuclear lysis. Both buffers contain magnesium to ensure the integrity of contaminating ribosomes and facilitate their later removal. Both also have a relatively low

[5] D. S. Pederson, M. Venkatesan, F. Thoma, and R. T. Simpson, *Proc. Natl. Acad. Sci. U.S.A.* **83,** 7206 (1986).

ionic strength so as not to dissociate chromatin components. Finally, both buffers lack detergents to prevent mitochondrial lysis.

The method described below is for the isolation of plasmid chromatin from about 25 g wet weight cells, which is the amount of yeast obtained from a 10-liter culture grown to an $A_{600 \, nm}$ of 4. A culture of SC-3 transformants, growing in synthetic, defined media, is still in log-phase growth at this density provided it has been aerated vigorously and the pH maintained at 5.5 by addition of ammonium hydroxide. The 25 g wet weight cells contains roughly 0.1 mg of TRP1ARS1 DNA and 10 mg genomic DNA. Since the overall yield of plasmid chromatin is typically about 10%, one isolation on this scale will yield about 10 μg plasmid DNA, 2 μg of each core histone, and about 0.5 μg of a nonhistone protein of 50,000 Mr, should it be present at a stoichiometry of one molecule per plasmid. These protein amounts are readily visible on gels stained by a variety of methods.

The method can be scaled to smaller cultures provided that cell mass to buffer amount ratios are kept constant. Scaling the isolation to larger cultures is limited first by the nucleus isolation procedure. Percoll gradients, performed in 250-ml bottles, have been used for isolating nuclei[5] but do not work as well as those performed in 50-ml tubes (described below). Efforts to form and use Percoll gradients in 1-liter bottles were unsuccessful. Finally, because working with very large cultures lengthens the nucleus isolation procedure, more plasmid chromatin is lost through slow leaching from nuclei prior to the elution step. As a compromise, it appears possible to quick-freeze the nuclear extract at −80°, after it has been passed through a Sephacryl S-300 column, and to purify chromatin further after material from two or more nuclear isolations has been accumulated. By working rapidly so that the first density gradient is begun at the end of day 1, it is possible to complete the purification in 4–5 days.

Reagents

Buffer A
80 mM KCl
5 mM MgCl$_2$
10 mM 2-(N-morpholino)ethanesulfonic acid, pH 6.3 with NaOH
1 mM EGTA
0.5 mM spermidine-HCl
Buffer B
200 mM NaCl
5 mM MgCl$_2$
10 mM piperazine-N,N'-bis(2-ethanesulfonic acid), pH 7.3
0.5 mM EGTA

Additions
  M: 0.1% (v/v) 2-mercaptoethanol
  S: 1 M sorbitol
  F: 18% (w/v) Ficoll, average M_r 400,000 (Pharmacia, Piscataway, NJ)
  I: 1 mM phenylmethylsulfonyl fluoride (PMSF) (from 250× stock in ethanol),
     1 mM iodoacetate (from 1000× stock in ethanol), and
     5 μg/ml pepstatin A (from 1000× stock in ethanol)

Cell Growth and Spheroplast Preparation

Yeast containing the plasmid are grown either in flasks with vigorous shaking or in fermentors at 30° in 2% dextrose and 0.67% nitrogen base without amino acids, supplemented with whatever other nutritional factors the particular *S. cerevisiae* strain requires. For SC-3 (*matα trp1 his3 ura3 gal2 gal10 cir⁰*) containing the unmodified TRP1ARS1 plasmid (Fig. 1A), which we used for the methodologic development, 0.002% each uracil and histidine is added. Cultures are grown to late log phase. This corresponds to $A_{600\,nm} = 1–1.5$ for a flask culture and 4–5 for fermentor cultures grown as described above.

Cultures are cooled to room temperature, and cells are harvested using a Beckman JA-10 or Sorvall GS3 rotor for flask cultures (3500 rpm for 5 min) and a continuous-flow Sharples centrifuge or Millipore tangential flow filtration apparatus for fermentor cultures. Wash 25 g of cells by suspension in 375 ml of buffer A plus M and pellet as above. The pellet is resuspended in 175 ml of buffer A plus M and S, and 100 mg of zymolyase 60,000 (Miles, Elkhart, IN), previously suspended in 10 ml buffer A plus S and M, is added. The suspension is incubated at 30° with gentle rotation, sufficient to keep the yeast from settling. The extent of spheroplasting is assayed by diluting 5 μl of the suspension into 1 ml of 1% sodium dodecyl sulfate (SDS) and measuring optical density at 600 nm. The incubation is continued until this value is 5–10% of the initial value (usually 30–60 min).

The suspension is then diluted with 200 ml of buffer A plus S and centrifuged as above. If plasmid transcriptional properties are to be investigated, spheroplasts are suspended in growth media plus 1 M sorbitol and allowed to recover at 30° with gentle shaking for 1 hr. I is then added and cells collected by centrifugation as above. Alternatively, spheroplasts are gently resuspended in 375 ml of buffer A plus S and I and recentrifuged as above. All procedures that follow are performed at 0–4°.

Nuclear Isolation and Plasmid Elution

Spheroplasts are suspended in 375 ml of buffer A plus F and I and stirred for 15–30 min. A few strokes of homogenization with a motor-driven Teflon pestle serve to complete lysis of the spheroplasts. The lysate is then centrifuged for 20 min at 11,000 rpm in a Sorvall GSA rotor. Most of the mitochrondria fail to pellet through this buffer, as assessed by electrophoresis of DNA from the supernatant.

Sixteen 34-ml gradients of 33% Percoll in buffer A plus S are preformed up to 24 hr in advance by centrifugation for 80 min at 15,000 rpm in a Beckman JA-20 or Sorvall SS-34 rotor. The spheroplast lysate pellet is resuspended in 75 ml of buffer A plus I, and 6-ml aliquots are layered onto the gradients. These are then centrifuged in a Sorvall HB4 rotor at 7500 rpm or a Sorvall HS4 rotor at 7000 rpm for 15 min; the rotor is allowed to coast to a stop. Much of the material distributes into two bands. The upper band is usually about 40% (16 ml) into the gradient and contains half the nuclei and 5–10% of the original cell mass. The bulkier lower band is usually about 80% (32 ml) into the gradient. If this lower band is collected and recentrifuged on another Percoll gradient, an upper band again forms, suggesting that nuclei previously trapped in the debris of the lower band have been released.

Nuclei in the upper band seem undamaged, judging by the following criteria: they are not sticky; they support run-on transcription reactions; and they contain undegraded DNA and histones and intact chromatin, as assessed by digestion with micrococcal nuclease. The upper band nuclei are collected with a pipette, diluted with 2 volumes of buffer A plus $0.1 \times$ I, and centrifuged for 5 min at 7500 rpm in the JA-10 or GSA rotor. Nuclei are washed by resuspension in 50 ml buffer A plus $0.1 \times$ I and centrifugation for 5 min at 7500 rpm in an HB4 rotor.

Plasmid chromatin is eluted by suspending nuclei in 8 ml of nuclear elution buffer consisting of buffer B plus M and $0.1 \times$ I (without iodoacetate) and incubating with occasional gentle mixing on ice for 75 min. The nuclei are then pelleted by centrifugation for 5 min at 7500 rpm in an HB4 rotor. Resuspension, incubation, and centrifugation are repeated 2 more times. The three extracts are combined and clarified by centrifugation at 11,000 rpm for 10 min in the HB4 rotor.

Purification of Plasmid Chromatin

Up to 50 ml of nuclear extract is applied to a 5×25 cm column of Sephacryl S-300 equilibrated with nuclear elution buffer and eluted with the same buffer. Material eluting in the void volume is made 40% (w/v) Nycodenz (Nyegaard, Oslo) by slow addition of 100% (w/v) Nycodenz.

The mixture is loaded into centrifuge tubes, filled to the top with 40% (w/v) Nycodenz plus 0.6 × nuclear elution buffer or mineral oil if necessary, and centrifuged for 36 hr at 50,000 rpm in a VTi50 rotor. Plasmid chromatin bands near the center of the gradient at a density of about 1.2 g/ml. The gradient is fractionated into about 20 fractions by any standard method. Eighty-microliter aliquots of each fraction are mixed with 40 μl of 8.1 M NH$_4$OAc/50 mM HNa$_3$EDTA and precipitated with 240 μl ethanol. The pellet is rinsed with 95% ethanol, air-dried, suspended in 10 μl of 16 mM NaOAc/1 mM EDTA/4 mM Tris containing 50 μg/ml ribonuclease A, and incubated at 37° for 30 min. Samples are mixed with 2 μl of 36% glycerol/0.6% SDS/75 mM HNa$_3$EDTA/0.12% bromphenol blue, heated at 60° for 5 min, and analyzed by agarose gel electrophoresis. Fractions containing plasmid chromatin are pooled, loaded into centrifuge tubes for a VTi65.2 rotor, and centrifuged for 18 hr at 50,000 rpm. These gradients are fractionated, and 20-μl aliquots are analyzed for plasmid chromatin as above. The plasmid-containing fractions are pooled.

To remove Nycodenz, up to 6 ml of pooled plasmid chromatin is applied to a 2.5 × 12 cm column of Sephacryl S-300 equilibrated with 0.6 × nuclear elution buffer and eluted with the same buffer. Material eluting in the void volume is loaded onto 0.4 – 1.0 M linear sucrose gradients in 0.6 × buffer B. Typically, material from 25 g of cells is in a volume of less than 4 ml and is loaded onto a single 35-ml gradient. Gradients are centrifuged at 50,000 rpm in a VTi50 rotor for 80 min. The rotor is braked to 10,000 rpm and allowed to coast to a stop. After fractionating the velocity gradient, fractions containing plasmid chromatin are identified by agarose gel electrophoresis of 40-μl aliquots as above, except for omission of the ribonuclease treatment. Figure 2 shows such an agarose gel analysis of the sucrose gradient. Note the absence of contaminating RNA and high molecular weight genomic or mitochondrial DNA. Most of the plasmid DNA is supercoiled, indicating a lack of nuclease degradation of the nucleic acid during the purification. Selected fractions are pooled and either used immediately for analyses or quick-frozen at −80°.

Comments

When examined by electron microscopy, most plasmid chromatin molecules are full circles with morphologically normal nucleosomes. Figure 3 is a montage of six selected molecules. Analysis of a large number of chromatin molecules showed that the number of nucleosomes per molecule peaked at seven with some skewing to lower numbers.[5] This skewing, as well as the tailing of the sucrose gradient peak to smaller sedimentation values (Fig. 2), indicates that some nucleosome dissociation does take place during the purification procedure.

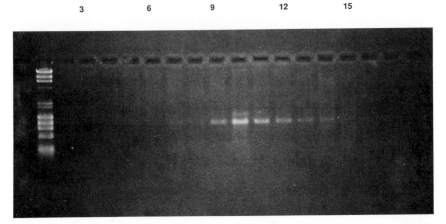

Fig. 2. Sucrose gradient fractionation of TRP1ARS1 plasmid chromatin. The final sucrose gradient in the purification was collected in 20 fractions, and total DNA from numbers 2 to 17 were analyzed by agarose gel electrophoresis and ethidium bromide staining. Sedimentation was from right to left across the gel. Size standards (far left lane) are lambda and ϕX174 RF DNA cut with HindIII and HaeIII, respectively. Most of the plasmid DNA migrates as covalently closed circles; the trace of more slowly migrating DNA in fractions 10 and 11 is nicked plasmid DNA.

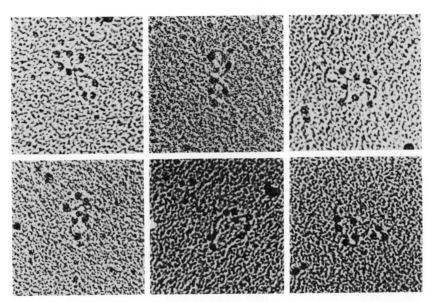

Fig. 3. Electron micrographs of purified plasmid chromatin. A montage of six molecules visualized using electron microscopy is shown. Analysis of the number of nucleosomes per plasmid chromatin molecule in a larger sample showed the average to be 6.5.[5]

Protein – Nucleic Acid Affinity Method

The protein – nucleic acid affinity method is based on the well-characterized, high-affinity interaction between the *E. coli lac* operator and the *lac* repressor protein.[6] The interaction has been used successfully in purification of a DNA-binding protein from crude extracts.[7] A segment of recombinant DNA, containing both the recognition sequence for the protein of interest and the *lac* operator, was incubated in the extract and the complex with the DNA-binding protein rescued from the extract using repressor – operator affinity.[7] We have modified this approach to purify *in vivo* assembled chromatin complexes. Using DNA constructions similar to that shown in Fig. 1C and containing the *lac* operator in a nucleosome-free region of the TRP1ARS1 plasmid,[4] we have transformed yeast to allow chromatin assembly *in vivo*. Spheroplasts are lysed in a buffer that allows leakage of the plasmid chromatin from nuclei. Debris is removed by centrifugation, and a *lac* repressor – β-galactosidase fusion protein, which retains operator- and inducer-binding properties,[8] is added to the supernatant. Plasmid chromatin bound to the fusion protein is precipitated by the addition of anti-β-galactosidase antibodies immobilized on beads. After washing, addition of isopropylthio-β-D-galactopyranoside (IPTG) releases the bound chromatin from the beads into the supernatant. The rapid and gentle purification yields a 20 – 25% recovery of chromatin that is intact as judged by the criteria described in the introduction.

Reagents and Materials

Yeast wash buffer (YWB)
  40 mM potassium phosphate, pH 7.5
  1 M sorbitol
  0.5 mM PMSF (freshly added from 1000× solution in dimethyl sulfoxide)
Elution and binding buffer (EBB)
  150 mM NaCl
  10 mM MgCl$_2$
  10 mM Tris-Cl, pH 7.4

Protein Reagents

The repressor – β-galactosidase fusion protein is produced by *E. coli* strain BMH 72-19-1[8] and purified exactly as described by Fowler and Zabin.[9] The purified protein is stored in aliquots at −80°.

[6] A. D. Riggs, H. Suzuki, and S. Bourgeois, *J. Mol. Biol.* **48,** 67 (1970).
[7] D. Levens and P. M. Howley, *Mol. Cell. Biol.* **5,** 2307 (1985).
[8] B. Muller-Hill and J. Kania, *Nature (London)* **249,** 561 (1974).
[9] A. V. Fowler and I. Zabin, *J. Biol. Chem.* **258,** 14354 (1983).

Rabbit anti-β-galactosidase antibodies (Cappel Laboratories, Malvern PA) are repurified by binding to *E. coli* β-galactosidase attached to cyanogen bromide-activated Sepharose prepared according to the procedures recommended by the manufacturer (Pharmacia). The antibodies are eluted with 0.1 M citric acid, pH 2.2, neutralized with NaOH, dialyzed against phosphate-buffered saline (PBS, 150 mM NaCl, 15 mM Na$_2$HPO$_4$, 4 mM KH$_2$PO$_4$), and frozen in aliquots at $-80°$.

Immunobeads coated with goat anti-rabbit IgG antibody (Bio-Rad Laboratories, Richmond, CA) are suspended according to the manufacturer's instructions except at 5-fold higher concentration.

Preliminary Titration of Protein Reagents

The amount of Immunobeads required to bind a given amount of purified rabbit anti-β-galactosidase can be calculated from the titer determined for each lot by the manufacturer. The amount of Immunobead–anti-β-galactosidase complex required to precipitate a given amount of *lac* repressor–β-galactosidase fusion protein is determined by incubating increasing amounts of fusion protein with preformed, washed antibody–bead complex in PBS for 1 hr at 4°. After centrifugation, the supernatant is assayed for β-galactosidase activity[10] and the equivalence point determined. Finally, the amount of fusion protein needed to bind a given amount of plasmid is assayed using a segment of linear DNA containing the *lac* operator, increasing amounts of fusion protein, and sufficient antibody–Immunobead complex to precipitate the largest amount of fusion protein being tested, using the absorption procedure detailed below. The amount of plasmid DNA remaining in the supernatant and the amount released from the bead complex by IPTG are analyzed by agarose gel electrophoresis and staining with ethidium bromide, using standard gel electrophoretic methods. See below for further considerations on the amount of fusion protein to use for optimal purification.

Growth of Yeast Culture

Saccharomyces cerevisiae containing the plasmid of interest are grown to late log phase, as described above. The steps that follow are based on a 1-liter culture. Proportional scale-up is possible; alterations of centrifugation times and equipment may be required.

---

[10] Worthington Manual, Worthington Biochemical Corp., Freehold, New Jersey.

Preparation of Spheroplasts and Nuclear Eluate

Yeast cells are collected by centrifugation in the GSA or JA-10 rotor for 5 min at 5000 rpm and 4°, washed with 100 ml cold water into one bottle, and recentrifuged. Cells are suspended in 50 ml room temperature YWB containing 20 mM 2-mercaptoethanol, and 0.5 ml zymolyase suspension (10 mg/ml in water) is added. Spheroplasts are allowed to form during a 30-min incubation at 30° rotating at 50 rpm. The $A_{600\,nm}$ of an aliquot diluted into 1% SDS should drop to 10% of the original value. Spheroplasts are centrifuged at 3400 rpm for 5 min at 4° in a 50-ml plastic conical centrifuge tube using a swinging-bucket rotor. Spheroplasts are resuspended gently using a 25-ml pipette in 30 ml of YWB at 4° and recentrifuged as above. The pellet is resuspended in 5 ml of EBB containing 0.5 mM PMSF at 4° and incubated 15 min on ice. Lysis of spheroplasts is completed by 2–5 strokes of homogenization in a motor-driven Teflon–glass homogenizer operated at the slowest speed that keeps the pestle moving. The lysate is allowed to stand on ice for 2–4 hr with occasional mixing to elute plasmid chromatin from nuclei. The mixture is then transferred to 1.5-ml Eppendorf tubes and centrifuged for 20 min at 4° in a microfuge. Supernatants are transferred to fresh tubes and kept on ice. It is convenient to perform the subsequent steps on 1-ml aliquots in Eppendorf microcentrifuge tubes.

Preparation of Antibody–Immunobead Complex

A 1-ml aliquot of Immunobeads is washed twice with PBS and resuspended in 1 ml of PBS; a 10-sec centrifugation in a microfuge at room temperature is sufficient to pellet the beads in this and all steps following. Bovine serum albumin, 1 μl of a 10 mg/ml solution in water per 100 μl of bead suspension, is added together with an appropriate amount of affinity-purified anti-β-galactosidase (typically 5–10 μl of protein at a concentration of 0.5 mg/ml per 100 μl of bead suspension). The mixture is incubated on ice for 2 hr with occasional mixing. Following absorption of antibody, the beads are washed with 1 ml EBB twice and resuspended in the original volumn of EBB.

Plasmid Binding and Elution

Sufficient *lac* repressor–β-galactosidase fusion protein is added to 1-ml aliquots of the nuclear eluate to bind the expected plasmid chromatin. A liter of cells grown as above contains about 3 μg of plasmid DNA. If recovery were 100% in the nuclear eluate, this would give about 600 ng per 1-ml aliquot. Typically, we use 20 μl of *lac* repressor–β-galactosidase fusion protein at 0.5 mg/ml for each milliliter of eluate, an amount of fusion protein sufficient to bind 200 ng of naked plasmid DNA. The fusion protein and eluate are incubated for 15 min on ice. Immunobead–anti-β-galactosidase complex, 200 μl per milliliter of nuclear eluate, is added in 4 aliquots over a 30-min period, and the mixture is incubated for an additional 1–1.5 hr on ice with occasional mixing.

The mixture is centrifuged, the supernatant removed, and the beads washed rapidly twice with 200 μl EBB at 4°. The bead complex is washed once with 200 μl EBB containing 25–50 μg/ml poly[d(A-T)], and the supernatant removed completely. Beads are then suspended in 100–200 μl 10 mM Tris-Cl, pH 7.5, 1 mM EDTA, 150 mM NaCl, 2 mM IPTG and incubated for 10 min on ice. The mixture is then centrifuged for 30 sec in a microfuge, and the supernatant, containing plasmid chromatin, removed to use for further studies. Figure 4 shows that the plasmid chromatin may be separated from virtually all the RNA and probably 99% of genomic

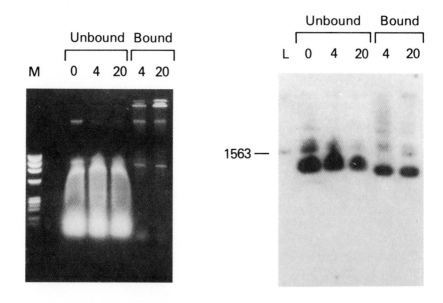

FIG. 4. Plasmid chromatin in a crude nuclear eluate was bound to the affinity matrix in the presence of increasing amounts of repressor–β-galactosidase fusion protein (4 and 20 μl at 0.5 mg/ml). After washing, the bound material was eluted with IPTG. Total nucleic acid in the bound and unbound fractions was electrophoresed in a 1% agarose gel and stained with ethidium bromide (left). The gel was blotted to nitrocellulose and hybridized with a TRP1ARS1 ^{32}P-labeled DNA probe (right), and autoradiography was performed. Lane M contains size standards as in Fig. 2. Lane L contains linearized AT9 DNA.

DNA in a single purification step. Note that plasmid chromatin is almost entirely in the form of intact supercoiled molecules.

Comments

The binding of the fusion protein to operator-containing plasmid chromatin has been successfully performed at concentrations of NaCl between 20 and 200 mM and in the presence or absence of $MgCl_2$. NaCl concentrations of 50–100 mM increase the difference between the affinities of the repressor for operator- versus nonoperator-containing DNA.[11] However, it is likely that the buffer conditions for both binding and release of operator-containing plasmid chromatin could be varied substantially depending on the intended use of the purified chromatin (e.g., structural studies of chromatin or identification of specific proteins bound *in vivo*). Furthermore, inhibitors of nucleases and proteases such as EGTA, aprotinin, and pepstatin do not interfere with the binding–elution protocol described above.

It is important not to use excess fusion protein since the affinity of repressor for nonoperator-containing DNA is high and any excess repressor, over that amount necessary to bind operator DNA, will bind to contaminating genomic or mitochrondrial DNA. For this reason, gentle lysis of spheroplasts is suggested to avoid damage to nuclei, which can permit genomic DNA to leach into the nuclear eluate. In typical preparations, 20–50% of the purified chromatin DNA is yeast genomic DNA. This DNA represents at most 1% of yeast chromosomal DNA since plasmid DNA is also about 1% of the total DNA present in a yeast cell. The genomic DNA could be eliminated by preparing intact nuclei using Percoll gradients, as outlined in the biochemical method above.

The structure of plasmid chromatin prepared by the protein–nucleic acid affinity method appears similar to that prepared by the conventional biochemical method[5] and to that of partially purified plasmid chromatin previously described.[4] Direct hybridization of a micrococcal nuclease-digested sample of chromatin as well as analysis of topological properties of chromatin treated with topoisomerase I suggest that nucleosome dissociation does not occur during the affinity purification. A small, but variable (< 10%), amount of nicked, circular DNA can be seen by gel electrophoretic analysis of the DNA in purified plasmid chromatin. Although topological studies have generally been performed shortly after chromatin isolation, storage of plasmid chromatin for up to a week at 4° has not altered the topological properties of the material, suggesting that chromatin is stable for at least that period of time.

[11] S.-Y. Lin and A. D. Riggs, *J. Mol. Biol.* **72,** 671 (1972).

Conclusions

Strengths and Weaknesses

Each of the methods has advantages and disadvantages. The conventional protocol is lengthy, raising concern over proteolysis or dissociation of chromatin components. It does, however, lead to material that appears to be biochemically pure. The affinity method is more rapid but does not yield totally clean chromatin after a single stage of purification. The affinity method also requires preparation of the protein reagents and titration of them one against another, a time-consuming process. It is likely that a combination of the two, either preparation of clean nuclei followed by the affinity method or the affinity method followed by one of the several purification steps of the conventional method, will turn out to be the approach of choice.

A choice must be based in part on the nature of the experiment. For example, if one were interested in identifying a *trans*-acting factor involved in control of expression of a unique gene, then the affinity method would be preferred. Even with a 50% contamination by genomic chromatin, the molar contribution of any specific nonhistone associated with genomic chromatin to the total is vanishingly small while a specific protein associated with plasmid chromatin should be a major component of the purified material. On the other hand, if one were interested in the histones associated with a particular gene, the conventional method (or a combination of affinity with one step of the conventional method) would be required since genomic and plasmid chromatin contribute to the histone pool in proportion to their content in the preparation.

Prospects and Cautions

We present methods that have allowed us to purify a plasmid containing unique *S. cerevisiae* DNA sequences as chromatin, assembled *in vivo*. Our hope is that these methods will allow advances in understanding the structure–function relationships involved in regulation of transcription of eukaryotic genes. At this time, that is still only a hope. Several cautions have entered into our selection of the first genes we plan to study; we think it appropriate to share some of them with readers interested in these methods.

A major concern is the homogeneity of the plasmid population, both functionally and structurally. In the case of the *propositus* plasmid, we do not know the fraction of the molecules that are transcribing the TRP1 gene. This is not of concern to us since we use this gene and its neighboring ARS as a vector, not an object for investigation, per se. Run-on transcription experiments indicate that the elution procedure used in the biochemical methods does not select against those molecules which are being transcribed. We do not know whether this is the case for replicating molecules.

For regulated genes, it is important to establish that regulation occurs when the gene is present as an amplified episomal element. Studies of mRNA abundance and decay rates should be carried out in parallel with studies of plasmid copy number. The possibility of titrating out regulatory factors by increasing the copy number of a regulated gene must be kept in mind. In this vein, it should be noted that during the spheroplasting procedure yeast cells are without nutrients for about half a cell cycle. Alterations in transcriptional metabolism occur during this starvation. We now routinely allow a recovery period after spheroplast formation (A. Dranginis, unpublished observations). The spheroplasts are cultured for 1 hr with slow shaking at 30° in growth medium containing 1 M sorbitol prior to lysis or nuclear preparation. Using this approach and estimates of mRNA decay rates, early results suggest that at least 50% (and perhaps all) of HSP26 genes present as amplified episomal elements are transcribed at a rate comparable to that of the single-copy genomic gene after heat shock.

We emphasize the importance of understanding the chromatin structure and functional DNA anatomy of the plasmid prior to making constructions to be used for study of protein–DNA interactions. For example, in construction of the *lac* operator-containing plasmids, we inserted the operator sequence at two sites: the *Eco*RI site in the nucleosome-free region at the 5′ end of the TRP1 gene and the *Bam*HI site located partially in and partially between two positioned nucleosomes. The short DNA segment allowed efficient interactions with the *lac* repressor–β-galactosidase fusion protein in the former location but not in the latter. Possible interference with function of either ARS or promoter sequences should be kept in mind. For short *cis*-acting sequences inserted to rescue *trans*-acting factors, tandem repeats may help to ensure that the sequence of interest is exposed to interact with solution components.

[4] Affinity Isolation of Replicating Simian Virus 40 Chromosomes

By Timothy M. Herman

Replicating SV40 chromosomes represent a good model system for the investigation of eukaryotic DNA and chromatin replication.[1] However, it has always been difficult to separate replicating SV40 chromosomes from

[1] M. L. DePamphilis and P. M. Wassarman, *in* "Organization and Replication of Viral DNA" (A. Kaplan, ed.), pp. 37–114. CRC Press, Boca Raton, Florida, 1982.

Copyright © 1989 by Academic Press, Inc.
All rights of reproduction in any form reserved.

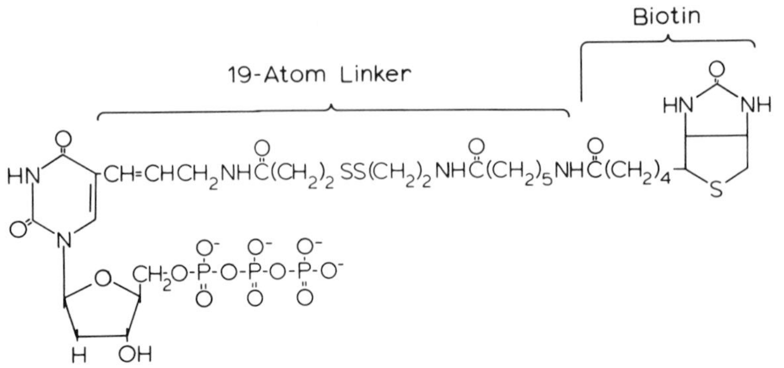

FIG. 1. Structure of the chemically cleavable biotinylated nucleotide analog Bio-19-SS-dUTP.

mature, nonreplicating chromosomes. Recently it has become possible to isolate replicating SV40 chromosomes by an affinity chromatography procedure utilizing the chemically cleavable biotinylated nucleotide Bio-19-SS-dUTP.[2]* Bio-19-SS-dUTP contains a chemically cleavable disulfide bond in the 19-atom linker arm joining biotin to the pyrimidine base uracil (Fig. 1). This biotinylated nucleotide is first incorporated into replicating SV40 chromosomes during a brief pulse-label *in vitro*. The replicating SV40 chromosomes are then separated from the mature chromosomes by affinity chromatography using streptavidin and biotin–cellulose.

This affinity purification procedure is illustrated in Fig. 2. The methods involved are presented in three sections. First, synthesis of the chemically cleavable biotinylated nucleotide Bio-19-SS-dUTP is described. Second, the *in vitro* DNA replication reaction used to affinity-label replicating SV40 chromosomes is described. Third, procedures to affinity-isolate the replicating SV40 chromosomes are outlined.

Synthesis of Bio-19-SS-dUTP

Bio-19-SS-dUTP is synthesized in a three-step procedure adapted from Langer *et al.*[3] Briefly, dUTP is first mercurated at the 5 position and then reacted with allylamine to form allylamine-dUTP. Allylamine-dUTP is then purified by ion-exchange chromatography and reacted with suc-

---

* Bio-19-SS-dUTP,   5-[(*N*-biotin-amido)hexanoamido-ethyl-1,3-dithiopropionyl-3-aminoallyl]-2′-deoxyuridine 5′-triphosphate.

[2] T. M. Herman, E. Lefever, and M. Shimkus, *Anal. Biochem.* **156**, 48 (1986).

[3] P. R. Langer, A. A. Waldrop, and D. C. Ward, *Proc. Natl. Acad. Sci. U.S.A.* **78**, 6633 (1981).

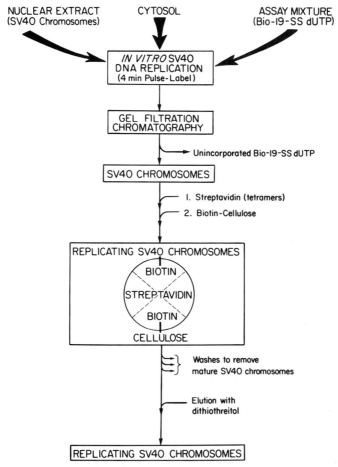

FIG. 2. Schematic illustration of the procedure to affinity-isolate replicating SV40 chromosomes.

cinimido-2-[6-(biotin-amido) hexanoamido] ethyl-1-1,3′-dithiopropionate (NHS-LC-SS-biotin) to form the final biotinylated nucleotide. Bio-19-SS-dUTP is purified by ion-exchange chromatography followed by reversed-phase HPLC. The procedures for this synthesis and purification are given below. Examples of the chromatographic procedures described here have been reported.[4] A more detailed description of this synthesis will be published in a future volume in this series.[5]

[4] M. L. Shimkus, P. Guaglianone, and T. M. Herman, *DNA* **5**, 247 (1986).
[5] T. M. Herman, this series, submitted for publication.

Synthesis of 5-Mercurated dUTP (Hg-dUTP)

1. Dissolve 50 mg (89 μmol) of deoxyuridine 5'-triphosphate (Sigma, St. Louis, MO) in 10 ml of 0.1 M sodium acetate, pH 6.0.
2. Add 142 mg mercury(II) acetate (Alfa) to the dUTP solution and heat at 50° for 4 hr.
3. Cool reaction on ice and add 34 mg LiCl.
4. Remove the precipitated $HgCl_2$ by extracting the solution 6 times with an equal volume of ice-cold ethyl acetate.
5. Precipitate the Hg-dUTP by adding 4 volumes of absolute ethanol and cooling to $-15°$ for 1 hr. Collect the precipitate by centrifugation and wash twice with ice-cold absolute ethanol and once with cold diethyl ether.
6. Dissolve the mercurated nucleotide in 0.1 M sodium acetate, pH 5.0, and adjust to a concentration of 20 mM based on the millimolar extinction coefficient of 10.2 at 260 nm.

Synthesis of 5-(3-Amino)allyldeoxyuridine Triphosphate (AA-dUTP)

1. Prepare a fresh solution of 2.0 M allylamine by adding 1.5 ml of 13.3 M allylamine (Aldrich, Milwaukee, WI) slowly to 8.5 ml of ice-cold 2 N glacial acetic acid. Neutralize the solution with 5.0 N NaOH.
2. Add 480 μl (960 μmol) of the 2.0 M allylamine solution to 80 μmol of Hg-dUTP (4 ml of a 20 mM solution in 0.1 M sodium acetate, pH 5.0) and 80 μmol of K_2PdCl_4 (Aldrich, 0.64 ml of a 125 mM solution in H_2O).
3. Incubate reaction for 18 hr at room temperature.
4. Filter the reaction to remove the black precipitate, and dilute the filtrate with 5 volumes of H_2O.
5. Load the diluted filtrate onto a 35-ml DEAE-Sephadex A-25 column (1 $\times$ 45 cm) equilibrated with 0.1 M sodium acetate, pH 5.5. Wash with 150 ml of the equilibration buffer before eluting the AA-dUTP with a 180-ml linear gradient from 0.1 M sodium acetate, pH 5.5, to 0.6 M sodium acetate, pH 8.5. Monitor the absorbance of column fractions at 260 nm.
6. Pool fractions containing AA-dUTP (absorbance at 288 nm ~1.6 times absorbance at 260 nm) and precipitate by adding 3 volumes of absolute ethanol. Collect precipitate by centrifugation and dissolve in 3 ml of 0.1 M sodium borate, pH 8.5. Determine the concentration of AA-dUTP using a millimolar extinction coefficient of 7.1 at 288 nm.

Synthesis of Bio-19-SS-dUTP

1. Add 14.5 μmol of sulfo-NHS-LC-SS-biotin to 14.5 μmol of AA-dUTP in 2.7 ml of 0.1 M sodium borate, pH 8.5. Incubate reaction mixture for 2 hr at room temperature.
2. Apply reaction to a 35-ml DEAE-Sephadex A-25 column (1.7 × 21.5 cm) equilibrated in 0.1 M triethylamine carbonate, pH 7.5. Wash column with 2 volumes of equilibration buffer and then elute Bio-19-SS-dUTP with a 180-ml linear gradient of 0.1–0.9 M triethylamine carbonate, pH 7.5. Monitor the absorbance of column fractions at 288 nm.
3. Pool fractions containing Bio-19-SS-dUTP (last major peak to elute between 0.8 and 0.9 M triethylamine) and rotary-evaporate to dryness with several additions of ethanol.
4. Dissolve Bio-19-SS-dUTP in 2.0 ml of 50 mM triethylamine carbonate, pH 7.0, in preparation for its final purification by reversed-phase HPLC.
5. Inject 200-μl aliquots of the partially purified Bio-19-SS-dUTP onto a Bio-Sil ODS-5S column (250 × 4 mm, Bio-Rad, Richmond, CA) equilibrated with 50 mM triethylamine carbonate, pH 7.0, 15% acetonitrile. Bio-19-SS-dUTP is recovered with a retention time of approximately 10 min (flow rate 1.0 ml/min) in this isocratic system. It is the last major peak to elute from the column. Monitor the absorbance of the eluant at 288 nm.
6. Concentrate the Bio-19-SS-dUTP by rotary evaporation and dissolve in 10 mM Tris-HCl, pH 7.5. Store nucleotide at −20°.

Incorporation of Bio-19-SS-dUTP into Replicating SV40
Chromosomes

Bio-19-SS-dUTP is incorporated into replicating SV40 chromosomes in the *in vitro* DNA replication system of Su and DePamphilis.[6] In this system, replicating SV40 chromosomes that have initiated DNA replication *in vivo* continue replication *in vitro*. The system is composed of three components: (1) a nuclear extract containing the replicating SV40 chromosomes, (2) cytosol prepared from uninfected cells, and (3) an assay mixture containing appropriate salts, buffer, nucleotides, an ATP-regenerating system, and Bio-19-SS-dUTP.

[6] R. Su and M. L. DePamphilis, *J. Virol.* **28,** 53 (1978).

Growth of Cells and Infection with SV40

1. Grow CV1 cells (an African green monkey kidney cell line) to confluency in 15-cm plastic culture dishes in Dulbecco's modified Eagle medium (DMEM, 4500 mg glucose/liter) supplemented with 10% fetal calf serum.

2. Pour media from each dish and add SV40 stock (strain Rh 911) diluted to a final volume of 0.2 ml in TS buffer (20 mM Tris, pH 7.4, 137 mM NaCl, 5 mM KCl, 1.0 mM CaCl$_2$, and 0.5 mM MgCl$_2$). The SV40 stock is prepared as described by DePamphilis *et al.*[7] Sufficient virus is used to result in the maximum rate of SV40 DNA synthesis at 36 hr postinfection. This corresponds to 20–40 PFU (plaque-forming units) per cell.

3. Return dishes to CO$_2$ incubator for 1 hr to allow virus to adsorb to cells. Dishes should be agitated frequently during this time to keep the small volume of virus distributed evenly over the cells. After 1 hr, add 20 ml of DMEM supplemented with 20% fetal calf serum to each dish and continue incubation. At 24 hr postinfection, add 2 μCi [^{3}H]thymidine to each dish of cells to radiolabel the mature, nonreplicating SV40 chromosomes. At 36-hr postinfection, the cells are used to prepare the nuclear extract.

Preparation of Nuclear Extract

The nuclear extract is a hypotonic extract of nuclei of SV40-infected cells. It contains both mature, nonreplicating SV40 chromosomes and replicating SV40 chromosomes capable of continuing DNA synthesis *in vitro*. The procedure begins with CV1 cells 36-hr postinfection with SV40, as described above. All steps should be performed in a cold room with ice-cold buffers and glassware. The procedure follows closely that described by Su and DePamphilis[6] with one notable exception: dithiothreitol is eliminated from the hypotonic extraction buffer used in the terminal stages of the procedure in order to preserve the disulfide bond in Bio-19-SS-dUTP.

1. Pour media from dishes into a Clorox solution to inactivate the SV40 virus.

2. Wash each dish 2 times with 15-ml aliquots of TS buffer followed by three 15-ml aliquots of hypotonic buffer (10 mM HEPES, pH 7.8, 5 mM potassium acetate, 0.5 mM magnesium acetate, 2.5 mM sodium bisulfite, and 0.5 mM dithiothreitol).

[7] M. L. DePamphilis, P. Beard, and P. Berg, *J. Biol. Chem.* **250**, 4340 (1975).

3. Wash each dish with 15 ml hypotonic buffer minus dithiothreitol. Each plate should be tilted at a sharp angle to allow residual buffer to drain from the cells. Aspirate as much of the hypotonic buffer as possible from the drained dishes. This step is important to minimize the loss of SV40 chromosomes from nuclei when the cells are initially disrupted.

4. Scrape cells from each dish and transfer to a 7-ml Dounce homogenizer (Kontes Glass Co.). Lyse cells with five strokes with a tight-fitting B pestle.

5. Transfer lysed cells immediately to a prechilled 15-ml Corex centrifuge tube and pellet nuclei by centrifugation at 3000 g for 5 min at 2°.

6. Aspirate the supernatant from the tube and discard. Resuspend the nuclei in 80 μl of hypotonic buffer minus dithiothreitol for every 15-cm dish of SV40-infected cells.

7. Incubate nuclear suspension on ice for 1 hr with frequent, vigorous mixing using a vortex mixing device.

8. Pellet nuclei by centrifugation at 10,000 g for 10 min at 2°.

9. Carefully remove the supernatant. This supernatant contains both mature and replicating SV40 chromosomes and is referred to as the nuclear extract.

Preparation of Cytosol

1. Grow uninfected CV1 cells to approximately 90% confluency. At this time, remove the conditioned media and replace with 20 ml of fresh DMEM supplemented with 10% fetal calf serum. Incubate dishes for an additional 15 hr.

2. Pour media from dishes and wash 3 times with 15-ml aliquots of cold TS buffer followed by three aliquots of hypotonic buffer supplemented to 50 mM potassium acetate and 0.25 M sucrose.

3. Scrape cells from the dish and transfer to a 15-ml Corex centrifuge tube.

4. Pellet cells by centrifugation at 2000 g for 3 min, and resuspend in 0.1 ml of hypotonic buffer minus dithiothreitol per 15-cm dish. Incubate cells for 10 min on ice.

5. Transfer swollen cells to a 7-ml Dounce homogenizer and lyse cells with five strokes using a tight-fitting B pestle.

6. Transfer lysed cells to a 15-ml Corex centrifuge tube and pellet nuclei by centrifugation at 10,000 g for 10 min at 2°.

7. Remove supernatant from the pelleted nuclei and centrifuge a second time at 100,000 g for 1 hr at 2°. This final supernatant is known

as cytosol. It is either used immediately or stored in 100-μl aliquots at $-70°$.

Preparation of Assay Mix

1. Prepare the following 5× assay mixture:

    200 mM HEPES, pH 7.8
    230 mM potassium acetate
    25 mM MgCl$_2$
    10 mM ATP
    250 μM GTP, CTP, and UTP
    250 μM dATP and dGTP
    0.5 μM dCTP
    250 μM Bio-19-SS-dUTP
    100 μg/ml pyruvate kinase (supplied in glycerol, Boehringer-Mannheim Biochemicals, Indianapolis, IN)
    21 mM phosphenolpyruvate

The assay mixture contains the required buffer, salts, nucleotides, and an ATP-regenerating system. The concentration of dCTP is reduced to 0.5 μM to increase the incorporation of [^{32}P]dCTP into DNA. TTP is eliminated from the assay mix and replaced with Bio-19-SS-dUTP. It is important to note that both the nuclear extract and the cytosol contain endogenous deoxynucleotides (both dCTP and TTP). Therefore, it is not possible to accurately calculate the amount of DNA synthesized in this *in vitro* reaction.

The in Vitro DNA Replication Reaction

Individual reactions are constructed on ice as follows:

1. Dry 5 μCi of [^{32}P]dCTP in a siliconized glass tube.
2. Add 100 μl cytosol and gently mix to resuspend the [^{32}P]dCTP.
3. Add 100 μl nuclear extract and mix gently.
4. Add 50 μl of 5× assay mixture and mix gently. The order of addition of the cytosol, nuclear extract, and assay mix is important to prevent precipitation of the SV40 chromosomes.
5. Transfer reaction to a 30° water bath and incubate for 4 min.
6. Stop reaction by placing on ice and adding EDTA to a final concentration of 20 mM. DNA synthesis normally continues in this *in vitro* reaction for 20–30 min. However, the 4-min pulse label, described above, is sufficient to effeciently label the replicating SV40 chromosomes.

7. Remove unincorporated nucleotides from the reaction by chromatographing the terminated reaction on a 10-ml Sephadex G-50 column equilibrated in TE buffer (10 mM Tris, pH 7.5, 1 mM EDTA). Measure the radioactivity in 2-μl aliquots of each fraction (0.1 ml) and pool fractions containing SV40 chromosomes. This step is necessary to remove the excess, unincorporated Bio-19-SS-dUTP, which would otherwise interfere with the subsequent affinity chromatography.

Affinity Purification of Replicating SV40 Chromosomes

The replicating SV40 chromosomes are further purified by affinity chromatography in a streptavidin/biotin–cellulose affinity system. The biotin–cellulose is synthesized by a modification of the procedure described by Porath.[8]

Synthesis of Biotin–Cellulose

1. Dissolve 30 mg of sodium borohydride in 15 ml 1.0 N NaOH.
2. Add 750 mg of cellulose powder (Whatman, Clifton, NJ; CC31, microgranular) and 15 ml 1,4-butanediol diglycidyl ether (Aldrich). Stir gently for 30 min at room temperature. Longer incubation will result in cross-linking of the cellulose.
3. Filter the activated cellulose and wash with approximately 200 ml deionized water. Dry the activated cellulose in the filter apparatus.
4. Prepare a solution of 4.5 ml of 0.2 M sodium carbonate, pH 11.0, and 70 mg biotin hydrazide (Calbiochem-Behring, San Diego, CA). This solution will appear as a flocculent slurry.
5. Add the dried cellulose prepared in step 3 to the solution prepared in step 4. Incubate overnight at 45° with gentle stirring.
6. Filter the biotin–cellulose and wash with 200-ml aliquots of (1) deionized water, (2) 1.0 M NaCl, and (3) deionized water.
7. Resuspend the biotin–cellulose in 30 ml 1.0 M Tris, pH 7.5, and stir gently for 1–2 hr.
8. Filter the biotin–cellulose and wash with 400 ml of TE buffer and with 200 ml TEN$_{50}$ buffer (10 mM Tris, pH 7.5, 0.1 mM EDTA, 50 mM NaCl).
9. Resuspend the biotin–cellulose in 5.0 ml TEN$_{50}$ buffer. Store at 5° until needed.

[8] J. Porath, this series, Vol. 34, p. 13.

This procedure results in the immobilization of 1.0–2.0 nmol of biotin per 50 μl of the final slurry. The degree of modification can be determined using the fluorometric assay for biotin described by Lin and Kirsch.[9]

Streptavidin/Biotin–Cellulose Affinity Chromatography

Streptavidin is a 60-kDa protein composed of four identical subunits. Each subunit is capable of binding biotin with high affinity.[10] Streptavidin is superior to avidin in this application in that it does not bind DNA nonspecifically in solutions of low ionic strength. In addition, when Bio-19-SS-dUTP is bound by streptavidin, the disulfide bond in the linker arm of the nucleotide remains susceptible to reduction by dithiothreitol.[2] In this procedure, an excess of streptavidin tetramers is added to the mixture of SV40 chromosomes and allowed to bind to the incorporated biotinylated nucleotides. This complex is then added to an excess of biotin–cellulose (~ 100-fold of biotin–cellulose versus streptavidin monomer). One of the three remaining biotin-binding sites of the complex binds to the biotin–cellulose and immoblilizes the replicating SV40 chromosomes. In the final step, the replicating chromosomes are eluted from the biotin–cellulose following reduction of the disulfide bond of Bio-19-SS-dUTP by dithiothreitol.

1. Add 5 μl of streptavidin (200 μg/ml) to the pooled Sephadex G-50 column fractions containing the SV40 chromosomes. Incubate at room temperature for 20 min to allow the streptavidin tetramers to bind to the incorporated Bio-19-SS-dUTP.
2. Wash 100 μl of biotin–cellulose (1.0–2.0 nmol immobilized biotin) with 500 μl of TEN$_{50}$. This is conveniently done using a microfilter unit (Schleicher & Schuell, Keene, NH; SS009/0) equipped with a 10-μm hydrophobic filter (Millipore, Bedford, MA; LCWP 025 00). After thorough mixing of the biotin–cellulose with the TEN$_{50}$, centrifugation of the filter unit for 2 min at 1000 g is sufficient to remove the buffer, leaving the washed biotin–cellulose as a loosely packed pellet in the filter unit.
3. Add the streptavidin–SV40 chromosome mixture to the washed biotin–cellulose prepared in step 2. Resuspend the biotin–cellulose by gentle vortex mixing. Incubate for 20 min at room temperature to allow the remaining biotin binding sites on the streptavidin to bind to biotin–cellulose.
4. Centrifuge as before to remove the buffer, soluble proteins, and

[9] H. J. Lin and J. F. Kirsch, *Anal. Biochem.* **81**, 442 (1977).
[10] N. M. Green, *Biochem. J.* **89**, 585 (1963).

nonbound SV40 chromosomes. Measure the radioactivity in an aliquot of the eluted solution to determine the fraction of ^{3}H-labeled mature and ^{32}P-labeled replicating SV40 chromosomes that failed to bind to the biotin–cellulose.

5. Wash the biotin–cellulose with three successive 200-μl aliquots of TEN$_{50}$. The buffer is added to the filter unit, mixed gently to resuspend the biotin–cellulose, and centrifuged immediately. The radioactivity in an aliquot of each wash is measured to determine the SV40 chromosomes removed by this washing procedure.

6. Elute the bound, replicating SV40 chromosomes by incubating the biotin–cellulose with three successive 200-μl aliquots of TEN$_{50}$ supplemented with 50 mM dithiothreitol. The pH of this elution buffer is carefully adjusted to 8.5 to increase the rate of disulfinde-bond reduction by dithiothreitol. Following addition of the elution buffer, the filter unit is gently vortexed to resuspend the biotin–cellulose and incubated at room temperature for 10 min prior to centrifugation.

7. Following the last wash, transfer the biotin–cellulose resin to a scintillation vial and measure the radioactivity to determine the fraction of SV40 chromosomes remaining bound to the affinity resin.

A typical result of the affinity purification procedure described above is shown in Table I. Recovery of 66% of the ^{32}P-labeled replicating SV40 chromosomes, labeled with Bio-19-SS-dUTP, was made in the dithiothreitol washes of the biotin–cellulose column. The remaining replicating chromosomes either failed to bind to the biotin–cellulose (18.8%) or

TABLE I
Affinity Purification of Replicating SV40 Chromosomes

Biotin–labeled nucleotide[a]	Chromosome	Flow-through plus TEN$_{50}$ washes	Dithiothreitol washes	Biotin–cellulose
Bio-4-dUTP	^{3}H-labeled mature	93.2	3.3	3.4
	^{32}P-labeled replicating	8.5	3.6	87.8
Bio-19-SS-dUTP	^{3}H-labeled mature	95.8	2.8	1.4
	^{32}P-labeled replicating	18.8	66.6	14.6

[a] Replicating SV40 chromosomes were pulse-labeled for 4 min with [^{32}P]dCTP and either Bio-19-SS-dUTP or the noncleavable analog Bio-4-dUTP. The chromosomes were then subjected to streptavidin/biotin–cellulose affinity chromatography. The percentage of mature (^{3}H-labeled) and replicating (^{32}P-labeled) SV40 chromosomes eluting in (1) the flow-through plus three successive TEN$_{50}$ washes, (2) three washes with TEN$_{50}$ plus 50 mM dithiothreitol, and (3) that remaining bound to the biotin–cellulose resin following the last dithiothreitol wash is indicated.

remained bound to the affinity resin after the last dithiothreitol wash (14.6%). In contrast, approximately 96% of the ^{3}H-labeled nonreplicating chromosomes failed to bind to the biotin–cellulose. The usefulness of the chemically cleavable Bio-19-SS-dUTP in this application is further demonstrated by comparing these results with those obtained with the noncleavable analog, Bio-4-dUTP {5-[(N-biotinyl)-3-aminoallyl]-2'-deoxyuridine 5'-triphosphate}. In this case, approximately 88% of the replicating chromosomes remained bound to the biotin–cellulose following the last dithiothreitol wash. Thus, the chemically cleavable analog Bio-19-SS-dUTP provides a new approach to the isolation of a population of replicating SV40 chromosomes.

Acknowledgments

I wish to thank Rolf Knippers and Hans Stahl for their help in the initial experiments that led to the development of this procedure. This work was supported by National Science Foundation Grant PCM 8316167.

Section II

Chromatin and Nucleosomes: Analytical Techniques

[5] Sucrose Gradient Techniques and Applications to Nucleosome Structure

By HANS NOLL and MARKUS NOLL

Introduction

Separation and analysis of macromolecules by banding in density gradients is a powerful and versatile tool that in the past 25 years has led to the discovery of new structures and the solution of problems for which no other methods were readily available. Examples are the discovery and characterization of ribosomes[1] and polysomes,[2-5] nucleosomes and polynucleosomes,[6] initiation complexes of DNA replication,[7] of RNA transcription,[8] splicing,[9] and translation,[10,11] and the classical demonstration of the mechanism of semiconservative DNA replication.[12] In recent years, newly developed methods based on molecular sieving in gels by electrophoresis or by filtration at low or high pressure have to a large extent displaced centrifugation. In many applications, however, centrifugation in density gradients, when practiced competently, remains the method of choice. Unfortunately, the potential of the method has not been fully realized in practice, partly out of ignorance of the underlying principles and partly because of inadequate analytical instrumentation. It is one of the paradoxes of the scientific marketplace that sophisticated microprocessor-controlled centrifuges are standard equipment in most laboratories while the equipment for making and analyzing gradients consists in most places of primitive contraptions pieced together from laboratory scrap. The results are the same as if an astronomer would fit his powerful telescope with an inexpensive box camera to examine distant galaxies. In the following, we discuss those aspects of zone velocity centrifugation in sucrose

---

[1] K. McQuillen *et al., Proc. Natl. Acad. Sci. U.S.A.* **45,** 1437 (1959); A. Tissières *et al., Proc. Natl. Acad. Sci. U.S.A.* **46,** 1450 (1960).

[2] A. Gierer, *J. Mol. Biol.* **6,** 148 (1963).

[3] J. R. Warner *et al., Proc. Natl. Acad. Sci. U.S.A.* **79,** 122 (1963).

[4] F. O. Wettstein *et al., Nature (London)* **197,** 430 (1963).

[5] T. Staehelin *et al., Nature (London)* **199,** 865 (1963).

[6] M. Noll, *Nature (London)* **251,** 249 (1974).

[7] H. Maki and A. Kornberg, *Proc. Natl. Acad. Sci. U.S.A.* **84,** 4389 (1987).

[8] Y. Lorch *et al., Cell* **49,** 203 (1987).

[9] A. L. Beyer *et al., Cell* **11,** 127 (1977); Y. N. Osheim *et al., Cell* **43,** 143 (1985); P. J. Grabowski *et al., Cell* **42,** 345 (1985); P. Frendewey and W. Keller, *Cell* **42,** 355 (1985).

[10] M. Noll and H. Noll, *Nature (London) New Biol.* **238,** 225 (1972).

[11] M. H. Schreier and T. Staehelin, *Nature (London) New Biol.* **242,** 35 (1973).

[12] M. Meselson and F. W. Stahl, *Proc. Natl. Acad. Sci. U.S.A.* **44,** 671 (1958).

METHODS IN ENZYMOLOGY, VOL. 170

Copyright © 1989 by Academic Press, Inc.
All rights of reproduction in any form reserved.

gradients that are essential for obtaining optimal results. This information will then furnish the criteria to identify those applications in which this technique is the method of choice. Finally, we give practical examples that illustrate various aspects of the method.

There are two ways of separating macromolecules in density gradients: Centrifugation to isopycnic equilibrium causes the molecules to band according to their density and regardless of their molecular mass, while centrifugation through a gradient of much lesser density than the molecules under study resolves mixtures into bands that move toward the bottom at rates determined by the mass, density, and shape of the molecules. In the following the discussion is limited to zone velocity centrifugation in sucrose gradients.

General Principles

Theory of Sedimentation

Solute molecules exposed to a centrifugal field accelerate until the opposing forces of friction and buoyancy equal the driving force. At a given temperature, the rate at which a particle of density D_p sediments through a medium of density D_m and viscosity η_m in a centrifugal field $\omega^2 r$ is

$$dr/dt = s_{20,w}(\omega^2/A)(D_p - D_m)r/\eta_m \tag{1}$$

The term $A = (D_p - D_{20,w})/\eta_{20,w}$ normalizes measurements with respect to temperature and the viscosity of water at 20°; it is constant for any given D_p. The sedimentation coefficient $s_{20,w}$ is a characteristic of each molecule and, for convenience, is expressed in units of 10^{-13} sec (Svedberg, S). Because s lumps together contributions from such different and often divergent properties as molecular mass, density, and shape, the information it provides is rather ambiguous. However, if any two of these parameters are kept constant within a series, the change in s will reflect the change in one particular property only and thus disclose very precise information about the molecules of interest. According to Eq. (1), the rate at which particles sediment in water increases with the square of the angular velocity; at constant revolutions per minute the rate increases with the distance r from the center of rotation. The frictional resistance depends on the viscosity of the medium and the shape of the molecule: for spheres it is given by Stokes' law, for less compact shapes calculations are more complicated. For particles of similar shape and density the rate of sedimentation depends only on their mass. Larger particles sediment faster because their relative frictional resistance is smaller. The relationship between molecular weight and s for molecules of similar shape and density may be represented as

$$M_r = as^b \tag{2}$$

in which a and b are empirical constants. For spherical particles $b = 1.5$, and since most macromolecules deviate at least somewhat from the shape of a perfect sphere, an exponent of 1.5 must be regarded as a lower limit. The large number of globular proteins give satisfactory fits for $a = 6000$ and $b = 1.64$:

$$M_{r \text{ globular proteins}} = 6000s^{1.64} \tag{2a}$$

while random coils of nucleic acid have been found to obey[13]

$$M_{r \text{ nucleic acids}} = 1100s^{2.2} \tag{2b}$$

Values for b may be determined experimentally from the slope of the straight line obtained when the molecular masses corresponding to the members of a homologous series are plotted against their s values in a log–log plot:

$$\log M_r = \log a + b \log s \tag{3}$$

Conversely, this method has been used to show that s values of particles conforming to this relationship are multiples of a basic repeat unit, as in the case of polysomes[5,14,15] and polynucleosomes.[16] Thus, if the molecular mass of the oligomer M_n is a multiple of n of the monomer M_1, it follows from Eq. (2) that $M_n/M_1 = n = (s_n/s_1)^b$ and

$$\log n = b(\log s_n - \log s_1) \tag{3a}$$

or

$$\log s_n = \log s_1 + 1/b \log n \tag{3b}$$

It should be clear from these considerations that the shape of a molecule has a profound influence on its sedimentation rate. It follows that molecules whose shape is dependent on the environment may sediment at widely different rates. Examples are polyelectrolytes like nucleic acids that unfold if the negative charges are not neutralized by suitable counterions in the solution. A particularly dramatic example is the unfolding of ribosomes after removal of Mg^{2+} with EDTA: the subunits whose RNA are still associated with ribosomal proteins now sediment with the much slower rates of their ribosomal RNA.[17] Hence, changes in sedimentation rate as a

[13] A. Gierer, Z. Naturforsch. B (London) 13B, 788 (1958).
[14] T. Staehelin et al., Nature (London) 201, 264 (1964).
[15] H. Noll, in "Techniques in Protein Biosynthesis" (P. N. Campbell and J. R. Sargent, eds.) Vol. 2, pp. 101–179. Academic Press, London, 1969.
[16] M. Noll and R. D. Kornberg, J. Mol. Biol. 109, 393 (1977).
[17] R. F. Gesteland, J. Mol. Biol. 18, 356 (1966).

function of the ionic environment may often provide valuable information on secondary and tertiary structure, especially when combined with biological data. The potential of the method for the structural analysis of supramolecular complexes and structures is far from being fully exploited. On the other hand, because of this dependence of s values on shape, it is imperative to define the ionic conditions precisely if meaningful comparisons are to be made.

In centrifugation with the analytical centrifuge equipped with schlieren optics, the measured s values are usually extrapolated to zero concentration because the intrinsic viscosity of the high local solute concentration makes a large contribution to the viscosity of the aqueous medium. In sucrose gradients this factor is negligible because of the high viscosity of the sucrose medium. The change in viscosity along the gradient introduces other complications that have been eliminated by the introduction of isokinetic gradients; these aspects are discussed in subsequent sections.

The new dimension that the introduction of sucrose gradients has brought to ultracentrifugation is the ability to separate into discrete bands or zones a mixture of macromolecules and to recover them for further analysis. Without a concentration gradient, the fluid column supporting the material applied to the top would not remain stable during centrifugation. This is because some of the radial lines of force intersect with the walls of the tube, and thus the solvent molecules reflected at the wall would set up a convection current. Even so, some of the solute molecules are lost or retarded at the tube wall, an effect that was found to be dependent on the amount of input, the molecular composition, the distance of migration, and effective mass of the particle.[18]

This wall effect has been studied with a mixture of vacant 70 S ribosomes ("tight couples") and pure 30 S subunits at various input levels. Figure 1A shows the actual gradients and Fig. 1B the calibration curves obtained after comparing the inputs with the recoveries computed by integration of the corresponding peak areas. Liquid columns, into which a soluble gradient has been introduced, are so remarkably stable against mixing because the density gradient restricts the movement of the solvent molecules in the direction of the axis of the gradient. This feature, together with the principle that the density gradient is maintained by the gravitational field, is exploited in the technique of the reorienting gradients. In this technique the buckets, instead of swinging out with the centrifugal force, remain fixed in the original vertical position. During acceleration and deceleration, the gradient reorients itself so as to remain aligned with the field of force. As a result, during centrifugation the gradient, now

---

[18] L. M. Noll, Ph.D. thesis, Northwestern Univ., 1972.

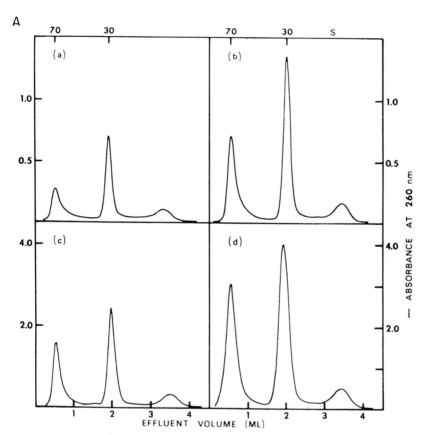

FIG. 1. (A) Wall effect. Various amounts of a mixture of pure vacant ribosome couples (A_{260} = 6.0) and pure 30 S subunits (A_{260} = 6.6) were analyzed on 3.4-ml isokinetic sucrose gradients (C_t = 5%, C_R = 28.8%, V_m = 3.7 ml) after 1.25 hr at 60K rpm and 4°: (a) 0.025 ml, (b) 0.050 ml, (c) 0.100 ml, and (d) 0.200 ml. The pronounced trailing edge of the 70 S peak results from particles traveling along the wall of the tube. (From Noll.[18]) (B, p. 60) Calibration curves for correction of the wall effect. In analogous experiments as described in the legend to Fig. 1A, the loss during centrifugation for various periods (i.e., distance traveled in tube), particles, and inputs was obtained from a comparison of the measured input with the output computed by integration of the corresponding peak areas. The recovery is plotted as a function of the input and the other parameters: 30 S subunits traveled the full (□) or half the distance (●); 50 S subunits traveled the full distance (+); 70 S particles traveled the full (△) or half the distance (○). Full distance is reached under the conditions described in the legend to Fig. 1A. (From Noll.[18])

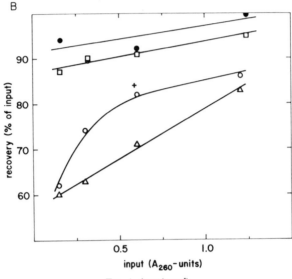

FIG. 1. *(continued)*

perpendicular to the axis of the tube, is compressed according to the ratio of tube length to diameter with a corresponding shortening of the sedimentation path.

One of the most successful applications of sucrose gradients has been the study of macromolecular interactions such as the formation of multicomponent complexes. A dramatic example is the elucidation of the individual steps in the assembly of prokaryotic and eukaryotic ribosomal initiation complexes.[10,11,19] One of the complications encountered here is shifts in the equilibrium of multicomponent complexes produced by the high hydrostatic pressures generated at high angular velocities.[20-23] This may result in the dissociation of the complexes or one of the components. Dissociation usually produces trailings of the lighter component and a final position of the heavier component at a distance intermediate between that of its free and associated forms.[20,23] In cases where the change of free energy would favor pressure-induced association, the effect would be difficult to detect, and no examples have been reported to our knowledge. In general, however, sucrose gradient analysis is a sensitive analytical tool to detect

[19] M. Noll and H. Noll, *J. Mol. Biol.* **89,** 447 (1974).
[20] M. Noll *et al., J. Mol. Biol.* **75,** 281 (1973).
[21] B. Hapke and H. Noll, *J. Mol. Biol.* **105,** 97 (1976).
[22] M. Noll and H. Noll, *J. Mol. Biol.* **105,** 111 (1976).
[23] A. A. Infante and R. Baierlein, *Proc. Natl. Acad. Sci. U.S.A.* **68,** 1780 (1971).

biological complexes because it will not normally disrupt complexes in the biological range of stability ($> 10^{-5}$ mol/liter).

Resolution

An important factor determining resolution is the band width which in turn depends on the molecular weight and concentration of the macromolecules in the sample, the volume and overall density of the sample layer, and the shape of the gradient. To maintain hydrostatic stability, it is necessary to prevent local density inversions, which occur whenever the trailing edge of a band is allowed to reach a higher density than its leading edge.

In a theoretical treatment of this subject, Berman has shown that a band that is initially stable can become unstable during migration.[24] In general, the following rules apply. Increasing the slope of the gradient causes band sharpening since the viscous drag rises enormously with increasing sucrose concentration. This band compression may reduce the band capacity (i.e., the maximum mass of particles that can be carried in a stable band of a given width) to such an extent that density inversion occurs. For this reason, resolution as a rule suffers less when, in order to accommodate a given load, the top zone is increased rather than the sample concentration. Poor resolution is almost always the result of overloading. On the other hand, astonishingly sharp bands can be obtained with very large loads if the volume of the sample layer is expanded and the slope of the gradient increases. Thus, well-resolved polysome patterns were obtained after layering a volume of 1.5 ml containing 10 mg of material over a 3.5-ml gradient.

If the top zones are wider than about 0.25 ml/cm^2, they should be stabilized against convection by means of a gradient prepared by gradually diluting a small volume of sucrose solution with the sample solution. The starting concentration of this sucrose solution must be equal to or slightly less than that of the top of the gradient, and, as it is diluted with sample, it must be carefully layered over the gradient. Thus, the sample is introduced in the form of an inverted gradient. (However, for stability the total density of the top zone must still decrease toward the top.) This measure also protects against density inversion, which would be inevitable if all the particles of a large top zone were allowed to accumulate near the viscosity barrier at the interface between gradient and top zone. In a quick empirical test for the stability of the top layer, a small test tube is filled with sucrose equal in concentration to that of the top of the gradient (C_t). A drop of the solution to be analyzed is allowed to fall into it from about 1 cm above the

[24] S. Berman, *Natl. Cancer Inst. Monogr.* **21**, 41 (1966).

surface. It should bounce back and float; if it sinks, it will be too dense for the top layer of the gradient.

Another limitation of resolution is imposed by the fact that in the most commonly used gradients and rotors, the molecules slow down as they move away from the top layer because the viscous drag of the medium increases more rapidly than the centrifugal force. As a result, the separation between two components fails to improve with sedimentation beyond one-third to one-half of the entire length of the tube, and the potential resolution offered by the available path length remains largely unexploited. To remedy this situation and at the same time facilitate the determination of sedimentation coefficients, isokinetic sucrose gradients were introduced[29] which were calculated to keep the sedimentation rate of particles of a given density constant throughout the length of the tube.

Of practical interest is the question concerning the limit of resolving power defined by the ability to distinguish two components as separate peaks. There is no simple answer to this question because it depends on many variables such as experimental techniques, the number and molecular mass and relative proportions of the components present, input, shape of the gradient, volume of the top layer, and interaction between molecules. However, the following general rules apply. The peak widths are determined by diffusion and the width of the top layer. Diffusion in turn is dependent on time of centrifugation, molecular mass, temperature, and viscosity of the medium. Because the viscosity of sucrose solutions itself is strongly temperature dependent, lowering the temperature restricts diffusion and sharpens peaks by the amplified effects of reducing thermal energy and increasing viscosity. Increasing the rate of spinning reduces the centrifugation time by the square of the angular velocity [Eq. (1)] and thus is nearly always desirable. Reducing diffusion exponentially by linear increases of the sucrose concentration, while very effective, has to be paid for by a corresponding reduction in sedimentation rate. For no better reasons than habit, most workers use linear 15–30% sucrose gradients. However, because high sucrose concentrations and long centrifugation times sometimes are a nuisance, we have routinely used 5% top concentrations and isokinetic gradients for most of our applications.

The influence of the width of the top layer on band broadening is minimal as long as a certain value is not exceeded. Studies with ribosomes revealed that in the 4-ml tubes of the 60K rpm swinging bucket rotors (inside diameter ~9 mm) no loss of resolution was observed with top layers of up to 0.100 ml at 4° and 5% sucrose top concentrations. At 0.200 ml some band broadening was noticeable (Fig. 1A, d).

The assessment of the true effect of these parameters on peak width is dependent on our ability to measure the peaks without introducing experimental band broadening. In practice, this condition is often not fulfilled. If the peaks are measured by the discontinuous collection of fractions, the true peak shape is recorded only if the size of the fractions is small relative to the peak width. In most cases this requires the collection of at least 30 fractions. If the peaks can be detected optically, determination of the true peak shape is approximated by continuous spectrophotometric recording of the gradient as it is pumped out through a flow cell.[5,25] But here again the condition applies that the volume of the flow cell be small relative to the volume of the peak. In addition, the flow must be nonturbulent to prevent excessive mixing and optical noise. Continuous recording, of course, is not practical when the concentration of the molecules of interest must be determined by slower methods, such as measurement of radioactivity, enzymatic activity, or some sort of bioassay. This point is illustrated by the usually considerably broader distribution of the radioactivity as compared to absorbancy when the same peak is measured by the two methods. Very small differences in sedimentation rate can be detected by mixing two components after labeling each with a different isotope,[26] a useful device to analyze the effect of modifications on the macromolecules under study, such as processing, binding of smaller components, or conformational transitions.

Separation of the components also depends, of course, on the distance of travel, provided their initial sedimentation rate does not slow down as it does in most linear gradients. However, the advantage gained by the use of longer tubes is partially offset by a corresponding peak broadening.

Foremost among the factors influencing resolution that cannot be controlled experimentally is the molecular mass. The diffusion-dependent half-width of peaks decreases very significantly as the molecular mass increases. Thus, particles of large mass and minimal individual variations, such as certain viruses, sediment as extremely sharp peaks.

To have a reasonably realistic idea of the resolution that can be expected is of considerable practical importance. Experience has shown that with the techniques and instrumentation described in this chapter, components which differ by 10% in mass and about 7% by sedimentation rate can in most cases be recorded as two distinct, though overlapping, peaks. The estimate for the mass difference assumes, of course, that the molecules have similar density and shape, otherwise their s values may be equal or widely different. This is illustrated by polysome patterns in which up to 11 peaks are resolved (Fig. 2). This estimate of resolving power makes sucrose gradient analysis attractive for many applications, especially in the higher

[25] H. Noll, *Anal. Biochem.* **27**, 130 (1969).
[26] H. Noll *et al., Virology* **18**, 154 (1962).

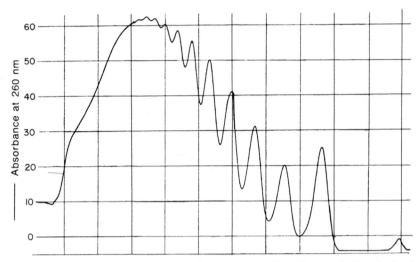

FIG. 2. Sedimentation pattern of mouse liver polysomes. Centrifugation in 4.8-ml tubes in a Spinco rotor SW-65 was for 15 min at 65K rpm and 8° in a nonisokinetic sucrose gradient (concave exponential, $C_R = 0$, $C_m = 1.3\ M$). The scale on the ordinate is linear with concentration and the numbers indicate percentage of full scale. [From H. Noll, *Anal. Biochem.* (Ref. 25), with permission.]

molecular mass ranges where alternative methods are scarce. Other examples are the separation of normal and small heads of phages,[27] the separation of phage particles having extended tail fibers from those in which they are retracted,* or the resolution of linear and circular forms of single-stranded, ϕX174 DNA (Fig. 3).[28] The more than doubled rate of sedimentation observed with the double-stranded RFI form reflects the doubling in mass and a much more compact shape. From such considerations, many other applications may be predicted, for example, the separation of the entire set of human metaphase chromosomes.

Determination of Sedimentation Coefficients

The determination of sedimentation coefficients in sucrose gradients has been largely a matter of guesswork. Very often, grossly erroneous values were derived by linear inter- or extrapolation relative to an internal standard. In order to appreciate the salient points of this problem, a short discussion of some of the quantitative aspects is necessary.

* E. Kellenberger, personal communication.

[27] A. H. Doermann *et al., J. Virol.* **12,** 374 (1973).
[28] M. L. Bayne and L. B. Dumas, *Anal. Biochem.* **91,** 432 (1978).

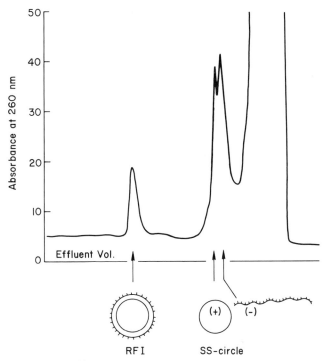

FIG. 3. Separation of ϕX174 DNA in an isokinetic alkaline sucrose gradient: 43 S replicative intermediate (denatured RFI), 18 S and 16 S circular and open single-stranded forms. The scale on the ordinate is linear with concentration and the numbers indicate percentage of full scale. (From B. Hapke and H. Noll, unpublished results; see also Ref. 28.)

In order to determine the sedimentation coefficient of a particle from the distance traveled in a given time, we have to integrate Eq. (1) unless the sedimentation velocity is constant as in the isokinetic gradients. For molecules sedimenting in an aqueous solution, the integration is simple since the force acting on the molecules is proportional to their distance from the rotor center. However, when Eq. (1) is applied to the sedimentation kinetics in a sucrose gradient, the integration becomes complicated, since D_m and η_m are no longer constant but functions of the changing sucrose concentration. To complicate matters further, the change in the medium might alter the shape and density of the sedimenting molecules, e.g., osmotic particles might shrink.

In the case of arbitrary, nonisokinetic gradients, the sedimentation rate is not a linear or other simple mathematical function of time. Hence integration has to resort to complicated mathematical approximation procedures. However, all of these difficulties are avoided by a simple empirical calibration method of determining sedimentation coefficients. A general method that gives fairly accurate values and is applicable to all particles of equal density and osmotic behavior and to any given rotor type and gradient is described below. Although this method requires a valid internal standard of known sedimentation coefficient, it is, unlike the mathematical methods, not dependent on the validity of whatever assumptions are made about the hydrodynamic behavior of the molecules under investigation.

After choosing a certain rotor type and gradient, we calibrate the gradient with a standard which must have the same density as the particles whose sedimentation coefficient we wish to determine. This is done by measuring the distance traveled by the standard as a function of centrifugation time, using sufficient different time intervals to cover the entire length of the tube. We then plot the distance traveled (d) or the effluent volume corresponding to the band center against the product obtained by multiplying the known sedimentation constant (s) with the experimentally chosen centrifugation time (t) and the square of the angular velocity (ω^2).[15]

$$d = st\omega^2 \tag{4}$$

The resulting calibration curve, reproduced in Fig. 4, is characteristic for a given gradient and rotor and, as already mentioned, for molecules of the same density. It usually exhibits a convex curvature that, for a given gradient, becomes less convex with increasing particle density. The s value of the unknown is obtained from the graph by finding the value $st\omega^2$ on the abscissa that corresponds to the experimentally determined distance of migration d_x. Since both the centrifugation time t and centrifugal force ω^2 are known, s can be immediately computed. Sedimentation coefficients of polysome peaks determined by this method were in close agreement with the values obtained in the analytical ultracentrifuge.[15]

As already pointed out, the necessity of constructing empirical calibration curves or the use of complicated mathematical integrations for the determination of s values is eliminated by isokinetic gradients, which, in addition, offer better resolution. To produce constant sedimentation rates over the entire tube length, it is necessary to construct a gradient in which the driving force acting on the particles is exactly compensated by an equivalent increase in the opposing forces of viscous drag and buoyancy. An implicit and attractive feature of such gradients is that the s values of the peaks are proportional to their distance from the top. Similarly, the peak positions, that is, the distance traveled from the top at a given rotor speed, will be proportional to the centrifugation time and hence conveniently predictable.[29] How such gradients are derived and produced by simple devices is described in the following sections.

[29] H. Noll, *Nature (London)* **215**, 360 (1967).

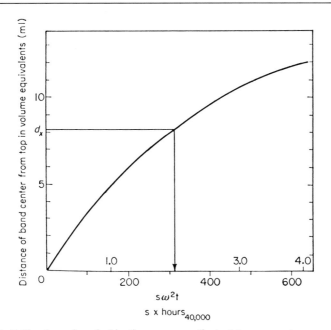

FIG. 4. Calibration of nonisokinetic sucrose gradients. Measurements were carried out with an IEC SB-283 rotor, using *Escherichia coli* ribosomes as calibration standard. The curve describes sedimentation in a convex exponential gradient ($C_t = 0.3\ M$, $C_R = 1.4\ M$, $V_m = 10$ ml). The upper scale of the abscissa represents $s\omega^2t$. The lower scale has been calibrated in units of 10^{-13} sec (Svedberg) at a rotor speed of 40K rpm. Determination of the s value of an unknown ribosomal peak: according to the calibration curve, the distance of sedimentation d_x observed after 2 hr of centrifugation at 30K rpm corresponds to a value on the abscissa of 2.0 (upper scale). Hence $s\omega^2t = 2.0$, where $\omega = 2\pi \times$ rotations/sec, or

$$s = \frac{2}{(2\pi \times 500)^2 \times 7200}\ \text{sec} = 280 \times 10^{-13}\ \text{sec} \qquad \text{or} \qquad 280\ \text{S}$$

If we use the lower scale, which is calibrated for a rotor speed of 40K rpm, a correction for the lower rotor speed of 30K rpm must be made. Thus, 2 hr at 30K rpm is equivalent to 1.125 hr at 40K rpm $[2 \times (30\text{K})^2/(40\text{K})^2]$, and hence $s \times 1.125 = 316$ and $s = 316/1.125 = 280$. [From H. Noll, *in* "Techniques in Protein Biosynthesis" (Ref. 15), with permission.]

Isokinetic Gradients

The equation for isokinetic sedimentation is easily derived by demanding that the second term in Eq. (1), $r(D_p - D_m)/\eta_m$, must remain constant with increasing r.[29] If the initial conditions are defined by the distance r_t of the meniscus from the center of rotation, by the choice of the sucrose concentration C_t at the top of the gradient, and the corresponding values of D_t and η_t, it follows from the constant velocity condition that at any distance r in the gradient

$$\eta_m(r)/\eta_t = r[D_p - D_m(r)]/r_t[D_p - D_t] \tag{5}$$

If we now insert into Eq. (5) the functions $\eta^*(C)$, $D^*(C)$ describing viscosity and density as functions of the sucrose concentrations, we obtain the desired isokinetic gradient

$$\eta^*(C) = \frac{\eta_t}{r_t(D_p - D_t)} r[D_p - D^*(C)] \tag{6}$$

Equation (6) is transcendental and cannot be solved explicitly for C. The isokinetic gradient described by the function $C = f(r)$ may be evaluated by computer, using empirical expressions for the viscosity and density functions. The shape of the gradient is convex and can be approximated by the type of exponential gradient generated by a simple constant volume mixing device consisting of mixing chamber and burette (see below). The parameters for making isokinetic gradients corresponding to most commercial swinging bucket rotors, top concentrations, and temperatures have been determined by computer and published.[15,49] In collaboration with David Calhoun, we have developed a program that can be applied to any rotor (see Appendix).

Apart from the errors introduced in preparing the sucrose solutions, the accuracy of isokinetic gradients depends largely on the equipment used to produce the gradients, the reproducibility of the volume in the gradient tubes, and the uniformity of the tube geometry. Errors may also result from variations in the wall thickness of the plastic centrifuge tubes and from the deformation of tubes and supporting titanium buckets induced by the extremely high forces generated at angular velocities in the 40K to 60K rpm range. Nevertheless, tests with ribosomes and polysomes, using different rotors and performed in different laboratories, showed that the expected constancy of sedimentation velocity held true over the entire tube length provided the isokinetic gradient for the correct particle density was selected. As predicted, the separation between peaks continued to increase throughout the whole length of the tube in contrast to the piling up of the components toward the bottom observed in conventional gradients.

The results of the computer program listed in the Appendix (Table A.I) have been tested with 30 S ribosome subunits from *Escherichia coli*. A constant input was centrifuged for time intervals ranging from 20 to 120 min (Fig. 5A) and the distance of sedimentation measured. When plotted against centrifugation time, no deviation from linearity was observed (Fig. 5B). Note the increase in peak width with centrifugation time discussed above (Fig. 5A).

The use of vertical rotors for the separation of macromolecules in

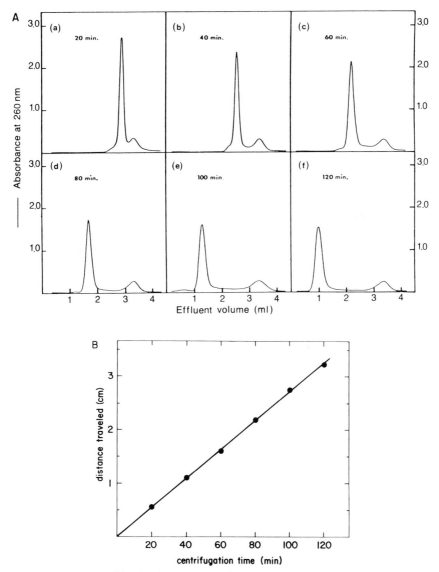

FIG. 5. (A) Test of isokinetic sucrose gradients determined by a computer program. Unwashed 30 S subunits (0.56 A_{260} units) in 0.100 ml buffer E (50 mM HEPES, pH 7.6, 50 mM NH$_4$Ac, 10 mM MgAc$_2$, 10 mM 2-mercaptoethanol) were centrifuged at 60K rpm and 4° for the periods indicated. The isokinetic gradients were the same as described in the legend to Fig. 1A. (From Noll.[18]) (B) Constant velocity of particles sedimenting in isokinetic sucrose gradients. A plot of the data obtained from the experiment illustrated in Fig. 5A is shown. The distance of the 30 S peak from the origin was determined by taking as origin the center of the absorbance peak of 2-mercaptoethanol at the top of the gradient. (From Noll.[18])

isokinetic gradients has been described.[30] In this case, the volume is no longer proportional to the long axis of the tube but is determined by the more complicated geometry produced by the reorientation. Taking this into consideration, the values of the gradient volume $[V = f(r)]$ as a function of the sucrose concentration (C) were obtained from Eq. (6) and, together with the new volume function,[30] used to program a microprocessor-controlled gradient mixer. Procedure and apparatus were tested with polysomes and oligonucleosomes from rat liver.[30] The results were nearly equivalent to similar preparative separations in large swinging-bucket rotors. The advantages are the relatively high capacity and short centrifugation times resulting from the short sedimentation path. However, without numerical solutions or a commercial model of the gradient mixer with the cognate program, the method is of little practical usefulness.

Instrumentation

The equipment needed for experiments discussed in this chapter include (1) a high-speed preparative ultracentrifuge and rotors, (2) devices for preparing sucrose gradients, (3) equipment to remove the gradient from the tube for analysis (tube-puncturing apparatus and pumps), (4) UV-transparent flow-through cells for the spectrophotometric monitoring of protein and nucleic acid bands, (5) fraction collectors, and (6) information storage and processing equipment.

Ultracentrifugation

The most significant event in recent ultracentrifuge technology is the development of high-speed, titanium swinging-bucket rotors. Since the sedimentation rate increases with the square of the angular velocity (ω^2), the new rotors have shortened centrifugation times about 4-fold. This means that a large class of biological particles requires centrifugation times of no longer than 1–2 hr, and instead of one set of analyses in a working day, up to four or more are now possible.

An important feature of the new high-speed rotors is the relatively small volume ($\cong 4$ ml) of the buckets dictated by the necessity to withstand the extreme stress at these enormous centrifugal forces. Because the techniques of collection and analysis widely used for the large volume gradients (12–40 ml) were inadequate for the analysis of 4-ml gradients, it was generally assumed that the resolution obtained in these gradients was intrinsically inferior. Systematic tests with proper techniques and equipment, however, have shown that the resolution is not significantly lower in 4-ml tubes and is adequate for almost all applications. An exception is the

---

[30] H. Merkel and V. Muth, *Biol. Chem. Hoppe-Seyler* **366,** 1097 (1985).

case in which maximal separation of two very close bands is required; in this case, separation is improved by the use of an isokinetic gradient on a longer tube to increase the distance of travel. The use of smaller samples in the 4-ml tubes is both a requirement and an advantage because valuable biological materials that often take many weeks to prepare can be used for a much larger number of experiments, a benefit that analytical instrumentation has always been striving for. Clearly, an instrument that saves time very quickly pays for itself.

For nearly all analytical applications, the 60K rpm rotors with six buckets holding 4-ml tubes are the state-of-the-art equipment. For preparative work with larger inputs (10–1000 mg), high-capacity rotors capable of spinning six 35-ml tubes at speeds of up to 25K rpm are the equipment of choice. However, the cylindrical tube is clearly not the ideal tool to handle large inputs with optimal resolution because the loading capacity is limited by the relatively small volume of the top layer. Much superior are the zonal rotors that can handle much larger volumes at comparable speeds and resolution. The six-bucket rotors holding 12- to 14-ml tubes have a much more limited use.

Modern ultracentrifuges accepting swinging-bucket rotors capable of 60K rpm are all adequate. The preprogrammable microprocessor-controlled instruments are particularly well suited for work with isokinetic gradients because they can be set to a predetermined value of $\omega^2 t$. Hence, the exact position of a known standard can be predicted without having to estimate the contributions of acceleration and deceleration.

Preparation of Gradients

There are three basic types of gradients that represent simple mathematical functions and are easy to prepare with ordinary laboratory equipment, namely, (1) linear, (2) convex exponential, and (3) concave exponential gradients (Fig. 6). All three types of gradients can be produced by continuously adding a sucrose solution of concentration C_R from a reservoir to a mixing chamber containing sucrose solution of concentration C_v, while at the same time displacing the mixture into the gradient tube (Fig. 7). The initial concentration in the mixing chamber ($C_v = C_i$) changes as a function of the volume added and gradually approaches that in the reservoir ($C_v \rightarrow C_R$).

The concentration of sucrose in the mixing chamber changes in a linear fashion as a function of the volume (dV) added from the reservoir if the volume of the mixture, V_m, contracts by the same amount dV_m (Fig. 8). This process is described by the differential equation:

$$\frac{dC_v}{dV} = \frac{C_R - C_v}{V_m} \tag{7}$$

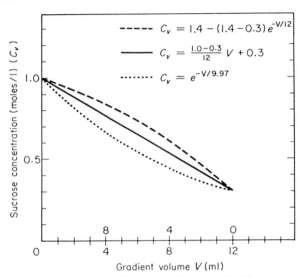

FIG. 6. Most frequently used types of gradients: linear (straight line), convex exponential (dashed), and concave exponential (dotted). The equations are represented by numerical examples for a 12-ml gradient tube. The scales on the abscissa are antiparallel because in the preparation of convex exponential gradients, the lowest concentration enters the tube first (upper scale), while the highest concentration is added first in the case of concave exponential gradients. [From H. Noll, *in* "Techniques in Protein Biosynthesis" (Ref. 15), with permission.]

Integrating and taking into account that $dV = -dV_m$

$$\ln|C_v - C_R| + K = \ln V_m \qquad (8)$$

The value of the integration constant is obtained from the initial conditions, $V_m = V_i$ and $C_v = C_i$,

$$K = \ln \frac{V_i}{|C_i - C_R|}$$

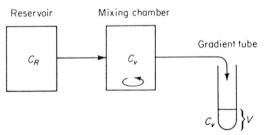

FIG. 7. Flow diagram for the preparation of linear and exponential gradients. [From H. Noll, *in* "Techniques in Protein Biosynthesis" (Ref. 15), with permission.]

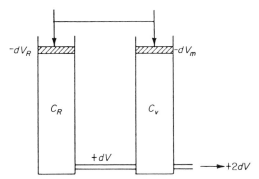

FIG. 8. Diagram illustrating the derivation of the differential equation describing the preparation of linear gradients. [From H. Noll, *in* "Techniques in Protein Biosynthesis" (Ref. 15), with permission.]

and inserted into Eq. (8):

$$\ln|C_v - C_R| + \ln \frac{V_i}{|C_i - C_R|} = \ln V_m$$

Solving for C_v, we obtain

$$C_v = C_R + (C_i - C_R)(V_m/V_i) \tag{9}$$

Equation (9) expresses the concentration in the gradient as a function of the volume remaining in the mixing chamber. It is more convenient, however, to express the concentration as a function of the volume V accumulating in the gradient tube. Thus, since $V_m = V_i - V/2$, we rewrite Eq. (9) so that*

$$C_v = C_R + (C_i - C_R)(1 - V/2V_i)$$

$$C_v = \frac{C_R - C_i}{2V_i}V + C_i \tag{10}$$

The initial concentration in the mixing chamber, C_i, may be either larger or smaller than C_R, the concentration of the solution added from the reservoir. If $C_R > C_i$, then $C_i = C_t$ (concentration at top of gradient), whereas if $C_R < C_i$, $C_i = C_b$ (concentration at bottom of gradient).

* Equation (10) applies strictly only if the volumes in the two vessels in Fig. 9 change at the same rate as, for example, in gradient machines consisting of closed vessels in which the liquid is displaced by means of two pistons (Fig. 8). If two open communicating cylinders are used (Fig. 9), $V_R \neq V_m$ at equilibrium because of the differences in hydrostatic pressure of sucrose solutions of different density. Equation (10) will then take the form

$$C_v = \frac{C_R - C_i}{V_g}V + C_i \tag{10a}$$

in which V_g is the final gradient volume.

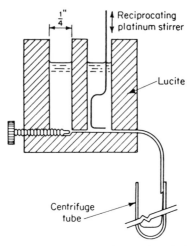

FIG. 9. Linear gradient maker. [From R. J. Britten and R. B. Roberts, *Science* (Ref. 31), with permission.]

If, in contrast to the case of linear gradients treated above, the mixing volume is kept constant during the procedure, an exponential gradient is obtained because as the concentration in the mixing chamber rises (or falls) each successive addition of the more (or less) concentrated sucrose solution causes a smaller amount of change in the mixture. The general form of the differential equation describing exponential gradients is identical to Eq. (7) given above for linear gradients. The integrated form differs, however, because the mixing volume V_m remains constant:

$$\ln|C_v - C_R| + K = -V/V_m \tag{11}$$

Equation (11) describes the concentration of the solution in the mixing chamber as a function of the volume V added from the reservoir or, since $V_m = $ constant, as a function of the volume accumulating in the gradient. Hence, the initial conditions, $V = 0$ and $C_v = C_i$, and the value of the integration constant, $K = -\ln|C_i - C_R|$, inserted into Eq. (11) give

$$\ln\left|\frac{C_v - C_R}{C_i - C_R}\right| = -\frac{V}{V_m}, \quad (C_v - C_R)/(C_i - C_R) = e^{-V/V_m},$$
$$C_v = C_R + (C_i - C_R)e^{-V/V_m} \tag{12}$$

Linear Gradients. The simplest device for producing linear gradients consists of two cylindrical vessels connected by a short piece of tubing at the bottom (Fig. 9) as described by Britten and Roberts.[31] The connection

[31] R. J. Britten and R. B. Roberts, *Science* **131**, 32 (1960).

between the two vessels can be opened or closed by means of a clamp or stopcock. One of the two vessels serves as a reservoir; the other is a mixing chamber fitted with a magnetic stirring bar and an outlet. The two vessels are filled with sucrose solutions of the balanced volumes V_t and V_b, corresponding to the desired top and bottom concentrations of the gradient. The gradient is formed by drawing fluid from the outlet of the mixing chamber. There are two methods for carrying out this procedure, depending on whether the heavy or the light sucrose is put into the mixing chamber.

Method 1: Heavy sucrose in mixing chamber, light sucrose in reservoir. In this arrangement the concentration of the sucrose solution drawn from the mixing chamber decreases at a linear rate. To prevent undesirable turbulence, it is necessary to put the tip of the outlet in contact with the inner wall of the gradient tube near the top so that the fluid is allowed to flow down the side as the tube is filled. This method is not feasible with water-repellent polyallomer tubes because their surface is not wettable.

Method 2: Heavy sucrose in reservoir, light sucrose in mixing chamber. In this method the tip of the outlet, consisting of a capillary pipette, must touch the bottom of the gradient tube near the center so that the light sucrose introduced is allowed to float to the top as fluid of increasing density is fed in. At the end of the procedure, the capillary pipette is carefully pulled out through the gradient. This method is suited for tubes made of water-repellent plastic.

The rate of flow is controlled either by a thumb screw or by the diameter of the outlet opening and should not be too rapid, otherwise it would cause turbulence. In the second procedure, flow rates of up to 2 ml/min are possible, whereas in the first method, slower rates of about 1 ml/min are mandatory to prevent the stream, flowing down the side of the tube, from stirring up the gradient as it hits the surface. This tendency is reduced by tilting the gradient tube.

The hydrostatic pressure of the two fluid columns must be equal so that when the connection is opened there will be no flow in either direction. Balanced volumes (V_t, V_b) corresponding to the top and bottom concentrations are calculated from their densities (d_t, d_b)

$$V_t d_t = V_b d_b \qquad (13)$$

The densities are taken from Fig. 10 in which the density is plotted as a function of the sucrose concentration. The desired gradient volume V_g combines the two unknown volumes of the solutions in the reservoir and mixing chamber:

$$V_g = V_t + V_b \qquad (14)$$

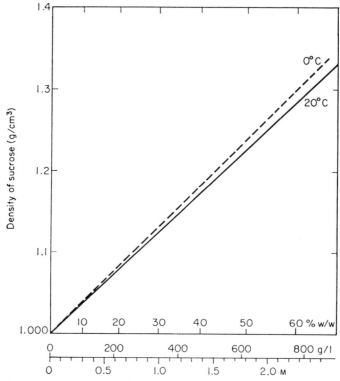

FIG. 10. Density of sucrose solutions as a function of concentration. [From H. Noll, *in* "Techniques of Protein Biosynthesis" (Ref. 15), with permission.]

Combining Eqs. (13) and (14) and solving for V_b, we obtain

$$V_b = \frac{V_g d_t}{d_b + d_t} \tag{15}$$

Since V_g is given, we immediately obtain V_t from Eq. (14).

If gradients are made by letting the gradient maker empty itself, a small portion will usually remain in the mixing chamber and reservoir depending on the particular construction of the gradient maker. To calibrate the device for correct delivery volumes, the volume V_r of this residue must be determined and included in the total volume to be added to the gradient maker:

$$V'_g = V_g + V_r \tag{16}$$

Evidently, the correction depends on whether the light or the heavy sucrose is delivered last. If the desired top and bottom concentrations of the gradient are C_t and C_b, the corrected concentrations (C_t', C_b') and volumes (V_t', V_b') to be added to the gradient maker are computed as follows:

a. Light sucrose in mixing chamber:

$$C_t' = C_t; \qquad C_b' = C_t + (V_g'/V_g)(C_b - C_t)$$
$$V_b' = (V_g' d_t)/(d_b' + d_t); \qquad V_t' = V_g' - V_b'$$

b. Heavy sucrose in mixing chamber:

$$C_t' = C_b - (V_g'/V_t)(C_b - C_t); \qquad C_b' = C_b$$
$$V_b' = (V_g' d_t)/(d_b + d_t'); \qquad V_t' = V_g' - V_b'$$

Example: we wish to prepare 28 ml of a $0.3-1.0\ M$ gradient. The gradient maker has a retention volume of 3.2 ml.

a. Light sucrose in mixing chamber (introduction to bottom of tube):

$$C_t' = C_t = 0.3\ M; \qquad C_b' = 0.3 + (31.2/28)(1.0 - 0.3) = 1.08\ M$$
$$V_b' = (31.2 \times 1.04)/(1.14 + 1.04) = 14.9\ ml$$
$$V_t' = 31.2 - 14.9 = 16.3\ ml$$

b. Heavy sucrose in mixing chamber (introduction to top of tube):

$$C_b' = C_b = 1.0\ M; \qquad C_t' = 1.0 - (31.2/28)(1.0 - 0.3) = 0.22\ M$$
$$V_b' = (31.2 \times 1.04)/(1.14 + 1.03) = 15.0\ ml$$
$$V_t' = 31.2 - 15.0 = 16.2\ ml$$

A variety of linear gradient makers are available commercially. Some make use of the principle of a double-barreled syringe pump instead of the hydrostatic pressure. Perfectly adequate linear gradient makers, of the type shown in Fig. 9, can be constructed easily by gluing the barrels of disposable plastic syringes to a support plate. An important condition, not always observed in commercial products, is that the connection between the two vessels be slightly raised from the bottom to prevent the fluid entering from the reservoir to tunnel through the mixing chamber without adequate mixing.

Exponential Gradients. Concave or convex exponential gradients are formed depending on whether the solution introduced into a mixing chamber of constant volume has a lower or higher concentration than the solution initially present in the mixing chamber. Thus, the same device can be used for making both types of gradients. A simple apparatus consists of a burette connected through an air-tight rubber stopper to a mixing flask fitted with magnetic stirring bar and outlet tube through which the gradient solution is displaced into the centrifuge tube as a corresponding fluid volume is added to the mixing flask (Fig. 11).

A semiautomatic gradient maker that produces an entire set of simultaneous and identical gradients has been described earlier.[15] This apparatus makes use of a multichannel roller pump that withdraws a desired volume from the mixing chamber and distributes it to a set of parallel tubes by means of a manifold. The pump that we have found most satisfactory for this purpose is the DESAGA STA-Multipurpose pump. The use of a precision pump pulling the fluid out of the mixing chamber ensures greatest uniformity and reproducibility of the gradients. The more primitive, burette-operated device, shown in Fig. 11, appears to give satisfactory reproducibility when used for the preparation of gradients in the 3–15 ml range.[29] However, as pointed out by Henderson, if high precision is required such as in the preparation of isokinetic gradients, certain difficulties are encountered in the preparation of gradients of larger volume (20–50 ml). Apparently, the change in hydrostatic pressure in the burette and the compressibility of the air in the space above the fluid level of the mixing chamber cause changes in the flow rates that make it difficult to keep the gradient volume constant within the required limits. Henderson recommends an improved mixing chamber in which the air space has been eliminated (Fig. 12).

The best commercial instrument is the MICO gradient maker, which combines into an integrated unit the DESAGA pump and an air-tight mixing chamber with a fully automatic dispenser set (Fig. 13). The volume of the mixing chamber (V_m) is determined by the position of a piston and can be varied so as to cover the entire volume range required in preparing six simultaneous isokinetic gradients for all commercial centrifuge tubes (20–200 ml). The end of a piece of rubber tubing attached to an outlet at the top of the mixing chamber is dropped into an open vessel serving as a reservoir. Thus, any fluid pumped out through the manifold, connected by tubing to an opening on the side near the bottom of the mixing chamber, is immediately replaced by aspiration from the reservoir.

The dispensing device used to fill the centrifuge tubes consists of six pieces of thin stainless steel tubing mounted in a horizontal bar that slides up and down along two vertical steel posts on ruler bearings. In order to fill the tube, the bar is lowered to a preset position so that the tips of the delivery tubing nearly touch the bottom of the centrifuge tubes. At the positions corresponding to three of the six centrifuge tubes, the bar also contains adjustable sensing pins. A voltage applied between sensing pin and delivery tubing is used to trigger a switch that shuts off the pump as soon as the fluid level reaches the tip of the sensing pin and thus closes the circuit. The tip of each sensing pin is adjusted in height so as to correspond

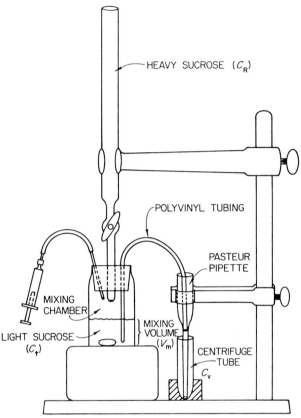

FIG. 11. Exponential gradient maker. The burette contains sucrose solution of concentration C_R and is connected through an air-tight rubber stopper to a mixing flask with magnetic stirring bar. A thin piece of polyvinyl tubing connects the mixing flask with the gradient tube. A small syringe is attached to the mixing chamber through a short piece of tubing to adjust in a reproducible way the amount of pressure at the beginning. The mixing flask has been filled with a sucrose solution of a concentration C_t equal to that desired at the top (or bottom) of the gradient. The procedure is started by increasing the air pressure in the mixing chamber with the syringe until the solution has risen to a fixed point in the ascending portion of the outlet tubing. The tubing between the syringe and mixing flask is then clamped off. The stopcock of the burette is now opened to add the heavy (or light) sucrose solution with rapid stirring. As the volume V added from the burette increases, a corresponding volume of the mixture is displaced into the gradient tube. The end of the outlet tubing from the mixing chamber is inserted into a capillary pipette whose tip must touch the bottom of the gradient tube so that the light sucrose introduced first is allowed to float to the top as solution of increasing density is fed in. At the end of the procedure, the capillary pipette is carefully pulled out through the gradient. For the preparation of isokinetic gradients, the parameters V_m, C_R, and C_t have to be chosen according to rotor dimensions and particle density. (See Appendix.) [From H. Noll, *Nature (London)* (Ref. 29), with permission.]

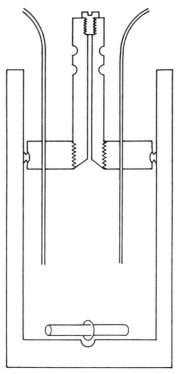

FIG. 12. Cross section of mixing chamber recommended by Henderson. [From A. R. Henderson, *Anal. Biochem.* **27**, 315 (1969), with permission.]

to the volumes of different sets of centrifuge tubes. The sensing pin corresponding to the tube size selected is "activated" by a three-way toggle switch. A plexiglass rack for each set of tubes slides into position on a track below the dispensing device. Its position can be exchanged with that of a collecting tray.

The operation is fully automatic. To prime the system, the mixing chamber (normally filled with distilled water) is filled with sucrose solution of the top concentration (C_t) from the reservoir. The reservoir is then filled with the heavy sucrose concentration taken from the isokinetic gradient table. The pump is now started by pushing the "prime" button on the front panel. It will run for the duration of the prime cycle (37 sec) necessary to displace the dead volume into the collecting tray and to bring the beginning of the gradient to the tip of the dispensing tubes. (Pump speed and connecting tubing are calibrated to give an interval of exactly 37 sec, the time constant of the electronic timer.) After inserting the rack with the

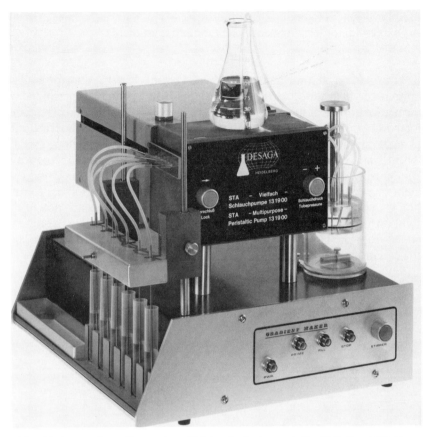

Fig. 13. Automatic gradient maker of Molecular Instruments Co. (P.O. Box 1652, Evanston, IL 60201). (Courtesy Molecular Instruments Co.)

centrifuge tubes and lowering the dispensing bar, the "fill" button is pressed. The pump now fills the gradient tubes until it is shut off by contact of the meniscus with the sensing pin.

Reproducibility of the volumes dispensed within the set of six tubes is remarkably constant: it should be within ±1% (about ±0.5 mm in the position of the meniscus in 4-ml tubes), thus eliminating the need to weigh the tubes. If larger variations occur, the tubing is worn and should be replaced.

Concave exponential gradients. Concave exponential gradients are formed if the initial concentration in the mixing chamber is larger than in the reservoir. Consequently, the concentration of the solution at the bottom of the gradient tube is given by the initial concentration (C_b) in the mixing vessel. Introducing the condition $C_i = C_b > C_R$ into Eq. (12) yields the expression describing concave exponential gradients.

$$C_v = C_R + (C_b - C_R)e^{-V/V_m}$$

It is evident that for a given gradient volume $V = V_g$, and for a given bottom concentration C_b, the top concentration $C_v = C_t$ is determined by the choice of the parameters C_R and V_m. Conversely, for any chosen pair of top and bottom concentrations, there is a family of gradients differing only by their curvature (Fig. 14). Thus, the shape of the curve is determined by the mixing volume: decreasing the mixing volume increases the concave curvature, whereas increasing the mixing volume tends to make the gradient more nearly linear. It is also evident that if the same top concentration is to be reached with a larger mixing volume, the concentration of the solution in the reservoir (C_R) has to be reduced. In practice, the diluent added from the burette is often buffer without sucrose so that $C_R = 0$ and the equation takes the form

$$C_v = C_b\, e^{-V/V_m}$$

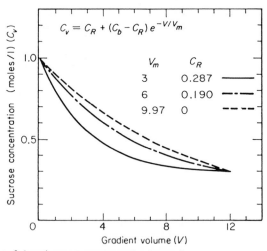

FIG. 14. Effect of changing parameters on the shape of concave exponential gradients with common top and bottom concentrations. With increasing mixing volume, the gradients become flatter. In the example chosen, the mixing volume cannot be increased beyond 9.97 ml if a top concentration of 0.3 M is desired because the sucrose concentration in the reservoir cannot become less than 0. Hence, of all concave exponential 12-ml gradients with a 0.3 M top and 1.0 M bottom concentration, the one represented by the top curve is the shallowest possible. [From H. Noll, in "Techniques in Protein Biosynthesis" (Ref. 15), with permission.]

In this case, V_m is fixed for a given top and bottom concentration, and hence the shape of the gradient is no longer a matter of choice. As already mentioned in the discussion of linear gradients, the solution has to be introduced from the top of the gradient tube because the sucrose concentration is decreasing.

Convex exponential gradients. Convex exponential gradients are formed if the initial concentration in the mixing chamber is smaller than in the reservoir. Thus, the solution first introduced into the gradient tube will form the top concentration, $C_i = C_t < C_R$, and

$$C_v = C_R + (C_t - C_R) e^{-V/V_m}$$

Similar conditions apply as in the discussion of concave exponential gradients. For a given gradient volume V_g and top concentration C_t, the bottom concentration is determined by two variables, namely, the sucrose concentration in the reservoir, C_R, and the volume in the mixing vessel, V_m. Thus, we can produce the same bottom concentration with different combinations of these two variables. It is easy to see that for a given set of top and bottom concentrations, the shape of the gradient is determined by the choice of these two variables. Decreasing the mixing volume increases the convex curvature of the gradients, whereas increasing the mixing volume tends to flatten the gradients (Fig. 15). At the same time, the reservoir concentration has to be decreased or increased so that the stipulated bot-

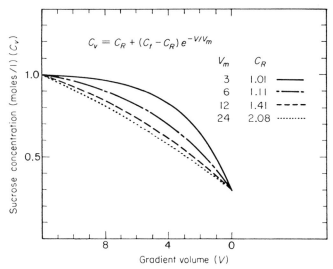

FIG. 15. Effect of increasing mixing volume on the shape of convex exponential gradients. [From H. Noll, *in* "Techniques in Protein Biosynthesis" (Ref. 15), with permission.]

tom concentration can be reached. Since the sucrose concentration increases during filling of the gradient tube, the outlet of the mixing vessel has to be introduced to the bottom of the gradient tube.

Analysis of Gradients

To exploit the analytical potential of zone velocity centrifugation, the methods applied to detect the components in the gradient should match the resolving power of the sedimentation process. This is rarely the case with discontinuous sampling methods which, in addition to the tedium, are subject to the cumulative errors from the many individual manipulations of counting drops and measuring the absorbancy of each fraction. Automatic scanning methods with suitable instrumentation are not only much more rapid and convenient but, in addition, are capable of reproducing the distribution of the macromolecules in the gradient without loss of resolution. In view of these advantages, the costs for a high-resolution gradient analyzing system are moderate, since relatively inexpensive components can be combined with high-performance spectrophotometric equipment that is standard in most laboratories. Since these components are easily disconnected at the end of an experiment, the spectrophotometer remains available for its normal routine use.

An overall view of an early version of such a system is depicted in Fig. 16. The centrifuge tube is seated in a puncturing device mounted on a spectrophotometer in such a way that the puncturing needle is close to the cuvette compartment. After puncturing the tube, the gradient is displaced through a flow cell in the spectrophotometer by pumping distilled water onto the top of the gradient tube. The gradient drains at the negative hydrostatic pressure and constant rate maintained by a pulse-free syringe pump. Negative pressure is necessary because the tube would drain too fast and at a variable rate at atmospheric pressure. Both the flow cell and the tube perforator must be precision engineered in order to meet the stringent requirements of reproducibility and high resolution. The perforator and the turbulence-free flow cell of the Molecular Instruments Company (Evanston, IL), described below, fulfill this condition.

Perforator. The MICO perforator is designed to puncture the gradient tubes through the bottom in a precise and reproducible manner while the contents remain under an air-tight seal (Fig. 17). Reproducibility is ensured by an adjustable needle setting and a level that elevates the needle assembly by an exact distance. During the puncturing operation, air is prevented from entering the gradient by a liquid seal of heavy sucrose.

For most applications, displacement of the gradient through the needle at the bottom (i.e., downward) is preferred because it avoids contamination

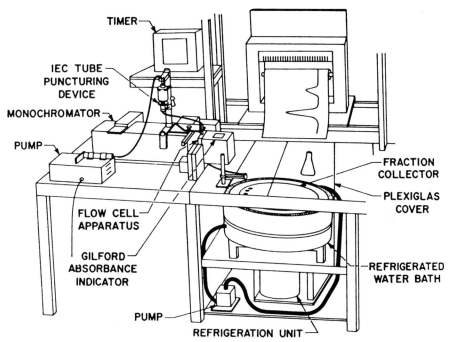

TIMER

IEC TUBE
PUNCTURING
DEVICE

MONOCHROMATOR

PUMP

FRACTION
COLLECTOR

PLEXIGLAS
COVER

FLOW CELL
APPARATUS

GILFORD
ABSORBANCE
INDICATOR

REFRIGERATED
WATER BATH

PUMP

REFRIGERATION UNIT

FIG. 16. Overall view of the high-resolution gradient analyzing system. [From H. Noll, *Anal. Biochem.* (Ref. 25), with permission.]

of the tubing and flow cell with the low molecular components of the top layer that most often contains a large background of absorbancy and/or radioactivity. If the contents are displaced through the top opening (upward), trailing from the top layer owing to laminar flow would obscure bands in the gradient that have low absorbancy and/or radioactivity. Moreover, pumping heavy sucrose solutions is messy and difficult to do at highly reproducible rates.

Other devices on the market introduce needles into the gradient and collect the fluid at atmospheric pressure by suction. In this case, the needle either draws the fluid from the top and is lowered as the meniscus recedes, or the gradient is collected from the bottom by inserting the needle through the gradient so as to place the tip near the bottom. There are several objections to this procedure: (1) suction requires a roller pump between centrifuge tube and delivery tip of the outlet tubing with the risk of mixing and loss of resolution; (2) collection from the top is objectionable for the reasons discussed above; and (3) collection from the bottom by introducing a needle through the gradient causes contamination of the lower regions with material from the top layer as well as some disturbance of the bands.

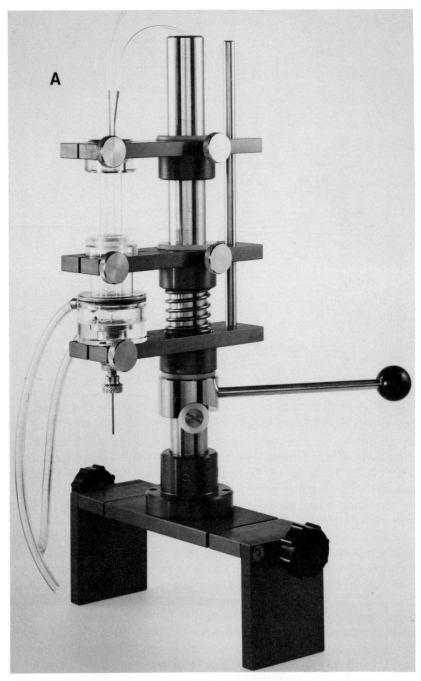

F<small>IG</small>. 17. Perforator of the Molecular Instruments Co. (P.O. Box 1652, Evanston, IL 60201). (A) Two clamps slide on the center post, the upper holds the top cap and the lower the tube holder in position with set screws. The puncturing needle is adjusted in height by a

B

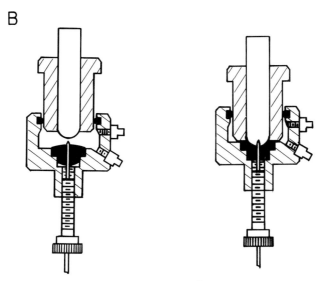

FIG.17. *(continued)*

set screw so that drainage holes of the needle are barely above the rubber seal after puncturing (Fig. 17B). To displace the gradient, distilled water is pumped through an inlet after remaining air has been displaced through the air escape tube. (Courtesy Molecular Instruments Co.). (B) View of puncturing chamber before and after puncturing.

Additional disadvantages of collecting upward are the reversal of the gradient in the descending portions of the tubing and the resulting risk of mixing.

Thus, for most applications, downward collection after puncturing the bottom of the tube is an absolute necessity for obtaining high resolution and sensitivity. This in turn makes it necessary to prevent air from entering the tube during puncturing. This is accomplished by an air-tight top cap that seals the tube with an O ring and by means of a liquid seal at the bottom, a feature found only in the MICO perforator (Fig. 17). In all other devices, puncturing remains a rather hazardous procedure with results that are not reproducible enough for the computerized data processing possible with the MICO system.

Turbulence-Free Flow Cell. In order to monitor the bands of the gradient without loss of resolution and at optimal sensitivity, the volume element in the light path should be smaller than the band width, and the band should pass through with minimal disturbance. Because high resolution and high sensitivity have the conflicting requirements of small cell volume and long light path, a compromise must be reached. Since the signal can be boosted much more efficiently by electronic amplification than by increasing the light path, however, the signal-to-noise ratio is the limiting factor. Optical noise resulting from turbulence is a serious problem with density gradients because even slight disturbances of the laminar flow will produce local changes of refractive index (schlieren), which give rise to fluctuations in the energy of the transmitted light beam. Amplified by the photomultiplier tube, these fluctuations appear as noise on the strip-chart recordings.

Most commercial flow cells are designed for chromatography and hence are not well suited for gradient analysis, especially if they entail a reversal of flow (U-shaped flow path). The MICO turbulence-free flow cell has been specially designed for optimal performance in gradient analysis. It consists of two injection-molded longitudinal sections made from Delrin and held together by eight screws (Fig. 18). The interior cavity is hydrodynamically streamlined; it opens gradually to a 4.1 mm light path between the detachable quartz windows and is completely smooth with no edge or other discontinuities that could produce turbulence. This is illustrated in the sections through the cell in the direction of flow and across (Fig. 19). Around the circular quartz window, the inside walls of the cell taper out into a knife edge that joins the quartz surface tangentially. The volume in

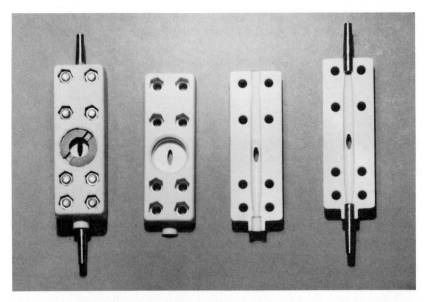

FIG. 18. Turbulence-free flow cell of Molecular Instruments Co. (U.S. Patent No. 3,728,032). Left, assembled cell; middle, one-half viewed from outside or from inside after removal of quartz windows and nipples; right, inside view of cell with nipples connecting to inlet and outlet tubing. (Courtesy Molecular Instruments Co.)

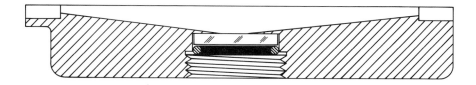

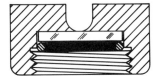

FIG. 19. Cross sections of flow cell shown in Fig. 18 to illustrate step-free contact of windows with streamlined interior surface. (Courtesy Molecular Instruments Co.)

the light path is kept to about 15 μl by an ellipsoid cross section with the long axis (4.1 mm) perpendicular to the direction of flow and parallel to the light beam. The overall outside dimensions of the flow cell are those of a standard quartz cuvette (36 × 12 × 12 mm) and thus permit it to be inserted into standard flow cell or cuvette holders.

Monitoring band separation without significant loss of resolution is no problem as most flow cells have a straight flow-through path if the band volume is severalfold larger than the volume of the cell. This is normally the case with centrifuge tubes of 25 mm diameter (25–35 ml). As the diameter is reduced to 14–16 mm (12–17 ml), some loss of resolution is likely to occur, and with further reduction (3.5–5 ml), both a small cell volume and nonturbulent flow become critical.

Illustrations of the resolution that can be achieved using 3.6-ml gradients and the MICO turbulence-free flow cell are the sedimentation patterns of reticulocyte polysomes (Fig. 20), the separation of the circular plus strand from the linear minus strand of replicative ϕX174 DNA (Fig. 3), and the resolution of the prokaryotic 38 S initiation complex consisting of fMet-tRNA, 30 S subunit, and R17 RNA (Fig. 21) from the free components (30 S subunit and 27 S R17 RNA).[10] In earlier experiments that had failed to separate these latter components, the formation of such a complex had to be inferred from the presence of radioactive fMet-tRNA in the 30 S region. With simply these data, however, one could not have distinguished between the alternative possibilities of fMet-tRNA binding to either 30 S

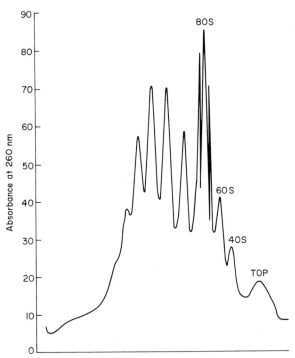

FIG. 20. Sedimentation pattern of rabbit reticulocyte polysomes monitored with the MICO turbulence-free flow cell. A_{260} full scale = 0.5. Convex exponential gradient (3.5 ml); $C_t = 0.5\ M$, $V_m = 2.1$ ml, $C_R = 1.5\ M$. Centrifugation for 0.7 hr at 60K rpm and 4°. [From H. Noll, *Vierteljahrsschr. Naturforsch. Ges. Zuerich* **116**, 377 (1971), with permission.]

subunits or 27 S RNA alone. In fact, it was later shown that fMet-tRNA can bind to 30 S subunits in the absence of R17 RNA.[19] This example further illustrates the power of combining radioactivity measurements with absorbancy scanning at high resolution. The conclusion that the postulated

FIG. 21. Resolution of the 38 S initiation complex from its free components by sedimentation in an isokinetic sucrose gradient. The incubation mixtures contained all the purified components for initiation, except as indicated in the individual graphs. In experiment (e), IF-3 was the only initiation factor added. In (g) the messenger had been treated with formaldehyde. In (h) both GTP and IF-1 plus IF-2 were omitted, and a tRNA mixture, free of GTP, charged with [³H]Met, and formylated, was used; in all others, formylated pure fMet-tRNA from Oak Ridge was the only tRNA added. In (g) the centrifugation time was reduced from 1.7 to 1.2 hr to display the polysomes resulting from the attachment of several 30 S subunits to the formaldehyde-unfolded R17 RNA messenger. [From *Nature (London) New Biol.* (Ref. 10), with permission.]

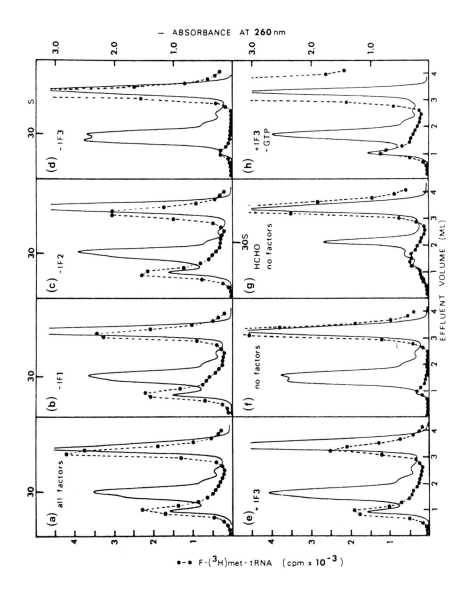

ternary complex had formed was compelling because the fMet-tRNA radioactivity coincided with the new 38 S absorbance peak and because binding of fMet-tRNA to the 30 S subunit alone does not cause a measurable shift of the latter's sedimentation rate. Moreover, by a quantitative evaluation of the 38 S peak area with respect to both radioactivity and absorbancy, it was possible to show that the molar ratios of the three components were close to unity. By the same technique, it was found that conversion of the initiation complex to a 70 S complex required only 50 S subunits and GTP but no initiation factors. Again it was possible to identify the mature initiation complex because it could be resolved optically as a 76 S peak from the vacant 70 S peak because the fMet-tRNA radioactivity coincided with the 76 S not the 70 S peak.[19]

If turbulence-free flow is critical for resolution in monitoring the small volumes of the high-speed gradients, it is even more important for sensitivity, which, as we have seen, is limited largely by the amount of electronic amplification permitted by the signal-to-noise ratio. Optical noise produced by mixing in the cell may be reduced by keeping the flow rate below a critical value which in turn depends on the cell's design. Experiments on the effect of flow rate on resolution with the MICO cell revealed that no significant gain in resolution was observed when the flow rate was reduced to below 0.5–0.6 ml/min for 3.6-ml gradients. The corresponding values are 1.2 and 2.4 ml for 12–16 and 25–35 ml gradients. Testing the sensitivity of the MICO cell with decreasing inputs of R17 RNA sedimenting through 3.6-ml isokinetic sucrose gradients, it was possible to detect a band with as little as 100 ng of RNA input (Fig. 22) at a flow rate of 0.6 ml and a spectrophotometer setting of $0.07A_{260}$ full scale (the maximum amplification factor of the spectrophotometer used for these experiments). Considering the still favorable signal-to-noise ratio, further reduction to 20 or even 10 ng might be possible with greater amplification, for example, by expanding the scale factor of the strip-chart recorder. Thus, with the improvements of turbulence-free flow, the sensitivity can be pushed into regions normally requiring radioactive labeling.

The flow cell is connected to the puncturing needle with silicon rubber tubing, which enters the cuvette compartment of the spectrophotometer through a hole in the cover. In many modern spectrophotometers there are preexisting holes at the top and bottom of the cuvette compartment for flow cell tubings; if not, small holes can easily be drilled for that purpose. It might be necessary to seal the tubing at the entrance to the compartment with about an inch of black insulating tape to prevent stray light from reaching the photomultiplier tube. In any event, the tubing from the puncturing needle to the collecting tube must always go down because any reversal of the gradient would cause mixing.

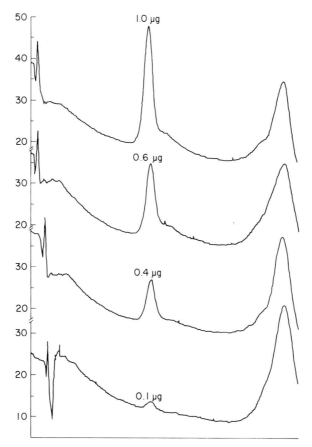

FIG. 22. Test of sensitivity of the MICO flow cell by UV scanning of gradients containing decreasing inputs of R17 RNA ranging from 1.0 to 0.1 μg.

Spectrophotometer. Most convenient for monitoring sucrose gradients are UV spectrophotometers equipped with a monochromatic light system, as are nearly all commercial spectrophotometers. Important requirements are good stabilization of the energy source (usually a deuterium lamp) and sufficient energy at the far UV to be able to monitor in the 225–210–nm range where proteins absorb strongly (1.0 $A_{220} \simeq 100$ μg/ml). Unless the protein inputs are in the milligram range, we recommend monitoring at 220 nm or even lower wavelengths rather than at the traditional 275–280 nm peak associated with aromatic residues. Nucleic acid bands are of course scanned at 260 nm.

Pump. A precision syringe pump is needed to displace the gradient through the flow cell by pumping distilled water onto the gradient at constant speed and under negative pressure. The syringe should hold at least 50 ml, preferably 200 ml. Plastic syringes have the advantages of not breaking and smoother piston action. If the piston advance is not completely smooth and shows the slightest tendency of stepping, the pressure differential produces optical noise which may reach unacceptable levels at high amplification. Reproducibility of the pumping speeds is important and naturally easier to achieve with a series of defined gear ratios rather than a continuously variable drive mechanism. The speeds should range from 0.5 to at least 2.5 ml/min. The drive should be equipped with a clutch so as to permit instantaneous starting and stopping. Without a clutch, the considerable inertia of the system will cause the piston to accelerate and decelerate after switching the power on and off. As a result, the fluid volume delivered per time unit is not constant at the beginning and end of the gradient.

Fraction Collectors. The collection of fractions is necessary when continuous optical scanning is not possible (1) because of the lack of equipment (flow cell, pump, etc.), (2) because the distribution of the molecules in the gradient has to be determined by radioassay methods, or (3) because the fractions must be preserved for further analysis and study.

Determination of sedimentation profiles by collecting equal volume fractions with any time-operated fraction collector presents no problem if the volume of the fraction is large relative to drop size. Drops carried over to the next fraction or lost during changing will cause but minor variations in the fraction volumes collected. However, this is no longer true if the 3.5–4 ml gradients of the high-performance rotors are to be analyzed. In order to obtain acceptable resolution, about 30 fractions will have to be collected, which corresponds to about 0.15 ml per fraction or about 3 drops. Thus, if a drop misses its tube and is added to the next, the first tube will have 0.1 ml and the next 0.2, a difference of 100%. Switching the fraction collector from time to drop counting is also not the answer because the drop size varies with the sucrose concentration, and it is difficult to correlate the volumes collected with the time-dependent travel of the strip-chart recording when optical scanning is carried out at the same time. One solution is to collect fractions by hand by touching the tip of the delivery tubing to the side of the test tube and switching to the next tube according to a stopwatch. As no drops are produced, the method, although tedious, is independent of drop size.

Automation of this procedure is possible if dilution of the samples is acceptable. Thus, an arrangement was designed in which a timer-operated dispensing head flushed the delivery tip with a preset volume of water immediately before collection is advanced to the next tube.[15] This method was found to be extremely useful for the radioassay of gradients when combined with the automatic addition of scintillation fluid. These features have been incorporated into the commercial radiograd fraction collector of the Molecular Instruments Company. A schematic representation of the dispensing head is shown in Fig. 23. It consists of a Teflon block with a vertical bore that holds the needle connected to the tubing from the gradient tube. The lower portion of the needle is surrounded coaxially by stainless steel tubing connected through the rinse channel to the pump delivering distilled water. A separate channel and spout are provided for the pump delivering the reagent. During collection of a gradient for radioassay, a preset volume of scintillation fluid is added to each vial up to a maximum of nine 1-ml shots spaced 0.5 sec apart. The turntable advances immediately after addition of the last shot.

The automated collection of fractions of exactly the same size and the automatic addition of scintillation fluid ensures high resolution and reproducibility, a fact that is reflected in the extraordinary smoothness of the radioactivity curves (Fig. 21). An equally impressive advantage of the automated operation is the enormous saving of time: it takes about 1 hr to scan six gradients for absorbancy and have them ready for counting, distributed into 180 scintillation vials filled with scintillator. To perform the same operation manually would take 3–5 hr.

Operation and Automation. The components described above perform with the precision and reproducibility that is required for integration into a fully automated system. Crucial are the preparation of the gradients, the standardized puncturing of the tubes, and the displacement through the

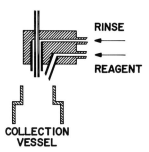

FIG. 23. Schematic of the dispensing head of the MICO radiograd fraction collector. Drops forming at the orifice of the center tube (which is connected to the flow cell and puncturing needle) are washed down by 1 ml of distilled water dispensed through the jacket of the rinse channel immediately before the fraction collector advances to the next vial. Liquid scintillation fluid is dispensed at the same time through the reagent spout. [From H. Noll, *Vierteljahrsschr. Naturforsch. Ges. Zuerich* **116**, 377 (1971), with permission.]

spectrophotometer. The operational arrangement of the components to function as an integrated automated system is shown in the block diagram of Fig. 24.

The process begins with the automated preparation of six identical gradients. After centrifugation, the gradients are analyzed serially. The tube is inserted into the perforator and closed by a top cap held in place by a clamp (Fig. 17). The top cap has an inlet for distilled water from the syringe pump and an air escape tube. The pump is turned on to fill the top of the gradient tube with distilled water, which displaces the air through the short silicon tubing of the air escape. The pump is turned off when the water has risen into the air escape tubing. The flow cell is then flushed out with a syringe filled with sucrose slightly heavier than the bottom concentration of the gradient. The syringe is connected to the outlet tubing below the flow cell compartment of the spectrophotometer (Fig. 16). The heavy sucrose is pushed through the flow cell and on through the puncturing needle until the puncturing needle is flooded. If slight and repeated pressure on the syringe plunger no longer causes a response of the recorder pen, all air has been flushed out of the flow cell. The tube is now punctured slowly and smoothly to prevent mixing by turning the lever that elevates the needle chamber. The air escape tube is kept open to allow equalization of pressure and then closed with a hemostat. If the syringe is then disconnected (leaving the needle attached to the outlet tubing), no fluid will drain from the gradient, unless there is a leak at the top cap or the flow cell. Collection is now started by clutching-in the pump and turning on the chart recorder. After the gradient volume has passed through, the pump clutch is disengaged.

The start of the gradient is usually marked by a blip caused by the passage of the heavy sucrose/gradient interface through the flow cell. Un-

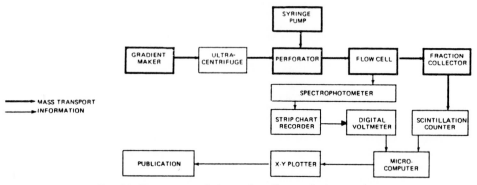

FIG. 24. Components of advanced gradient analyzing system.

less there is already low molecular material present, it is useful to mark the top of the gradient with an absorbance marker (e.g., DNA because of slow diffusion) added after centrifugation but before puncturing. For data processing and analysis, we recommend recording the output of the spectrophotometer via digital converter and microcomputer onto cassette tapes and/or floppy or hard disks. The gradients can then be reproduced by an X–Y plotter in any format desired and without loss of information. Programs may be written that integrate all the peak areas, resolve overlapping peaks into their components, determine s values, etc. In addition, we have greatly benefited from writing a program that plots the radioactivity corresponding to each fraction after typing into the computer the radioactivity values from the scintillation counter. Storing gradients in this form has the advantage that both x and y coordinates, as well as the overall size of the graph, can be chosen at will (for example, to meet specific requirements for publication) while preserving the smooth curves of the original absorbance peaks. Similarly, when plotting the radioactivity into the gradient graph, our program automatically computes the scale factor by asking for the counts per minute value corresponding to 100% ordinate. For example, if the highest counts per minute for any fraction is 3600, we would enter 4000. In some cases, this value may be chosen so as to harmonize with certain absorbance peaks. In any case, the ability to print out in a few minutes and in publishable form the information desired is an enormous saving of time and trouble when compared to the tortuous methods used without this computerized data processing. With the type of equipment described and the system summarized in Fig. 16, we have routinely processed six gradient tubes (one rotor load) in about 1 hr, 20 min with respect to recording of absorbance and collecting of 6×30 fractions for radioactivity measurement (including addition of scintillation fluid). Thus, at the end of 1 hr, the scintillation vials were ready for counting except for the addition of the screw caps, which still had to be done by hand.

Applications*

Discovery of Polysomes and Nucleosomes

A classical application of analysis by zone velocity sedimentation in sucrose gradients is the examination of nucleoprotein complexes suspected of forming repeat structures. By this method the repeat structure of poly-

---

* This section is not intended to be a review of nucleosome structure but rather an illustration of how sucrose gradients as both a preparative and analytical tool contributed to the solution of problems encountered in the study of nucleosomes.

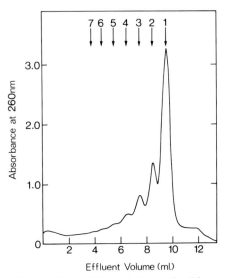

FIG. 25. Sucrose gradient analysis of digested chromatin. Digested chromatin was prepared from rat liver nuclei (1.5×10^8/ml) by incubation with micrococcal nuclease (2 min at 37°, 150 units/ml) followed by pelleting of the nuclei at low speed, resuspension in 0.2 mM EDTA, and homogenization in a Waring blendor. The homogenate was centrifuged for 30 min at 12,000 g, and the chromatin supernatant (5.5 A_{280} units) was layered on a 11.5-ml isokinetic sucrose gradient ($C_t = 5\%$, $C_R = 28.8\%$, $V_m = 9.2$ ml) in 0.2 mM EDTA, pH 7, and centrifuged for 12.75 hr at 28K rpm and 4° in a Beckman SW40 rotor. The absorbancy was recorded by passing the effluent through a MICO flow cell in a spectrophotometer at 260 nm. [From M. Noll, *Nature (London)* (Ref. 6), with permission.]

somes was first discovered[3-5] and, by an exactly analogous experiment, that of nucleosomes 11 years later.[6] Yet, while it was the biological activity of protein synthesis that prompted a closer look at the rapidly sedimenting ribosome structures, the impetus for the experiments with chromatin came from purely structural considerations. Thus, Kornberg, integrating existing data,[32,33] had proposed a fairly precise model for a chromatin repeat unit in which about 200 base pairs (bp) of DNA was associated with two each of the histones H2A, H2B, H3, and H4 and one H1.[34]

Convincing evidence required the demonstration that the postulated complex actually existed as a soluble particle in both its monomeric form and as higher multiples. This was accomplished by zone velocity centrifugation of the soluble chromatin after mild digestion of nuclei with micrococcal nuclease (Fig. 25).[6] As in the case of the polysomes, the seven peaks

[32] D. R. Hewish and L. A. Burgoyne, *Biochem. Biophys. Res. Commun.* **52**, 504 (1973).
[33] R. D. Kornberg and J. O. Thomas, *Science* **184**, 865 (1974).
[34] R. D. Kornberg, *Science* **184**, 868 (1974).

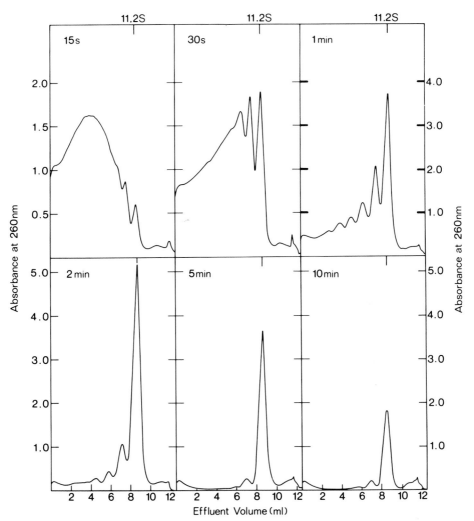

FIG. 26. Time course of micrococcal nuclease digestion: sedimentation analysis of chromatin fragments. Two milliliters of rat liver nuclei (1.5×10^8/ml) was digested with micrococcal nuclease (450 U/ml) and then spun down (5 min at 500 g). The pellet was suspended in 2 ml of 1 mM NaEDTA (pH 7) to lyse the nuclei. After centrifugation for 5 min at 5000 g, the precipitate was discarded, and samples (0.45 ml) of the supernatant were applied to 11.5-ml isokinetic sucrose gradients (as described in the legend to Fig. 25) and centrifuged for 14.25 hr at 30K rpm in a Beckman SW40 rotor at 4°. Analysis of sucrose gradients was as in the legend to Fig. 25. [From M. Noll and R. D. Kornberg, *J. Mol. Biol.* (Ref. 36), with permission.]

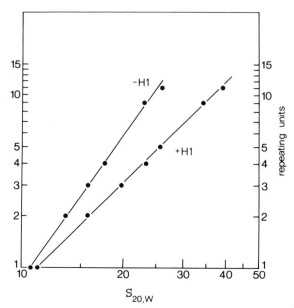

FIG. 27. Log–log plot of number of nucleosome repeating units against sedimentation coefficients of chromatin fragments with and without H1. Chromatin fragments containing 1, 2, 3, 4, and 5 repeating units were prepared as described by Finch et al.[35] Mixtures of chromatin fragments containing DNA of 7 to 11 times 200 bp and 8 to 14 times 200 bp were obtained as cuts from an isokinetic sucrose gradient of a micrococcal digest of nuclei, and the size distributions of the DNA were determined by polyacrylamide gel electrophoresis with HincII restriction fragments of φX174 DNA as calibration markers. Sedimentation coefficients, before and after the removal of H1, were determined on 36-ml isokinetic sucrose gradients. Analysis of histone compositions by sodium dodecyl sulfate–18% polyacrylamide gel electrophoresis and densitometry[48] showed that the removal of H1 was complete but that 8% of the H2B was removed as well. [From M. Noll and R. D. Kornberg, J. Mol. Biol. (Ref. 36), with permission.]

revealed by the strip-chart recording of the continuous optical scan would have been difficult to discover by the discontinuous assay of fractions. The faster sedimenting peaks were shown to be multiples of the 11 S component because the DNA extracted from the isolated peak fractions corresponded in length to multiples of 200 bp.[35] In addition, 11 S monomers were generated at the expense of larger aggregates as digestion progressed (Fig. 26). Finally, it could be shown that when the s values of the monomer and oligomeric peaks were plotted against subunit number (or total molecular mass) in a log–log plot according to Eq. (3a), a straight line was obtained (Fig. 27).[36] Analogous tests had previously been applied in characterizing polysomes.[4,5]

[35] J. T. Finch et al., Proc. Natl. Acad. Sci. U.S.A. 72, 3320 (1975).
[36] M. Noll and R. D. Kornberg, J. Mol. Biol. 109, 393 (1977).

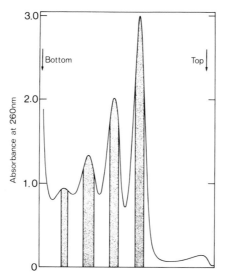

FIG. 28. Purification of oligonucleosomes by separation on preparative isokinetic sucrose gradients. Chromatin (A_{260} = 46) was prepared[46] by digestion of rat liver nuclei (1.5×10^8/ ml) with micrococcal nuclease (110 units/ml, 2 min at 37°). The nuclei were spun down and lysed in one-half the original volume of 0.2 mM EDTA. After centrifugation for 5 min at 4000 g, 0.8 ml was layered on each of five 36.3-ml isokinetic sucrose gradients, as described in the text. [From Finch *et al.*, *Proc. Natl. Acad. Sci. U.S.A.* (Ref. 35), with permission.]

Characterization of the Nucleosome as Chromatin Repeat Unit by Sucrose Gradient Analysis, Electron Microscopy, and Gel Electrophoresis

Further details of the structure of polynucleosomes or chromatin fibers were revealed by combining purification by preparative sucrose gradient fractionation with electron microscopy and electrophoretic analysis of the extracted DNA.[35,37] After mild digestion, chromatin was resolved into monomers, dimers, trimers, and tetramers by two cycles of preparative ultracentrifugation in sucrose gradients. The results are summarized in Figs. 28 and 29. Figure 28 shows the separation in the first gradient, and Fig. 29A shows the four sedimentation patterns resulting from the further purification through a second cycle of gradients of the four collected peak fractions reproduced in Fig. 28. In Fig. 29B are photographs of the polyacrylamide gel electrophoresis (PAGE) patterns of the DNA extracted from the purified peak fractions (Fig. 29A), and Fig. 29C shows electron micrographs of the corresponding fractions. The importance of the purification

[37] M. Noll, *in* "Organization and Expression of Chromosomes" (V. G. Allfrey, E. K. F. Bautz, B. J. McCarthy, R. T. Schimke, and A. Tissières, eds.), pp. 239–252. Dahlem Konferenzen, Berlin; 1976.

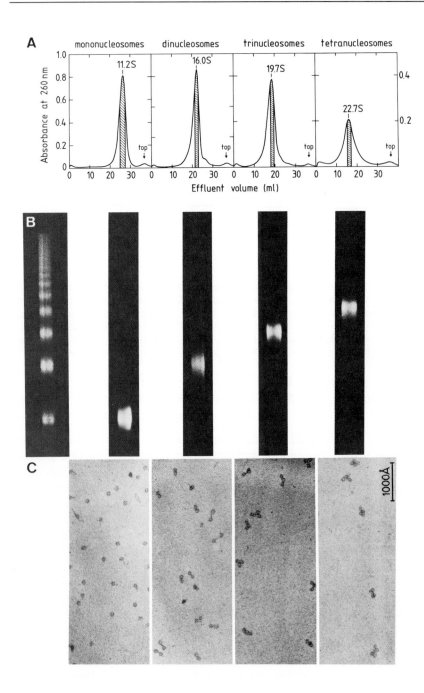

by sucrose gradients in this experiment and others, still to be discussed, is that this method alone was capable of providing components of uniform composition and precise physical chemical definition. This is reflected in the fact that the cognate DNA fragments have an average length corresponding to multiples of 200 bp and, in the electron micrographs, show uniformly shaped, nearly spherical particles that are in close contact and whose numbers are exactly determined by the peak position, one for the monomer, two for the dimer peak, etc.[35]

Other preparations of nucleosomes had yielded electron micrographs and DNA fragments that were much more variable. While some electron micrographs of chromatin were similar to those in Fig. 29C, a form seen frequently in earlier micrographs was that of a string of beads separated by three to four bead diameters, or about 24 nm. The DNA content of these beads, which were obtained after more extensive digestion with micrococcal nuclease and without purification in sucrose gradients, was often as low as 110–120 bp. Since the chromatin fragments purified on sucrose gradients were prepared under the mildest possible conditions, a strong inference was that the picture of the densely packed beads in Fig. 29C represented the state of native chromatin more closely than the widely separated beads on a string. The packing ratio (the ratio of the extended length of DNA to the corresponding length of fiber) is about 7 in the "native" as compared to 2 in the stretched conformation. These results raised the question about the nature of the changes responsible for the wide spacing of the beads seen in some electron micrographs. As we shall see below, sucrose gradient analysis was again essential for the solution of this problem.

The variable configuration of the subunits, shown by the electron micrographs of Fig. 29C, implies a flexible linkage between subunits. When viewed together, these results are consistent with the model of a chain of identical and closely spaced, nearly spherical subunits linked by a flexible region[34] in which the DNA is preferentially cleaved by micrococcal nuclease.

Before turning to an analysis of the transition from the wide to the close

---

FIG. 29. Characterization of oligonucleosomes by (A) sedimentation, (B) DNA size, and (C) electron microscopy. Sedimentation patterns of nucleosomes, dinucleosomes, trinucleosomes, and tetranucleosomes derived by a second cycle of centrifugation from the corresponding effluent fractions of the gradient shown in Fig. 28 (shaded columns). Samples of the peak fractions (shaded columns in A) were analyzed for DNA by electrophoresis (B) and electron microscopy (C) after staining with uranyl acetate.[35] [(A) From unpublished results of M. Noll;[39] (B) and (C) from Finch et al., Proc. Natl. Acad. Sci. U.S.A. (Ref. 35), with permission.]

bead spacing, we have to describe the experimental details of the double sucrose gradient purification used in the experiment above. This technique has proved useful before; the first time was in experiments establishing the length of mRNA per ribosome in a polysome and the coding ratio, i.e., the number of amino acids added to the nascent chain as a ribosome traverses the segment of mRNA extending center to center from one ribosome to the next.[38] In that case, fractions corresponding to polysome peaks and ranging from monomers to tetradecamers were collected and treated with sodium dodecyl sulfate (SDS) to dissociate ^{32}P-labeled mRNA, which was then analyzed for size on a second set of sucrose gradients.

To prepare defined nucleosome oligomers for biochemical and electron microscopy studies, relatively large quantities of chromatin (0.8 ml of $A_{260} = 46$, or about 4 mg) were loaded onto each of five 36.3-ml isokinetic sucrose gradients ($C_t = 5\%$, $C_R = 28.8\%$, and $V_m = 33$ ml) and centrifuged for 22 hr at 27K rpm and 4° in a Beckman SW27 rotor.[35] Fractions corresponding to the peaks from monomer to tetramer were collected as indicated by the vertical columns in Fig. 28, pooled, and dialyzed to remove the sucrose. As apparent from the absorbancy profile of the first gradient in Fig. 28, there is still considerable overlap, especially of the faster peaks. The purity of each peak was determined densitometrically after electrophoresis of the extracted DNA in polyacrylamide gels and found to be 95, 87, 78, and 67% for monomer, dimer, trimer, and tetramer fractions. A 3-ml portion of each dialyzed peak sample was again loaded onto an identical sucrose gradient for further purification. The profiles of the second gradients are shown in Fig. 29A.[39] Densitometric analysis of the corresponding DNA patterns in Fig. 29B showed that cross-contamination in the peak fractions 1–4 of the second gradients (Fig. 29A) had been reduced to less than 5%. The view that the highly purified nucleosomes described above were in the native state was confirmed by the finding that they contained all five histones in the expected stoichiometry.[36,37]

Nucleosome Fine Structure

Nature of Core Particle and Structural Role of H1. A number of other important issues of chromatin structure could only be settled by combining biochemical analysis with the preparative and analytical technique of sucrose gradients. Good examples are the problems related to the nature of the core particles and the structural role of H1. Thus, when the DNA fragments associated with purified 11 S particles after extensive treatment

---

[38] T. Staehelin *et al., Nature (London)* **201**, 264 (1964).
[39] M. Noll, unpublished results, 1974.

of nuclei with micrococcal nuclease were analyzed by electrophoresis, the smallest fragment was 146 bp. Such a "nucleosome core" sediments at 10.6 S as compared to 11.2 S of the intact nucleosome associated with the 200-bp repeat of DNA. Analysis of the core particles for protein by SDS–PAGE revealed the absence of H1, while all other histones were present[36] in the stoichiometry predicted by the octamer core model.[34]

When the time course of digestion by micrococcal nuclease was followed by DNA and protein analysis of the monomer particles, it was found that the reaction paused after approximately 40 bp was removed from the ends.[36] Further digestion of this intermediary 166-bp particle resulted in the loss of H1 and another 20 bp, implying that H1 and the DNA protected by it are more tightly associated with the core particle than the DNA removed initially from the ends. By contrast, H1-depleted chromatin is degraded to core particles nearly 10 times faster and without the formation of a 166-bp intermediate. H1-less polynucleosomes sediment much more slowly than their native counterparts. This is dramatically illustrated in a log–log plot of the number of subunits versus their s values (Fig. 27) according to Eq. (3a). Both sets of s values fall onto straight lines with slopes of 1.92 for the native and 2.78 for H1-depleted fibers.[36] As discussed in General Principles, the slope reflects the change in frictional resistance and approaches a lower value of 1.5 for spherical particles. Hence the shift from 1.92 to 2.78 must be the result of an enormous shape change, as illustrated in Fig. 30 by the transition from the compact structure of packing ratio 7 (**II**) to the open "denatured" form of packing ratio 2 (**I**).[35] This transition is most likely amplified by increased electrostatic repulsion of negative charges previously neutralized by H1 and the low ionic strength

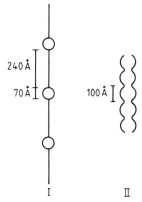

FIG. 30. Schematic models proposed for chromatin structure. [From Finch *et al., Proc. Natl. Acad. Sci. U.S.A.* (Ref. 35), with permission.]

of the sucrose solution (1 mM Na EDTA). Consequently, the DNA between the core particles is expected to be in a greatly extended state.

It should be pointed out that if the change in sedimentation rate between the two forms were merely the result of a change in mass, rather than conformation, the slopes of the two straight lines in Fig. 27 should be the same, i.e., we would have two parallel lines intersecting the abscissa at the s values corresponding to their monomers (10.6 and 11.2 S), as in the case of polysomes of 70 and 80 S ribosomes (Fig. 31). The relatively small difference in sedimentation rate between the 200-bp native monomer and the 146-bp core particle contrasts sharply with the large difference between the corresponding oligomeric forms as, for example, 37 S for the native and 24 S for the H1-depleted decamer. Actually, the 10.6 S core particle sediments more rapidly than would be expected from a mass loss of about 57 kDa or 22% which, according to Fig. 27, would produce a 9.7 S particle if no shape change had occurred. This finding is consistent with the generation of a more compact particle by the removal of the less compact structures that are open to the attack of micrococcal nuclease.

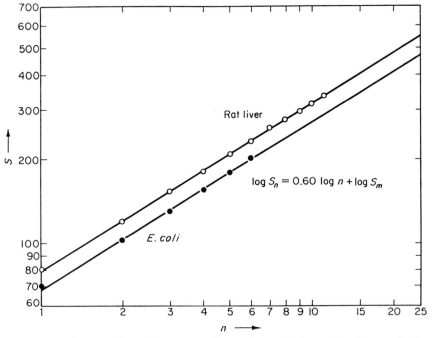

FIG. 31. Linear relationship between sedimentation coefficient and polysome size in log–log plot. [From H. Noll, *in* "Techniques in Protein Biosynthesis" (Ref. 15), with permission.]

Internal Structure of Core Particles Revealed by DNase I Sensitivity.
Experiments with pancreatic nuclease also produced a conversion of
polynucleosomes into single 11 S nucleosomes.[40] This was expected because DNA associated with a histone octamer would be less vulnerable to
enzymatic attack than the DNA of the linker region. However, the DNA of
these monomers, as well as that of higher oligomers isolated from sucrose
gradients, was extensively nicked in a pattern corresponding to sensitive
sites, spaced 10 nucleotides apart.[40] It could be further shown by the
electrophoretic resolution of the fragments into doublets in nondenaturing
gels and by the band shifts produced in denaturing gels after trimming with
S_1 nuclease that the DNase I cleavage sites on opposite strands are staggered by two bases. The generally high resolution required to establish the
details of fine structure discussed above, that is, the ability to resolve the
doublets resulting from staggered cleavage sites, demands careful control of
the experimental conditions. As pointed out previously, purification of the
11 S nucleosomes on sucrose gradients after digestion was essential in these
experiments to prevent contamination with degradation products not specific for nucleosome structure.[40]

In all of the examples discussed so far, isolated nuclei were treated first
with the various nucleases and the nucleosomes from solubilized nuclei
purified subsequently on sucrose gradients. By contrast, in the following
example, the enzymatic treatment with DNase I was performed on sucrose
gradient-purified 146-bp core particles prepared by micrococcal nuclease.
The 5' ends of the two DNA strands in these particles had been labeled
with ^{32}P by the use of polynucleotide kinase in order to determine the size
distribution of the fragments starting at the same fixed point at the end of
the tightly bound 146 bp. At the same time, combining such an experiment with a kinetic analysis of the fragmentation reaction should give
information on such important questions as to whether the DNA is exposed to DNase I every 10 bases, whether the probability of nicks is
uniform or variable over the length of the DNA of the particle, and how
the nicks are located with respect to each other on the two strands.[40]

Below we describe the labeling of the 140-bp nucleosomes.* To ensure
that the labeling procedure did not alter the native structure of the core
particles, the particles were reisolated by a second purification through a
sucrose gradient prior to the analysis of the nicking pattern. As evident

---

* When these experiments were performed, the best estimates for the core particle DNA
length was 140 bp[36] and for the DNase I fragments an average of multiples of 10 nucleotides
implying an uncertainty of ±0.5 nucleotides.[41] Later measurements gave a value of 146 and
10.4 nucleotides, respectively.[42]

[40] M. Noll, *J. Mol. Biol.* **116**, 49 (1977).

from Fig. 32, the core particle sedimented as a homogeneous peak of about 11 S that coincides with all of the radioactivity in the gradient. The autoradiographs of the electrophoretic separations in denaturing gels showed a series of fragments of 11, 21, 31, . . . nucleotides cut from the 5′ end according to $10n + 1$ and another series 14, 24, 34, . . . cut according to $10n + 4$, implying that they correspond to cleavage sites on opposite strands.* Since it has already been established that cleavage sites on opposite strands are staggered by two nucleotides, the new finding would be consistent with a model of the 146-bp core* in which the 5′ ends are protruding by three nucleotides (Fig. 33). The results of the kinetic analysis are also shown in Fig. 33 by indicating the cleavage sites with arrows whose length is proportional to the nicking probabilities. The distribution of these probabilities appears to reflect a 2-fold symmetry of the nucleosome; sites that are about 10, 20, 40, and 50 bp from the ends are most susceptible to pancreatic DNase, whereas a 40-bp region, in the middle of the 146 bp, and sites around 30 bp from the ends exhibit a much lower accessibility. The results are consistent with a model in which the DNA is wound helically around the outside of the histone core (Fig. 34B) with a full turn of about 80 bp.[40]

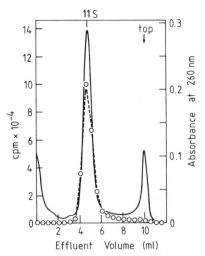

FIG. 32. Sedimentation analysis of 140-base pair nucleosomes labeled at the 5′ ends. The 11 S peak of ^{32}P-labeled 140-bp nucleosomes was collected, and a portion of it was analyzed together with 0.27 A_{260} units of unlabeled 140-bp subunits in an 11-ml isokinetic sucrose gradient. Centrifugation was at 4° for 12 hr at 40K rpm in a Beckman SW41 rotor. Fractions (0.5 ml) were counted by Cerenkov counting in a Packard Tri-Carb scintillation counter. Circles, ^{32}P radioactivity; straight line, absorbance at 260 nm. [From M. Noll, *J. Mol. Biol.* (Ref. 40), with permission.]

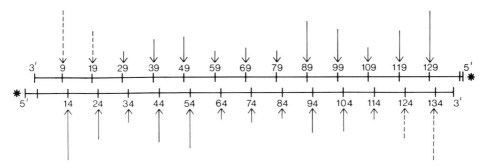

FIG. 33. Probability distribution of nicks at sites exposed to pancreatic DNase in the 140-base pair nucleosome. The accessibility of the DNA in the 140-bp nucleosome, as reflected by the nicking probabilities P_i, is illustrated schematically by the lengths of arrows. The location of the sites exposed to pancreatic DNase is indicated by their distance (number of bases) from the 5' ends of the 140 bp. It is assumed that P_i at the sites 19, 124 and 9, 134 (broken arrows) are the same as at the sites 119, 24 and 129, 14. [From M. Noll, *J. Mol. Biol.* (Ref. 40), with permission.]

In such a model, sites most accessible to pancreatic DNase would be located next to each other separated only by the pitch of the DNA helix. Thus, as nicks every 10 bases are the consequence of the pitch of the double helix, the predominant band of 80 bases would be created by the pitch of the DNA helix in the nucleosome. Two such groups of sites would be present in the 146-bp nucleosome (sites 9, 19, 89, and 99, and 39, 49, 119, and 129, plus corresponding sites on the opposite strand in Fig. 33) separated by three-eighths of a turn. This might reflect an important feature of the nucleosome that could play a crucial part as the site of DNA–protein interaction in higher orders of folding of the chromatin fiber or as binding sites of regulatory proteins.[40] These end-labeling experiments, combined with kinetic measurements of DNase I action, were later repeated and similar conclusions reached, although the results differed in minor details.[43] The more precise determination of the fragment lengths and their distribution revealed variations in the spacing of cleavage sites and the periodicity of fragment lengths that did not allow an unambiguous interpretation beyond the level of resolution indicated by the previous work, summarized in the model of Fig. 34B, because "the exact interpretation of any digestion data will be rather speculative until much more is learned about nucleosome structure."[42]

[41] M. Noll, *Nucleic Acids Res,* **1,** 1573 (1974).
[42] A. Prunell *et al., Science* **204,** 855 (1979).
[43] L. C. Lutter, *J. Mol. Biol.* **124,** 391 (1978).

Labeling of the 140-base pair nucleosomes. Nucleosomes with a reduced DNA size of 140 bp* were produced by degradation from the ends with micrococcal nuclease as described by Noll and Kornberg.[36] The nucleosomes (0.79 A_{260} units) were incubated in 0.40 ml of 50 mM Tris-HCl (pH 8.1), 10 mM MgCl$_2$, 2 mM dithiothreitol with 110 nmol [γ-^{32}P]ATP (145 Ci/mmol) and polynucleotide kinase for 1 hr at 37° to label the 5' ends. Labeled nucleosomes were separated from unreacted [γ-^{32}P]ATP on a 6-ml Sephadex G-50 column equilibrated in 5 mM Tris-HCl (pH 7.4); 70% of the 5' ends were labeled. Peak fractions eluted at the void volume were pooled, loaded together with 0.25 A_{260} units of unlabeled 140-bp nucleosomes on 11-ml isokinetic sucrose gradients, and centrifuged at 4° for 15 hr at 40K rpm in a Beckman SW41 rotor. This 11 S peak was collected and used for digestion with pancreatic DNase. Figure 32 shows that all the radioactivity of this material sediments as 11 S subunits.

Structural Model Derived from Electron Microscopy of Nucleosome Arcs and Cylinders Obtained from Highly Purified Core Particles. The last application of sucrose gradients to be described is the purification of 10.6 S core particles in sufficient quantity and purity for crystallization studies. In this procedure (for details, see below), rat liver nuclei were extensively digested with micrococcal nuclease (to optimize the yield of 146-bp core particles), lysed, and the clarified supernatant loaded onto preparative sucrose gradients for isolation of the 10.6 S peak. This material was concentrated by precipitation with MgCl$_2$, and the material, redissolved in high salt (0.6 M NaCl) to remove traces of H1 and nonhistone proteins, was spun into a pellet through a cushion of 1 M sucrose. The pellet was dissolved and dialyzed against the crystallization buffer.

Both hexagonal crystals and two-dimensional hexagonal associations of nucleosomes were obtained. Most frequently, however, two new types of ordered structures were observed: the cylinders and the arclike arrays of nucleosomes. A detailed examination of these structures revealed interactions between and within nucleosomes, which may be important in maintaining higher orders of chromatin structure of the nucleosome well suited for its multiple physiological tasks (repression, activation, replication).[45] The following picture emerged from a quantitative analysis of electron micrographs: The overall structure of the nucleosome core resembles a wedge-shaped, short cylinder with a mean height of about 60 Å and a

---

* When these experiments were performed, the best estimates for the core particle DNA length was 140 bp[36] and for the DNase I fragments an average of multiples of 10 nucleotides[41] implying an uncertainty of ±0.5 nucleotides. Later measurements gave a value of 146 and 10.4 nucleotides, respectively.[42]

[44] J. T. Finch *et al., Nature (London)* **269,** 29 (1977).
[45] J. Dubochet and M. Noll, *Science* **202,** 280 (1978).

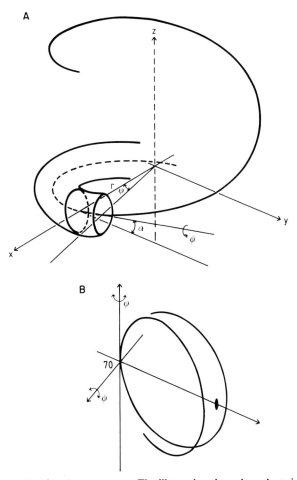

FIG. 34. (A) Helix of nucleosome cores. The illustration shows how the twist angle ψ and the wedge angle ϕ of a nucleosome core are related to the radius r and the pitch angle α of the nucleosomal helix. (B) Model of conformational variability of the nucleosome core. To explain the conformational variability of the nucleosome, a pivot is postulated in the region of the 70th base pair. The two axes of the pivot define with the presumptive dyad of the nucleosome core a Cartesian coordinate system with its origin at the 70th base pair. Rotations around the two axes of the pivot result in changes of the wedge angle ϕ and the twist angle ψ of the nucleosome core. The digaram shows only the 140 bp of the nucleosome core, which have been assumed to form nearly two loops on the outside of the histone octamer.[40,44] [From J. Dubochet and M. Noll, *Science* (Ref. 45), with permission.]

diameter of approximately 110 Å.[44,45] The existence of nucleosome arcs and cylinders indicates that this basic structure of the nucleosome core is not rigid but may show large changes in shape.[45] The model and its conformational variability are shown in Fig. 34.

The 140 bp* form about two loops in two nearly parallel planes. A dyad or pseudodyad passes between the two planes through the 70th base pair where the two loops are joined. Only this region of the DNA that is common to both loops must remain fixed during a shape change of the nucleosomes. The conformational variability is explained by rotations around this pivot (at the origin in Fig. 34B). Thus, change of the wedge angle ϕ alters the angle between the planes of the two loops, which involves a torsion of the DNA at the pivot. A change of the twist angle ψ, on the other hand, is generated by a rotation around an axis nearly perpendicular to these planes. By this movement, the DNA is bent (or kinked) or straightened in the region of the pivot. Another feature of this model is that the largest movements, for a given change of wedge and twist angle, occur opposite to the pivot at the surface of the nucleosome. Thus, the extreme displacements are 26 Å if the change of the twist angle is maximal.[45]

The large shape changes and the nearly continuous spectrum of conformations are unusual and possibly related to the physiological role of the nucleosome. Since the organization of a particular chromatin region depends on its role at a specific time (e.g., during its activation, transcription, replication, polytenization, or recombination), the remarkable conformational variability of the nucleosome could provide the freedom necessary for such a dynamic structure.

Large-scale preparation and analysis of nucleosome cores. Rat liver nuclei were prepared as described.[32,46] Nuclei (1.5×10^8/ml) were digested with micrococcal nuclease (60 units/ml) in 0.34 M sucrose, buffer A (15 mM Tris-HCl, pH 7.4, 60 mM KCl, 15 mM NaCl, 15 mM 2-mercaptoethanol, 0.5 mM spermidine, 0.15 mM spermine), 0.5 mM phenylmethylsulfonyl fluoride (PMSF), and 1 mM CaCl$_2$ for 20 min at 37°. Digestion was stopped by the addition of one-twentieth volume of 0.1 M EDTA, pH 7. The nuclei were sedimented and then lysed in one-third of the original volume of 1 mM EDTA, pH 7. After centrifugation for 5 min at 2000 rpm (Sorval GLC-1), the supernatant was removed and layered on isokinetic sucrose gradients.[35] The analysis of such a gradient after centrifugation (Beckman SW27 rotor) at 4° for 36.5 hr at 26K rpm is shown in the top half of Fig. 35.

---

* When these experiments were performed, the best estimates for the core particle DNA length was 140 bp[36] and for the DNase I fragments an average of multiples of 10 nucleotides implying an uncertainty of ±0.5 nucleotides.[41] Later measurements gave a value of 146 and 10.4 nucleotides, respectively.[42]

[46] M. Noll et al., Science 187, 1203 (1975).

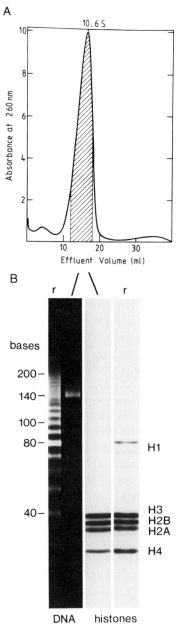

FIG. 35. Large-scale preparation of highly purified nucleosome cores. (Top) Core particles, prepared as described in the text, were purified on a sucrose gradient, and the fractions corresponding to the shaded portion of the 10.6 S peak were concentrated by precipitation with MgCl$_2$ and used for crystallization. (Bottom) Analysis for DNA (left) and protein (right) of these nucleosomes after 5 weeks of association at room temperature. [From J. Dubochet and M. Noll, *Science* (Ref. 45), with permission.]

The material of the monosome peak was collected and pooled from several gradients. The nucleosomes were precipitated with 10 mM $MgCl_2$. After centrifugation for 20 min at 10K rpm (Sorval SS-34 rotor), the nucleosomal pellet was dissolved in 10 mM EDTA, pH 7 ($A_{260} = 80$). Traces of H1 and most nonhistone proteins were washed off the nucleosomes by the addition of two volumes of 0.9 M NaCl, 15 mM Tris-HCl, pH 7.4, 0.75 mM PMSF, and subsequent centrifugation through a 0.5-ml cushion of 1 M sucrose, 0.6 M NaCl, 10 mM Tris-HCl, pH 7.4, 0.5 mM PMSF (Beckman SW50.1 rotor) for 49.5 hr at 50K rpm. The nucleosomes were dissolved at a concentration of $A_{260} = 188$ in 10 mM sodium cacodylate, pH 7.0, 0.2 mM PMSF, equilibrated against this buffer, and used for crystallization or association experiments. Analysis of the DNA and protein components of these nucleosomes after five weeks of association at room temperature is shown in Fig. 35 at bottom left and right, respectively. The DNAs of a DNase I digest of chromatin[41] and histones of native chromatin[46] are shown for reference (r) in the left and right lanes, respectively. DNA analysis was performed in an 8% polyacrylamide slab gel,[47] and histone analysis in an SDS–18% polyacrylamide slab gel.[48]

Appendix

As explained in the Instrumentation section under Preparation of Gradients, convex exponential gradients that are isokinetic may be prepared by adding heavy sucrose of concentration C_R from a reservoir to a constant volume (V_m) mixing chamber containing sucrose of an initial concentration C_t corresponding to the top of the gradient. Once C_t is chosen, the isokinetic gradient is determined by the parameters C_R, V_m, D_p, rotor dimensions, and temperature. Tables giving C_R and V_m for many different rotors, C_t, and temperatures of centrifugation have been published.[15,49] These tables, while generally adequate, do not take into consideration variations in the level to which the centrifuges are filled. In collaboration with David Calhoun of G. D. Searle, we have developed a computer program that provides the same C_R for all rotors by selection of the V_m corresponding to the rotor or centrifuge chosen. The mixing volume V_m is calculated from the known diameter ($2r$) of the tube and the distance R_i of the meniscus from the center of rotation:

$$V_m = 4.02 R_i r^2 \qquad (A.1)$$

[47] T. Maniatis et al., Biochemistry **14**, 3787 (1975).

[48] J. O. Thomas and R. D. Kornberg, Proc. Natl. Acad. Sci. U.S.A. **72**, 2626 (1975); J. O. Thomas and R. D. Kornberg, FEBS Lett. **58**, 353 (1975).

[49] K. S. McCarty et al., Anal. Biochem. **24**, 314 (1968); K. S. McCarty et al., Anal. Biochem. **61**, 165 (1974).

TABLE A.I
ISOKINETIC TABLE

C_t (%)	1.1	1.2	1.3	1.4	1.5	1.6	1.7	1.8	1.9
					C_R[% sucrose (w/v)] for D_p				
4°									
5	19.2 (22)	27.5 (10)	31.5 (6)	34.1 (5)	35.9 (4)	36.8 (3)	37.8 (3)	37.8 (2)	38.7 (2)
10	21.2 (35)	30.3 (13)	34.6 (8)	37.2 (6)	38.9 (5)	39.8 (4)	40.6 (3)	40.6 (3)	41.5 (2)
20	24.6 (138)	36.6 (25)	41.1 (14)	43.4 (9)	45.0 (7)	45.7 (6)	46.5 (5)	46.5 (4)	47.3 (4)
30	—	42.6 (56)	47.6 (25)	49.8 (16)	51.2 (12)	51.9 (10)	52.5 (8)	53.2 (7)	53.2 (6)
40	—	48.0 (172)	54.9 (53)	57.0 (32)	58.2 (22)	58.8 (17)	59.4 (14)	59.4 (12)	60.0 (10)
22°									
5	19.9 (13)	29.5 (6)	34.0 (4)	36.7 (3)	38.5 (2)	39.4 (2)	40.3 (2)	41.3 (1)	42.2 (1)
10	22.1 (20)	32.2 (8)	37.0 (5)	39.6 (3)	41.3 (3)	42.2 (2)	43.1 (2)	44.0 (2)	44.0 (1)
20	25.5 (67)	37.9 (14)	43.2 (8)	45.5 (5)	47.1 (4)	47.9 (3)	48.7 (3)	49.4 (2)	50.2 (2)
30	—	43.8 (29)	49.6 (14)	52.3 (9)	53.7 (7)	54.4 (5)	55.1 (4)	55.8 (4)	55.8 (3)
40	—	49.0 (79)	56.7 (27)	59.2 (16)	60.4 (11)	61.6 (9)	61.6 (7)	62.2 (6)	62.8 (5)

If a gradient maker is used that feeds six tubes simultaneously from one mixing chamber (Fig. 13), Eq. (A.1) has to be multiplied by 6:

$$V_{\mathrm{m}} = 24.1 R_{\mathrm{i}} r^2 \tag{A.2}$$

The constants R_{i} and r can easily be measured or taken from the manufacturer's manual, which usually lists the distance from the upper edge of the tube to the center of rotation, so that we have to add the few millimeters from the meniscus of the gradient (before adding the top layer) to the edge of the tube.

In Table A.I, the sucrose concentrations (w/v) in the reservoir (C_{R}) are listed for particle densities, D_{p}, ranging from 1.1. to 1.9 g/cm^3 and for top concentrations (C_{t}) covering the useful span from 5 to 40%. The values are given for two temperatures (4° and 22°). The numbers in parentheses are very rough estimates of the relative sedimentation rate in arbitrary units. It is immediately evident that C_{R} depends strongly on particle density, somewhat less on C_{t}, and very little on the temperature. Hence, temperature control is not very critical for maintaining isokinetic conditions; however, it strongly influences the rate of sedimentation. Plotting C_{R} against either C_{t} or D_{p} further shows that for intermediary values not tabulated, linear interpolations are permissible. The same applies to the values for temperature.

Acknowledgments

This work was supported by Grant No. 6M38984-01 of the National Institutes of Health (H.N.) and No. 3.348-0.86 of the Swiss National Science Foundation and by the Kanton Basel (M.N.). Hans Noll is a lifetime Career Professor of the American Cancer Society.

[6] Electrophoretic Analyses of Nucleosomes and Other Protein–DNA Complexes

By Sue-Ying Huang and William T. Garrard

Introduction

Electrophoresis has proven to be a powerful resolving technique for both analytical and preparative biochemical experiments. Electrophoretic resolution of ribosome and polysome components was successfully

Copyright © 1989 by Academic Press, Inc.
All rights of reproduction in any form reserved.

achieved in 1969 by Dahlberg *et al.*[1] Thus, following the discovery of the subunit structure of chromatin in 1974, we decided to develop one- and two-dimensional electrophoretic techniques to address questions of nucleosome heterogeneity, structure, and function.[2] Independently, two other groups reported the successful electrophoretic separation of nucleosomes as early as 1976.[3,4] The techniques described here are routinely used in the authors' laboratory and represent the evolution of methods over a 10-year period. These procedures led to the discovery and solution of nucleosome heterogeneity,[5-7] the elucidation of independent binding sites on nucleosomes for histone H1 and HMG proteins,[5,6] the description of nuclease-mediated processing events between poly- and mononucleosomes,[7] the determination of the protein composition of nucleosomes in specific genes[8,9] or methylated DNA domains,[10] and the localization of the nucleosomal origins of a spectrum of cross-linked chromosomal proteins.[11,12] These techniques are also directly applicable to the gel retardation assay to study the interaction of any sequence-specific binding protein with DNA.[13] In addition, many of the methods described here may be useful for the electrophoretic analyses of ribonucleoprotein complexes.

This chapter discusses the following topics: (1) one-dimensional electrophoretic resolution of nucleosomes; (2) enzymatic preparation of chromatin and chromatin fractionation; (3) two-dimensional electrophoretic resolution of DNA and detection of specific sequences; (4) two-dimensional electrophoretic resolution of proteins using Triton–acid–urea gels; and (5) further applications for electrophoretic analyses.

[1] A. E. Dahlberg, C. W. Dingman, and A. C. Peacock, *J. Mol. Biol.* **41,** 139 (1969).

[2] R. D. Todd and W. T. Garrard, *J. Biol. Chem.* **252,** 4729 (1977).

[3] A. Varshavsky, V. V. Bakayev, and G. P. Georgiev, *Nucleic Acids Res.* **3,** 477 (1976).

[4] A. L. Olins, R. D. Carlson, E. B. Wright, and D. E. Olins, *Nucleic Acids Res.* **3,** 3271 (1976).

[5] P. P. Nelson, S. C. Albright, J. M. Wiseman, and W. T. Garrard, *J. Biol. Chem.* **254,** 11751 (1979).

[6] S. C. Albright, J. M. Wiseman, R. A. Lange, and W. T. Garrard, *J. Biol. Chem.* **255,** 3673 (1980).

[7] R. D. Todd and W. T. Garrard, *J. Biol. Chem.* **254,** 3074 (1979).

[8] A. H. Davis, T. L. Reudelhuber, and W. T. Garrard, *J. Mol. Biol.* **167,** 133 (1983).

[9] S.-Y. Huang, M. B. Barnard, M. Xu, S.-I. Matsui, S. M. Rose, and W. T. Garrard, *Proc. Natl. Acad. Sci. U.S.A.* **83,** 3738 (1986).

[10] D. J. Ball, D. S. Gross, and W. T. Garrard, *Proc. Natl. Acad. Sci. U.S.A.* **80,** 5490 (1983).

[11] T. L. Reudelhuber, T. Boulikas, and W. T. Garrard, *J. Biol. Chem.* **255,** 4511 (1980).

[12] T. Boulikas, J. M. Wiseman, and W. T. Garrard, *Proc. Natl. Acad. Sci. U.S.A.* **77,** 127 (1980).

[13] D. M. Crothers, *Nature (London)* **325,** 464 (1987).

One-Dimensional Electrophoretic Resolution of Nucleosomes

Background

The procedure recommended here for one-dimensional electrophoretic separation of nucleosomes cannot be underemphasized as it forms the basis for all further analyses employing second-dimensional gels. The system consists of a low ionic strength, EDTA-containing buffer and a 3.5% polyacrylamide–0.5% agarose gel that contains 30% glycerol. We arrived at this by systematically testing the effects of ionic strength, pH, gel composition, and glycerol concentration on particle resolution and the extent of sample entry into gels.

The low ionic strength, EDTA-containing buffer eliminates selective precipitation of chromatin subfractions that occurs in the presence of divalent metals and/or at moderate ionic strengths. The low ionic strength is also necessary to reduce histone H1 and HMG protein dissociation and exchange,[6,14] while the EDTA inhibits the activity of traces of micrococcal nuclease and endogenous nucleases that may remain weakly bound to nucleosomes even during electrophoresis. The presence of 30% glycerol reduces the dielectric constant of the medium and further stabilizes weak protein–DNA interactions[7] (see also Refs. 15–17). The presence of agarose stabilizes gels that contain low concentrations of polyacrylamide (e.g., 3.5%), allowing optimal resolution of monosomes while still maintaining reasonable resolution of larger oligomers. It has been our experience that higher percentage polyacrylamide gels are less desirable; nucleosome bands are broader and less discrete because DNA length heterogeneity contributes more to nucleoprotein mobility under these conditions.

Theoretically, electrophoresis should fractionate nucleosomes based on their net negative charge, mass, and conformation. The molecular basis of electrophoretic separation of nucleosomes under the conditions described here has been unequivocally established to be due primarily to particle mass, but conformational effects are also detectable.[6] Native mononucleosomes are resolved into distinct electrophoretic forms that possess well-defined major protein compositions, whose electrophoretic fractionation primarily depends on whether histone H1 and/or HMG proteins are associated with DNA–histone octamer complexes.[5,6] Slight variation in particle mobility can also occur owing to the phasing frame of associated

[14] F. Caron and J. O. Thomas, *J. Mol. Biol.* **146**, 513 (1981).
[15] K. Gekko and S. N. Timasheff, *Biochemistry* **20**, 4667 (1981).
[16] K. Gekko and S. N. Timasheff, *Biochemistry* **20**, 4677 (1981).
[17] G. C. Na and S. N. Timasheff, *J. Mol. Biol.* **151**, 165 (1981).
[18] W. Linxweiler and W. Horz, *Nucleic Acids Res.* **12**, 9395 (1984).

DNA.[18] Other conditions of electrophoresis that employ a higher ionic strength buffer have provided evidence for separation of nucleosomes based on their content of acetylated histones[19] and are necessary to study the cooperative binding of HMG proteins to nucleosomes.[20]

Reagents

Acrylamide and bisacrylamide (Bio-Rad Laboratories, Richmond, CA)

Agarose (Bio-Rad)

Ammonium persulfate (Bio-Rad)

TEMED (Bio-Rad)

Glycerol (Sigma, St. Louis, MO)

Stock Solutions

Solution A, 62.5× nucleoprotein gel electrophoresis buffer: 400 mM Tris base, 200 mM sodium acetate, 20 mM disodium EDTA, pH adjusted to 8.5 with acetic acid (store at 4°)

Solution B, 5.7× acrylamide gel solution: 20% acrylamide, 1% bisacrylamide (store at 4° in a dark container)

Solution C, 1.3× acrylamide plug solution: 20% acrylamide, 1% bisacrylamide, 40% glycerol (store at 4° in a dark container)

Solution D, 5000× gel staining solution: 5 mg/ml ethidium bromide (store in light-shielded container at room temperature)

Preparation of Gel

The following description applies to the formation and running of a vertical slab gel 16 × 14 × 0.3 cm using the apparatus shown in Fig. 1.[21] The recipes and running conditions presented here and in following sections should be appropriately modified to accommodate other apparatuses. First form a 15% polyacrylamide plug at the gel bottom to prevent the potential sliding out of the running gel during electrophoresis. Mix at room temperature 5.7 ml solution C, 0.12 ml solution A, 1.7 ml water, 15 μl TEMED, and 75 μl 20% (w/v) ammonium persulfate (freshly prepared).

Rapidly pour this solution into the bottom of the unit.

After polymerization (usually <5 min), cast the running gel on top of the supporting plug as follows. Weigh out 28.4 g of glycerol into a 250-ml

[19] J. Bode, M. M. Gomez-Lira, and H. Schroter, *Eur. J. Biochem.* **130**, 437 (1983).
[20] A. E. Paton, E. Wilkinson-Singley, and D. E. Olins, *J. Biol. Chem.* **258**, 13221 (1983).
[21] It has been our experience that the 0.3 cm thickness gives best resolution when combined with 0.8-cm-wide wells. We have been less successful employing thinner gels with equivalent or narrower sample wells.

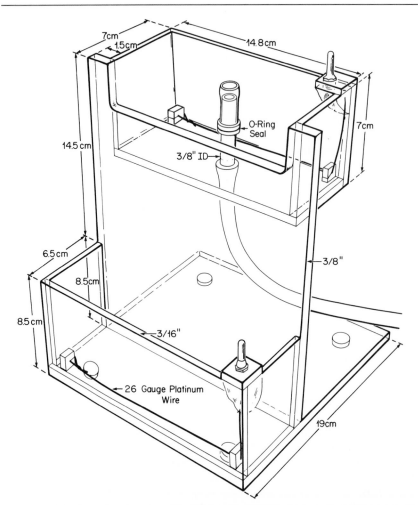

FIG. 1. One-dimensional vertical slab gel apparatus. During the run, electrophoresis buffer is pumped from the lower tray to the upper tray. The upper tray has an adjustable O-ring drain that provides a gravity feed to redistribute buffer to the lower chamber.

Erlenmeyer flask and pipette into this flask 2.0 ml water, 13.1 ml solution B, and 1.2 ml solution A. Mix and place in a 55° water bath. Prepare a second 250-ml Erlenmeyer flask containing 375 mg agarose and 18.75 ml water. Melt the agarose using a microwave oven, and place this second flask also at 55°. After the solutions in both flasks become equilibrated to 55°, pour the flask containing acrylamide into the agarose solution and mix. Then add 60 μl TEMED and mix. Finally, add 60 μl of 20% ammonium persulfate (freshly prepared), mix, and rapidly pour the solution into the gel casting unit on top of the polyacrylamide plug. Submerge a 10-tooth Teflon comb with 0.8-cm-wide sample wells (Bio-Rad) into the gel top. After polymerization (usually at least 1 hr at room temperature), the sample well comb is removed, and the wells are cleaned of bits, rinsed, and finally overlayed with 1.6% solution A, 30% glycerol. Assemble the casted gel in the running apparatus and fill the upper and lower buffer trays with 1.6% solution A. Equilibrate the gel to 4° in a cold room. Take care to avoid air bubbles at the gel bottom.

Sample Loading and Electrophoresis

Immediately before loading samples, remove the overlay solution and rinse the sample wells with tray buffer using a Pasteur pipette. Nucleoprotein samples should be in 50% glycerol–1 mM EDTA, 0.02% bromphenol blue, pH 7.4, at a concentration of 1–5 mg/ml (as DNA). (The tracking dye is optional). For preparation of the $S1$ and $S2$ chromatin fractions, see the later section in this chapter entitled, "Enzymatic Preparation of Chromatin and Chromatin Fractionation." For optimal resolution, the amount loaded depends on the number of different electrophoretic components. For a sample containing mainly mononucleosomes, 5–15 μg (as DNA) per lane gives the best resolution ($S1$ chromatin). For a sample containing a spectrum of N-mers, 30 μg (as DNA) works well ($S2$ chromatin). Loading volumes of about 20 μl per lane are recommended. Capillary pipettes with a mechanical micropipettor provide one convenient way for sample loading.

Connect the power leads with the positive pole at the bottom. Running time is 110 mA-hr at constant current (overnight, e.g., 12 hr at 9 mA) at 4°. An outer lane containing 10 μl of 50% glycerol, 0.1% bromphenol blue tracking dye is beneficial to follow the progress of electrophoresis; the dye runs about 1–2 cm into the gel plug under these conditions, while mononucleosomes run about 9 cm. *Note:* It is imperative to recirculate the buffer between the upper and lower trays during electrophoresis at a rate of about 10 ml/min because pH gradients rapidly form.

Gel Staining, Fluorography, and Autoradiography

Immediately after the run, carefully remove the gel from between the glass plates and peel off the 15% acrylamide plug. Place the gel in a plastic tray, and add 1 liter of ethidium bromide staining solution consisting of 16 ml solution A, 0.2 ml solution D and 984 ml water. Shake the gel gently for about 45 min at room temperature. Although gels can be directly visualized under UV light and photographed at this point, for best results

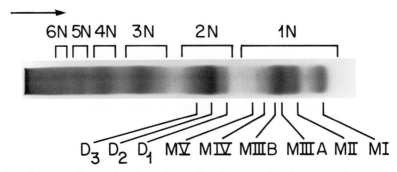

FIG. 2. Nomenclature of different electrophoretic forms of nucleosomes. Electrophoretic separation of a partial micrococcal nuclease digest of mammalian chromatin (the *S2* chromatin fraction). A negative print of an ethidium bromide-stained gel is shown. The positions of mono- and polynucleosomes are indicated (1N–6N) as well as different electrophoretic classes of disomes (D_1–D_3) and monosomes (MI–MV).

destain about 45 min with 1.6% solution A. Short-wave UV transillumination (Ultraviolet Products, Inc., San Gabriel, CA) gives the greatest sensitivity for visualization but can cause damage to components if further analyses are desired. For photography with a Polaroid MP4 camera using type 55 film, we use *f*4.5, 10–30 sec exposures, and a Kodak Wratten yellow gelatin filter No. 15.

For fluorography, soak the gel in 250 ml of liquid EN³HANCE [New England Nuclear (NEN), Boston, MA] for 1 hr at room temperature with agitation. After soaking the gel in 2% glycerol for an additional 1 hr, dry the gel onto Whatman 3 MM paper using a slab gel dryer (Bio-Rad) set below 60°. For autoradiography, omit the impregnation with fluor. Exposures against preflashed Kodak XAR-5 film are performed at −70°, using a DuPont Cronex Lightning-Plus AG intensifying screen when necessary.[22,23]

Nomenclature and Properties of Different Electrophoretic Forms of Nucleosomes

Figure 2 illustrates the pattern obtained upon electrophoretic separation of a partial micrococcal nuclease digest of mammalian chromatin (*S2* chromatin). Dinucleosomes (2N) are resolved into three overlapping electrophoretic forms, termed D_1, D_2, and D_3,[24] while mononucleosomes

[22] R. A. Laskey and A. D. Mills, *Eur. J. Biochem.* **56**, 335 (1975).
[23] R. Swanstrom and P. R. Shank, *Anal. Biochem.* **86**, 184 (1978).
[24] S. C. Albright, P. P. Nelson, and W. T. Garrard, *J. Biol. Chem.* **254**, 1065 (1979).

(1N) are resolved into five partially overlapping electrophoretic species, termed MI through MV. These mononucleosome classes were initially defined by two-dimensional electrophoresis, which is necessary to achieve their complete resolution (Fig. 3).[2] The subunit structures of major components within each of these classes are known (Table I). Defined subsets of MII to MV can be reconstituted by reassociating pure histone H1 and/or HMG17 to stripped nucleosomes (Fig. 4). By doubling the length of gels and the running time, even linker-trimmed mononucleosome class MI

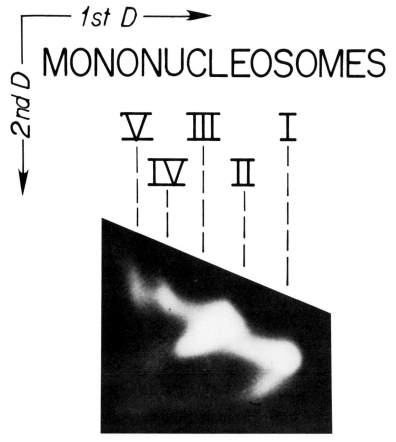

FIG. 3. Two-dimensional map of mononucleosomes. DNA fragments of electrophoretically separated mononucleosomes were displayed by electrophoresis in the presence of 0.1% sodium dodecyl sulfate (SDS) in the second dimension. Ethidium bromide was used for staining. The positions of different electrophoretic forms of mononucleosomes (MI–MV) are indicated. (Reproduced, with permission, from Ref. 7.)

TABLE I
PROPERTIES OF DIFFERENT ELECTROPHORETIC FORMS OF NUCLEOSOMES[a]

Nucleosome class and electrophoretic form nomenclature	R_f^b	Mass proportion within class[c] (%)	DNA length range (bp)	Protein composition of major components within class	
				Moles HMG14 or 17 per particle	Moles histone H1 per particle
Disomes					
D_3	0.57	20	350–390	ND[d]	2
D_2	0.62	55	320–370	ND	2
D_1	0.68	25	295–330	ND	1
Monosomes					
MV	0.76	<2	190–220	2	1
MIV	0.81	<2	185–210	1	1
MIIIB	0.84	<10	145–185	2	0
MIIIA	0.86	60–90	160–210	0	1
MII	0.91	<10	140–185	1	0
MI	1.00	<10	140–175	0	0

[a] Data pertain to nucleosomes of mammalian cells (see Refs. 2, 5–7, and 24).
[b] Naked DNA has the following R_f values (relative to MI as 1.00) for mono-, di-, tri-, and tetranucleosomal fragments, respectively: 1.33, 1.00, 0.79, and 0.68 (see Fig. 5).
[c] The mass proportion changes during digestion owing to nuclease-mediated interconversion of different electrophoretic forms.[7]
[d] ND, Not determined.

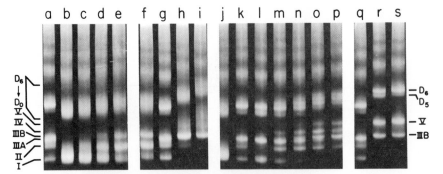

FIG. 4. Electrophoretic patterns of reconstituted nucleosome components. The S2 chromatin fraction was depleted of histone H1 and nonhistone proteins and reassociated with either purified HMG17 or purified histone H1 (or both) by step-gradient dialysis. Resulting reconstitutes were separated by electrophoresis, and gels were stained with ethidium bromide. Samples: a, g, k, and q, nondepleted control chromatin; b and j, depleted chromatin; c–f, h, and i, depleted chromatin reassociated with 0.25, 0.50, 0.75, 1.0, 2.0, and 3.0 mol of HMG17/mol of nucleosome, respectively; l–p, r, and s, depleted chromatin reassociated with 0.7 mol of histone H1/mol of nucleosome in the presence of 0, 0.25, 0.50, 0.75, 1.0, 2.0, and 3.0 mol of HMG-17/mol of nucleosome, respectively. MI to MV and D_0 to D_6 refer to different electrophoretic forms of mono- and dinucleosomes, respectively. (Reproduced, with permission, from Ref. 6.)

(nucleosome core particles) can be fractionated into three components, termed MIa, MIb, and MIc. Second-dimensional protein gel analysis reveals that these components possess 0, 1, and 2 copies of ubiquitinated histone molecules per octamer, respectively (see Fig. 8).[9,25] Besides these analyses, the assembly of restriction fragments into nucleosomes can also be assayed by gel electrophoresis (Fig. 5) (Ball and Garrard, unpublished results, 1984).

Comments on Nomenclature

It should be emphasized that even though the subunit structures of major components within each mononucleosome class are known, each class is compositionally heterogeneous. For example, the binding of any protein of approximately 12,000 molecular weight to MI nucleosomes would be predicted to generate MII nucleosomes. For this reason, it is misleading to designate MII as MN_{HMG} as have Varshavsky and co-workers.[26]

Preparative Gels: Electroeluting and Concentrating Nucleosomes

A slab gel is cast as described above, but the sample well comb is inserted upside down so as to generate a single, broad sample well. $S2$ chromatin is loaded at a density of $60-80 \mu g/cm$ for 0.3-cm-thick gels. After electrophoresis, cut strips (<1 cm wide) from the gel center and outside edges and stain with ethidium bromide. Cut gels on glass plates using a knife with an even up and down motion. Store the remaining gel pieces at 4° in Saran Wrap. Do not stain these with ethidium bromide because native nucleosomes are disrupted by this intercalating dye.[27] Under UV light, visualize and record the distances migrated of the desired nucleosome components. In a cold room, place the stored, unstained preparative gel pieces on top of a thin-layer chromatography (TLC) plate containing fluorescent indicator dye covered with Saran Wrap. Using a hand-held, long-wave UV light, locate the desired nucleosome components by fluorescent quenching of the underlying indicator dye. Verify these positions using the distances migrated that had been recorded by the ethidium bromide staining of the gel strips.

Cut out the desired segments using a razor blade, and place them in dialysis bags with an equal volume of 1.6% solution A. Drop these bags in an empty submarine horizontal slab gel electrophoresis unit filled with

[25] L. Levinger and A. Varshavsky, *Proc. Natl. Acad. Sci. U.S.A.* **77**, 3244 (1980).
[26] J. Barsoum, L. Levinger, and A. Varshavsky, *J. Biol. Chem.* **257**, 5274 (1982).
[27] C. T. McMurray and K. E. van Holde, *Proc. Natl. Acad. Sci. U.S.A.* **83**, 8472 (1986).

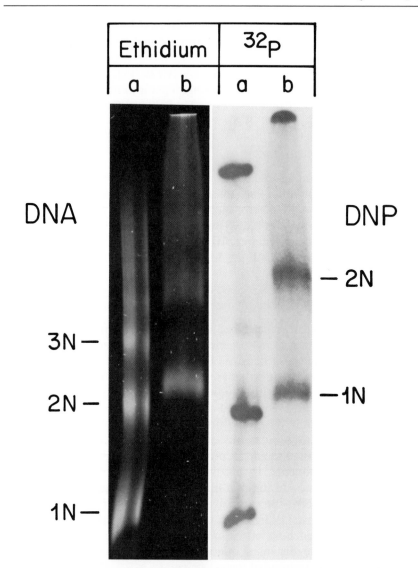

FIG. 5. Electrophoretic assay for nucleosome assembly. pBR322 was cut with *Bam*HI, *Sph*I, and *Eco*RI, yielding fragments of 187, 377, and 3797 base pairs. The fragments were end-labeled with ^{32}P and mixed with a 5000-fold excess of *S2* chromatin that had been depleted of histone H1 and nonhistone proteins. This material was then subjected to 2.5 *M* NaCl, followed by step-gradient dialysis to low ionic strength (see Ref. 3). Lane a, purified nucleosomal DNA mixed with the ^{32}P-labeled restriction fragments; lane b, reconstituted chromatin. DNP, Deoxyribonucleoprotein.

1.6% solution A at 4°. Electroelute at 4° for 1–2 hr at 50 mA with buffer recirculation. The progress can be monitored under long-wave UV light using as an indicator a corresponding piece from the ethidium bromide-stained gel placed in a separate dialysis bag. Reverse the current for 1 min. Remove the eluted nucleosomes, centrifuge at low speed to remove debris, and concentrate sevenfold by dialysis for 2 hr at 4° against 65% glycerol, 1% solution A. Samples can be stored at −20° without freezing. Alternatively, samples can be concentrated using a Minicon-B15 concentrator at 4° prior to glycerol addition.

Comments on Electroeluted Nucleosome Stability

This procedure works extremely well for isolation of MI nucleosomes (core particles). Analytical gels performed on this material reveal that not even traces of histone octamers have dissociated from the DNA, since no free DNA band is observable. However, isolated MII and MIII tend to be less stable; analytical gels performed on these components reveal partial conversion to MI, indicating accessory protein loss.

Enzymatic Preparation of Chromatin and Chromatin Fractionation

Reagents

Micrococcal nuclease (Cooper Biomedical, Malvern, PA)

Solutions

Solution E, digestion buffer: 300 mM sucrose, 50 mM triethanolamine, 25 mM KCl, 5 mM sodium butyrate, 4 mM MgCl$_2$, 1 mM CaCl$_2$, 1% thiodiglycol, 1 mM iodoacetamide, 1 mM phenylmethylsulfonyl fluoride (PMSF) (pH 7.4 at 37°) (store at 4°) (the iodoacetamide and PMSF are added just before use; PMSF is first put into solution at 100 mM in 2-propanol)

Solution F, micrococcal nuclease: dissolve at 20,000 units/ml in solution E minus butyrate, iodoacetamide, and PMSF (store frozen at −20°)

Solution G, nuclei lysing solution: 2 mM EDTA, pH 7.4, at 4° (adjust the pH of the free acid with tetrapropylammonium hydroxide)

Method

Nuclei are isolated from fresh tissues or exponential cultured cells by a modification of the Blobel and Potter method, as reported in Ref. 28, but

[28] W. T. Garrard and R. Hancock, *Methods Cell Biol.* **17**, 27 (1978).

with supplementation of all solutions with 5 mM sodium butyrate, 1 mM iodoacetamide, 1 mM PMSF, and 1% thiodiglycol (to inhibit histone deacetylases, isopeptidases, proteases, and chemical oxidation, respectively). We avoid the use of nonionic detergents, which often cause undesired nuclear lysis. When nuclei from liver are required, owing to very high endogenous nucleases, it is necessary to use an alternative procedure for isolation[29] and for digestion,[8] but supplementation with the above inhibitors is recommended.

The freshly isolated nuclei are washed twice in solution E at 4° by low-speed centrifugation and are suspended in the same buffer at a concentration of about 1 – 5 mg/ml (as DNA). (DNA is roughly quantitated by taking A_{260} readings in 1 N NaOH using 1 mg as 27 absorbance units.) Preincubate the nuclei at 37° for 5 min and then add solution F to yield 10, 20, 30, and 40 units of micrococcal nuclease per milligram DNA. Incubate 10 min at 37° and then rapidly cool the samples in ice. (Although about 30 sec of continued digestion occurs upon cooling, the addition of EDTA to stop digestion can cause undesired nuclear lysis or selective leakage of nucleosome components.) The degree of digestion under these conditions should convert between 5 and 20% of the DNA to acid-soluble oligonucleotides, yielding between 15 and 60% of the mass fraction of the sample as mononucleosomes.

After 10 min on ice, centrifuge the samples for 10 min at 13,000 g at 4° and collect the supernatant, termed the *S1* chromatin fraction. Wipe the tubes free of residual solution, and add a small amount of solution G. Disperse the pellet gently with a Pasteur pipette, first without suction so as to break up large aggregates during nuclear lysis, and next with suction to solubilize the sample. Add additional solution G to bring the final DNA concentration to about 4 mg/ml and continue to solubilize the sample using suction into the pipette. (Lysis at higher DNA concentrations is less efficient.) After 10 min on ice, centrifuge as above and collect the supernatant, termed the *S2* chromatin fraction. The *S2* fraction is ready for electrophoretic analysis after addition to an equal volume of glycerol. This fraction consists of mono- and polynucleosomes, whereas the *S1* fraction consists primarily of mononucleosomes.

For electrophoretic analysis of the *S1* chromatin fraction, samples should be dialyzed at 4° against solution G after addition of one-tenth volume of 100 mM EDTA, pH 7.4. Direct dialysis without EDTA addition can also be used; under these conditions, excess linker DNA is trimmed because the residual micrococcal nuclease takes about 0.5 – 1 hr to be

[29] P. G. Wallace, D. R. Hewish, M. M. Venning, and L. A. Burgoyne, *Biochem. J.* **125**, 47 (1971).

inhibited by the slow entry of the EDTA solution. The dialyzed *S1* fraction is concentrated using a Minicon-B15 concentrator at 4° prior to addition of an equal volume of glycerol to prepare for electrophoretic sample loading. Samples can also be concentrated by dialysis against 65% glycerol, 1% solution A (see above). While we do not recommend storing samples, the material can be stored at −20° without freezing because of the presence of glycerol. Freezing samples is not advisable because salt gradients transiently form that result in the disassociation and rebinding of chromosomal proteins.

Comments on Chromatin Fractionation

Figure 6A shows the flow chart for generating *S1* and *S2* chromatin fractions from MPC-11 mouse plasmacytoma cell nuclei. The bulk of the nuclear DNA (85%) resides in the *S2* chromatin fraction. SDS–gel electrophoresis reveals that the *S1* fraction is enriched in nonhistone proteins and depleted in histone H1 relative to the proteins present either in the *S2* fraction or in total unfractionated nuclei (T). Figure 6B shows that both potentially active ($V_{\lambda 1}$) and active (C_{κ}, rDNA) sequences fractionate to the *S1* and *P* chromatin fractions, whereas transcriptionally inactive sequences appear primarily in the *S2* fraction [$d(CA)_n$, β-globin, satellite] (see also Refs. 8–10 and 30–32). Thus, since specific sequences partition uniquely during chromatin fractionation, an obvious necessary prelude to any electrophoretic analysis of nucleosomes that package specific genes is to determine which chromatin fraction contains the sequence under investigation.

Second-Dimensional Electrophoretic Resolution of DNA and Detection of Specific Sequences

Reagents

Proteinase K (Boehringer-Mannheim Biochemicals, Indianapolis, IN)
Sodium dodecyl sulfate (Bio-Rad)

Stock Solutions

Solution H, 10× DNA gel electrophoresis buffer: 360 mM Tris base, 300 mM monosodium phosphate, 100 mM disodium EDTA, pH adjusted to 7.8 with hydrochloric acid (store at room temperature)

---

[30] W. T. Garrard, P. N. Cockerill, D. W. Hunting, D. McDaniel-Gerwiz, C. Szent-Gyorgyi, M. Xu, and D. S. Gross, *in* "Chromosomes and Chromatin" (K. W. Adolph, ed.), pp. 133–178. CRC Press, Boca Raton, Florida, 1988.

[31] S. M. Rose and W. T. Garrard, *J. Biol. Chem.* **259**, 8534 (1984).

[32] M. Xu, M. B. Barnard, S. M. Rose, P. N. Cockerill, S.-Y. Huang, and W. T. Garrard, *J. Biol. Chem.* **261**, 3838 (1986).

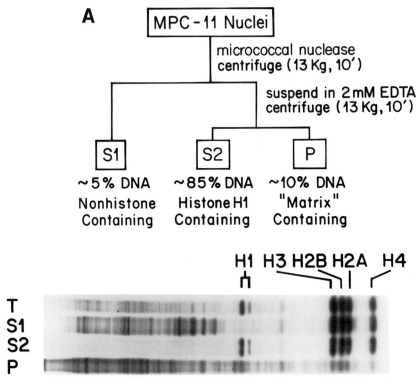

FIG. 6. Procedure for chromatin fractionation, and distributions of DNA, proteins, and specific sequences. (A) The flow diagram illustrates the mass proportions of bulk, acid-insoluble DNA that are partitioned to the different chromatin fractions. MPC-11 plasmacytoma cells were used as the source of nuclei. (Reprinted, with permission, from Ref. 30.) (B, p. 131) Dot-blot hybridization analyses of specific sequences within DNA samples purified from the indicated MPC-11 chromatin fractions and unfractionated nuclei (T) (see Refs. 8–10, 31, and 32). (Reproduced, with permission, from Ref. 9.)

Solution I, 10% sodium dodecyl sulfate (store at room temperature)
Solution J, $10\times$ DNA transfer buffer: 400 mM Tris base, 200 mM sodium acetate, 10 mM disodium EDTA, pH adjusted to 7.4 with acetic acid

Second-Dimensional SDS DNA Gels

Form a 5% polyacrylamide, 0.1% SDS DNA gel, $16 \times 14 \times 0.3$ cm, as follows. Mix at room temperature 25 ml solution B, 10 ml solution H, 1 ml solution I, 64 ml water, 0.5 ml 20% ammonium persulfate (freshly

B

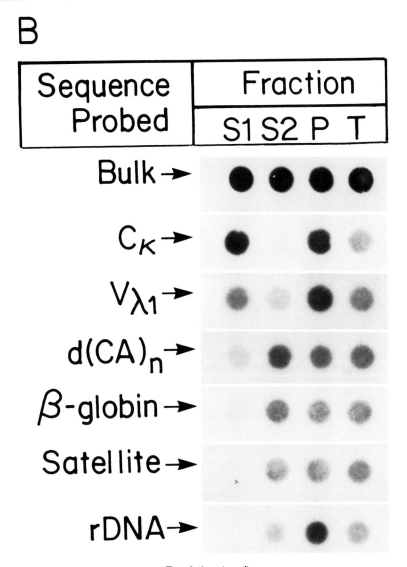

FIG. 6. *(continued)*

prepared), and 0.2 ml TEMED. Fill the gel mold up to 2 cm from the top and gently overlay with water so as to create an even surface across the top of the gel. Polymerization usually takes less than 1 hr (note that a 15% polyacrylamide gel plug is not necessary here). Assemble the cast slab gel into the running apparatus and fill the lower tray with electrophoresis buffer (10% solution H, 1% solution I), taking care to remove any air bubbles from the gel bottom.

Ethidium bromide-stained one-dimensional nucleosome gels are visualized under long-wave UV light, and the desired 0.8-cm-wide sample lanes are cut out on a glass plate using a knife by pressing down with an even motion. Excise the center 0.5-cm portion, and avoid taking the outer edges of the sample lane, which are often distorted. Trim the length to 11 cm. This yields a strip, 0.3 cm thick, 0.5 cm wide, and 11 cm long. Soak the strip 1 hr at room temperature in 10% solution H, 10% solution I.

Prepare to polymerize the one-dimensional gel strip in place across the top of the second-dimensional DNA gel as follows. Remove the water overlay from the second-dimensional gel. Mix at room temperature 5 ml solution H, 10 ml solution B, 0.5 ml solution I, 1.0 ml 5 mg/ml solution of proteinase K[33] (freshly dissolved), 33 ml water, 50 μl TEMED, and 0.3 ml 20% ammonium persulfate (freshly prepared). Pour this solution across the top of the second-dimensional DNA gel and quickly insert the gel strip in place, forcing out air bubbles during the process. Place the strip so that the top end of the one-dimensional gel abuts one side of the second-dimensional gel. A second-dimensional standard sample well may be conveniently formed at this stage by inserting a 0.8-cm-wide, 0.3-cm-thick, 5-cm-long spacer in between the bottom end of the one-dimensional gel and the side of the gel apparatus.

To prepare the standard, mix a 5 mg/ml solution of proteinase K[33] with the same chromatin sample that was separated in the one-dimensional gel with 10% solution H, 10% solution I such that the final concentration of DNA is 0.75 mg/ml and the protease, 100 μg/ml. Incubate at 37° for 30 min. Add an equal volume of 50% glycerol, 0.1% bromphenol blue.

Fill the upper tray with electrophoresis buffer (10% solution H, 1% solution I). Load 20 μl of the standard into the sample well (7.5 μg as DNA). Connect power leads with the positive pole at the bottom and run at room temperature overnight at constant current with buffer recirculation until the tracking dye runs to 3 cm above the gel bottom, approximately 600 mA-hr (e.g., 19 hr at 30 mA).

Gel Staining, DNA Transfer, Fluorography, and Nucleic Acid Hybridization

After termination of electrophoresis, carefully place the fragile gel in 50% methanol. Shake gently at room temperature for at least 4 hr, chang-

---

[33] The presence of proteinase K eliminates quenching of ethidium bromide fluorescence within the mononucleosome region by digesting histone H1.

ing the 50% methanol solution at least once. This removes the SDS, which severely interferes with ethidium bromide staining. Gels can be conveniently stored for several days in this solution at room temperature because the DNA becomes fixed. To stain the DNA with ethidium bromide, rehydrate the shrunken gel in 10-ml solution H, 0.2 ml solution D, 990 ml water. After 1–2 hr with shaking at room temperature, destain in 1% solution H for 1 hr and photograph under long-wave UV light (to prevent DNA damage), as described above for one-dimensional nucleosome gels. If the DNA is radioactively labeled, gels may be prepared for fluorography, dried, and exposed, as already described for nucleosome gels.

To transfer the DNA electrophoretically to a positively charged nylon membrane, first soak the ethidium bromide-stained gel in 10% solution J at room temperature for 30 min. Then boil the gel in fresh 10% solution J for 15 min in a large beaker or in a glass tray within a microwave oven. This step denatures the DNA and is used in place of base treatment, which causes partial depolymerization of the polyacrylamide and excessive gel swelling during the subsequent electrophoretic transfer of DNA. Soak the boiled gel and precut Zeta-Probe membrane (Bio-Rad) separately in 10% solution J at 4° for at least 30 min. In a cold room, assemble the membrane against the gel in an E-C apparatus electrophoretic transfer cassette. Sandwich the gel and membrane between Whatman 3 MM paper and $\frac{1}{2}$-inch sponges that have been equilibrated in cold 10% solution J. Take great care to avoid air bubbles between the gel and membrane. Assemble the cassette in the E-C Corporation (St. Petersburg, FL) electroblotter that contains 6 liters of cold 10% solution J. Electrophoretically transfer toward the positive pole with buffer recirculation and stirring of each buffer compartment (to reduce pH gradients) using a power setting of 300 mA for the first 30 min and 600–1000 mA for the next 6 hr at 4° (see also Refs. 34–37). If problems with air bubbles on the blots persist, then the electroblotter can be modified to eliminate this artifact.[37] After transfer, bake the membrane at 80° for 2 hr in a vacuum oven.

It is particularly useful to be able to visualize the electroeluted DNA that actually becomes bound to the membrane. For this purpose, we recommend performing analyses on samples containing uniformly labeled DNA. Cultured cells are conveniently labeled by growth for several generations in medium containing 1 μCi/ml [³H]thymidine plus 5 μg/ml unlabeled thymidine (more label kills cells). To visualize the labeled DNA that

[34] E. J. Stellwag and A. E. Dahlberg, *Nucleic Acids Res.* **8**, 299 (1980).
[35] M. Bittner, P. Kupferer, and C. F. Morris, *Anal. Biochem.* **102**, 460 (1980).
[36] D. S. Gross, S.-Y. Huang, and W. T. Garrard, *J. Mol. Biol.* **183**, 251 (1985).
[37] C. Szent-Gyorgyi, D. B. Finkelstein, and W. T. Garrard, *J. Mol. Biol.* **193**, 71 (1987).

was transferred to the membrane by fluorography, first evenly spray the blot with a generous coat of NEN EN³HANCE spray and air-dry in the hood at least 1 hr. Then, for alignment purposes, mark the edges of the membrane with dots of phosphorescent indicator ink (UltEmit autoradiography marker, NEN). Expose to preflashed film as described above for nucleosome electrophoresis.[22] Fluor is removed by washing the membrane 4 times with 100% acetone. Traces of RNA are then removed (to eliminate misinterpretation of subsequent hybridization patterns) by soaking the membrane for 1 hr at 37° in 0.3 N sodium hydroxide. The membrane is then extensively rinsed in distilled water and is remarked in the same positions with the phosphorescent indicator ink. The blot is prehybrized and hybridized with a ^{32}P-labeled probe, as described elsewhere,[36] but 1% SDS is also included in all solutions (see also Ref. 37). Exposure against preflashed film with an intensifying screen is as described above for nucleosome electrophoresis.[22,23] If extensive ^{3}H radioactivity is also present on the blot, place a piece of black paper between the membrane and the film. This, however, reduces the ^{32}P signal about 50% and normally is not necessary. Exposures to detect single-copy mammalian gene sequences take up to 14 days at −70°.[9] X-Ray films are superimposed with fluorograms by using the alignment dots; this allows for absolute identification of hybridizing nucleosome components. One should also take into account that nylon membranes expand approximately 2.5% following fluorography as a result of subsequent treatments.

Comments on Hybridization Mapping Technique

Figure 7 shows a fluorogram displaying the [^{3}H]thymidine-labeled DNA of a second-dimensional DNA gel that was transferred to Zeta-Probe membrane. For purposes of subsequent hybridization, we have found that positively charged nylon membranes like Zeta-Probe yield much higher signals than does diazotized paper. Even so, some enrichment of single-copy mammalian gene sequences is necessary to readily detect them in nucleosome-sized fragments. This is possible for active genes, which are enriched by chromatin fractionation in the *S1* fraction (Fig. 6B).[9] It should be stressed that the localization of a hybridization signal to a specific electrophoretic form of mononucleosome does not prove that the particles carrying the specific sequence have the same protein composition as the bulk DNA within that class. It is a correlation, and each electrophoretic class probably contains a family of different protein compositions. Additional lines of evidence are necessary to prove the association of specific proteins with specific sequences (see below).

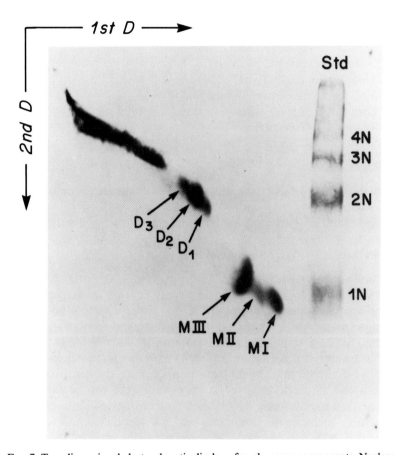

FIG. 7. Two-dimensional electrophoretic display of nucleosome components. Nucleoprotein complexes of the S2 chromatin fraction of [³H]thymidine-labeled cultured cells were electrophoretically separated in the first dimension, and DNA was displayed by a second dimension of electrophoresis in the presence of SDS. A standard derived from the same chromatin sample was included at the right side of the second-dimensional gel. DNA was transferred to Zeta-Probe membrane, and the blot was fluorographed. Note the alignment spots near the corners made by fluorographic indicator ink. Different electrophoretic forms of mono- and dinucleosomes are indicated, as well as nucleosome oligomers. Subsequent hybridization with a ³²P-labeled probe, followed by superimposition of X-ray films through the alignment dots, permits the unequivocal assignment of hybridizing components to nucleosome classes.

Second-Dimensional Electrophoretic Resolution of Proteins Using
Triton – Acid – Urea Gels

Reagents

Protamine sulfate (Sigma)
2-Mercaptoethanol (Bio-Rad)
Thioglycolic acid (Sigma)
2-Mercaptoethylamine (Sigma)
Urea (ultra-pure) (Bethesda Research Laboratories, Bethesda, MD)
Triton X-100 (Sigma)
Serva Blue R (Serva, Heidelberg)

Stock Solutions

Solution K: 30% acrylamide, 0.2% bisacrylamide (store in dark container at 4°)
Solution L: 10% Triton X-100 (store at room temperature)

Method (see also Refs. 6, 9, and 38 – 40)

Form a 12% polyacrylamide, 8 M urea, 6 mM Triton X-100 gel, 16 ×
14 × 0.3 cm, as follows. Mix in a 200-ml beaker, using gentle heating
under hot tap water or on a stirring hot plate, 32 ml solution K, 3 ml
solution L, 4 ml glacial acetic acid, 38.4 g urea, and 13 ml water. After the
solution has cooled to room temperature, add 0.4 ml TEMED and 0.48 ml
of 10% ammonium persulfate (freshly prepared). (Final volume should be
80 ml.) Fill the gel mold up to 2 cm from the top, and gently overlay with
water so as to create an even surface across the top of the gel. After
polymerization, assemble the second-dimensional gel into the running
apparatus. To the lower buffer tray add 5% acetic acid and to the upper
buffer tray add 30% glycerol, 5% acetic acid (the glycerol reduces excess
swelling of the gel during preelectrophoresis). Be sure to remove all traces
of agarose (if this was used to seal the plates) from regions which would be
in contact with the lower tray buffer. Agarose decomposes to deleterious
polyanionic species on exposure to the electrophoresis buffer. Connect
power leads with the positive pole at the top, and perform preelectrophoresis at room temperature, without buffer recirculation, for 16 hr at 125 V
(constant voltage). Immediately before the second-dimensional run, per-

[38] A. Zweidler, *Methods Cell Biol.* **17**, 223 (1978).
[39] R. G. Richards and B. R. Shaw, *Anal. Biochem.* **121**, 69 (1982).
[40] N. L. Bafus, S. C. Albright, R. D. Todd, and W. T. Garrard, *J. Biol. Chem.* **253**, 2568
(1978).

form a second preelectrophoresis at 125 V for 1 hr at room temperature after replacing upper and lower tray buffers with 5% acetic acid and overlaying the sample well chamber with 2 ml 2 M 2-mercaptoethylamine, 5% acetic acid. Then replace the tray buffer with fresh 5% acetic acid.

For best results, we recommend fluorographic analysis of labeled proteins. For this purpose, samples are prepared from cultured cells that have been labeled for 2 hr with 10 μCi/ml [^{3}H]lysine in lysine-free medium.[9] This level of isotope kills the cells, and thus longer labeling times are not advisable. If it is not possible to label cells and one must use the relatively insensitive Coomassie blue protein-staining procedure, it is then necessary to run preparative one-dimensional nucleosome gels by loading approximately 0.8 mg chromatin (as DNA) per centimeter of gel width on the 0.3-cm-thick slabs. After electrophoresis, 1.5-cm-wide strips are cut from the preparative gels parallel to the direction of electrophoresis and trimmed to the exact second-dimensional gel width (~ 13 cm). Soak the one-dimensional gel strip for 45 min at room temperature in freshly prepared 8 M urea, 5% acetic acid, 5% 2-mercaptoethanol, 2.5% thioglycolic acid. Insert the soaked strip into the well across the top of the second-dimensional gel so that no air bubbles are present and there is a tight fit at both sides. Overlay the gel strip with 2 ml of the above soaking solution supplemented with 1% protamine sulfate. If the sides of the strip are not flush against the spacers, this solution will undesirably collect in flanking pockets. To avoid this, fill these gaps with 30% glycerol and 5% acetic acid prior to overlaying with the protamine solution. Perform electrophoresis at 125 V, constant current, at room temperature *without* buffer recirculation for 17 hr (2160 V/hr). Note that the protamine solution stepwise displaces the histones as it electrophoreses through the one-dimensional gel strip and concomitantly leaves DNA as a precipitate.[39] This permits the use of wide, one-dimensional gel strips because of the stacking effect.

Gel Staining, Photography, and Fluorography

After the run, gels are stained in 2% Amido black 10B, 45% methanol, 9% acetic acid for 1 hr at room temperature and destained in 10% 2-propanol, 1% acetic acid with gentle shaking. (Be careful to avoid contamination with the lower tray buffer since it contains the electrophoresed protamine.) Addition of wool yarn to the destaining solution facilitates dye removal. Gels are restained with 0.06% Serva Blue R, 45% methanol, 9% acetic acid for 1 hr at room temperature and destained with 5% methanol, 7.5% acetic acid with shaking at room temperature. We have found this two-step staining procedure to be necessary to best fix the proteins in these gels since it prevents "bubble effects" in the protein bands. Gels are photographed with a Polaroid MP4 camera and type 55 film using fluorescent light transillumination, a Kodak Wratten yellow gelatin filter No. 15, and exposures for 1 sec at $f12.5$ to $f22$. Gels are impregnated with EN³HANCE, dried, and exposed to X-ray film,[22,23] as outlined above under "one-dimensional electrophoretic resolution of nucleosomes."

Comments on Second-Dimensional Protein Gels

Figure 8 shows an example of second-dimensional Triton–acid–urea gel electrophoresis performed on nucleosome core particles that have been resolved on a long one-dimensional nucleoprotein gel. Alignment with a second-dimensional DNA gel, run in parallel, reveals that components a, b, and c differ in their content of ubiquitinated histones H2A and H2B (uH2A and uH2B) and histone H2A.X but not in the other histone variants or their acetylated counterparts (see also Ref. 9). The high resolving power of these Triton–acid–urea gels is to be appreciated. However, oxidation artifacts are possible in this system since even conversion of methionine to its sulfoxide derivative in proteins will lead to the appearance of new bands. Some sources of Triton X-100 have oxidizing agent contaminants that can lead to this artifact.[38] Therefore, for many applications, simply running acid–urea gels is adequate. This is performed essentially identically to that described here but with a different second-dimensional gel (see also Refs. 6 and 38–40). In addition, second-dimensional SDS gels can also be readily performed,[2] but their resolving power is minimal compared to the urea gel systems. (Acid–urea gels, plus or minus Triton X-100, can resolve posttranslationally modified forms of proteins, whereas only acid–urea–Triton gels can separate proteins differing by even a single neutral amino acid substitution.[38])

Further Applications for Electrophoretic Analyses

We have described techniques to (1) determine the overall electrophoretic patterns of various nucleoprotein species; (2) analyze the bulk DNA lengths of such species by performing a second dimension of electrophoresis in the presence of SDS; (3) determine the position of specific sequences in such two-dimensional displays after DNA transfer and nucleic acid hybridization; and (4) analyze the bulk protein compositions of such species by performing a second dimension of electrophoresis in Triton–acid–urea gels. For purposes of completeness, we list below six additional studies that can be performed by related methodologies.

Precursor–Product Relationships. Electrophoretically separated mono- and polynucleosomes can be digested *in situ* within the gel prior to per-

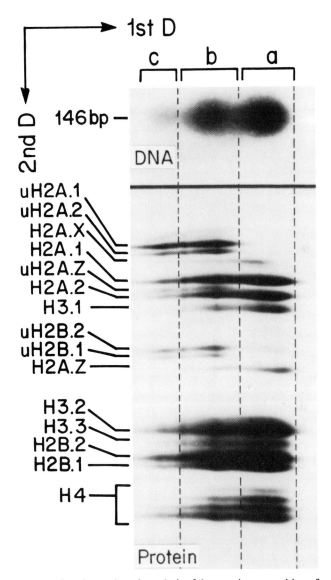

FIG. 8. High-resolution electrophoretic analysis of the protein composition of nucleosome core particles. Chromatin from the *S1* fraction of [³H]lysine-labeled mouse cells was trimmed with micrococcal nuclease, extracted with 0.35 *M* NaCl, and resolved on a 35-cm-long, 3.5% acrylamide–0.5% agarose–30% glycerol nucleoprotein gel (1st D). DNA and proteins were separately resolved by performing second-dimensional electrophoresis (2nd D) using 5% acrylamide–0.1% SDS and 12% acrylamide–6 m*M* Triton X-100–8 *M* urea gels, respectively. The DNA gel reveals that MI is fractionated into three components (labeled a, b, and c). Histone variants and their ubiquitinated counterparts are indicated for the fluorogram of the protein gel. (Reproduced, with permission, from Ref. 9.)

forming electrophoresis in the second dimension to display either DNA or nucleoprotein complexes. This is accomplished by incorporating reversibly inactivated micrococcal nuclease within the first-dimensional gel matrix and running buffer. The enzyme is reactivated after electrophoresis by exposing gels to Ca^{2+}. Since the nuclease is positively charged at neutral pH while chromatin is negatively charged, enzyme protein is maintained within the gel during electrophoresis by attaching a dialysis bag containing the enzyme to the anode end of gel tubes. We have termed this general procedure "chromatin fingerprinting."[2,5-7] Since we have found that the enzyme tends to "stick" adversely to nucleosomes in gels containing glycerol, we now recommend running gels without glycerol when using this procedure.

Protein Exchange and Stability of Nucleoprotein Complexes. The stabilities and extent of protein exchange between various protein–DNA complexes can be readily ascertained by repeating nucleoprotein gel electrophoresis in a second dimension. Nucleoprotein complexes lying above the diagonal have bound additional proteins donated by the nucleoproteins lying below the diagonal (see Fig. 10 of Ref. 6).

Analyses of Protein–Protein Interactions by Cross-Linking. By employing a variety of protein–protein cross-linking reagents prior to nucleoprotein electrophoresis in the first dimension, the nucleosomal origins of the cross-linked products can be readily determined by running second-dimensional protein gels. For example, cross-linked proteins resulting from internucleosomal contacts can be readily distinguished from those resulting from intranucleosomal interactions. Furthermore, specific electrophoretic forms of mononucleosomes may generate specific cross-linked proteins owing to their unique protein composition.[11,12]

Location of Methylated DNA Sequences. We have developed a procedure, which employs an analog affinity column that contains cytidine and 5-methyluridine, to purify highly specific polyclonal antibodies against 5-methylcytosine. These antibodies can be used to immunologically probe DNA transferred from nucleoprotein gels to diazotized paper.[10]

Immunological and Enzymatic Location of Specific Proteins. Second-dimensional SDS protein gels performed on electrophoretically separated nucleoprotein species[2] can be transferred to nitrocellulose and probed with specific antibodies against chromosomal proteins using existing techniques.[41] Similar experiments employing Triton–acid–urea gels should be possible after soaking such gels in SDS transfer buffer. Nucleosomes that contain both a specific sequence and ubiquitinated histones can be identi-

---

[41] H. Towbin and J. Gordon, *J. Immunol. Methods* **55,** 297 (1984).

fied by employing isopeptidase and hybridization mapping to determine whether the particles shift their mobility after enzymatic deubiquitination.[9] Furthermore, previous studies have shown that nucleosomes complexed with IgG electrophoretically migrate as discrete bands with reduced mobilities.[42] Therefore, antibodies against specific proteins can be used to shift the mobilities of nucleosomes. Second-dimensional SDS DNA gels will reveal components that have bound one or two antibody molecules per particle as new diagonal lines, shifted in mobility from those nucleoprotein species that do not bind antibodies. Such displays can be hybridized after DNA transfer with specific gene probes as outlined above.

Localization of other enzyme–nucleoprotein complexes should be possible by soaking nucleoprotein gels in the appropriate reaction mixture containing a radioactive substrate (e.g., assays for RNA polymerase, DNA polymerase, and histone acetyltransferases). Thorough removal of unincorporated substrates followed by fluorography, either before or after second-dimensional electrophoresis, will reveal the specific reaction products. Related techniques have been used to locate *Escherichia coli* RNA polymerase in gels,[43] and poly(ADP)-ribose polymerase[44] (NAD^+ ADP-ribosyltransferase).

One-Dimensional Hybridization Mapping and Analysis of Nucleosome Unfolding. We previously reported a technique for transferring DNA from nucleoprotein gels to diazotized paper that employed *in situ* histone removal using a cationic detergent.[45] It should now be possible to replace this procedure with direct electrophoretic transfer of the native nucleoprotein complexes to Zeta-Probe membrane using the low ionic strength nucleosome electrophoresis buffer. Such blots could then be soaked in base,[46] treated with proteases, and processed for nucleic acid hybridization as discussed above. This procedure should be particularly well suited to study nucleosome core particle unfolding patterns in response to increasing urea concentrations in 5% acrylamide gels containing transverse urea gradients (0 to 8 M) (Lynch and Garrard, unpublished results, 1982). DNA transfers from such gels can be hybridized with selected specific gene probes to address biological questions related to alterations in the conformational stability of nucleosomes.

[42] C. S. M. Tahowidin and M. Bustin, *Biochemistry* **19**, 4387 (1980).

[43] L. D. Beckman and C. D. Frenkel, *Nucleic Acids Res.* **3**, 1727 (1976).

[44] C. P. Giri, M. H. P. West, M. L. Ramirez, and M. Smulson, *Biochemistry* **17**, 3501 (1978).

[45] T. L. Reudelhuber, D. J. Ball, A. H. Davis, and W. T. Garrard, *Nucleic Acids Res.* **10**, 1311 (1982).

[46] K. C. Reed and D. A. Mann, *Nucleic Acids Res.* **13**, 7207 (1985).

Acknowledgments

We are indebted to the following colleagues from this laboratory for important contributions in perfecting the techniques described here: Richard Todd, Stephen Albright, Phyllis Nelson, Tim Reudelhuber, Alan Davis, Teni Boulikas, Dorothy Ball, David Gross, Stephen Rose, Peter Cockerill, Darel Hunting, Christopher Szent-Gyorgyi, Charles Lynch, and Mary Barnard. We thank Veronica Blasquez, Barbara Fishel, and Christopher Szent-Gyorgyi for review and Marie Rotondi and Susan Alexander for the preparation of the manuscript. This work has been supported by Grants GM22201, GM29935, GM25829, and GM31689 from the National Institutes of Health and Grant I-823 from The Robert A. Welch Foundation.

[7] Electron Microscopy of Chromatin

By José M. Sogo and Fritz Thoma

Introduction

The fundamental problem in electron microscopy of chromatin concerns specimen preparation. In the first part of this chapter, we describe a method for visualizing soluble chromatin as a nucleoprotein complex. This allows one to obtain information about the three-dimensional structure of chromatin complexes, in particular about the presence or absence of nucleosomes, and the folding or unfolding of nucleosomes and chromatin fibers. This information can be used to support interpretations of biophysical and biochemical data. In the second part, we describe the use of DNA cross-linking reagents (psoralens) as probes of different chromatin structures. Since the accessibility of DNA to psoralen is restricted by various proteins bound to DNA (e.g., histone octamers in nucleosomes), this method allows one to distinguish between the presence or absence of nucleosomes on the DNA. It is particularly useful for studying nucleosome arrangements on actively transcribed and replicated DNA and on inactive DNA, provided that these DNAs can be identified in the electron microscope (e.g., extrachromosomal ribosomal DNAs, viral DNAs, plasmid DNA).

Standard Protocol for Soluble Chromatin

In order to apply conventional transmission electron microscopy to chromatin samples, the specimen must be prepared in a way that enables it to withstand both high vacuum and electron bombardment. Chromatin has to be released from nuclei, immobilized on a surface, stained, dehydrated, and dried. Since all these steps may change structural features of

Copyright © 1989 by Academic Press, Inc.
All rights of reproduction in any form reserved.

the specimen, they must be carefully controlled. It is important to keep chromatin in solution as long as possible. This allows one to apply methods independent of electron microscopy. From the moment the material is immobilized on a support grid, controls can be performed by electron microscopy only. The most powerful approach is a comparative analysis of similar samples where just one parameter (such as protein composition or ionic strength) is changed, while the material is still in solution, and all other steps are the same.

Preparation of Soluble Chromatin

A gentle procedure to obtain soluble chromatin fragments involves limited digestion of nuclear DNA by endogenous or exogenous nucleases (staphylococcal nuclease) and release of chromatin fragments by disruption of the nuclear structure (see Kornberg et al., this volume [1]). Rat liver nuclei are isolated according to Hewish and Burgoyne.[1] The nuclei (~ 2 – 3 mg DNA/ml) are adjusted to 1 mM CaCl$_2$, preincubated at 37° for 5 min, and digested with staphylococcal nuclease (15 units/ml nuclei) at 37° for an optimal time to obtain long chromatin fragments (1 – 2 min, as determined in a pilot digest). The reaction is terminated by addition of 2 mM EDTA on ice. The nuclei are pelleted, resuspended in cold lysis buffer [1 mM triethanolamine-chloride (TEACl), pH 7, 0.2 mM EDTA] for 10 – 30 min, and nuclear debris and insoluble chromatin are pelleted. The supernatant contains the "soluble chromatin" as well as other nuclear components that are not necessarily chromatin constituents.

For further purification of chromatin and removal of nonhistone components or histone H1, fractionation on sucrose gradients (Noll and Noll, this volume [5]), which contain 5 mM TEACl, pH 7, 0.2 mM EDTA, and various concentrations of NaCl, is recommended. At 10 mM NaCl, loosely bound or nonbound components are sufficiently removed, whereas 300 – 350 mM NaCl is required to remove nonhistone proteins. At 500 mM NaCl, histone H1 is removed in addition to nonhistone protein.[2]

Preparation of Carbon-Coated Grids (C Grids)

Materials and Equipment

High Vacuum Evaporator BA 360 M (Balzers) with electron gun, quartz crystal, thin film monitor QSG 201, and control unit EVM 052

---

[1] D. R. Hewish and L. A. Burgoyne, *Biochem. Biophys. Res. Commun.* **52,** 504 (1973).

[2] F. Thoma and T. Koller, *J. Mol. Biol.* **149,** 709 (1981).

Carbon, platinum (90%)–carbon (10%), and mica (Balzers Union A.G., FL-9496 Balzers)

Procedure

1. Lay a sheet of mica (with a freshly cleaved surface facing up) at a distance of about 12 cm below the carbon evaporator gun. Place the quartz close to the mica. Cover mica with the shutter. At a vacuum of about 5×10^{-6} torr, preheat the filament. When the evaporation rate is constant, open shutter and deposit about 60–70 Hz (measured with the thin film monitor) carbon on the mica. Film thickness is approximately 6–7 nm.
2. The mica (carbon face up) is placed on a wet filter paper in a petri dish and processed within the next hour.
3. Copper grids are spread on a filter paper, rinsed with one or two droplets of a Scotch solution (20–30 cm Scotch tape in 50 ml chloroform; cellophane support does not dissolve) and air-dried. The tape adhesive will keep the carbon film attached.
4. A filter paper with the approximate size of the carbon film is submerged in water and placed on a supporting wire mesh stand. Copper grids are placed on the filter paper.
5. The carbon-coated mica is lowered into the water at an angle of approximately 45 degrees (carbon side up) until the carbon film is completely released and floating on the water surface.
6. The carbon film is placed on the grids by carefully lowering the water level using an aspirator.
7. The paper with the carbon-coated grids is placed on another filter paper in a glass petri dish and baked for 3–4 min at 160°. These "C grids" are stored at room temperature and used within the next 2–15 hr.

Preparation of Alcian Blue-Coated Carbon Grids (ABC Grids[3])

1. A 0.2% stock solution of Alcian blue 8GX (Serva, Heidelberg) in 3% acetic acid is diluted in redistilled water to make 0.002% Alcian blue and filtered through a 0.22 μm filter unit (Millipore, Bedford, MA).
2. Carbon grids (old grids can be used, too) are floated carbon side down for 5 min on 0.002% Alcian blue at room temperature. The excess of Alcian blue is washed off by floating on redistilled water for 10 min. The grids are blotted dry on filter paper and used within the next hour.

---

[3] P. Labhart and T. Koller, *Eur. J. Cell Biol.* **24,** 309 (1981).

Preparation of Soluble Chromatin on C or ABC Grids

Reagents

Salt series buffers (pH 7)

	Buffer				
Type	"1"	"10"	"40"	"60"	"100"
TEACl (mM)	1	5	5	5	5
EDTA (mM)	0.2	0.2	0.2	0.2	0.2
NaCl (mM)	0	10	40	60	100

TEACl (pH 7) stock: 0.5 M triethanolamine adjusted with HCl to pH 7

BAC stock: 0.2% BAC (benzyldimethylalkylammonium chloride; *n*-alkyl mixture: $C_{12}H_{25}$, 60%; $C_{14}H_{29}$, 40%; Bayer, Leverkusen, FRG) in redistilled water, kept at room temperature

Adsorption buffer: 1 or 5 mM TEACl (pH 7), 0.2 mM EDTA, 2×10^{-4}% BAC

Glutaraldehyde: 25% (Merck, Darmstadt, FRG), kept at 4° under nitrogen

Dialysis bags: Spectrapore, boiled in EDTA, rinsed, and stored in redistilled water

Procedure

1. Chromatin ($\sim 20-100$ μg DNA/ml) is dialyzed extensively against one of the above salt series buffers at 4°.
2. Chromatin is fixed by dialysis for at least 16 hr at 4° after addition of 0.1% glutaraldehyde to the dialysis buffer. The fixed chromatin can be stored at 4° for weeks. (Do not freeze!)
3. The fixed chromatin samples are diluted into adsorption buffer to make 1 μg DNA/ml for extended chromatin ("1", "10") or up to 5 μg DNA/ml for more compact chromatin (higher salt). Leave at room temperature for 15–30 min.
4. Attach C grids (or ABC grids) with carbon side up on a sheet of Parafilm (American Can Co., Greenwich, CT) by pressing the edge with the tips of tweezers.
5. Place a droplet of chromatin (5–20 μl) carefully on a grid. Chromatin is allowed to adsorb for 5 min (longer adsorption time is not critical).
6. Wash off excess salt by floating on redistilled water for 10 min.

7. Dehydrate the sample by dipping into 98% (v/v) ethanol for about 3 sec.
8. Place the grid on filter paper and air-dry.
9. For contrast enhancement, the grids are placed on a rotary stage about 12 cm from a platinum–carbon evaporation gun. The angle between the beam center and the carbon film plane is 7°. Evaporation of platinum–carbon is at a pressure below 10^{-4} torr. Evaporation is measured by the quartz perpendicular to the beam. Either 500 Hz (C grids) or 750 Hz (ABC grids) provides sufficient contrast to visualize DNA and chromatin.
10. Electron microscopy is performed at 75–100 kV and a magnification of about 20,000 [routinely checked using calibration grids (2160 lines per mm, Balzers Union)]. Resolution of shadowed specimens is approximately 3 nm as defined by the distance between platinum clusters.

Discussion

Solubilization of Chromatin

1. Soluble chromatin should be kept at 4° and processed as quickly as possible. Freezing and pelleting of chromatin damages the structure. Between 60 and 250 mM NaCl, aggregation of chromatin might occur. Aggregated chromatin does not efficiently adsorb on C and ABC grids. Electron microscopy under these conditions might lead to a selection of only soluble material. Furthermore, exposure of chromatin to high salt and/or temperature (37°) might lead to rearrangement of proteins and sliding of nucleosomes (in particular, for H1-depleted chromatin).

2. Structural properties of chromatin, such as condensation, depend on its protein composition (Fig. 1). During the digestion with nucleases and lysis in low salt, chromosomal proteins might be lost or nonchromosomal proteins might be trapped.

3. Nuclease digestion releases only a subpopulation of the nuclear chromatin. Lysis in low-salt buffers yields extended chromatin (nucleosome filaments). The fibers that are obtained by addition of salt are therefore reconstituted chromatin fibers. Their structure, however, appears to be similar to the structure in nuclei and chromosomes as estimated from X-ray diffraction data.[4] Lysis in higher salt (80 mM NaCl) might select for only the most soluble material, whereas insoluble and less soluble material remain precipitated (or trapped) in the nuclear debris. High salt also promotes redistribution of histone H1.

[4] J. Widom and A. Klug, *Cell* **43**, 207 (1985).

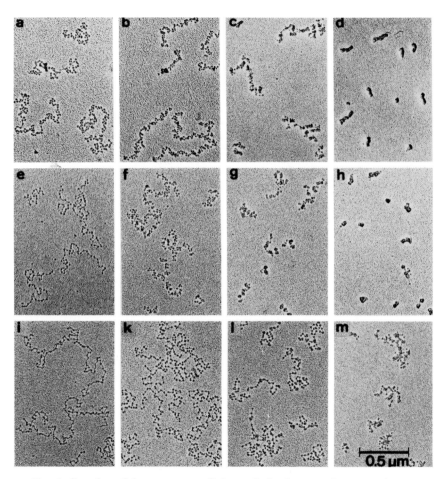

FIG. 1. Overview of the appearance of chromatin in electron microscopy. Appearance depends on protein composition and ionic conditions. Soluble rat liver chromatin (a–d), nonhistone-depleted chromatin (e–h), and chromatin depleted of nonhistone components and histone H1 (i–m) were dialyzed against "1" (a, e, i), "10" (b, f, k), "40" (c, g, l), and "100" (d, h, m) at pH 7 and prepared for electron microscopy as described in the text. The overview shows that the appearance at a given condition is homogeneous: only one type of fibers or filaments is observed. The different morphologies are described in the text and summarized in Table I. [Reproduced from Thoma and Koller, *J. Mol. Biol.* **149**, 709 (1981), with permission.]

4. High pH (≥ 9) is frequently recommended to promote lysis of nuclei, for example, in the spreading procedures of Miller and Beatty.[5] High pH, however, disturbs higher order chromatin structures and promotes unfolding of nucleosomes, especially in H1-depleted chromatin at very low ionic strength.[6] We are not yet sure to what extent these pH-dependent structural changes are reversible (see also below).

Fixation. Adsorption, washing, staining, and dehydration are steps where strong forces act on the structures. Components not firmly bound or structures not firmly stabilized might fall apart or rearrange. When unfixed chromatin is prepared according to the standard procedure on C grids, chromatin fibers and nucleosomes are not observed and unravelled filaments appear, presumably DNA with proteins still attached (Fig. 2a and 2b). To overcome this problem, glutaraldehyde fixation was used. Since glutaraldehyde preferentially cross-links proteins, it presumably stabilizes nucleosomal and supranucleosomal structures. Buffers that might react with the fixative have to be avoided. Furthermore, since the buffering capacity of $1-5$ mM TEACl (pH 7) is low and glutaraldehyde or CO_2 uptake from the air might lower the pH, the pH of the fixation buffers should be routinely checked.

For interpretation of the data, it is important to realize that fixation can shift an "unfolding–folding" equilibrium in a structure toward the more compact, folded state. Furthermore, chromatin fixed in an extended state (low ionic strength) can still condense if ionic conditions are appropriately changed, but chromatin fixed in a condensed state (high salt) is less likely to unfold when the ionic strength is decreased. Therefore, the adsorption buffer should never be higher in ionic strength than the fixation buffer.[7]

Adsorption. Conditions must be found that allow firm attachment of the specimen to the support film so that the structure does not change in subsequent steps. "Good" adsorption is defined according to the following criteria: at a chromatin concentration of about $1-2$ μg DNA/ml adsorption buffer, the whole grid should be uniformly covered with chromatin fibers. Chromatin that is homogeneous by biochemical assays should also result in a homogeneous population as viewed in the electron microscope (Fig. 1). Stretched fibers indicate insufficient adhesion points and therefore should not be used for structural interpretations.

In $1-5$ mM TEACl, pH 7, and 0.2 mM EDTA, chromatin adsorbs well only on carbon films if more than 10 mM NaCl is present. If one wants to

[5] O. L. Miller and B. R. Beatty, *Science* **164**, 955 (1969).
[6] P. Labhart, F. Thoma, and T. Koller, *J. Cell Biol.* **25**, 19 (1981).
[7] F. Thoma, T. Koller, and A. Klug, *J. Cell Biol.* **83**, 403 (1979).

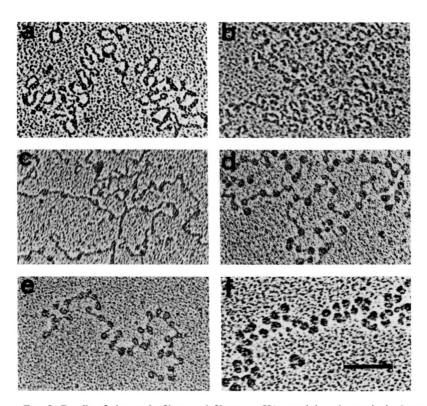

FIG. 2. Details of chromatin fibers and filaments. H1-containing chromatin is shown which was not fixed (a, b) or fixed with glutaraldehyde (e, f) in "1" (a, e) or "10" (b, f). H1-depleted chromatin is shown fixed in "1" (c) or "10" (d). Unfixed chromatin adsorbed to C grids appears as DNA-like filaments and does not show a salt-dependent condensation. This is consistent with unfolding or partial disintegration of chromatin. H1-depleted chromatin fixed in "1" (c) or "10" (d) shows unfolded and folded nucleosomes, respectively. This ionic strength-dependent folding–unfolding transition is reversible. Note that the linker DNA enters and leaves the nucleosomal beads at opposite sides (d) which results in the typical beads-on-a-string appearance. H1-containing chromatin in "1" (e) shows nucleosomal beads separated by linker DNA. The linker enters and leaves on the same side, which results in the typical open zigzag appearance of the nucleosome filament. H1 prevents unfolding of the nucleosomes under these conditions. H1-containing chromatin in "10" (f) shows nucleosomes in close contact (closed zigzag fiber). This fiber is observed in H1-containing chromatin only and reflects a first "higher order structure." Bar, 100 nm. [Reproduced from Thoma, Koller, and Klug, *J. Cell Biol.* **83,** 403 (1979), with permission of The Rockefeller Press.]

visualize chromatin at lower ionic strength, a minimum of $2 \times 10^{-4}\%$ BAC has to be added to provide sufficient adsorption. Since the quality of carbon films is variable in different hands and laboratories, procedures have been established to coat the surface with positively charged molecules such as amylamine,[8] polylysine,[9] or Alcian blue.[3] However, it is not known to what extent these molecules interact with the specimen and alter its structure. Unfixed chromatin adsorbed to ABC grids or amylamine-coated grids does not unfold and disintegrate to the same extent as on C grids. Major disadvantages of coated grids are (1) a significantly higher background, which frequently requires positive staining of DNA prior to metal shadowing, and (2) patchwise coating of the grids, which might lead to different structures in the different patches, in particular with unfixed chromatin, and promote misinterpretations.

The use of BAC in solution instead of coated grids has the advantage that the structural influence of BAC can be checked by methods other than electron microscopy. Furthermore, instead of using a "standing droplet" adsorption procedure as described above, the detergent properties of BAC allow one to spread chromatin from a mica ramp on a hypophase of water similar to spreading of DNA[10] (see below). The chromatin is then picked by the support grid from the water surface. This procedure might be useful if length measurements of chromatin or other protein–DNA complexes are required. However, spreading of a fragile nucleoprotein complex might lead to artificially stretched material and disruption of the structures.

Washing, Dehydration, and Drying. One way to test whether chromatin samples are firmly attached to the carbon film is to wash the specimens on 1–5 mM MgCl$_2$ solutions before dehydration in ethanol. Chromatin that was fixed in an extended conformation at very low ionic strength and firmly absorbed to the grid should not condense by this wash treatment.[7]

Ethanol dehydration of DNA and chromatin may induce folded forms, which, in the case of chromatin, could easily be mistaken for 30-nm chromatin fibers. To control for dehydration artifacts, specimens can be freeze-dried by rapid freezing in liquid nitrogen and slowly drying *in vacuo* on a cold stage. We have noted no significant difference between ethanol dehydrated, air-dried, and freeze-dried specimens.[7]

Staining and/or Metal Shadowing. Biological specimens scatter electrons poorly and do not create sufficient contrast for conventional electron microscopy. The specimen, therefore, is contrasted by heavy metals either

[8] J. Dubochet, M. Ducommun, M. Zollinger, and E. Kellenberger, *J. Ultrastruct. Res.* **35**, 147 (1971).

[9] R. C. Williams, *Proc. Natl. Acad. Sci. U.S.A.* **74**, 2311 (1977).

[10] J. J. Vollenweider, J. M. Sogo, and T. Koller, *Proc. Natl. Acad. Sci. U.S.A.* **72**, 83 (1975).

by staining and/or by shadowing. There are fundamental differences in the interpretation of stained and shadowed specimens: Whereas shadowed specimens provide a projection of the metal deposited on the surface, electron micrographs of stained specimens represent a projection not only of stain bound on the outside but also of stain that has penetrated the structure.

Positive staining with uranyl acetate. Positive staining with uranyl acetate is done after adsorption but prior to washing on water. Chromatin samples adsorbed on C grids are floated for about 1–2 min on a freshly prepared and filtered 1% uranyl acetate solution (in water). Positive staining of the chromatin is obtained by uranyl dioxide binding to the DNA. Precaution is necessary for interpretation of positively stained structures: (1) Since chromatin structure is very sensitive to divalent cations, binding of uranyl dioxide (a divalent cation) might alter the local structure, and (2) selective binding to one component emphasizes a particular aspect of the structure (e.g., DNA), whereas others (e.g., proteins) might not be detected.

Negative staining with uranyl acetate. After adsorption of the chromatin to the supporting carbon grid, the grid surface is rinsed with 3 droplets of uranyl acetate (1% uranyl acetate in water, freshly filtered), and excess liquid is removed with filter paper. It is important to keep in mind that the success of this technique depends heavily on the surface properties of the specimen support films. Only chromatin that is not firmly adsorbed to the grid (high concentrations of chromatin) is negatively stained. Under these conditions, chromatin frequently appears as 10-nm-wide, continuous filaments ("nucleofilaments" [11]). It has been suggested that nucleosomes in these filaments are packed face to face and stabilized by the surrounding uranyl.[7]

Platinum–carbon shadowing. Contrast enhancement is achieved by evaporation and deposition of heavy metals on the surface of the specimen and support film *in vacuo.* In contrast to positive and negative staining in solution, which are based on the specific chemical interactions of substrate and stain, metal deposition is unlikely to alter the structure by chemical modification. Metal deposition is not necessarily random either, however, since preferential nucleation sites on the substrate and background might lead to preferential build up of platinum kernels and hence to "decoration" of a particular structure. Too much metal deposition will mask structural details. If length measurements are performed, shrinkage of the specimen because of dehydration and the thickness of the metal cast has to be taken into account. The diameter of nucleosomes (11 nm) in rotary shadowed specimens can be as large as 16–20 nm.

[11] J. T. Finch and A. Klug, *Proc. Natl. Acad. Sci. U.S.A.* **73,** 1897 (1976).

Metal shadowing from a fixed angle can be used to determine the height of the specimen. Owing to the flexibility of grids, the angles cannot be exactly reproduced. Therefore, latex beads of known diameters should be coprepared as internal standards. This can be done by floating the grids after the washing step on a suspension of latex beads (about 10^{12} beads/ml water; diameter 90 nm, Agar Scientific Ltd., Stansted Essex, UK).

Interpretation of Electron Micrographs. The appearance of chromatin depends on the protein composition, ionic strength, ion composition, and pH. The morphologies can be divided into several classes (Table I, Figs. 1 and 2).

1. Unravelled nucleosomes are observed in chromatin depleted of H1 and nonhistone proteins at very low ionic strength and pH 7. These morphologies are even more pronounced at pH 9 or 10. In solution, these nucleosomes appear to be extended to such a degree that cross-linking of the proteins by the fixative does not result in a folded nucleosome [Figs. 1i, 2c, 3 (black arrows)].

2. Beads on a string is a typical appearance for chromatin at very low ionic strength and high pH or for H1-depleted chromatin above 5 mM monovalent salt. The linker DNA enters and leaves the nucleosomal beads more or less on opposite sides (Figs. 1e, 1k, and 2d). In these structures, the ends of the nucleosomal DNA might be partially unfolded, with or without attached core histones.

TABLE I

APPEARANCE OF SOLUBLE RAT LIVER CHROMATIN AS A FUNCTION OF THE PROTEIN COMPOSITION, IONIC STRENGTH, AND pH[a]

Histone H1	+	+	+	−	−
NHCP	+	+	−	−	−
pH	7	10	7	7	10
Unravelled nucleosomes	—[b]	—	—	1	1
Beads on a string	—	1	1	5–40	5–40
Open zigzag	1	—	5–10	—	—
Closed zigzag	5–10	—	10	—	—
Compact fibers	60	—	60	—	—
Disrupted, ill-defined fibers	20–50	>20	20–50	—	—
Tangles	—	—	—	100	100

[a] NHCP are nonhistone chromatin components that can be removed in 350 mM NaCl. The numbers indicate approximate salt concentrations (millimolar NaCl), as described in the text. An overview of the pH 7 results is presented in Fig. 1.

[b] Structure not observed under the conditions of presence (+) or absence (−) of histone H1 and NHCP at the pH specified.

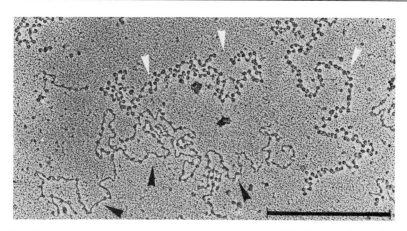

FIG. 3. Mixing experiment. H1-containing chromatin and H1-depleted chromatin were separately fixed in "1" and mixed in the spreading buffer before adsorption. Nucleosome filaments (white arrows) and unfolded nucleosomes (black arrows) are clearly visible, coprepared on the same grid. This indicates that the different morphologies observed reflect a structure in solution and are not produced by preparation artifacts. Bar, 500 nm. [Reproduced from Thoma, Koller, and Klug, *J Cell Biol.* **83**, 403 (1979), with permission of The Rockefeller Press.]

3. Open zigzag is observed in H1-containing chromatin only. Nucleosomes are interpreted to consist of two turns of DNA sealed by histone H1, and the linker DNA is visible. In solution, the nucleosomes are separated by distances longer than the fixative glutaraldehyde. Adsorption of nucleosomes is frequently (but not exclusively) on their flat sides. Since the entry and exit sites of the DNA appear to be on the same side, the nucleosome filament appears as a zigzag filament. Although this appearance is indicative of the presence of histone H1 and allowed its localization, it does not allow one to quantitate the number of H1-containing nucleosomes and hence the interpretation of these morphologies remains qualitative [Figs. 1a, 2e, and 3 (white arrows)].

4. The closed zigzag is observed exclusively in H1-containing chromatin and represents a higher order structure of chromatin at low ionic strength.[12] Neighboring nucleosomes are in solution close enough to be cross-linked by glutaraldehyde, presumably through histone H1 (Figs. 1b, 1f, and 2f).

5. Compact (or 30 nm) fibers are stably cross-linked by the fixative. The overall shape of the fibers appears to be rather irregular. Nucleosomes

[12] F. Thoma and T. Koller, *Cell* **12**, 101 (1977).

can hardly be resolved in shadowed specimens or by negative staining (Fig. 1d and 1h). Cross-striations with a spacing of a nucleosomal diameter can frequently be observed, but their angle toward the fiber axis can hardly be determined owing to the irregularity of the fiber structure. Cross-striations do not necessarily indicate the helical ramp of a nucleosome filament folded into a compact structure.[7]

6. Disrupted or ill-defined fibers are characterized by discontinuous, fibrous features caused by destabilization at high pH or at moderate ionic strength (20–40 mM NaCl; Fig. 1c and 1g).

7. Tangles are fiberless, compact structures (Fig. 1m).

Alternative Methods and Applications. Measurements of scattered electrons by STEM (scanning transmission electron microscopy) provide an excellent method to calculate the mass and dimensions of individual chromatin fibers and subsequently to determine the local DNA packaging density. For this purpose, chromatin is adsorbed to ultrathin carbon film or ABC films as described above, but staining and metal shadowing is omitted.[13] Owing to the necessity of special equipment, this approach remains restricted to a few specialized laboratories.

The most popular method for spreading chromatin of whole cells or chromosomes was developed by Miller and Beatty[5] and used by Olins and Olins[14] to provide electron microscopic evidence for the existence of nucleosomes. The specimen is centrifuged through a sucrose cushion (containing fixative) on the support grid of an electron microscope. (A detailed protocol has been published.[15]) This method is designed for optimal dispersion of chromatin in order to measure transcriptional complexes rather than to keep chromatin fibers in an intact shape. A major limitation of whole cell (or nuclei) spreads is that only a minor fraction of the chromatin can actually be analyzed, which makes a direct correlation to biochemical data (e.g., protein composition) and biophysical data (e.g., sedimentation) difficult. In order to allow a comparison with soluble chromatin, the centrifugation method has been modified by introducing a step gradient. The first step is designed to wash the specimen and to strip proteins (e.g., histone H1). Chromatin is adjusted to the desired ionic strength during centrifugation through the second step, and it is fixed in a third step containing the same ionic strength and glutaraldehyde.[3]

[13] A. Engel, S. Sutterlin, and T. Koller, *Proc. Eur. Congr. Electron Microsc., 7th, The Hague* **2**, 548 (1980).

[14] A. L. Olins and D. W. Olins, *Science* **183**, 330 (1974).

[15] H. Zentgraf, C. -T. Bock, and M. Schrenk, *in* "Electron Microscopy in Molecular Biology" (J. Somerville and U. Scheer, eds.), p. 81. IRL Press, Oxford, 1987.

Psoralen Cross-Linking of Chromatin

The psoralen cross-linking technique of chromatin, originally developed by Hanson et al.,[16] has proven to be a powerful approach to study the structure of bulk chromatin as well as that of chromatin active in transcription or replication. Psoralen derivatives intercalate in the DNA helix and, after irradiation with UV light (366 nm), form covalent cross-links between pyrimidine bases of opposite strands.[17,18] In chromatin, photocross-linking occurs predominantly in the linker regions between nucleosomes. The nucleosomal and higher order structures do not appear to be significantly affected.[19] While many biochemical assays to study chromatin (e.g., nuclease digestions) inevitably destroy the DNA, psoralen cross-linking can be performed in nuclei, cell extracts, or in purified chromatin preparations without DNA degradation. Thus the DNA can be extracted in an intact form and analyzed by molecular biological techniques and by electron microscopy.

Cross-Linking

Reagents and Equipment

Psoralens are from HIR Associates Inc. (Berkeley, CA) or from Sigma (St. Louis, MO). Stock solutions are stored at 4° in the dark and stirred at room temperature for several hours before use until the psoralen is completely dissolved.

TMP (4,5′,8-Trimethylpsoralen), 0.2–1 mg/ml in ethanol or in dimethyl sulfoxide

8-MOP (8-Methoxypsoralen), 5 mg/ml in ethanol

AMT aminomethylpsoralen, 1 mg/ml in ethanol

HMT hydroxymethylpsoralen, 1 mg/ml in redistilled water

Light source for cross-linkings and photoreversal: Psoralen cross-linking is done by placing the samples on ice and irradiating with long-wavelength UV light of 366 nm (Model UVL-56 Blak-Ray Lamp, Ultraviolet Products, Inc., San Gabriel, CA). Photoreversal of the cross-linked DNA is carried out at room temperature with UV light of 254 nm (Camag, Muttenz, Switzerland).

Chromatin, nuclei, and nucleoli: Standard procedures were used for preparation (Table II). SV40 infected nuclei and SV40 chromosomes are extracted from TC7 or CV1 cells [3–5 P150 petri dishes

[16] C. V. Hanson, C.-K. J. Shen, and J. E. Hearst, *Science* **193**, 62 (1976).
[17] L. Musajo and G. Rodrighiero, *Photochem. Photobiol.* **11**, 27 (1970).
[18] R. S. Cole, *Biochim. Biophys. Acta* **217**, 30 (1970).
[19] A. Conconi, R. Losa, T. Koller, and J. M. Sogo, *J. Mol. Biol.* **178**, 920 (1984).

TABLE II
PARAMETERS OF DNA–DNA CROSS-LINKING IN CHROMATIN

Source	Preparation[a]	Buffer[b]	Concentration[c]	TMP[d]	Time[e] (hr)
Soluble rat liver chromatin	(1)	a, a'	0.1–0.3 mg/ml	$3 \times 0.05v$ $4 \times 0.05v$	4 4
Rat liver nuclei	(2)	b	$1-15\ A_{260}$	$4 \times 0.05v$	6
Friend cell nuclei	(3)	c	$25\ A_{260}$	$4 \times 0.05v$	6
TC7 cells	(4)	d	Confluent	$4 \times 0.01v*$	6
CV1 nuclei	(4)	e	$1-15\ A_{260}$	$4 \times 0.05v$	6
SV40 chromo-somes	(4)	f	$3-20\ \mu g/ml$	$3 \times 0.05v$	4
Physarum nucleoli	(5)	g	$10-15\ A_{260}$	$3 \times 0.05v$ $4 \times 0.05v$	4 4
Dictyostelium nucleoli	(6)	h	$10-50\ A_{260}$	$3 \times 0.05v$	3–4
Dictyostelium nuclei	(6)	h	$50-100\ A_{260}$	$3 \times 0.05v$	4–5
Dictyostelium cells	(6)	i	10^6-10^7	$3 \times 0.01v*$	4–5

[a] References for preparation of the material: (1) see text; (2) D. R. Hewish and L. A. Burgoyne, *Biochem. Biophys. Res. Commun.* **52**, 504 (1973); (3) W. F. Marzluff and R. C. C. Huang, *in* "Transcription and Translation: A Practical Approach" (B. D. Hames and S. J. Higgins, eds.), p. 89. IRL Press, Oxford, 1984; (4) W. DeBernardin, T. Koller, and J. M. Sogo, *J. Mol. Biol.* **191**, 469 (1986); (5) K. Behrens, T. Seebeck, and R. Braun, *in* "Biology of *Physarum* and *Didymium*" (H. C. Aldrick and J. W. Daniel, eds.), Vol. 2, p. 301. Princeton University Press, Princeton, New Jersey, 1982; (6) R. Widmer, S. Fuhrer, and R. W. Parish, *FEBS Lett.* **106**, 363 (1979), and (7) P. J. Ness, P. Labhart, E. Banz, T. Koller, and R. W. Parish, *J. Mol. Biol.* **166**, 361 (1983); Erratum: *J. Mol. Biol.* **169**, 639.

[b] Buffers and media used to dissolve or resuspend the material for cross-linking: (a) 5 m*M* TEACl (pH 7), 0.2 m*M* EDTA, NaCl from 0 to 700 m*M*; (a') 5 m*M* diethanolamine chloride (pH 10), 0.2 m*M* EDTA, NaCl from 0 to 700 m*M*. (b) Buffer A (2); (c) 25% glycerol, 5 m*M* magnesium acetate, 0.1 m*M* EDTA, 50 m*M* Tris (pH 8), 5 m*M* dithiothreitol (DTT); (d) Eagle's 10% fetal calf serum; (e) 10 m*M* HEPES–KOH, 5 m*M* KCl, 0.5 m*M* DTT, 1 m*M* EDTA, 0.25 m*M* phenylmethylsulfonyl fluoride (PMSF); (f) pooled from sucrose gradients in 10 m*M* HEPES–KOH (pH 7.8), 5 m*M* KCl, 0.5 m*M* DTT, 1 m*M* EDTA; (g) 20 m*M* Tris-HCl (pH 7.6), 0.25 m*M* PMSF; (h) 20 m*M* Tris-HCl (pH 7.6); (i) HL-5.

[c] $\mu g/ml = \mu g$ DNA/ml; A_{260}, absorbance at 260 nm measured in 1 *M* NaOH.

[d] v, Sample volume; *, 1 mg/ml TMP in ethanol or dimethyl sulfoxide.

[e] Total time of irradiation at 366 nm in the dark on ice.

(15 cm diameter) or 7–10 P100 petri dishes]. SV40 chromosomes are further purified on sucrose gradients and pooled.[20]

Protocol for DNA–DNA Cross-Linking. Formation of DNA–DNA cross-links in chromatin is performed as follows (details are summarized in Table II).

[20] W. DeBernardin, T. Koller, and J. M. Sogo, *J. Mol. Biol.* **191**, 469 (1986).

1. A volume of sample (v; from 0.3 to 1 ml) is placed in an open petri dish (3.5 cm diameter) on ice. Alternatively, samples can be put in polypropylene tubes (Falcon No. 2059), which are placed horizontally on ice.
2. The light source (366 nm) is mounted at a distance of 7–8 cm above the petri dishes.
3. $0.05v$ TMP (0.2 mg/ml) is added to the sample. The petri dish is gently swirled to ensure solubility of the psoralen. After addition of psoralen, all manipulations are done in the dark or at dimmed light.
4. After 10 min, the samples are irradiated for 1 hr at 366 nm.
5. Addition of psoralen and cross-linking is repeated as indicated in Table II.

Protocol for Cross-Linking of DNA–RNA in Addition to DNA–DNA. Since 8-MOP has a higher DNA–RNA cross-linking efficiency than TMP, irradiation is carried out in two steps using 8-MOP followed by TMP.[20] Although this procedure can also be used for other chromatin sources, it will be described for SV40 minichromosomes.

1. A volume of $0.05v$ of 8-MOP is added to the pooled SV40 chromosome fractions, the solution is mixed for 10 min, and the sample is irradiated for 1 hr. This procedure is repeated once.
2. This procedure is then repeated 3 times with $0.05v$ of TMP (0.2 mg/ml).

In the case of SV40-infected nuclei, 8-MOP is added twice at intervals of 1.5 hr and TMP 3 times at intervals of 1 hr. We do not recommend using AMP or HMP to visualize DNA–RNA cross-links in the electron microscope, since intramolecular cross-links in the RNA chain prevent its unfolding during spreading and consequently proper interpretation of the pictures.

Controls for Cross-Linking. Control experiments are performed on parallel samples by omitting irradiation or by adding ethanol or dimethyl sulfoxide instead of cross-linking solutions.

Photoreversal of Cross-Linking. The DNA samples (maximum 1 ml per petri dish of 3.5 cm diameter, 10–1000 μg DNA/ml in 20 mM Tris, 1 mM EDTA, pH 7.8) or agarose gels are placed 20 cm from the light source and irradiated for 1 hr at room temperature in the dark.

Purification of Cross-Linked DNA and DNA–RNA

1. Nuclei or chromatin is adjusted to 20 mM EDTA, 30 mM TEACl (pH 7.8), 0.5% sodium dodecyl sulfate (SDS).
2. Deproteinization is performed by incubation in the presence of

pronase (0.6 mg/ml) overnight at room temperature and then in the presence of proteinase K (0.2 mg/ml) at 65° for 5 hr. Alternatively, twice proteinase K may be used with incubation times of 5 hr. During incubations, the tubes are wrapped in aluminum foil to prevent additional cross-linking.

3. After treatment with proteinase, nuclear DNA is sheared by forcing it 10 times up and down through a 21-gauge syringe needle followed by 5 times up and down through a 23-gauge needle.

4. After extraction with phenol, phenol–chloroform–isoamyl alcohol, and chloroform–isoamyl alcohol, the nucleic acids are precipitated with ethanol and dissolved in 30 mM TEACl (pH 7.8), 20 mM EDTA, 10 mM NaCl.

Preparation of EBC Grids[21]

Carbon grids (C grids) are prepared as described earlier with the following modifications: (1) Evaporation of carbon is at a pressure between 7 and 1×10^{-5} torr. (2) C grids are not baked and can be stored on filter paper for several weeks. (3) Coating with ethidium bromide is done by floating the C grids (carbon side down) for 15–20 min on a droplet containing 30 μg ethidium bromide/ml of redistilled water. The excess of ethidium bromide is removed with filter paper, and the grids are air-dried. (4) The EBC grids should be used within several hours.

It is important to keep in mind: (1) High vacuum during carbon evaporation or aging of EBC grids reduces the number of DNA molecules adsorbed by the BAC monolayer technique. (2) In only 30–40% of the different batches of EBC grids will the nucleic acids be sufficiently unfolded. (3) Alternatively, ABC grids or polylysine grids can be used, but the higher background from these grids makes it difficult to follow single-stranded DNA molecules.

Spreading of Cross-Linked DNA

The DNA is completely denatured and spread in a monolayer of BAC on the surface of an aqueous hypophase.[10]

Stock Solutions

TEN: 30 mM TEACl (pH 7.8), 20 mM EDTA, 10 mM NaCl
BAC stock: 0.2% (w/v) BAC in deionized formamide (can be kept at room temperature for at least 1 year)

---

[21] J. M. Sogo, P. Rodeno, T. Koller, E. Vinuela, and M. Salas, *Nucleic Acids Res.* **7**, 107 (1979).

BAC spreading solution: mix 9 volumes TEN and 1 volume BAC stock immediately before use

Uranyl acetate stock: 5 mM uranyl acetate, 5 mM HCl in water

Uranyl acetate staining solution (freshly prepared before use): 10-times diluted stock solution in ethanol

Formamide, deionized; glyoxal (30%, Merck)

Procedure

1. To prepare the hypophase, a petri dish (13.5 cm diameter) is filled with redistilled water. A freshly cleaved sheet of mica (5 × 2 cm) is inserted as a ramp. Graphite powder is dusted onto the water surface.
2. Psoralen-cross-linked DNA is denatured in a buffer containing 70% formamide and 0.4 M glyoxal[22] by sequential addition of 1 μl DNA (20–100 μg/ml) in TEN, 2 μl TEN, 9.3 μl formamide, and 1 μl glyoxal. This mixture is incubated at 37° for 25–30 min and then immediately cooled on ice.
3. Five microliters TEN is added to dilute formamide concentration to 50%, followed by 1 μl BAC spreading solution. The solution is gently mixed to avoid foaming.
4. A 5-μl droplet is carefully applied to the mica ramp and allowed to roll down and spread on the water surface. The resultant film has a diameter of several centimeters.
5. The DNA–BAC film is then picked up close to the graphite powder border or to the mica ramp by the EBC grids.
6. The grids are submerged for 15–20 sec in uranyl acetate staining solution, rinsed twice in 98% ethanol, and air-dried on filter paper (carbon side up).
7. The grids are rotary shadowed with 700–800 Hz platinum–carbon at an angle of 3°, as described above.

Additional Comments. The lengths of single-stranded or double-stranded molecules can vary with different grids. Therefore, it is important to include size markers for double-stranded DNA (e.g., completely cross-linked plasmid DNA) and for single-stranded DNA (e.g., ϕX174 DNA) directly into the spreading solution. Since no washing is performed, the amount of formamide that remains together with the deposited uranyl gives an excellent contrast, which is important for visualizing single-stranded filaments. Washing of the grids in water for 10–20 sec prior to

[22] T. R. Cech and M. L. Pardue, *Proc. Natl. Acad. Sci. U.S.A.* **73**, 2644 (1976).

staining reduces the background; however, prolonged washing might result in a loss of DNA molecules.

Enhancement of Denatured Regions by Binding Single-Strand Binding Proteins

We have used single-strand binding protein (gene 32 protein of bacteriophage T4) to increase the contrast of single-stranded regions. This allows the detection of single-stranded regions smaller than 40–50 bp,[23] which cannot be detected by using the technique described above.[24]

1. The cross-linked DNA is incubated for 90 min at 37° in a buffer containing 1 M glyoxal, 20 mM sodium phosphate (pH 7.0).
2. The samples are precipitated with 2 volumes of ethanol, washed twice with 70% (v/v) ethanol, and redissolved in 1 mM phosphate buffer (pH 7.0).
3. The glyoxylated DNA (20–100 μg/ml) is heated to 90° for 10 min and immediately placed on ice. Protein p32 is then added (ratio p32/DNA = 10), and the sample is incubated at 37° for 4 hr.
4. The complexes are fixed by addition of 1% glutaraldehyde (buffered in 10 mM phosphate, pH 7.8) to 0.1% end concentration and incubation for 15 min at 37°.
5. The complexes are purified on a column of Sephadex CL-2B (4 × 0.5 cm), equilibrated with 0.01% glutaraldehyde, 30 mM TEACl (pH 7.8).[25]
6. The complexed DNA samples are diluted to a final concentration of 1–10 μg/ml with 30 mM TEACl, and formamide is added to a final concentration of 30% (w/w).
7. The DNA–protein complexes are spread onto redistilled water by the BAC method as described above.

Discussion

Owing to the large amount of proteins bound to the DNA in chromatin, as well as its compact state, details of chromatin structure might be obstructed when chromatin is visualized as a nucleoprotein complex. Although the psoralen cross-linking assay is an indirect approach, it is well established that DNA in nucleosomes is not cross-linked and appears as

[23] J. M. Sogo, A. Stasiak, W. DeBernardin, R. Losa, and T. Koller, in "Electron Microscopy in Molecular Biology" (J. Sommerville and U. Scheer, eds.), p. 61. IRL Press, Oxford, 1987.
[24] R. Portmann, W. Schaffner, and M. Birnstiel, Nature (London) **264**, 31 (1976).
[25] H. Delius and B. Koller, J. Mol. Biol. **142**, 247 (1980).

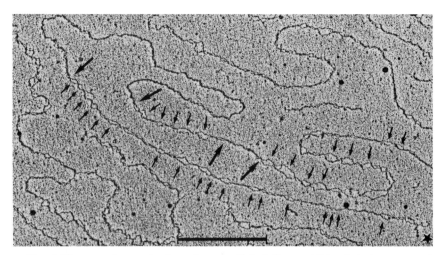

Fig. 4. Electron micrograph of psoralen-cross-linked ribosomal DNA in isolated nucleoli of *Dictyostelium discoideum*. Only half of the palindrome is shown. The star indicates one end of the palindrome. The coding region and putative regulatory 5' region are visible as duplex DNA (between large arrows) and hence intensely cross-linked. The inactive chromatin regions are organized in single-stranded bubbles reflecting the presence of nucleosomes (small arrows). Bar, 2 kb.

single-stranded bubbles of about 160 base pairs (bp) in denatured samples. It is therefore a sensitive method for detecting the linear arrangement of nucleosomes along DNA and the presence or absence of nucleosomes in bulk, as well as in actively transcribing or replicating chromatin. Furthermore, it allows one to obtain micrographs of the nucleosomal arrangement along several kilobases of chromatin (e.g., 95 kb ribosomal DNA of *Dictyostelium discoideum*.[26]) The use of restriction enzymes, gel electrophoresis, blotting, and hybridization procedures with cross-linked DNA in combination with electron microscopy makes psoralen cross-linking a powerful method.[26,27]

An example of cross-linked ribosomal DNA is shown in Fig. 4. Regions where regularly spaced, single-stranded bubbles are visible (smaller arrows) are interpreted to correspond to the terminal and central inactive domains organized in nucleosomes. The heavily cross-linked stretches (between large arrows, Fig. 4) with the appearance of duplex DNA support the notion that this region is free of normal nucleosomes. These duplex DNA

[26] J. M. Sogo, P. J. Ness, R. M. Widmer, R. W. Parish, and T. Koller, *J. Mol. Biol.* **178,** 897 (1984).

[27] R. Lucchini, U. Pauli, R. Braun, T. Koller, and J. M. Sogo, *J. Mol. Biol.* **196,** 829 (1987).

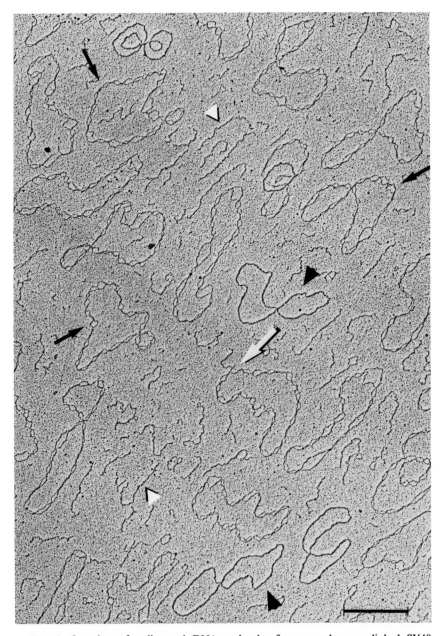

FIG. 5. Overview of well-spread DNA molecules from psoralen-cross-linked SV40 minichromosomes. The distribution is rather uniform, and although the number of molecules per area is high, they are not in contact with each other. Most of the circular molecules are perfectly unfolded or do not have more than one random overlapping point. Owing to the heavy contrast and the low background, the difference between double-stranded DNA (arrowheads) and single-stranded DNA (arrows) is clearly visible. RNA molecules are also well unfolded (white arrowheads). Bar, 1 kb. (Courtesy W. DeBernardin.)

stretches correspond to the known positions of the coding sequences (long duplex regions) and the putative regulatory 5′ regions (short duplex regions) where topoisomerase activity has been described.[26,28]

Figure 5 shows a view of cross-linked SV40 chromosomal DNA. Most of the circular molecules (arrows) are organized in single-stranded bubbles and are presumably inactive. Covalently closed circular DNA appears as duplex DNA (arrowheads). Nicking by DNase I digestion in presence of ethidium bromide or mechanical nicking by handling is required to visualize the single-stranded bubbles. When some sections in the same circular molecule are supercoiled, but others are relaxed, they can be unambiguously distinguished by the psoralen technique. In SV40 replicating intermediates (Fig. 6a), the unreplicated parental stem remains topologically fixed with negative superhelical turns, whereas the two newly replicated sections are topologically relaxed. After introducing a nick, the bubbles are again visible in the parental stem (Fig. 6b). Frequently, an asymmetrical structure at the replication fork is clearly visible with one double-stranded branch (arrows) and one single-stranded branch (arrowheads). Our interpretation is that the double-stranded branch comprises the leading (or forward) strand and the lagging (or retrograde) strand. The partially single-stranded branch is interpreted as the position where the discontinuous DNA synthesis occurs.

By using a combination of 8-MOP and TMP, it was possible to cross-link nascent RNA strands to the DNA template in transcribing chromatin. Figure 5 (white arrow) and Fig. 6c show SV40 DNA organized in regularly spaced single-stranded bubbles with one attached single-stranded tail. It is our opinion that these single-stranded tails represent RNA for the following reasons: (1) they disappear on incubation with RNase; (2) no tails were found to exceed the length of half the SV40 template; (3) in *Dictyostelium* ribosomal genes, several single-stranded tails were found attached in the coding region, which increased in length going toward the 3′ end of the gene (Fig. 2 in Ref. 20); (4) when such DNA–RNA complexes were cut with restriction enzymes near the transcription start sites, the length of the template DNA was similar to the length of the RNA.

Psoralen cross-linking can also be a powerful tool for detecting conformational changes in bulk chromatin. Bulk H1-containing chromatin has the appearance of single-stranded bubbles after cross-linking, which is independent of the ionic strength ($1 - 100$ mM) and pH ($7 - 10$).[29] In H1-depleted chromatin, nucleosomes are known to unfold partially at very low ionic strength and pH 7 (Figs. 2c and 1i). Complete unfolding is obtained

[28] P. J. Ness, R. W. Parish, and T. Koller, *J. Mol. Biol.* **188,** 287 (1986).
[29] R. Widmer, T. Koller, and J. M. Sogo, *NAR* **16,** 7013 (1988).

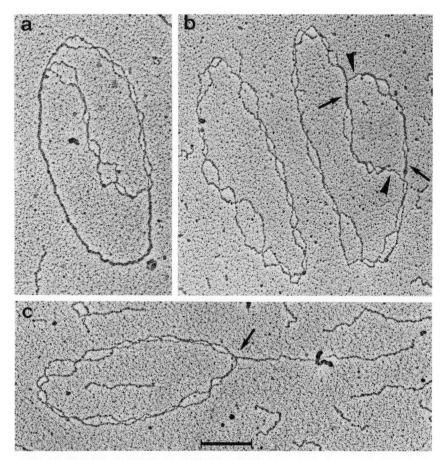

FIG. 6. DNA from psoralen-cross-linked replicating (a, b) and transcribing (c) SV40 minichromosomes. (a) The unreplicated parental strand is supercoiled and appears covalently closed in the micrograph. In the relaxed daughter strands, the appearance of the first single-stranded bubble after the replication fork is asymmetric. (b) Note the replication fork with the leading (arrows) and the lagging (arrowheads) strands. In this molecule, the parental strand is relaxed and reveals its nucleosomal bubbles. A putative inactive molecule is also shown. (c) A RNA molecule is cross-linked to SV40 DNA (arrow). Note that not all single-stranded bubbles have the same size. Some of them correspond to the size of two or three adjacent nucleosomes where the linker DNA was not cross-linked [for details, see J. M. Sogo, H. Stahl, T. Koller, and R. Knippers, *J. Mol. Biol.* **189**, 189 (1986)]. Bar, 0.5 kb.

at very low ionic strength and pH 10. Whereas H1-depleted chromatin cross-linked at very low ionic strength and pH 7 shows rather regularly distributed bubbles (Fig. 7a), the same chromatin cross-linked at very low ionic strength and pH 10 shows predominantly duplex DNA (Fig. 7b). The

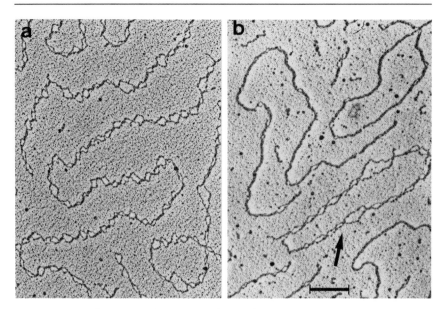

Fig. 7. DNA from H1-depleted chromatin (rat liver) was cross-linked with psoralen at very low ionic strength (a) in 5 m*M* TEACl (pH 7), 0.2 m*M* EDTA and (b) in 5 m*M* DEACl (pH 10), 0.2 m*M* EDTA. In order to demonstrate that spreading was done under optimal conditions, SV40 DNA from psoralen-cross-linked minichromosomes (arrow) was added to the spreading solution. Bar, 0.5 kb. (Courtesy R. M. Widmer.)

discriminating reactivity of psoralen in H1-depleted chromatin at different pH values indicates that this technique is a powerful way to analyze domains containing or lacking H1, especially in regions of ongoing transcription or replication.

Finally, it is important to note that since psoralen is permeable to the membranes, it is possible to perform the photoreaction in whole cells. This implies that, although being an indirect method, the psoralen cross-linking technique allows for the morphological study of chromatin with minimum manipulations.

Acknowledgments

We are grateful to Prof. T. Koller for his guidance and support. We thank Prof. M. Smerdon for discussions and comments on the manuscript and Mrs. Mayer-Rosa for the artwork. This work was supported by grants from the Swiss National Science Foundation (Nr. 3.202–085 to F. T. and Nr. 3.201.098 to J. M. S.) and from the Prof. Dr. Max Cloëtta Foundation (to F. T.).

[8] Electron Microscopy of Simian Virus 40 Minichromosomes

By PIERRE OUDET, PATRICK SCHULTZ, JEAN-CLAUDE HOMO, and PIERRE COLIN

Introduction

Electron microscopy has proven to be a very powerful technique in the characterization of protein–DNA interactions, providing direct structural information in cases where more precise and more laborious approaches, like X-ray diffraction of three-dimensional crystals, are not possible. The theoretical resolution of most electron microscopes is around 0.2–0.5 nm, but in practice resolution is usually limited to 2 nm on a biological specimen. This limitation is due mainly to the damage caused to the molecules during the preparation of the specimen and to the electron irradiation necessary for image formation. Two main techniques were developed in order to minimize such effects: staining of biological molecules adsorbed on charged grids with heavy salts and shadowing with tungsten particles, and direct observation of hydrated frozen samples.

Electron microscopic observation of chromatin fibers is a fast and direct method to analyze compaction of the DNA filament in nucleoprotein complexes and their modulation caused by modifications of the environment or interactions with regulatory factors. Direct observation of chromatin structure with electron microscope has, however, some limitations: at physiological ionic strength, the isolated polynucleosomal chain has the unfavorable characteristic of aggregating and precipitating, thus making the visualization and analysis of the substructures impossible. To determine the nature of the structures localized on a precise DNA sequence, it is necessary to obtain a fairly homogeneous population of well-spread structures, which can be analyzed statistically, in order to take into account variations introduced by the preparation. These two major problems are circumvented by the use of SV40 minichromosomes as a model system. The purified material can be obtained in sufficient quantity and is soluble in different ionic conditions even in the presence of divalent cations.

Specimen Preparation

Preparation of Electron Microscope Grids

Preparation of Grids. Grids with small squares (300–400 mesh nickel or copper grids; Fullam, Latham, NY) are generally used for the observa-

METHODS IN ENZYMOLOGY, VOL. 170   Copyright © 1989 by Academic Press, Inc.
All rights of reproduction in any form reserved.

tion of isolated macromolecules. The limited size of the squares increases the stability of the supporting film and permits the use of thin films. To favor sufficient adhesion of supporting films, the grids are cleaned before use by a 30-min wash in dichloro-1,2-ethane and then are rinsed in absolute ethanol. After drying, the grids are ready to be covered by the supporting carbon film.

Preparation of Supporting Carbon Films. Carbon films are generally used as the support for specimens in electron microscopy. Their preparation is easy and reproducible, and they have good stability under the electron beam in contrast to nitrocellulose or Formvar films. A thick carbon film (~20 nm thick) may be used for negatively stained or shadowed specimens that present a high contrast. For high-resolution or dark-field visualization of unstained or slightly positively stained specimens, it is necessary to use very thin films (2–5 nm thick). These thin films are fragile and need to be sustained by a thick, holey carbon film. The biological specimens are observed in the holes where they are supported only by the thin film.

Preparation of thick carbon film-coated grid

1. Mica sheets (4 × 4 cm; Pelanne Instruments, Toulouse, France) are carefully cleaved with forceps in order to obtain a clean and perfectly flat surface. The freshly cleaved mica sheets are immediately placed in an evaporator with their clean surface facing the carbon source.

2. The carbon is evaporated from a rod placed in an electron beam device (Edwards, Crawley, West Sussex, UK) under a vacuum of 10^{-6} to 10^{-7} torr. A good vacuum precludes the formation of a clean and homogenous carbon film. The mica sheets are placed at a distance approximately 20 cm from the source. The carbon evaporation has to be stable and regular. The mica sheets are protected by a retractable mask during the stabilization of the emission. The thickness of the evaporated carbon can be evaluated with a quartz instrument (Balzers Union AG, FL-9496, Balzers, Liechtenstein) or empirically by placing a folded piece of paper close to the mica sheet. A 10- to 20-nm-thick carbon film is produced when a slightly brown color is observed on the exposed paper surface.

3. The carbon-coated mica sheets can be stored in a petri dish for at least 2 months without alteration.

Preparation of thin, carbon-coated, holey carbon films. The method originally described by Fukami and Adachi[1] is generally used with some slightly modifications:

1. Standard 25 × 75 mm optical microscope glass slides are used to form a holey plastic film. The slides are first degreased and cleaned by

---

[1] A. Fukami and K. Adachi, *J. Electron Microsc.* **14**, 112 (1965).

boiling in a detergent solution (7×, full strength; Flow Laboratories, Ricksmansworth, UK), followed by a thorough overnight wash in flowing tap water. The cleaning and rinsing of the glass slides can be shortened by sonication for 15 and 30 min, respectively.

2. The slides are then made hydrophobic by a 20-min bath in a 0.1% aqueous solution of Osvan (Th. Christ AG, Aesch, Germany).

3. The slides are dipped in double-distilled water to remove excess detergent. Their hydrophobicity is attested by the absence of water droplets when the slides are pulled out of the water.

4. The dry slides are placed on a cold aluminum plate (Fig. 1a), about 2 cm thick, which is stored at −20°. Small water droplets will condense on

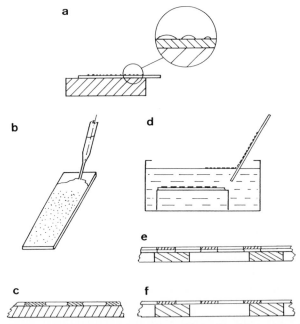

FIG. 1. Preparation of holey films. A clean glass slide is placed for a short time on a cold aluminum block, which between experiments is stored at −20° (a). A fine haze appears progressively on the slide owing to the condensation of water; after less than 1 min the droplets have a diameter of a few micrometers. The slide is removed, and the solution of Triafol is immediately applied (b), using a Pasteur pipette. A thick film of plastic is thus formed containing holes (c) whose diameters can be analyzed by optical phase-contrast microscopy. After complete drying, the film is floated on a surface of distilled water (d). Detachment from the glass slide is made easier by scratching the edges of the slide with forceps. The film is laid down on the grids simply by lowering the water level. A cross section of the grid after evaporation of a thick carbon film is represented (e). Dissolution of the Triafol results in a holey film (f).

the cooled slide and form a fine haze. This point is critical because the size and distribution of the final perforations are determined by the size of the water droplets. The condensation process is dependent on the partial water pressure of the atmosphere, and therefore the time of condensation varies (generally between 5 sec and 1 min).

5. When the condensed water layer is formed, the slide is immediately coated with Triafol (Balzers), a plastic solution made at a concentration of 1% (w/v) in ethyl acetate. The plastic solution is applied dropwise on the edge of the slide, which is tilted to allow the coating of the entire surface (Fig. 1b). Excess Triafol is removed by blotting an edge of the slide on filter paper. On evaporation of the solvent, a solid Triafol film is formed which is perforated at the places of the water droplets (Fig. 1c). The quality of the perforations is checked by light microscopy. The holes should be 0.1 – 5 μm in diameter, and they should be densely and homogeneously distributed. The size and distribution of the holes are generally optimal only on a portion of the slide, which can be delineated by a glass marker on the Triafol-free side of the slide.

6. The holey Triafol film is then transferred onto the electron microscope grids. As the film normally sticks on the glass slide, it is pealed off by a detergent. The Triafol-coated slides are dipped in a 0.2% aqueous solution of dioctyl sulfosuccinate (Sigma, St. Louis, MO) for 20 min. Excess detergent is removed by a 10-min immersion in a double-distilled water bath. The perforated Triafol film is then floated off the glass slide at the air–water interface of a 35-mm-deep petri dish (Fig. 1d). The floating is facilitated by scratching the Triafol film at the edges of the glass slide. A set of 50–75 precleaned electron microscope grids is immersed in the petri dish before floating of the Triafol film, and the set is placed on a piece of filter paper lying on a metallic mesh support (see Fig. 1d). The area covered by the grids should correspond to the preselected area of the perforated film. The floated Triafol film is deposited on the grids by lowering the water level with a water pump.

7. The Triafol film is unstable under the electron beam, where it melts and bubbles. To prevent this instability of the supporting film, it can be replaced by a thick carbon film. The filter paper on which the Triafol-coated grids are placed is introduced directly into the evaporator. A thick carbon layer is deposited over the Triafol film, as described in Preparation of Supporting Carbon Films (Fig. 1e). The Triafol film is later dissolved in vapors of the ethyl acetate. This is achieved by an overnight incubation of the filter paper, supporting the Triafol and carbon-coated grids, on 10–20 layers of filter paper soaked with ethyl acetate and placed in a petri dish (Fig. 1f).

8. The focalization and astigmatism corrections are generally on the

graininess of the image of the carbon film (phase-contrast image). However, in some cases, like the scanning transmission electron microscopy
(STEM) low-dose mode or for frozen hydrated specimens, this is not
possible. It may be helpful to deposit gold particles on the thick, perforated
carbon film, which can be done at this stage by evaporating a 2-cm-long,
0.1-mm-thick gold wire (Fullam) onto the carbon film (see Rotary Shadowing).

9. The grids, covered with a thick perforated carbon film, are then
covered with a thin carbon film evaporated on a freshly cleaved mica sheet
as described above. The adequate 2–5 nm thickness of the supporting film
corresponds to a barely perceptible brownish color on the folded piece of
paper, which generally cannot be observed from outside the evaporator.
The floating of such thin films may be difficult because of their fragility
and because they are barely visible at the surface of the water, even by
refringence. It is helpful to use a small device (Fig. 2a) that immerses the
carbon-coated mica automatically into the water with very little motion
and without shaking. The floating film can be followed visually by watch-
ing the displacement of small dust particles, which are often adsorbed on
the edge of the film.

Modifications of Surface Properties of Carbon Films

Nucleosomes must be absorbed on the supporting carbon film before
being further processed for observation by electron microscopy. The sur-

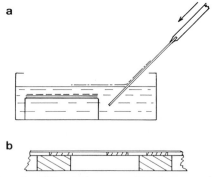

FIG. 2. Floating thin carbon films. Thin carbon films 2–5 nm thick, evaporated on mica
sheets, are very fragile and must be processed carefully; otherwise they split into small pieces
before being adsorbed on the electron microscopic grids. The carbon-coated mica sheet is
handled by forceps held by a motor-driven axis (5 cm/min). This allows a regular, slow
detachment of the carbon film on the water surface (a). The position of the film on the water
surface is visualized by the displacement of small dust particles. A cross section of the thin
carbon film supported by the thick holey film is represented in b.

face properties of the carbon film are essential for efficient adsorption of the nucleoprotein complex. The freshly evaporated carbon film is generally hydrophobic and slightly negatively charged[2,3] and can be used as such to visualize proteins, but it is not suitable to obtain fixation of well-spread nucleic acids.

In our laboratory we employ the glow-discharge method developed by Dubochet et al.[4] to modify the surface properties of the carbon film. The method consists of ionizing various gases in the presence of the carbon film, thus producing a charged surface. Two types of glows are frequently used, giving reproducible results: (1) The air glow discharge produces a hydrophilic and negatively charged grid.[2,3] The carbon film then appears to be suited for efficient adsorption of proteins and protein complexes and allows negative staining. (2) The amylamine glow is believed to provide a positively charged surface as indicated by the selective adsorption of negatively charged polystyrene spheres.[3] Nucleic acids as well as nucleoprotein complexes are efficiently adsorbed on amylamine glow-discharged grids.

The glow-discharge apparatus is composed of two separate bell jars and an electrical system that generates a potential variable between 10 and 1000 V (Fig. 3). Two separate vacuum bells are necessary in order to produce an amylamine-free atmosphere for the air glow. Amylamine is corrosive both for the rotary pump and for the operator. An external outlet is therefore advisable.

Experimental Procedure

1. The grids are placed on a glass slide and deposited onto the lower electrode in the glow-discharge bell.

2. The bell is pumped down by the rotary pump to 10^{-2} torr.

3. Amylamine is introduced at a constant flow until the vacuum reaches 10^{-1} torr, using a Teflon needle valve.

4. A potential of 500 V is then applied between the two electrodes, which ionizes the residual atmosphere and gives rise to a current of 50 μA. The glow produces a barely perceptible purple color.

5. After 30 sec of glow discharge, the ionization and pumping are stopped, and the atmospheric pressure is restored by an automatic air-inlet valve.

6. The grids remain active for about 20 min.

[2] J. Dubochet, M. Groom, and S. Mueller-Neuteboom, *Adv. Opt. Electron Microsc.* **8,** 107 (1981).

[3] C. L. F. Woodcock, L. L. Frado, G. R. Green, and L. Einck, *J. Microsc. (Oxford)* **121,** 211 (1981).

[4] J. Dubochet, M. Ducommun, M. Zollinger, and E. Kellenberger, *J. Ultrastruct. Res.* **35,** 147 (1971).

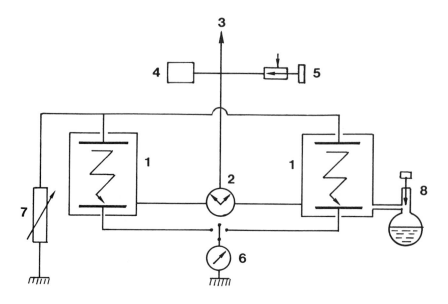

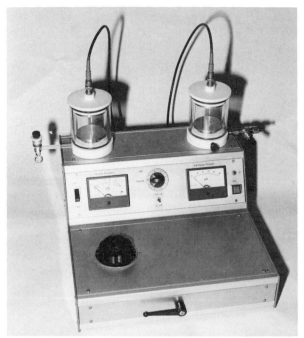

FIG. 3. Glow-discharge apparatus used to modify the surface properties of carbon films. The system is composed of two independent bell jars (1) allowing discharges in the presence of amylamine on one side or glow discharges in a clean, amylamine-free bell jar on the other side. The jars are connected to the vacuum pump (3) by a three-way valve (2). The vacuum is measured using a Pirani gauge (4) connected to the vacuum circuit. Air is introduced using

Adsorption and Staining Procedures

1. A 10-μl drop of nucleosomes or SV40 minichromosomes at a DNA concentration of 0.1–5 μg/ml is applied on the activated surface of the grid. The adsorption can be performed under various conditions: at low (10 mM Tris-HCl, pH 7.4, 0.2 mM EDTA) or at nearly physiological ionic strength (150 mM NaCl, 10 mM Tris-HCl, pH 7.4, 0.2 mM EDTA) and in the presence or absence of divalent cations. The presence of less than 5% sucrose, as is the case with material purified on sucrose gradients, does not interfere. The adsorption is more efficient at low ionic strength than at high ionic strength (200 mM NaCl). The charged grid is slightly hydrophobic when the drop of suspension is applied. After contact for 30 sec, however, the wettability of the grid is apparently modified, and a thin layer of suspension should remain tightly associated with the surface when the excess solution is pumped from the side of the grid using a clean filter paper. Under these conditions, maximum adsorption is obtained after 1 min of diffusion (P. Oudet, P. Schultz, J.-C. Homo, and P. Calin, unpublished observation).

2. The grid is rinsed twice with 10 μl of adsorption buffer or distilled water. The change in the ionic environment does not seem to affect the macromolecular structures, tightly adsorbed on the grids. For the observation of unstained specimens, the grid is blotted against a filter paper and air-dried.

3. Positive staining is obtained by a 30-sec application of 1% (w/v) uranyl acetate solution in water. This staining solution is quickly replaced by a 10-μl drop of double-distilled water for 10 sec and then the grid is blotted on a filter paper and air-dried. Negative staining is difficult to obtain when the grid has been glow discharged in the presence of amylamine. This is probably a consequence of the negative charge of the surface.

Rotary Shadowing

To enhance the contrast of the nucleoprotein complexes, and particularly of the nucleic acids that are sometimes difficult to follow with precision, the grids are rotary shadowed with platinum–tungsten before examination by electron microscopy. However, fine structural details are generally masked by this technique.

---

the air-inlet needle valve (5). Adjustable alternating current is applied to the two sets of electrodes by a voltage transformer 10 VA/0–1000 V (7) and followed by a ammeter 0.5 mA/0–100 μA (6). Introduction of amylamine from a small reservoir is adjusted using the amylamine-inlet needle valve (8). Detailed description of the apparatus and operating procedures is available on request from one of the authors (J.-C. Homo).

1. Place the grid on a rotary support, the side on which the specimen is adsorbed toward the source, at a distance of 10 cm and at an angle of 7° – 10°. Wrap a piece of platinum wire (5 cm long, 0.1 mm thick; Fullam) around a 1-mm-thick tungsten filament.

2. Carry out the platinum vaporization under a vacuum of 10^{-6} torr with the grid support rotating at 60 rpm. Increase the voltage until melting of the platinum is observed through dark glasses. Wait 5 sec and then double the voltage for 5 – 10 sec. The girds are now ready to be examined by electron microscopy.

Frozen Hydrated Suspensions of Nucleosomes

The structure of nucleosomes and chromatin is fragile; adsorption, staining, and dehydration are likely to modify the architecture of the nucleoprotein complexes. Methods were developed to preserve the full hydration of the specimen in the electron microscope.[5-8] The principle of the described method, now used routinely in several laboratories, is to freeze a thin layer of suspension quickly in order to vitrify the solvent. The biological specimen is included in a layer of noncrystalline ice and is transferred and kept in the electron microscope in a cold stage operating below − 170°. This method has the following advantages: (1) the aqueous environment of the specimen is conserved; (2) the specimen does not interact with the carbon surface; (3) the specimen may be observed at high concentration; (4) the ionic environment of the specimen may be controlled; and (5) the specimen is observed in the absence of stain by phase contrast, and all of its parts contribute equally to the image formation.

The usefulness and efficiency of this method were demonstrated by the observation of well-preserved periodic or nonperiodic structures.[8-10] The biological specimens must be maintained in a hydrated condition in order to keep the structures intact at high resolution (for a recent review, see Ref. 11). Unfortunately this approach has severe constraints: specific equipment, such as the cold-stage adaptation in the electron microscope, and careful manipulation and transfer into the microscope of a clean, thin film of amorphous ice are required; potential interactions of the material of

[5] K. A. Taylor and R. M. Glaeser, *Science* **186,** 1036 (1975).

[6] K. A. Taylor and R. M. Glaeser, *J. Ultrastruct. Res.* **55,** 448 (1976).

[7] J. Dubochet, J. Lepault, R. Freeman, J. A. Berriman, and J.-C. Homo, *J. Microsc. (Oxford)* **128,** 1219 (1982).

[8] J. Lepault, and T. Pitt, *EMBO J.* **3,** 101 (1984).

[9] M. Adrian, J. Dubochet, J. Lepault, and A. W. McDowall, *Nature (London)* **308,** 32 (1984).

[10] G. P. A. Vigers, R. A. Crowther, and B. M. F. Pearse, *EMBO J.* **5,** 529 (1986).

[11] J. Dubochet, M. Adrian, J. J. Chang, J.-C. Homo, J. Lepault, A. W. McDowall, and P. Schultz, *Q. Rev. Biophy.* **21,** 2, 129 (1988).

interest with the surface can affect the structures; and the direct interpretation of the contrast of the different structures is difficult owing to the high defocalization values used (1 – 5 μm).

Scanning Transmission Electron Microscopy of the Chromatin Structure

Because of the high efficiency of the electron detectors, STEM makes it possible routinely to obtain images of good quality. This characteristic is improved further by the digitalization of the images and the resulting improvement in the contrast of specific structures. STEM allows the visualization of unstained chromatin preparations: individual nucleosomes are recognized, and the number of electrons diffused by the nucleosome is counted. The linear relationship between this number, at a first approximation, and the mass of the biological material allows a direct evaluation of the mass of individual particles with the use of an internal standard (see Fig. 4). Despite the necessity to associate cryogenic and image-processing developments, the scanning transmission electron microscope is the instrument of choice to characterize macromolecular complexes.

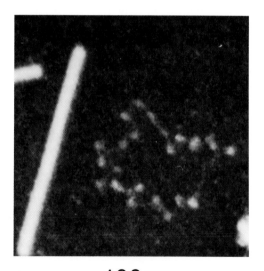

⊢————⊣ 100nm

FIG. 4. Mass determination of unstained minichromosomes. Together with TMV particles as an internal standard, minichromosomes are adsorbed in presence of low salt on amylamine-charged grids. This image originates from the signal on an annular detector of the scanning transmission electron microscope VG HB5. The clearly recognizable nucleosomes are connected by a thin strand. Counting of the scattered electrons allows a direct mass determination of the nucleosomes or of the whole minichromosome.

Observation of SV40 Minichromosomes and Nucleosomes

Minichromosomes Prepared in the Presence of Salt

The nucleoprotein complexes can be adsorbed on charged grids, as described above, in a salt environment similar to the physiological ionic strength. Under these conditions and after staining, the complexes are observed either as a compact structure 35–40 nm in diameter or as more or less unravelled or flattened structures 60–70 nm in diameter where nucleosomes are recognized (Fig. 5). Embedding of the structures in a negative staining prevents to some extent flattening of the structures.

Minichromosomes Prepared in Low Salt Conditions

In order to identify individuals nucleosome and to unravel the naturally compact structure of the minichromosomes and, more generally, of chromatin fibers, it is necessary to open the structure in low salt (less than 5–20 mM NaCl) and in the absence of divalent cations (less than 0.1 mM MgCl$_2$, for example) (Figs. 4 and 6). This compaction of the chromatin structure is due, at least in part, to the presence of the H1 histone and salts (for a more complete analysis and discussion, see Ref. 12). The low-salt conditions allow the visualization of individual nucleosomes connected by a short DNA filament. The nucleosomes appear rounded 10–13 nm in diameter, with an accumulation of stain in the center of some of the particles. The apparent length of the internucleosomal DNA depends on the time of incubation of the material in low salt: short exposures give rise to almost no visible internucleosomal DNA, while extensive dialysis against essentially no salt can induce the reversible loss of the nucleosomal structure.[13]

The observation of the nucleosomes as flat disks is presumably due to the preferential adsorption of the particles by their "flat surface" on the carbon film. In a hydrated and nonoriented state, nucleosomes, embedded in a film of amorphous ice, appear as flat disks 11–12 nm in diameter and 5–6 nm thick,[14] very similar to the structure of the nucleosomal core deduced from neutron diffraction and X-ray data (see Ref. 15 for a recent review).

[12] D. L. Bates, P. J. G. Butler, E. C. Pearson, and J. O. Thomas, *Eur. J. Biochem.* **119,** 469 (1981).

[13] P. Oudet, C. Spadafora, and P. Chambon, *Cold Spring Harbor Symp. Quant. Biol.* **42,** 301 (1977).

[14] J. Dubochet, M. Adrian, P. Schultz, and P. Oudet, *EMBO J.* **5,** 519 (1986).

[15] D. S. Pederson, F. Thoma, and R. T. Simpson, *Annu. Rev. Cell Biol.* **2,** 117 (1986).

Visualization of Subpopulations of SV40 Minichromosomes

The classic electron microscopic appearance of SV40 minichromosomes as 20–24 nucleosomes interconnected by short and variable DNA filaments represents the majority (75%) of the structures visualized from material purified 38–42 hr postinfection. One-fourth of the minichromo-

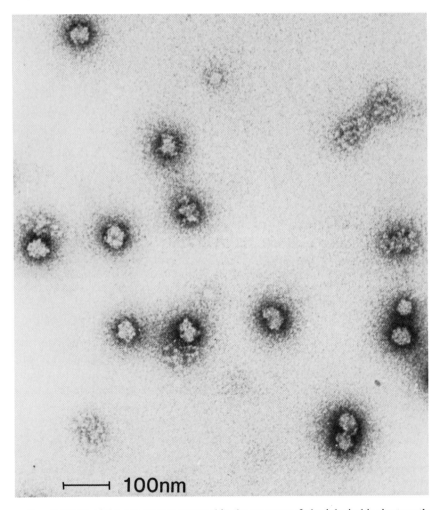

FIG. 5. SV40 minichromosomes prepared in the presence of physiological ionic strength. When adsorbed in the presence of 100–150 m*M* NaCl, the minichromosomes present a compact, ovoid structure where individual nucleosomes are barely seen. Some of the structures are flattened during the drying process, revealing the nucleosome arrangements. This appearance is not significantly modified by the presence of divalent cations.

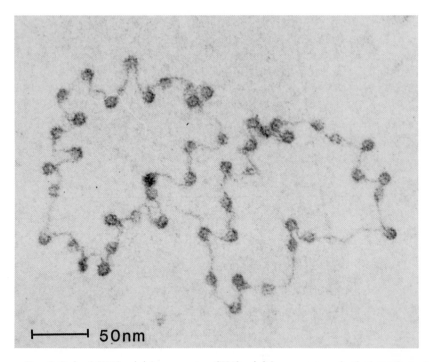

FIG. 6. Stained SV40 minichromosomes. SV40 minichromosomes adsorbed on charged grids in the presence of low salt concentrations (<20 mM NaCl) and in the absence of divalent cations show an open configuration. As seen on the right, some minichromosomes present a nucleosome-free region, or "gap," covering the origin of replication and the different initiation points for late and early transcription.

somes present a nucleosome-free region or gap precisely localized over the ORI region.[16-19] The biological role of this nucleosome-free region is still open to debate, but several possibilities have been proposed, such as attachment point to a specific nuclear structure or entry site for the transcriptional machinery.[20-23]

[16] S. Saragosti, G. Moyne, and M. Yaniv, Cell 20, 65 (1980).

[17] E. B. Jakobovits, S. Bratosin, and Y. Aloni, Nature (London) 285, 263 (1982).

[18] J. Jongstra, T. L. Reudelhuber, P. Oudet, C. Benoist, C. B. Chae, J. M. Jeltsch, D. J. Mathis, and P. Chambon, Nature (London) 307, 708 (1984).

[19] E. Weiss, C. Ruhlmann, and P. Oudet, Nucleic Acids Res. 14, 2054 (1986).

[20] L. C. Tack, P. M. Wassarman, and M. L. DePamphilis, J. Biol. Chem. 256, 8821 (1981).

[21] B. Wasylyk, C. Wasylyk, P. Augereau, and P. Chambon, Cell 32, 503 (1983).

[22] G. Moyne, F. Harper, S. Saragosti, and M. Yaniv, Cell 30, 123 (1982).

[23] R. Abulapia, A. Ben-Ze'ev, N. Hay, and Y. Aloni, J. Mol. Biol. 172, 467 (1984).

The SV40 transcriptional complexes represent a very low percentage (< 1%) of the extracted minichromosomes (for a recent discussion, see Ref. 19) as well as of the nuclear population. Such molecules can be identified by electron microscopy after *in vitro* elongation of the *in vivo* initiated complexes and complexation of the growing RNA with a single-strand binding protein like the product of the gene 32 of the T4 phage.[24] More extensive characterization of the structures involved in the transcriptional process requires preparations significantly enriched in such transcriptional complexes, allowing the necessary correlations between the biochemical and electron microscopic approaches. Replicative intermediates are characterized by an indirect approach after cross-linking of the accessible DNA regions with trimethylpsoralen and denaturation of the protected regions.[25]

Conclusions

The study of nucleosomal and chromatin organization using electron microscopy requires fairly homogeneous populations of molecules, which are available with the SV40 system. New approaches can be used to obtain episomal minichromatin structures, and the preparation of enriched subpopulations will allow the direct characterization of these structures at low resolution, but no other method could provide such information so directly. The development of methods improving the contrast and preservation of biological structures, allowing the detection of precise structures by immunomicroscopy, will reinforce the usefulness of electron microscopy in studying biological reactions.

New information on the structure of the minichromosomes will be available after image processing and correlation of electron micrographs obtained under conditions preserving as much of the native material as possible. This approach uses the benefit of the comparisons of individual images by reducing the noise introduction by the necessary manipulations and interaction with the electron beam. In this respect, the correlation with images obtained with the same material embedded in amorphous ice will enhance the visibility of statistically significant structural features.

Acknowledgments

We thank E. Weiss for critical reading of the manuscript and B. Boulay for the illustrations. This work was supported by grants from the INSERM, CNRS, and the French Ministère de la Recherche et de l'Enseignement Supérieur (Grand No. 86TO371).

[24] P. Gariglio, R. Llopis, P. Oudet, and P. Chambon, *J. Mol. Biol.* **131**, 75 (1979).
[25] J. M. Sogo, H. Stahl, T. Koller, and R. Knippers, *J. Mol. Biol.* **189**, (1986).

[9] Immunoelectron Microscopy of Nucleosomes

By CHRISTOPHER L. F. WOODCOCK

Introduction

In this chapter, the considerations involved in the binding of specific antibodies to nucleosomes and the visualizing of the resulting complexes with the electron microscope are discussed. The advantage of these methods, as compared to other uses of antibody probes, is that information is obtained at the level of the individual nucleosome. Neither immunofluorescence nor immunoelectron microscopy with thin sections provides the resolution needed to visualize individual nucleosomes. Among the applications of the technique are (1) localizing nucleosomal components; (2) detecting the presence (or absence) of a nucleosomal component on a particular stretch of chromatin; and (3) studying the accessibility to antibody of a nucleosomal component known to be present.

The methods discussed are based on practical experience with antihistone antibodies, but the same principles and precautions apply to work with other nucleosomal components including specific nucleotide sequences. There are several recent reviews of the general principles involved in immunoelectron microscopy, especially as applied to thin sections;[1-3] consequently, this chapter focuses on the particular requirements for chromatin.

In order to achieve resolution at the nucleosomal level with antihistone antibodies, there are a number of factors peculiar to chromatin that need to be taken into account. Compared with other antibody techniques, immunoelectron microscopy requires that a high proportion of antigen be complexed with antibody. For example, in an experiment in which an antihistone antibody is reacted with a group of contiguous nucleosomes, all of which are expected to contain the histone under study, the minimum satisfactory signal might be from 1 to 5 bound antibody molecules per 10 nucleosomes, depending on the aims of the experiment. If the ratio becomes much higher, the signal is lost in the background noise. In contrast, in a typical immunofluorescence experiment using the same antibody on

[1] M. A. Williams, in "Practical Methods in Electron Microscopy" (A. M. Glauert, ed.), Vol. 6(1), p. 1. North-Holland, Amsterdam, 1977.

[2] J. M. Polak and I. M. Varndell, eds., "Immunolabeling for Electron Microscopy." Elsevier, Amsterdam, 1984.

[3] I. M. Varndell and J. M. Polak, in "Electron Microscopy in Molecular Biology" (J. Sommerville and U. Scheer, eds.), p. 179. IRL Press, Oxford, 1987.

Copyright © 1989 by Academic Press, Inc.
All rights of reproduction in any form reserved.

isolated nuclei, the binding of 1 antibody per 1000 nucleosomes or more would probably provide a satisfactory fluorescence signal. Unfortunately, antibodies to histones tend to be of low affinity, requiring high antibody:antigen ratios in order to favor the formation of sufficient antibody–antigen complexes. Further, the complexes may have to be stabilized prior to separation from unbound reactants in order to prevent subsequent dissociation.

The salt concentration at which the binding reaction is carried out is often critical. The three factors that must be balanced here are the compaction state of the chromatin (except in cases where monomeric or small oligonucleosomes are employed), the possible migration of H1 histones (if such would affect the experiment), and the specificity of formation of antibody–antigen complexes. Most standard immunoassays are carried out in phosphate-buffered saline, with a monovalent ion concentration of about 150 mM. These conditions result in maximum compaction and possible precipitation of H1-containing chromatin[4] and are thus inappropriate for many applications. Lowering the salt concentration unfolds the chromatin so that individual nucleosomes can be resolved;[5] however, these conditions may promote the nonspecific binding of antibody, and controls designed to detect this type of artifact are essential.

Also, it is necessary to stain the final product of the reaction so that both antibody and antigen can be visualized simultaneously. This is primarily of concern in cases where unlabeled antibodies are used.

Preparation and Selection of Antibodies

The subject of preparation and selection of antibodies is treated extensively in other chapters of this volume[6,7] and is not considered in detail here. As mentioned above, the main requirement for electron microscopy is that the antibodies have a satisfactory affinity for the antigen under study, and it is advisable to screen all preparations for suitability. Although prior testing does not guarantee success, it does allow poorly reacting preparations to be recognized and excluded. Prescreening is also of use in selecting sera or monoclones for further purification of antibody. It should be stressed that the following recommendations apply only to testing the suitability of an antibody preparation for immunoelectron microscopy; other assays are needed to establish the specificity of the antibody for the immunogen.

[4] R. D. Kornberg, J. W. LaPointe, and Y. Lorch, this volume [1].
[5] F. Thoma, T. Koller, and A. Klug, *J. Cell Biol.* **83,** 408 (1979).
[6] M. Bustin, this volume [11].
[7] S. Muller and M. H. V. Van Regenmortel, this volume [12].

Prescreening

The most reliable method of predicting the suitability of an antibody preparation for immunoelectron microscopy is a simple immunofluorescence assay performed on isolated nuclei. Although nonquantitative, it has the advantages that binding conditions can be identical to those desired for electron microscopy and the conformational state of the antigen can be properly controlled. In the case of antihistone antibodies, the preparation may bind well to the immunogen but poorly or not at all to native nucleosomes because of the masking of antigenic sites. Assays that involve the adsorption of nucleosomes or chromatin to a solid substrate should be avoided because conformational changes during adsorption may expose sites that are normally inaccessible and give misleading results.

The use of isolated nuclei rather than whole cells or tissue sections is preferable for immunofluorescence prescreening. With isolated nuclei, problems of penetration and prefixation are avoided, and, in any case, nuclei will be needed for the subsequent preparation of nucleosomes or chromatin. If the immunoelectron microscopy reaction is to be carried out under physiological salt conditions, then the immunofluorescence experiment can be done in solution; for working with lower salt concentrations, the nuclei must be affixed to glass slides before equilibrating to the desired ionic strength, and the reaction completed on the slides.

A dilution series should first be run to establish the minimum concentration of antibody that yields a positive fluorescence signal. It is difficult to give a precise cutoff level of reactivity below which immunoelectron microscopy would be fruitless since the considerations depend greatly on the experiment and its aims. In the case of affinity-purified antibodies from animal sera or purified immunoglobins from monoclones, a strong positive reaction should be obtained with 100 μg/ml antibody (1 mg/ml immunoglobulin gives an absorbance of approximately 0.75 at 280 nm[8]).

Immunofluorescence screening should also include controls. In addition to the omission of the primary antibody, tests with purified preimmune or nonspecific primary antibody should be conducted. While these will detect some types of nonspecific antibody adhesion, they cannot cover the full spectrum of possible artifacts, nor can they substitute for controls at the ultrastructural level. Particularly troublesome are monoclonal antibodies for which the only rigorous control may be removal of the antigen from the chromatin preparation; this aspect is discussed below.

Immunofluorescence Conditions

The main goal of the fluorescence assay is to provide conditions that are as close as possible to those used in the subsequent electron microscopi-

---

[8] B. A. L. Hurn and S. M. Chantler, this series, Vol. 70 [5].

cal assay. This generally means careful selection of salt and buffer conditions, the avoidance of prefixation of the nuclei, and not using "blocking" agents such as bovine serum albumin or animal sera that are incompatible with staining for electron microscopy. Within these limitations, there are many possible variations, and the following protocols should be taken as guidelines only.

Suspension Assay. Nuclei are prepared and washed in 150 mM NaCl, 0.1 mM phenylmethylsulfonyl fluoride (PMSF), 10 mM triethanolamine-HCl (TEA) buffer, pH 7.2. They are then reacted with varying amounts of antibody (e.g., 10^6 nuclei with 2–20 μg antibody in 40 μl buffer) for 30 min on ice, washed several times in buffer by gentle centrifugation, resuspended in a suitable dilution of commercial fluorescein-conjugated second antibody (1 : 30 is optimal for the nonaffinity products from Cappel, Cooper Biomedical Inc., Malvern, PA 19355), incubated a further 30 min, washed again, resuspended in a small volume of buffer containing 50% glycerol and 5% propyl gallate, and placed on a microscope slide for examination.

The main precaution in this procedure is that centrifugation conditions should be mild enough to prevent the nuclei from adhering to each other. Although divalent cations or polyamines may better preserve nuclear integrity, they result in such compact nuclei that penetration of antibody may be impaired.

Slide Assay. Nuclei are prepared and washed in buffer as above, then spread over the surface of slides pretreated with either an aqueous solution of 1% Ficoll or 1 mg/ml polylysine (475,000 MW, Sigma Chemical Co., St. Louis, MO 63178), pH 8.2. Polylysine slides are washed 3 times in distilled water and dried before use, while Ficoll-treated slides are left unwashed. In the case of Ficoll slides, adhesion requires a further step after spreading the nuclear suspension in which the slides are dipped into 100% ethanol for 5 min, then acetone for 2 min, followed by air-drying. Nuclei adhere to polylysine slides without further treatment. After the adhesion step, slides are equilibrated in the desired ionic strength solution prepared by lowering the NaCl concentration of the standard buffer. They are then laid flat in a humid chamber, drops of primary antibody in the desired ionic strength buffer added, and the drops covered with a glass coverslip. After a 30-min incubation, the coverslips are carefully removed, and the slides are washed several times by immersion in buffer. Second antibody is applied as a droplet in the same way, and, after final washing, the same mounting medium is used as for the suspension assay.

These methods are clearly not as satisfactory as the suspension assay since the nuclei are exposed to conditions (ethanol, acetone, and air-drying or polylysine) that could alter conformations at the nucleosome level and give a false impression of the activity of the antibody. However, if the reaction is to be carried out under salt conditions that cause nuclei to swell, some disturbance of this sort seems unavoidable.

State of the Antibody

For most immunoelectron microscopy applications, the IgG class of antibody is preferred over IgM because of its much smaller size. However, IgM is much easier to visualize in the unlabeled form and, for some purposes, may be preferable.

It is important that the antibody preparation be free of extraneous macromolecules including nonspecific immunoglobulins that would otherwise saturate the system. This normally means that the starting animal serum, monoclone culture medium, or ascites fluid must be affinity purified. Methods for affinity purification are available in the literature.[9] In carrying out such a purification scheme, the most critical step is the release of the antibody from the affinity column or other substrate, which is achieved by exposure to extremes of pH, chaotropic reagents, or other potentially denaturing conditions. It is especially important to keep this step as short as possible, otherwise the preparation may contain such a high proportion of nonreactive antibody molecules as to defeat the experiment. Another unfortunate consequence of the affinity purification of polyclonal antibodies is that the most avid species of antibody are probably never released from the affinity column. Thus, for such polyclonal preparations, it is worth experimenting with highly purified IgG fractions that have not been affinity purified.

After purification, the antibody should ideally be concentrated to about 1 mg/ml in a low ionic strength buffer. This provides the most flexibility when designing the critical step in which the antibody–nucleosome complex is formed. Antibodies prepared in phosphate-buffered saline (PBS), for example, would have to be diluted 15-fold or dialyzed (with inevitable losses) if the reaction were to be carried out at 10 mM monovalent ions. For long-term storage, the concentrated antibody is divided into aliquots, frozen rapidly in liquid nitrogen, and kept at $-80°$ or below.

Subnucleosomal Localization

Since IgG molecules and nucleosomes are similar in size, only a very limited spatial resolution is possible with antibody methods. A second limitation arising from the featureless nature of the nucleosome disk (at the level of resolution readily attainable) can be partially overcome by using

---

[9] P. Cuatrecasas and C. B. Anfinsen, this series, Vol. 22 [31].

polynucleosomes where the linker DNA entry and exit sites provide a limited amount of positional information. Thus, it is feasible to determine the location of a nucleosomal component with respect to its proximity to the linker DNA entry – exit site. The procedure consists of the preparation of polynucleosomes, reaction with the antibody, fixation of antibody – antigen complexes, removal of unbound antibody, and staining for electron microscopy. To a large degree, the methods have been based on those developed for immunoelectron microscopy of ribosomal subunits.[10,11]

Preparation of Polynucleosomes

General methods for preparing nucleosomes are discussed elsewhere in this volume,[4] and only minor changes should be required to adapt such methods to the system under study. For immunolocalization, the minimum size of polynucleosome is the dinucleosome, and for many applications this unit is ideal. As the size of the polynucleosome increases, it becomes more difficult to display the structure adequately in the electron microscope, and antibodies that are bound to two nucleosomes may further complicate the analysis.[12] In order to use the linker as a positional marker, it is essential that the H1 class of histones remains in place during the isolation and purification procedure. Thus it is important to verify that the H1 content of the final polynucleosome preparation is similar to that of the starting nuclei. Also, of course, the component to be localized must not be displaced.

Preparation of Antibody – Nucleosome Complexes

Optimum conditions for antibody – antigen complex formation depend largely on the affinity of the antibody and accessibility of antigenic sites, factors that differ for each application. The following suggestions should therefore be taken as guidelines only.

In order to have sufficient material in high enough concentration after the separation of unbound antibody, it is necessary to start with at least 50 μg of chromatin DNA (1 absorbance unit at 260 nm) per experiment in a final volume not greater than 0.5 ml. This starting material is dialyzed, if necessary, into a reaction buffer containing the desired level of monovalent ions (as close to 100 mM as possible without precipitating the chromatin)

[10] J. A. Lake, J. Mol. Biol. 105, 131 (1976).

[11] J. A. Lake, in "Advanced Techniques in Biological Electron Microscopy II" (J. K. Koehler, ed.), p. 173. Springer-Verlag, Berlin, 1978.

[12] A. Mazen, G. DeMurcia, S. Bernard, J. Pouyet, and M. Champagne, Eur. J. Biochem. 127, 169 (1982).

and 0.2 mM EDTA, pH 7.0. Additional buffers with pH values up to 8.0 may be added if desired, but Tris and phosphate should be avoided. It is important to retain EDTA at this step to avoid reactivation of the micrococcal nuclease used to generate the polynucleosomes.[13] Antibody is then added at a ratio determined from the immunofluorescence screening. Experience with various antihistone antibodies has shown that chromatin DNA:antibody protein ratios of 0.5–5 are most likely to provide useful amounts of antibody complex. Complex formation seems to be favored by an initial incubation at 40° for 2 min, followed by 20 min on ice.

Fixation of Antibody–Nucleosome Complexes

Ideally, fixation of antibody–nucleosome complexes should be omitted, but this step may be needed to prevent weakly associating complexes from dissociating in the absence of a driving amount of free antibody. The danger of fixing at this stage is that nonspecific cross-links may be formed. In practice, however, controls in which the antigenic component is removed from dinucleosomes before adding antibody show a negligible amount of spurious antibody binding due to this effect.[14]

A mild but effective fixation is provided by adding formaldehyde (freshly generated from paraformaldehyde) to 1% for 15 min on ice, then glutaraldehyde to 0.6% for a further 15 min on ice. Both fixatives are prepared at 10-fold greater concentration in reaction buffer. Separation of unbound antibody is then carried out immediately and effectively removes excess fixatives.

Removal of Unbound Antibody

Detection of nucleosome–antibody complexes under conditions of antibody excess is impracticable, and removal or reduction of unbound antibody is recommended. Two methods developed for achieving this vary in the efficiency of antibody separation and in the quality of the final electron micrographic preparation. Sepharose 2B chromatography provides only a partial separation but has the advantages that it is rapid and adds no extra components. Sucrose gradient centrifugation, on the other hand, removes all unbound antibody but requires an overnight run, and the final sample now contains about 15% sucrose, which must be diluted for effective electron microscopy.

A 1-ml-bed-volume Sepharose 2B column is set up in a Pasteur pipette

[13] C. W. Carter, Jr. and L. F. Levinger, *Biochem. Biophys. Res. Commun.* **74**, 955 (1977).
[14] L.-L. Y. Frado, C. V. Mura, B. D. Stollar, and C. L. F. Woodcock, *J. Biol. Chem.* **258**, 11984 (1983).

and equilibrated with 50 mM NaCl, 1 mM EDTA, 50 mM Tricine buffer, pH 8.0. The antibody–polynucleosome mixture is added and eluted with buffer. Fractions (0.1 ml) are collected and checked by electron microscopy for the presence of polynucleosomes. This can be done rapidly by mixing 5 μl of each fraction with 50 μl of 50 mM NaCl and placing on a glow-discharged carbon film for 30 sec. The grid is then washed dropwise with 50 mM NaCl, stained with 5 drops of 2% (w/v) uranyl acetate, and further washed with 1 or 2 drops of water. The grid is then air-dried and observed with the electron microscope. Dark-field illumination provides optimum contrast. It should be noted that these conditions are used only to identify the nucleosome-containing fractions: antibodies will not normally be visible at this stage. The first fractions containing polynucleosomes will be enriched in antibody complexes and depleted of free antibody.

For sucrose gradient separation, 10–30% sucrose gradients are prepared containing 10 mM NaCl, 0.2 mM EDTA, 0.1 mM PMSF, pH 7.0. Approximately 1.0 absorbance unit (260 nm) of antibody–nucleosome mixture is layered on top, and the gradients centrifuged at 35,000 rpm for 16 hr (Beckman SW41 rotor). The gradients are then pumped through an absorbance monitor, and 0.2-ml fractions are collected. IgG molecules (7 S) are completely separated from nucleosomes (11 S) and polynucleosomes, but there is often no separation between free and antibody-bound polynucleosomes. In similar experiments in which ribosomal subunits were reacted with antibodies to specific ribosomal proteins, sucrose gradient centrifugation revealed additional, faster sedimenting material formed from subunits cross-linked by antibody molecules. These species proved to be the most valuable in mapping the positions of ribosomal proteins.[10] Most antihistone antibodies appear to be of too low affinity to form an appreciable yield of cross-linked species, but there are some reports of nucleosomes increasing in s value following antibody binding.[15,16]

Electron Microscopy

The sandwich technique, in which the specimens are sandwiched in stain between two carbon films, is the most satisfactory way of preparing antibody–nucleosome complexes for electron microscope.[11] The following slight modifications allow sandwich preparations to be made from as little as 100 μl of antibody–nucleosome complex.

A thin carbon film is evaporated onto freshly cleaved mica, and the

[15] D. Absolom and M. H. V. Van Regenmortel, *FEBS Lett.* **85,** 61 (1978).
[16] R. T. Simpson and M. Bustin, *Biochemistry* **15,** 4305 (1976).

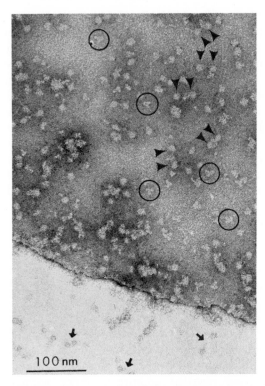

Fig. 1. Portion of a sandwich preparation of dinucleosomes reacted with an antihistone H5 antibody and partially purified by Sepharose 2B chromatography. At lower left, on the single layer of carbon, only dinucleosomes are seen (arrows), while in the double layer, both dinucleosomes and free IgG molecules (circled) are resolved. A few dinucleosomes with bound antibody are denoted (double arrowheads). In addition, a number of unidentifiable structures are usually present in the void volume of the column. Most of these are probably aggregates of IgG. (From Ref. 14.)

mica is cut into squares approximately 5 mm on a side. The nucleosome– antibody complex is diluted to an absorbance of 0.025 units (260 nm), and about 0.1 ml placed on a clean, white, glazed porcelain surface. Other surfaces where the carbon film can be monitored readily are also satisfactory, but materials such as Parafilm, which result in a highly curved meniscus, should be avoided. The mica fragment is lowered into the droplet so that the carbon floats on the surface but remains attached at one corner. Both mica and carbon are withdrawn from the droplet, and the carbon is then floated free onto a solution of 2% aqueous uranyl acetate. A bare 400-mesh copper grid, pretreated with 100% ethanol, is used to pick up the carbon in such a way that the carbon is folded upon itself, forming a double layer. This is best achieved by pushing the floating film against the side of the container (usually a porcelain depression plate) so that it is crumpled and thrown into folds. When such films are picked up and observed in the electron microscope, it is easy to recognize the many sandwiched areas and distinguish them from single layers (Fig. 1). The relatively thick stain layer is easily damaged by the electron beam, where-upon it assumes a coarse granular texture. Low-dose techniques should therefore be used if possible.

Analysis and Controls

Photomicrographs taken near the junction of single and double layers seem to provide the best areas for analysis (Fig. 1). Within the sandwiched regions it should be possible to recognize polynucleosomes and, in the case of Sepharose fractionation, free antibody. In contrast, in the single layers, only the nucleosomes are seen, since the immunoglobin molecules fail to pick up stain under these conditions. The presence of sucrose in the sample, even after considerable dilution, produces a coarser textured image, but it is nevertheless possible to distinguish free polynucleosomes from those with bound antibody (Fig. 2). Scoring of antibody position is done from enlarged images.

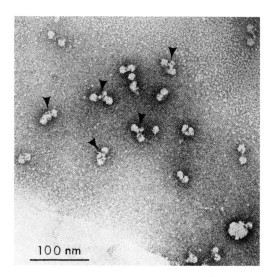

FIG. 2. Electron micrograph from a preparation similar to that shown in Fig. 1 except the complex was purified on a sucrose gradient. Free IgG molecules are absent, and the texture of the stain is much coarser owing to the presence of sucrose. Arrowheads denote dinucleosomes with putative bound antibodies. (From Ref. 14.)

Controls should include absence of antibody and, in the case of polyclonal antibodies, reaction with an equivalent concentration of normal IgG from the same animal. For monoclonal antibodies, normal mouse IgG or an unrelated monoclonal provide a partial control, but the unique nature of the molecule makes it impossible to exclude the possibility that binding to components other than the desired antigen is occurring. If it is possible to deplete the polynucleosome of the component under study, carrying out the reaction on the depleted material provides a rigorous control system.[14]

Indirect Localization Techniques

In many applications of immunoelectron microscopy, resolution at the subnucleosome level is not required, but the distribution or accessibility of a component at the chromatin fiber level is desired. In these cases, a different experimental approach may be taken, and a labeled second antibody used. In most instances, it is necessary to unfold the chromatin in question to the 10-nm fiber state in order to recognize the regions under study (e.g., replication forks[17] or transcriptionally active areas[18]) or in order to make the nucleosomes as accessible to antibody as possible. This requires exposure to low ionic strength solutions (< 10 mM monovalent ions) and brings with it the problem of increased nonspecific binding of antibody. The method of preparing the chromatin for electron microscopy is based on the "spreading" technique developed by Miller and colleagues,[19,20] and the centrifugation step in the method allows for the separation of unbound antibodies.

Preparation of Chromatin

Isolated nuclei are the usual starting material and must be resuspended in a solution in which they can be expanded to the 10-nm fiber step by simple dilution. Depending on the source of nuclei, this may require a final medium depleted of divalent cations or spermidine and spermine, agents that protect chromatin from decondensation even at high dilution. In a typical experiment with chicken erythrocytes,[21] a pellet of washed cells was

[17] S. L. McKnight and O. L. Miller, Jr., *Cell* **12**, 795 (1977).

[18] S. L. McKnight, M. Bustin, and O. L. Miller, Jr., *Cold Spring Harbor Symp. Quant. Biol.* **42**, 741 (1977).

[19] O. L. Miller, Jr. and A. H. Bakken, *Karolinska Symp. Res. Methods Reprod. Endocrinol. 5th*, p. 155 (1972).

[20] H. Zentgraf, C.-T. Bock, and M. Schenk, *in* "Electron Microscopy in Molecular Biology" (J. Somerville and U. Scheer, eds.), p. 81. IRL Press, Oxford, 1987.

[21] D. DiPadua Mathieu, C. V. Mura, L.-L. Y. Frado, C. L. F. Woodcock, and B. D. Stollar, *J. Cell Biol.* **91**, 135 (1981).

resuspended at a 1 : 1 (v/v) ratio in 150 mM NaCl, pH 8.0, and 5 μl of the suspension added to 1.5 ml water adjusted to pH 9.4 with sodium borate buffer. In this case, the cells and nuclei expand simultaneously, and the chromatin is better dispersed than if isolated nuclei were used. However, this is not always possible, and experiments to select the best conditions for spreading and observation should be completed before antibody binding is attempted. Other methods of dispersing chromatin are discussed in Ref. 20.

Antibody Binding

Antibody is added to a 25-μl aliquot of the expanded chromatin preparation and allowed to bind for 10 min at room temperature. Here it is important that the antibody not raise the salt concentration appreciably, or the chromatin will recondense. Again, the amount of antibody should be varied over a wide range, as indicated by prior immunofluorescence tests. There seems to be no simple way to remove the unbound first antibody at this stage, as is done with most immunological assays. All attempts to achieve this by centrifugation or other means have resulted in poorer results than if the second antibody is simply added to the mixture of chromatin and primary antibody. It is of course necessary in this case to add enough secondary antibody to saturate the primary antibody.

There are commercially available labeled second antibodies suitable for most types of primary antibody. In addition, protein A conjugates and streptavidin–biotin reagents should be satisfactory. Avidin itself should be avoided, however, since it binds nonspecifically to chromatin. Ferritin and colloidal gold are the most satisfactory labels, but with both, aggregation of the commercially obtainable product can be a problem. Monomeric ferritin conjugates can be separated from aggregates by passing the mixture over a Sepharose 2B column equilibrated with 10 mM NaCl, 0.2 mM EDTA, 1 mM TEA buffer, pH 8.0.[21] Aggregates appear in the void volume, and subsequent fractions containing the monomer can be selected by electron microscopy.

The labeled reagent is added, and incubation continued for a further 10 min. After this time, the mixture is carefully layered onto a cushion of 0.1 M sucrose containing 10% formaldehyde in a Lucite chamber 4 mm in diameter and 6 mm deep containing a glow-discharged carbon film at the bottom.[19,20] The chamber is centrifuged at about 1000 g for 5 min during which time the dispersed chromatin passes through the sucrose and is spread on the carbon film, leaving the unbound antibodies at the top. A critical step at this point in the procedure is to repeatedly remove the upper supernatant fluid from the chamber and replace it with distilled water.

Failure to do this rigorously results in an unacceptably high background of label on the grid.

Electron Microscopy

After removing unbound label and antibody, the chamber is filled completely with water and inverted, whereupon the grid can be removed, washed with distilled water, and stained with 2% (w/v) aqueous uranyl acetate. This results in a lower contrast than double staining with ethanolic uranyl acetate and phosphotungstic acid following a Photoflo rinse,[19] but it seems to be advantageous for the simultaneous visualization of nucleosomes and label.

Controls

The same considerations apply to controls as discussed in the section on subnucleosomal localization. In this case, however, it is usually not practicable to employ chromatin with the antigen under study removed. An alternative control for nonspecific antibody binding at low ionic strength is to use a different type of chromatin known (from other assays) not to react with the primary antibody.

[10] Use of Fluorescent Probes to Study Nucleosomes

By JOAN-RAMON DABAN and CHARLES R. CANTOR

Introduction

Fluorescence is the emission of photons that results from the relaxation of electronically excited molecules. At room temperature, most molecules are found in the lowest energy vibrational level of the ground singlet electronic state. The absorption of light of appropriate energy induces the transition from this level to an excited electronic state. Depending on the energy of the absorbed photon, a molecule can be excited to different vibrational levels of any excited state. However, vibrational relaxation and internal conversion processes rapidly relax the molecule to the lowest vibrational level of the first excited singlet state. In solution, these processes are generally completed in 10^{-12} sec and occur before fluorescence can take place. Thus, fluorescence emission is due to the relaxation of an excited molecule from the ground vibrational level of the first excited state to any vibrational level of the ground electronic state. This relaxation can also be produced by nonradiative processes, such as intersystem crossing

METHODS IN ENZYMOLOGY, VOL. 170
Copyright © 1989 by Academic Press, Inc.
All rights of reproduction in any form reserved.

(i.e., conversion into an excited triplet) or solvent quenching, which compete with fluorescence. In fact, the natural fluorescence of most biological molecules is very low, and extrinsic fluorophores with favorable emission properties must be attached to these molecules to apply fluorescence methods.

For many fluorophores, the lifetime of the first excited singlet state is of the order of 10 nsec. Consequently, if a process occurs with a rate equal to or greater than 10^8 sec^{-1}, it can modify the emission properties of the fluorescent chromophore. Various interesting processes can take place on this time scale, including solvent relaxation, collisions with solute molecules, rotational diffusion, formation of excimers, and nonradiative singlet–singlet energy transfer. Each of these processes can produce specific alterations of the spectral characteristics of the fluorophore, which usually are easily monitored using commercially available instruments. When a fluorophore is bound to a macromolecular system, its fluorescent properties are dependent, in addition, on the local environment and on the structural characteristics of the whole system. Thus, fluorescence can be used to study the static structure and dynamics of macromolecular assemblies. For an overview of this field, the interested reader is referred to general books on fluorescence spectroscopy.[1,2]

In the particular case of the nucleosome, intrinsic tyrosine fluorescence has been used to study the specific interactions between core histones[3] and structural transitions of nucleosome core particles.[4] This chapter, however, deals exclusively with studies on nucleosome structure carried out using extrinsic fluorescent labels. In particular, we consider extensively the methodology and results obtained using covalently attached labels. The specific location of the fluorescent labels has allowed a detailed investigation of particular regions of the nucleosome. These probes have also facilitated the study of the overall structure and stability, as well as many dynamic aspects, of this chromatin particle.

Labeling of Nucleosomes with Fluorescent Probes

The DNA structure in the nucleosome has been studied by taking advantage of the fluorescent properties of the noncovalent intercalative dye ethidium bromide.[5,6] However, the majority of the fluorescence stud-

[1] A. J. Pesce, C. G. Rosén, and T. L. Pasby, "Fluorescence Spectroscopy: An Introduction for Biology and Medicine." Dekker, New York, 1971.

[2] J. R. Lakowicz, "Principles of Fluorescence Spectroscopy." Plenum, New York, 1983.

[3] J. A. D'Anna and I. Isenberg, *Biochemistry* 13, 4992 (1974).

[4] L. J. Libertini and E. W. Small, *Biochemistry* 21, 3327 (1982).

[5] H. M. Wu, N. Dattagupta, M. Hogan, and D. M. Crothers, *Biochemistry* 19, 626 (1980).

[6] I. Ashikawa, K. Kinosita, A. Ikegami, Y. Nishimura, M. Tsuboi, K. Watanabe, K. Iso, and T. Nakano, *Biochemistry* 22, 6018 (1983).

TABLE I
COVALENT PROBES USED IN FLUORESCENCE STUDIES OF NUCLEOSOMES[a]

Structure	Abbreviated name	Covalent attachment to	Refs.[b]
	IAEDANS	Cys-110 of CE H3	7–11
		Cys-73 of SU H4	9
		Met-84 of CE H4	12, 13
	IAF	Cys-110 of CE H3	7, 8
		3' ends of core DNA	9
		Met-84 of CE H4	13
	NPM	Cys-110 of CE H3	14–18
		Single Cys of PP H3[c]	19
		Met-84 of CE H3[d]	20

[a] IAEDANS, 5-[(iodoacetamido)ethyl]aminonaphthalene-1-sulfonic acid; IAF, 5-(iodo-acetamido)fluorescein; NPM, N-(1-pyrene)maleimide; CE, chicken erythrocyte; SU, sea urchin; PP, *Physarum polycephalum;* H3, histone H3; H4, histone H4.
[b] Reference numbers refer to text footnotes.
[c] Iodoacetoxypyrene was used instead of NPM.
[d] 1-N-pyrenyliodoacetamide was used instead of NPM.

ies on the nucleosome structure have been carried out using covalent probes. As summarized in Table I, nucleosomes have been labeled covalently with the fluorescent dyes 5-[(iodoacetamido)ethyl]aminonaphthalene-1-sulfonic acid (IAEDANS),[7–13] 5-(iodoacetamido)fluorescein (IAF),[7–9,13] N-(1-pyrene)maleimide (NPM),[14–18] and other pyrene deriva-

[7] A. E. Dieterich, R. Axel, and C. R. Cantor, *Cold Spring Harbor Symp. Quant. Biol.* **42**, 199 (1978).
[8] A. E. Dieterich, R. Axel, and C. R. Cantor, *J. Mol. Biol.* **129**, 587 (1979).
[9] H. Eshaghpour, A. E. Dieterich, C. R. Cantor, and D. M. Crothers, *Biochemistry* **19**, 1797 (1980).
[10] A. E. Dieterich, H. Eshaghpour, D. M. Crothers, and C. R. Cantor, *Nucleic Acids Res.* **8**, 2475 (1980).
[11] A. E. Dieterich and C. R. Cantor, *Biopolymers* **20**, 111 (1981).
[12] D. G. Chung and P. N. Lewis, *Biochemistry* **24**, 8028 (1985).
[13] D. G. Chung and P. N. Lewis, *Biochemistry* **25**, 5036 (1986).
[14] M. Zama, P. N. Bryan, R. E. Harrington, A. L. Olins, and D. E. Olins, *Cold Spring Harbor Symp. Quant. Biol.* **42**, 31 (1978).
[15] M. Zama, D. E. Olins, B. Prescott, and G. J. Thomas, *Nucleic Acids Res.* **5**, 3881 (1978).
[16] J. R. Daban and C. R. Cantor, *J. Mol. Biol.* **156**, 749 (1982).
[17] J. R. Daban and C. R. Cantor, *J. Mol. Biol.* **156**, 771 (1982).

tives.[19,20] The dye pair IAEDANS (donor)–IAF (acceptor) is particularly convenient for energy-transfer experiments.[21] Furthermore, IAEDANS is an extremely sensitive environmental probe.[22] The pyrene derivative NPM is essentially nonfluorescent in aqueous solutions but forms intense fluorescent adducts with a long lifetime when it reacts with protein thiol groups.[23] The pyrene probes are very useful in fluorescence studies because they can form excited dimers (excimers) under appropriate conditions.[24]

In most studies, labels have been specifically attached to the single thiol group of chicken erythrocyte core histones, which is located at position 110 of the amino acid sequence of histone H3.[25] Since, at low and intermediate ionic strength, the reactivity of the two H3 thiol groups in the nucleosome is very low,[26] the labeling reaction has to be performed in the presence of denaturing agents and/or at high ionic strength. In the case of IAEDANS and IAF (Molecular Probes, Eugene, OR), the thiol groups of chicken erythrocyte nucleosome core particles can be labeled in the presence of 5 M urea, 2 M NaCl, pH 8, as described previously.[7,8] A similar procedure has been used to label Cys-73 of histone H4 from sea urchin sperm (*Arbacia lixula*) with IAF[9] and the single thiol group of *Physarum polycephalum* histone H3 with iodoacetoxypyrene.[19] In the case of the pyrene derivative NPM (Molecular Probes), the labeling of the thiol groups of chicken erythrocyte core particles can be achieved at 2 M NaCl, pH 7, in the absence of denaturing agents.[16] The single methionine residue of chicken erythrocyte histone H4, located at position 84 of its amino acid sequence, has been labeled with IAEDANS,[12,13] IAF,[13] and 1-N-pyrenyliodoacetamide,[20] following the procedure of Lewis.[27] The method of Eshaghpour *et al.*[28] can be used for the fluorescent labeling of the 3' ends of core DNA. In this case, 4-thiouridine is specifically added to the 3'-terminal hydroxyl groups of DNA in the presence of terminal deoxynucleotidyltransferase, and finally the resulting reactive thiol groups are modified with IAF.[9]

[18] P. Diaz and J. R. Daban, *Biochemistry* **25**, 7736 (1986).

[19] C. P. Prior, C. R. Cantor, E. M. Johnson, and V. G. Allfrey, *Cell* **20**, 597 (1980).

[20] D. G. Chung and P. N. Lewis, *Biochemistry* **25**, 2048 (1986).

[21] R. H. Fairclough and C. R. Cantor, this series, Vol. 48, p. 347.

[22] E. N. Hudson and G. Weber, *Biochemistry* **12**, 4154 (1973).

[23] J. K. Weltman, R. P. Szaro, A. R. Frackelton, R. M. Dowben, J. R. Bunting, and R. E. Cathou, *J. Biol. Chem.* **248**, 3173 (1973).

[24] T. Förster, *Angew. Chem. Int. Ed. Engl.* **8**, 333 (1969).

[25] W. F. Brandt and C. von Holt, *Eur. J. Biochem.* **46**, 419 (1974).

[26] N. T. N. Wong and E. P. M. Candido, *J. Biol. Chem.* **253**, 8263 (1978).

[27] P. N. Lewis, *Eur. J. Biochem.* **99**, 315 (1979).

[28] H. Eshaghpour, D. Söll, and D. M. Crothers, *Nucleic Acids Res.* **7**, 1485 (1979).

Once labeled, the nucleosomes are dialyzed extensively against 5 M urea, 2 M NaCl, reconstituted by stepwise dialysis against solutions of decreasing urea and salt concentrations, and finally purified by sedimentation in a sucrose gradient.[7,9] For energy-transfer experiments, adequate amounts of nucleosomes labeled independently with IAEDANS or IAF, or unlabeled, are mixed in the presence of 5 M urea, 2 M NaCl. In some experiments, the samples are prepared by mixing IAEDANS-labeled and unlabeled histones with the appropriate amount of core DNA, labeled with IAF or unlabeled. In both procedures, the nucleosomes are reconstituted by dialysis as described above. To carry out kinetic measurements of the nucleosome self-assembly reaction, however, NPM-labeled core particles, dissociated in 2 M NaCl, have to be reassociated by a rapid salt jump to 0.2 M NaCl.[16] In all cases, the integrity of the reconstituted particles should be tested by different methods, including analysis of histones and DNA, sedimentation, nucleoprotein gel electrophoresis, circular dichroism, and trypsin and DNase I digestions.[7–9,11,16,18]

To examine the specificity of the labeling, the resulting samples are analyzed on polyacrylamide gels. The fluorescence of the bands containing modified histones or DNA is visualized by UV illumination of the gel, prior to staining with dyes to detect proteins or DNA. The results obtained indicate that the labeling reactions considered above allow the specific modification of the reactive thiol groups of the different samples.[7,9,16] The extent of labeling can be determined spectrophotometrically by using previously published extinction coefficients for IAEDANS,[22] IAF,[7,9] and NPM.[23,29] The accurate quantitation of labeling with IAF is complicated because the extinction coefficient of this dye is very dependent on the environment and on ionic strength.[9] To facilitate the quantitation of labeling of the samples for energy-transfer experiments, it is particularly convenient to use a radioactive preparation of the donor.[21] For this purpose, ^{3}H-labeled IAEDANS can be synthesized as previously described.[30] This allows the determination of the amount of IAEDANS in double-labeled nucleosomes by scintillation counting. The results obtained indicate that, following the procedures considered above, the extent of labeling can be very high (89–99%) for IAEDANS- and IAF-labeled nucleosomes.[8] In the case of NPM-labeled nucleosomes, the degree of conversion of monomer to excimer can be used to estimate the NPM content in the nucleosome.[16] Our results indicate that the extent of labeling that can be attained with NPM is close to 100%.

[29] S. L. Betcher-Langue and S. S. Lehrer, *J. Biol. Chem.* **253**, 3757 (1978).
[30] K. H. Huang, R. H. Fairclough, and C. R. Cantor, *J. Mol. Biol.* **97**, 443 (1975).

Fluorescence Measurements

Steady-state fluorescence, quenching, and polarization studies can be performed with commercial spectrofluorometers equipped with monochromators to select excitation and emission wavelengths, preferably interfaced to a programmable calculator or a microcomputer to facilitate the measurements and further data analysis. For polarization measurements, polarizers are inserted in both the excitation and emission beams. Fluorescence lifetimes are measured using a single photon-counting appartus. The sample compartment can be conveniently maintained at a fixed temperature (generally, $20° - 25°$) with a circulating water bath attached to the cell holder. Dry-air purging is necessary to avoid vapor condensation on cell walls.

A sample concentration of the order of $10^{-7} M$ in nucleosomes is convenient to avoid inner filter effects of labeled samples. The excitation wavelength for IAEDANS- and NPM-labeled nucleosomes is 340 nm. However, for polarization studies of nucleosomes labeled with IAEDANS, the optimum excitation wavelength is 380 nm.[8] In kinetic experiments, the fluorescence intensity of IAEDANS-labeled nucleosomes is observed at 476 nm, and the fluorescence intensities of the monomer and excimer forms of NPM-labeled nucleosomes are measured at 375 and 460 nm, respectively.

For the determination of anisotropy and polarization, the fluorescence intensities, when the emission polarizer is oriented parallel and perpendicular to the vertically polarized excitation, have to be measured. However, to obtain the actual values of these intensities, it is necessary to determine the relative transmission efficiency (G) of the emission monochromator for vertically and horizontally polarized light.[2] The correction factor G is measured using horizontally polarized excitation, according to

$$G = I_{HV}/I_{HH} \tag{1}$$

where the subscripts indicate the orientation (H, horizontal; V, vertical) of the excitation and emission polarizers, respectively. Then, anisotropy (A) and polarization (P) are calculated as follows:

$$A = (I_{VV} - GI_{VH})/(I_{VV} + 2GI_{VH}) \tag{2}$$

$$P = (I_{VV} - GI_{VH})/(I_{VV} + GI_{VH}) \tag{3}$$

For energy-transfer studies, the quantum yield of the donor IAEDANS attached to the nucleosome core, under different conditions, can be determined by comparison with a standard.[7,21] The spectral overlap integral J_{DA} [see below, Eq. (18)] is obtained from the corrected emission spectrum of IAEDANS bound to the nucleosome and the corrected excitation spectrum and the extinction coefficient of the acceptor IAF in the nucleosome.[21] The efficiency of energy transfer between the donor and the acceptor can be calculated from the ratio ϕ_{DA}/ϕ_D [see below, Eq. (14)] where ϕ_{DA} and ϕ_D are, respectively, the quantum yields of the donor in the presence and absence of the acceptor. For the determination of this ratio, the fluorescence intensities of three different reconstituted nucleosome samples, containing, respectively, the donor, the acceptor, and the two dyes, are measured at regular intervals in the spectral overlap region (between 440 and 480 nm, in the case of the IAEDANS–IAF pair). The background intensity, determined using an unlabeled sample, has to be subtracted from all the measurements, and the values obtained have to be corrected for differences in concentration. The ratio ϕ_{DA}/ϕ_D is equal to the average value of $(F_{DA} - F_A)/F_D$, where F_D, F_A, and F_{DA} are the fluorescence intensities of the nucleosome cores labeled with the donor, the acceptor, and the donor–acceptor pair, respectively.

Applications of Fluorescence Spectroscopy to Study Nucleosome Structure and Dynamics

In the following sections, different experimental approaches that can be used to obtain structural and dynamic information from fluorescent-labeled nucleosomes are presented. The basic concepts of the different aspects of fluorescence spectroscopy involved in these studies are briefly considered. A summary of the results obtained so far using the different approaches is also presented in each section.

Environmental Effects on Emission Properties of Fluorescent Labels

The fluorescence spectrum of IAEDANS is highly sensitive to the polarity of the solvent.[22] This behavior is observed for many fluorophores. Generally it is found that increasing solvent polarity causes a progressive shift of the maximum of the emission spectrum to longer wavelengths (red shift). This is interpreted to be a consequence of the increased dipole moment of the excited fluorophore, which produces a reorganization (relaxation) of the polar solvent molecules before fluorescence emission can occur. Solvent relaxation gives rise to a decrease in the energy difference between the first excited singlet state and the ground state and, consequently, produces a red shift in the emission spectrum. For a detailed description of solvent effects, the reader is referred to the review by Brand and Gohlke.[31]

---

[31] L. Brand and J. R. Gohlke, *Annu. Rev. Biochem.* **41**, 843 (1972).

TABLE II
FLUORESCENT PROPERTIES OF IAEDANS–CORE NUCLEOSOMES[a]

| Sample | Parameters[b] | NaCl (M) | | | Refs.[c] |
		10^{-4}	10^{-2}	0.6	
IAEDANS-CE H3	λ_{max}	488	476	484	8
	F/F_I	0.65	1	0.35	
	$k_q/k_{q,I}$	4.3	1	11.6	
IAEDANS-CE H4[d]	F/F_I	1.05	1	0.84	12
	$k_q/k_{q,I}$	0.90	1	2.4	

[a] IAEDANS-CE H3, IAEDANS-labeled core nucleosomes at Cys-110 of CE histone H3; IAEDANS-CE H4, IAEDANS-labeled core nucleosomes at Met-84 of CE histone H4. More detailed definition of the abbreviations used is given in Table I.
[b] λ_{max}, wavelength of maximum emission; F/F_I, ratio of the observed fluorescence intensity at the indicated salt concentration to that found at intermediate ionic strength (10^{-2}–10^{-1} M NaCl); $k_q/k_{q,I}$, ratio of the acrylamide quenching rate constant [k_q, see Eq. (7)] observed at the indicated salt concentration to that found at intermediate ionic strength.
[c] Reference numbers refer to text footnotes.
[d] The 8-IAEDANS isomer[22] was used instead of 5-IAEDANS (Table I).

As an extension of the above considerations, it is found that when a sensitive fluorophore is bound to a macromolecular system, the position of its emission maximum is dependent on the polarity of its immediate environment.[32] In the particular case of IAEDANS-labeled core particles it is observed that, at intermediate ionic strength (from about 10^{-2} to 10^{-1} M NaCl), the wavelength of maximum emission is 476 nm.[8] This suggests that under these conditions the dye is in a hydrophobic environment, similar to 40–80% ethanol.[22] However, at low (10^{-4} M NaCl) and high ionic strength (0.6 M NaCl) the emission maxima are located at higher wavelengths (see Table II), indicating the existence of different conformational states in the nucleosome core particle.

The fluorescence quantum yield (ϕ) is the ratio of photons emitted to photons absorbed by the fluorophore. Since the rate of depopulation of the first excited singlet state is equal to the rate of fluorescence (k_f) plus the sum of the rates of all nonradiative processes ($\Sigma \, k_i$) that compete with fluorescence, ϕ can be defined as

$$\phi = k_f/(k_f + \Sigma \, k_i) \tag{4}$$

[32] L. Stryer, *J. Mol. Biol.* **13**, 482 (1965).

Often it is found that an emission maximum shift to the red is accompanied by a decrease in the quantum yield of the fluorophore, suggesting that a rise in the polarity of the environment enhances the rates of nonradiative processes that depopulate the first excited state.[31] In keeping with this, in the case of nucleosomes labeled with IAEDANS at Cys-110 of histone H3, it is observed that the fluorescence intensity is higher at intermediate ionic strength than at low or high salt concentrations (Table II). Thus, taken together, these modifications of the spectral parameters of fluorescent core nucleosomes indicate that the environment of IAEDANS in the nucleosome is more polar at low and high ionic strength than at intermediate salt concentrations.

In Table II, we have considered the two main salt-induced structural transitions observed for core nucleosomes labeled with IAEDANS at Cys-110 of histone H3. In addition, there is a minor conformational transition that occurs between 0.1 and 0.35 M NaCl,[8] which has not been indicated in Table II. Chung and Lewis,[12] using nucleosomes labeled at Met-84 of histone H4, have also detected several salt-induced transitions that produce small changes in emission properties. As observed with nucleosomes labeled at Cys-110 of histone H3, when the label is at Met-84 of histone H4 the main structural change is found between intermediate and high ionic strength. In the latter case, however, the transition is almost completed at 0.4 M NaCl, and the relative change in fluorescence intensity is smaller.

The equilibrium constants for the observed transitions can be calculated from the fluorescence data. The dependence of the apparent equilibrium constants on salt concentration allows the determination of the number of ions involved in these structural changes.[33] The results obtained with IAEDANS-labeled nucleosomes indicate that a total of 16 ions participate in the high-salt transition.[7,8] Only two to four ions are involved in the low-salt transition.[10] Furthermore, the observed changes in the fluorescence intensity can also be used to investigate the kinetics of these conformational changes. A detailed kinetic study of the low-salt transition has revealed that this process is complex and involves two distinct intermediate structures.[11]

Collisional Quenching of Fluorescent-Labeled Nucleosomes

In general, fluorescence quenching can be defined as a decrease in the quantum yield. Many processes can produce quenching. In particular, the quenching of fluorescence resulting from collisions with solute molecules is

[33] M. T. Record, T. M. Lohman, and P. Haseth, *J. Mol. Biol.* **107,** 145 (1976).

useful to study the relative exposure of extrinsic fluorophores in macromolecular systems.[34] This is called dynamic or collisional quenching, because the quencher diffuses and collides with the fluorophore during the lifetime of the excited state. In contrast, in static quenching a nonfluorescent complex is formed between the quencher and the fluorophore in the ground state. Consequently, whereas dynamic quenching competes with fluorescence in the relaxation of the excited state and produces a decrease in the measured excited state lifetime, static quenching does not alter this parameter.

If the lifetime of the first excited singlet state (τ), which is usually called fluorescence lifetime, is defined as the reciprocal of the sum of the rates of all the processes that can deexcite the fluorophore in the absence of solute quenchers, then, using the same symbols as in Eq. (4), one has

$$\tau = (k_f + \Sigma \, k_i)^{-1} \tag{5}$$

The rate of deexcitation owing to collisions with a quencher is $k_q[Q]$, where k_q is the bimolecular quenching rate constant and $[Q]$ the concentration of the quencher. Thus the fluorescence lifetime in the presence of the quencher (τ_q) can be defined as

$$\tau_q = (k_f + \Sigma \, k_i + k_q[Q])^{-1} \tag{6}$$

Combining these equations and using the definition of quantum yield given in Eq. (4), one obtains

$$\phi/\phi_q = \tau/\tau_q = 1 + k_q\tau[Q] \tag{7}$$

where ϕ_q is the fluorescence quantum yield in the presence of the quencher. Experimentally, taking into account that the fluorescence intensity observed at a given wavelength is expected to be linearly dependent on the quantum yield and that the value of τ can be measured in fluorescence decay experiments, it is possible to determine k_q by plotting F/F_q versus $[Q]$ (Stern–Volmer plot), where F and F_q are, respectively, the fluorescence intensities in the absence and presence of quencher. The high values of k_q found for fluorophores free in solution demonstrate that in these cases solute quenching is a diffusion-controlled process. When the fluorophore attached to a macromolecule is exposed to the medium, the observed values of k_q are close to values found for free-diffusing species. If the fluorophore is buried, however, the observed k_q is considerably smaller.[34]

A number of solutes can be used as collisional quenchers.[34] Using the nonionic quencher acrylamide[35] with nucleosome cores labeled with IAE-

[34] S. S. Lehrer and P. C. Leavis, this series, Vol. 49, p. 222.
[35] M. R. Eftink and C. A. Ghiron, *Biochemistry* 15, 672 (1976).

DANS at Cys-110 of histone H3, it is found that the experimental values of k_q determined as indicated above, are dependent on the ionic strength of the solution.[8] This indicates changes in the accessibility of IAEDANS to acrylamide quenching and hence suggests the existence of different structural states with distinct degrees of exposure of this dye to solvent. As expected, taking into account the results considered in the preceding section, the values of the quenching rate constant are higher at high and low ionic strength than at intermediate salt concentrations (see Table II). When IAEDANS is attached to Met-84 of histone H4, the relative changes of k_q are smaller, and the differences between 10^{-4} and 10^{-2} M NaCl are almost negligible.[12] In this case, however, a relatively small conformational transition is observed at 10^{-3} M NaCl. Again, independent of the position of the label, the main structural change occurs between intermediate and high ionic strength.

Fluorescence Polarization Studies

When a fluorescent sample is irradiated with polarized light, there is a preferential excitation (photoselection) of the fluorophores oriented with absorption transition dipoles parallel to the electric vector of the exciting beam. Thus, photoselection produces an anisotropic distribution of the excited molecules. Consequently, in a rigid system, for almost any relative orientation of the absorption and emission transition dipoles of the fluorophore, this distribution will produce polarized fluorescence; that is, the observed fluorescence anisotropy [A, Eq. (2)] and polarization [P, Eq. (3)] will not be zero. In contrast, if the excited fluorophores are mobile and can rotate and randomize their orientation during the lifetime of the first excited singlet state, A and P will be both zero. This is observed for small molecules in aqueous solutions, at room temperature. However, if the fluorophore is rigidly bound to a biological macromolecule, in general, its rotational diffusion is not negligible on the time scale of the fluorescence emission, but it is not so high as to produce the complete depolarization of the fluorescence. This situation is particularly interesting because the parameters that characterize the fluorescence polarization properties of these systems can give information about the size and shape of the macromolecules, and even about their conformational flexibility.[36]

In time-resolved fluorescence polarization experiments with fluorescent-labeled macromolecules, it is found that, as time progresses after the excitation with a short pulse of polarized light, there is a progressive

[36] C. R. Cantor and P. R. Schimmel, "Biophysical Chemistry," p. 433. Freeman, San Francisco, California, 1980.

depolarization of the fluorescence emission. This is due to the rotational diffusion of the molecule. In the case of a rigid spherical molecule, the time-resolved anisotropy decay [$A(t)$] is described by a single exponential:

$$A(t) = A_0\, e^{-t/\tau_R} \tag{8}$$

where A_0 is the anisotropy at $t = 0$ and τ_R the rotational correlation time of the molecule, which depends on the viscosity (η) and the absolute temperature (T) of the solution, as described by

$$\tau_R = \eta V/RT \tag{9}$$

where V is the hydrated molar volume of the molecule and R the gas constant. More complex anisotrophy decay curves are found if the macromolecular system is nonspherical. In these cases, the anisotrophy decay is described by a sum of exponential terms. A complex behavior is also observed if the fluorophore itself, or the part of the molecule in which it is held, can rotate independently of the rotational motion of the whole molecule. The reader interested in these questions is referred to a comprehensive article by Yguerabide.[37]

The equations describing the steady-state anisotropy and polarization can be derived from the equations of the time-resolved studies.[37] The steady-state values of A and P are related to the rotational correlation time, τ_R, and the fluorescence lifetime of the fluorophore, τ, according to the Perrin equations:

$$1/A = 1/A_0(1 + \tau/\tau_R) = 1/A_0(1 + \tau RT/\eta V) \tag{10}$$
$$(1/P - 1/3) = (1/P_0 - 1/3)(1 + \tau/\tau_R)$$
$$= (1/P_0 - 1/3)(1 + \tau RT/\eta V) \tag{11}$$

In these equations, A_0 and P_0 are the anisotropy and polarization that would be observed if the macromolecule cannot rotate during the lifetime of the excited state. The rotational correlation time has been expressed explicitly according to Eq. (9). Perrin plots of $1/A$ or $1/P$ versus $\tau RT/\eta$ are experimentally obtained by varying the temperature, the viscosity (by addition of sucrose or glycerol), or the lifetime of the dye (by addition of a quencher). These plots allow the determination of $1/A_0$ or $1/P_0$, V, and τ_R, provided that the fluorescence lifetime is known from fluorescence decay experiments. The experimental value of V represents the effective molar volume of the entire macromolecule, or, when the system shows segmental flexibility, this parameter corresponds to the molar volume of the region of the molecule where the dye is bound. If the value obtained for τ_R is

[37] J. Yguerabide, this series, Vol. 26, p. 498.

TABLE III
FLUORESCENCE POLARIZATION PROPERTIES OF FLUORESCENT-LABELED
NUCLEOSOMES[a]

Sample	Parameters[b]	Low	Intermediate	High	Refs.[d]
			Ionic strength[c]		
IAEDANS-CE H3	τ_R	26.4	145	6.2	8
	V	6.57	36.1	1.55	
IAEDANS-CE H4	τ_R	—	122[e]	25	12
	V	—	29.9[e]	6.1	
Ethidium-DNA	τ_R	—	170[f]	—	6

[a] Ethidium-DNA, core nucleosomes labeled with ethidium bromide; the other abbreviations are described in Tables I and II.
[b] τ_R, Rotational correlation time (nsec); V, effective molar volume (cm^3/mol $\times$ 10^{-4}). See Eqs. (8)–(11).
[c] Ionic strength: low, 0.1 mM Tris, 0.02 mM EDTA; intermediate, 10 mM Tris, 2 mM EDTA; high, 0.6 M NaCl.
[d] Reference numbers refer to text footnotes.
[e] Measurements were carried out in 50 mM NaCl.
[f] Measurements were carried out in 5 mM Tris, 0.2 mM EDTA.

larger than expected from the molecular weight, it is likely that the system is nonspherical. By the graphical method of Weber[38] it is possible to calculate the axial ratio of an equivalent ellipsoid of revolution.

Fluorescence polarization studies carried out with IAEDANS-labeled nucleosomes show significant changes of structure as a consequence of the variation of the ionic strength of the medium (see Table III). When the dye is attached to Cys-110 of histone H3, the values of τ_R and V found at low and high ionic strength are much lower than that observed at intermediate salt concentration.[8] These values are low enough to suggest that at low and high ionic strength the core nucleosome is flexible and allows the independent motion of the region that contains the Cys-110 of histone H3. The results obtained with core nucleosomes labeled with IAEDANS at Met-84 of histone H4 also indicate a significant increase in mobility when the ionic strength is raised to 0.6 M NaCl,[12] but in this case the observed decrease of τ_R and V is lower than that found when the dye is bound to histone H3. Independent of the position of the label, at intermediate ionic strength the nucleosome is more compact and shows a rotational correlation time higher than that expected for a spherical structure. The axial ratio obtained

[38] G. Weber, *Adv. Protein Chem.* **8**, 415 (1953).

for an oblate ellipsoid, according to the method of Weber,[38] is 3.3, which corresponds to dimensions of $130 \times 130 \times 39$ Å.[8]

Barkley and Zimm[39] have shown that the anisotropy decay of ethidium bromide intercalated in free DNA originates mainly by the torsional motion of DNA. This property of the noncovalent ethidium–DNA complexes can be used to investigate the internal motions of DNA in the nucleosome core particle. Ashikawa *et al.*[6] have performed time-resolved anisotropy measurements of core nucleosomes labeled with ethidium bromide where the ratio of dye molecules per base pair is in a range lower than that required to produce unfolding[5] or dissociation[40] of nucleosomes. The results obtained suggest that the DNA in the nucleosome core particle has a torsional rigidity similar to that of free DNA, indicating the existence of DNA flexibility even when the core particle adopts a compact conformation at intermediate ionic strength.

Excimer Formation in Pyrene-Labeled Core Nucleosomes

Dilute solutions of pyrene and other aromatic hydrocarbons show structured emission bands characteristic of the relaxation from the first excited singlet state. When the concentration of these solutions is increased, however, this band is progressively quenched and a new broad, structureless fluorescence band is found at a longer wavelength. Since the observed autoquenching obeys the normal equation of bimolecular collisional quenching [see Eq. (7)] and since these changes in the fluorescence spectra do not produce variations in the corresponding absorption spectra, these findings suggested[24] that the fluorophores can form dimers after the absorption of light. The excited dimers (usually called excimers) are formed according to the following reaction scheme:

$$P + P \xrightarrow{h\nu} P^* + P \rightarrow (PP)^* \tag{12}$$

where P and P* represent, respectively, the unexcited and excited monomer and (PP)* the excimer structure. The observed photoaddition does not produce permanent chemical changes: the typical monomer structure is restored after the excimer is relaxed by emission of light or by nonradiative processes. The transition to the ground state results in an unstable dimer that dissociates immediately. This is the cause of the broad, structureless band associated with this transition. Crystals of pyrene exhibit the typical fluorescence spectrum of excimer structures. In the crystal lattice, the pyrene molecules are arranged in parallel pairs with an interplanar distance

[39] M. D. Barkley and B. H. Zimm, *J. Chem. Phys.* **70**, 2991 (1979).
[40] C. T. McMurray and K. E. von Holde, *Proc. Natl. Acad. Sci. U.S.A.* **83**, 8472 (1986).

of 3.5 Å. More details about the photophysics of excimer structures can be found in reviews by Birks[41] and Stevens.[42]

The nucleosome core particle can be labeled with NPM at Cys-110 of histone H3.[14,16] The resulting particles containing two fluorescent pyrene rings show striking changes in the emission spectra as a function of the conditions used. As shown in Fig. 1, a decrease in NaCl concentration from 0.6 to 0.2 M produces the appearance of a band centered at about 460 nm,[16] which indicates the existence of intramolecular excimers of pyrene. The formation of intramolecular pyrene excimers has also been found in synthetic polymers[43] and in protein molecules.[29] This band is not observed when nucleosomes are in the presence of NaCl at concentrations higher than 0.6 M[14,16] but is found at low ionic strength.[14] Core particles can also be labeled with pyrene at Met-84 of histone H4.[20] In keeping with the results obtained with IAEDANS, it is observed that when this probe is attached to this position, the fluorescence changes induced by salt are less intense than those observed when the dye is bound to Cys-110 of histone H3.

Fluorescence quenching studies, carried out using the collisional quencher iodide, indicate that the NPM label attached to Cys-110 of histone H3 is less exposed to solvent in 0.2 M NaCl than in 0.6 M NaCl.[16] Time-resolved measurements show that in 0.2 M NaCl, the excimer fluorescence decays continuously after the end of the light pulse.[17] This indicates that in this conformational state, the formation of the excimer is completed in a few nanoseconds, probably because diffusion is not required for the encounter of the two pyrene rings. Taken together, these observations suggest that in 0.2 M NaCl, the nucleosome core particle is in a folded conformation in which the two pyrene rings, even in the ground state, are very close and buried inside the particle. The slight deformation of the excitation spectra, found when the emission wavelength is 460 nm,[14,16] confirms this interpretation. The crystallographic results of Richmond et al.[44] have shown that the heavy atom cluster tetrakis(acetoxymercuri)methane, attached to Cys-110 of histone H3, is located on the dyad axis of the core nucleosome, between the center of the particle and the central turn of the DNA. Thus, all these results suggest that the two pyrene rings that associate to produce excimers are also located in the same region of the nucleosome.

[41] J. B. Birks, "Photophysics of Aromatic Molecules," Chap. 7. Wiley, New York, 1970.
[42] B. Stevens, Adv. Photochem. 8, 161 (1971).
[43] A. E. C. Redpath and M. A. Winnik, Ann. N.Y. Acad. Sci. 366, 75 (1981).
[44] T. J. Richmond, J. T. Finch, B. Rushton, D. Rhodes, and A. Klug, Nature (London) 311, 532 (1984).

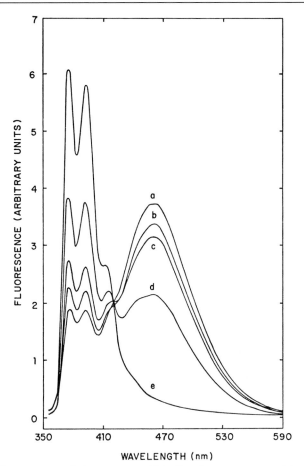

FIG. 1. Fluorescence emission spectra at 20° of nucleosome core particles labeled with NPM at Cys-110 of histone H3. The samples were prepared in 2 M NaCl and were salt jumped to the indicated final salt concentrations: (e) 0.6, (d) 0.5, (a) 0.4, (b) 0.3, and (c) 0.2 M NaCl. The sample solutions contained, in addition, 0.2 M EDTA, 10 mM Tricine, pH 7.0. The excitation wavelength was 340 nm. Note that the main structural change occurs between 0.6 and 0.4 M NaCl; between 0.4 and 0.2 M a minor structural change can also be seen. There is an isoemissive point at roughly 421 nm.

The properties of this interesting system can be used further to investigate different aspects of nucleosome structure and dynamics. Zama *et al.*[14,15] have shown that pH values lower than 4.5, urea at a concentration higher than 3 M, and various organic solvents produce the disappearance of the excimer band of core particles containing NPM. This sensitive probe for nucleosome conformation has also been used to investigate the role of histones and DNA in the stabilization of the nucleosome core at physiological ionic strength.[18] It has been found that extensive DNase I or trypsin digestion of NPM-labeled nucleosome cores in 0.2 M NaCl does not produce significant changes in excimer fluorescence, indicating that the covalent continuity of DNA and the basic amino-terminal regions of core histones are not required for the maintenance of the folded conformation.

This system has also been used to study the kinetics of nucleosome self-assembly reaction.[16] The results obtained indicate that the binding of core histones to DNA is a very fast process, which is followed by an intramolecular folding reaction with two first-order kinetic components that can be monitored by measuring the increase of excimer intensity (Table IV). The kinetics of the reaction of NPM-labeled histones H3,H4 with core DNA, compared with the kinetic results obtained with histones H2A,H2B, suggest that the intramolecular reaction is directed by histones H3,H4, which form an intermediate structure that in the presence of H2A,H2B produces the final nucleosome structure.[17]

Measurement of Internal Distances by Nonradiative Singlet–Singlet Energy Transfer

In principle, if the emission spectrum of a given chromophore (donor) overlaps the absorption spectrum of another chromophore (acceptor), fluo-

TABLE IV
KINETIC RESULTS OF SELF-ASSEMBLY REACTION OF NPM–CORE NUCLEOSOMES[a]

		Kinetic parameters[c]				
Sample	Conditions,[b] NaCl (M)	A_1	$\lambda_1 \times 10^{-1}$ (sec^{-1})	A_2	$\lambda_2 \times 10^{-3}$ (sec^{-1})	Refs.[d]
Nucleosome	2–0.2	0.62	0.93	0.38	6.7	16
cores	0.6–0.2	0.66	1.2	0.34	7.1	
Core histones plus DNA	0.2	0.60	0.70	0.40	6.8	16
(H3,H4) plus DNA	0.2	0.76	2.1	0.24	9.3	17
[(H3,H4) plus DNA] plus (H2A,H2B)	0.2	—	—	1	6.2	17

[a] All the samples contained NPM-labeled CE histone H3 (see Table I); DNA represents core DNA.

[b] The reaction conditions are as follows: 2–0.2 and 0.6–0.2, salt jump from 2.0 and 0.6 to 0.2 M NaCl; 0.2, direct mixing in 0.2 M NaCl.

[c] A_1, A_2, λ_1, and λ_2 are, respectively, the amplitudes and first-order rate constants of the two kinetic components, observed by measuring the increase of fluorescence intensity at 460 nm.

[d] Reference numbers refer to text footnotes.

rescence energy transfer can occur between them by a mechanism involving the emission and reabsorption of photons. In this trivial mechanism of energy transfer, no change in the fluorescence lifetime of the donor is expected. Nevertheless, this process is highly improbable in diluted solutions. A direct mechanism for energy transfer involves the coupling of the emission transition dipole of the donor to the absorption transition dipole of the acceptor. In this mechanism, the energy from the first excited singlet state of the donor is nonradiatively transferred to the acceptor, which becomes excited. This process of energy transfer competes with all the other processes that can deexcite the first excited singlet state of the donor. Therefore, if for a particular donor–acceptor pair the rate of singlet–singlet energy transfer is k_T, the efficiency (E) of the transfer can be defined as

$$E = k_T/(k_T + \tau_D^{-1})$$ (13)

where τ_D represents the fluorescence lifetime [see Eq. (5)] of the donor in the absence of the acceptor. Using the definition of quantum yield given in Eq. (4), the efficiency of the energy transfer can be expressed as a function of the quantum yield of the donor in the presence (ϕ_{DA}) and absence (ϕ_D) of the acceptor:

$$E = 1 - (\phi_{DA}/\phi_D)$$ (14)

Taking into account Eq. (5), the efficiency can be expressed as

$$E = 1 - (\tau_{DA}/\tau_D)$$ (15)

where τ_{DA} and τ_D are, respectively, the lifetimes of the donor in the presence and absence of the acceptor. A detailed description of the theoretical and experimental aspects of fluorescence energy transfer can be found in a previous review.[21]

For a given donor–acceptor pair, the rate of energy transfer depends on the distance between these molecules. This is an interesting feature of fluorescence spectroscopy because it allows the measurement of the distances between donor and acceptor fluorophores attached to macromolecular systems. According to Förster theory, the rate of energy transfer is given by

$$k_T = (1/\tau_D)(R_0/R)^6$$ (16)

where R is the distance between donor and acceptor and R_0 the critical distance at which energy transfer is 50% efficient [i.e., $k_T = \tau_D^{-1}$, according to Eq. (13)]. The critical distance can be determined experimentally (in angstroms), according to

$$R_0 = [(8.79 \times 10^{-5})n^{-4}\phi_D J_{DA}\kappa^2]^{1/6}$$ (17)

where n is the refractive index of the medium (1.4 for biological systems[21]), ϕ_D the quantum yield of the donor in the absence of acceptor, J_{DA} the spectral overlap integral, and κ^2 the dipole–dipole orientation factor. The spectral overlap integral is defined as

$$J_{DA} = \left[\int_0^\infty F_D(\lambda)\epsilon_A(\lambda)\lambda^4 \, d\lambda \right] / \left[\int_0^\infty F_D(\lambda) \, d\lambda \right] \qquad (18)$$

where $F_D(\lambda)$ is the fluorescence intensity of the donor in the absence of acceptor at wavelength λ (nanometers), and $\epsilon_A(\lambda)$ the molar extinction coefficient ($M^{-1}cm^{-1}$) of the acceptor at the same wavelength.

The geometric factor κ^2 takes into account the dependence of energy transfer on the relative orientation of the transition dipoles of the donor and acceptor chromophores. It is defined as

$$\kappa^2 = (\cos \theta_{DA} - 3 \cos \theta_D \cos \theta_A)^2 \qquad (19)$$

where θ_{DA} is the angle between donor emission and acceptor absorption transition dipoles, and θ_D and θ_A are, respectively, the angles between donor and acceptor dipoles and the vector joining the centers of the two chromophores. In theory, κ^2 may vary from 0 to 4. If both chromophores are free to rotate and randomize their orientations during the fluorescence lifetime of the donor, κ^2 approaches the limiting value of 2/3. When the two chromophores are more rigidly held, this value provides only an estimate of R_0. A method developed by Dale and Eisinger[45] allows the calculation of the actual range of possible κ^2 values for a given system. The half-angle (ψ) of the conical volume, within which a given chromophore is free to reorient, can be calculated from the limiting anisotropy [A_0 in Eq. (10)] determined from Perrin plots (see above).[21] The angle ψ has to be calculated for both donor and acceptor, and maximum and minimum values of κ^2 are determined using the graphical analysis of Dale and Eisinger.[45]

The distance R between donor and acceptor and the transfer efficiency E are related according to Eq. (20), which is obtained by combining Eqs. (13) and (16):

$$R = R_0(1/E - 1)^{1/6} \qquad (20)$$

If R_0 is calculated taking into account maximum and minimum values of κ^2 obtained as indicated above, Eq. (20) allows the determination of a range of distances. The transfer efficiency E can be obtained by measuring the additional emission (sensitized emission) of the acceptor produced by the presence of the donor.[21] Another procedure to determine E is measure-

[45] R. E. Dale and J. Eisinger, *Biopolymers* **13**, 1573 (1974).

ment of the donor quenching produced by the acceptor.[21] In this case, the efficiency is calculated according to Eq. (14) from the observed fluorescence intensity in the presence and absence of the acceptor, or according to Eq. (15) by direct observation of the fluorescence lifetimes τ_{DA} and τ_D. The lifetimes determined after the analysis of the fluorescence decay kinetics are independent of the stoichiometry of the labeling. However, when the other methods considered above are used, the observed values of the efficiency (E_{obs}) must be corrected for the acceptor stoichiometry. If the fraction of molecules with acceptor is f_A, the corrected values of E to be used in Eq. (20) are obtained by dividing E_{obs} by f_A. If the assembly contains more than one molecule of acceptor, the probabilities of the different species present in partially labeled samples must be considered for estimation of the true efficiency from the observed efficiency.[9]

The various intranucleosomal distances measured by energy transfer are shown in Table V. These results have been obtained by measuring the quenching of the IAEDANS fluorescence produced by IAF. The measurements of the efficiency of energy transfer have been carried out between 440 and 480 nm, because in this spectral region the fluorescence intensity of IAEDANS is high whereas the fluorescence of IAF is almost negligible (see Fig. 2). The distances shown in Table V are obtained with a value of 2/3 for the orientation factor κ^2. However, the range of distances, calculated considering the maximum and minimum values of κ^2, obtained as described above, does not substantially change the results shown in Table V.[8,9,13] In agreement with the results obtained with other fluorescence

TABLE V
ENERGY-TRANSFER MEASUREMENTS IN CORE NUCLEOSOMES[a]

Location of the donor (IAEDANS)–acceptor (IAF) pair	Distance measured (Å)[b]				
	Ionic strengths[c]			Urea	
	Low	Intermediate	High	(3–5 M)	Refs.[d]
Cys-110 CE H3	43–48	⩽30	⩾70	—	8
Met-84 CE H4	<20–35	<20	38	—	13
Cys-110 CE H3–3′ ends of DNA	50–58	50–53	54	47–55	9
Cys-73 SU H4–3′ ends of DNA	⩽31	⩽32	—	—	9

[a] Abbreviations are described in Tables I and II.

[b] The values of R have been obtained according to Eq. (20), assuming a value of 2/3 for the orientation factor κ^2.

[c] Ionic strength: low, 10^{-4}–10^{-3}; intermediate, 10^{-2}–10^{-1}; high, 0.6 M. See the original references for details about the composition of the buffers used.

[d] Reference numbers refer to text footnotes.

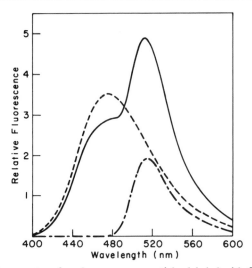

FIG. 2. Emission spectra of nucleosome core particles labeled with fluorescent dyes for energy-transfer experiments. All the samples were prepared at intermediate ionic strength. The scans correspond to nucleosomes labeled with IAEDANS at Cys-110 of histone H3 (- - -), labeled with IAF at 3′ ends of core DNA (—··—), and double-labeled with both IAEDANS and IAF at the indicated positions (——). The excitation wavelength was 340 nm. Note the negligible contribution of IAF to the emission spectra in the 400–480 nm region. The quenching of IAEDANS fluorescence produced by IAF, and the sensitized emission of IAF produced by IAEDANS, can also be clearly seen. [Reprinted, with permission, from Eshaghpour et al.[9] Copyright 1980 American Chemical Society.]

methods described in the preceding sections, the two main structural transitions induced by low and high salt concentrations are also clearly observed when IAEDANS and IAF are attached to the two copies of histone H3 in the nucleosome.[8] Less intense changes are observed when these dyes are bound to the two molecules of histone H4.[13] In this case, also in agreement with previous results, the main structural transition occurs between intermediate ionic strength and 0.35 M NaCl, and, in addition, a transition is detected at 1 mM NaCl. Energy-transfer studies carried out with core DNA, labeled at the 3′ ends with IAF,[9] indicate that the distance between this dye and IAEDANS attached to Cys-110 of histone H3 is 50–53 Å. This is in keeping with the central position of the NPM labels in the nucleosome core considered in the preceding section. The relatively short distance between the Cys-73 of sea urchin histone H4 and the 3′ ends of core DNA indicates that this group lies not too far from Cys-110 of histone H3. This is in agreement with the histone–DNA cross-linking results of Ebralidse and Mirzabekov[46] showing that the strong H4–DNA

[46] K. K. Ebralidse and A. D. Mirzabekov, *FEBS Lett.* **194**, 69 (1986).

binding sites overlap the binding sites of histone H3 in the central turn of core DNA. The most striking feature of these results is that the distance between the Cys-110 of histone H3 and the DNA ends remains essentially constant even in the presence of 0.6 M NaCl or 5 M urea.[9]

These results are consistent with the observations presented in the preceding sections and indicate again that the nucleosome is more compact at intermediate ionic strength than at low and high salt concentrations. The low and high salt structural transitions produce more intense changes in the fluorescence parameters when the labels are attached to Cys-110 of histone H3 than when they are bound to Met-84 of histone H4. This could indicate that Cys-110 of histone H3 is located in a more flexible region of the core particle. On the other hand, it could be that the presence of a label attached to histone H3 can produce alterations of the core particle structure. In fact, the high salt structural transition produces only a relatively small change in the sedimentation coefficient of the native nucleosome core.[47-49] However, the presence of the voluminous, heavy atom cluster, tetrakis(acetoxymercuri)methane,[50] attached at Cys-110 of histone H3, does not perturb the crystal structure of the nucleosome core particle.[44] Furthermore, the invariance of the distance between Cys-110 of histone H3 and the DNA ends considered above is consistent with the existence of contacts between this histone and the terminal region of core DNA as demonstrated in histone–DNA cross-linking experiments.[51] This suggests that the remarkable changes in the fluorescent parameters observed with fluorescent-labeled histone H3 are due to the flexibility of the core particle structure in the regions containing the DNA tails.

Concluding Remarks

The results obtained using fluorescence spectroscopy to study nucleosome core particles in solution have been summarized in the preceding sections. Many different aspects of this structure have been revealed using fluorescent labels attached to specific sites. It has been demonstrated that fluorescent labels are sensitive and versatile probes with which to investigate the static and dynamic properties of nucleosomes in solution.

Other useful applications of fluorescent labels in chromatin studies have been described. Recently, it has been found that the environmental

[47] H. Eisenberg and G. Felsenfeld, *J. Mol. Biol.* **150,** 537 (1981).
[48] T. D. Yager and K. E. van Holde, *J. Biol. Chem.* **259,** 4212 (1984).
[49] K. O. Greulich, J. Ausió, and H. Eisenberg, *J. Mol. Biol.* **186,** 167 (1985).
[50] J. J. Lipka, S. J. Lippart, and J. S. Wall, *Science* **206,** 1419 (1979).
[51] V. V. Shick, A. V. Belyavsky, S. G. Bavykin, and A. D. Mirzabekov, *J. Mol. Biol.* **139,** 491 (1980).

probe 1-anilinonaphthalene 8-sulfonate can be used for the rapid staining and quantitation of histones and histone peptides in polyacrylamide gels.[52-54] Furthermore, fluorescent labels can be useful for studying complex aspects of chromatin structure and function. In particular, fluorescent labels have been used to study the interaction of the nonhistone chromosomal proteins HMG 14 and 17 with nucleosomes[12] and the stability of the higher order structure of chromatin.[55] The histone distribution during chromatin replication[19] and the changes of chromatin structure associated with transcription[56] of *Physarum polycephalum* have also been investigated using fluorescent probes. All these reports suggest a wide application of these techniques in future analyses of nucleosomes and chromatin.

Acknowledgments

This work was supported in part by a grant from the Comisión Asesora de Investigación Cietífica y Técnica (2384-83), and a grant from the U.S. National Cancer Institute (CA 39782).

[52] J. R. Daban and A. M. Aragay, *Anal. Biochem.* **138**, 223 (1984).
[53] B. Piña, A. M. Aragay, P. Suau, and J. R. Daban, *Anal. Biochem.* **146**, 431 (1985).
[54] A. M. Aragay, P. Diaz, and J. R. Daban, *Electrophoresis* **6**, 527 (1985).
[55] I. Ashikawa, K. Kinosita, A. Ikegami, Y. Nishimura, and M. Tsuboi, *Biochemistry* **24**, 1291 (1985).
[56] C. P. Prior, C. R. Cantor, E. M. Johson, V. C. Littau, and V. G. Allfrey, *Cell* **34**, 1033 (1983).

[11] Preparation and Application of Immunological Probes for Nucleosomes

By MICHAEL BUSTIN

Introduction

Immunochemical reagents are versatile tools for a variety of studies on the structure and function of chromatin and chromosomes. The serological approach complements other physical and biochemical techniques used in the study of chromatin and nucleosomes and can provide information which is not readily obtainable in other ways. Antibodies are remarkable analytical reagents since they can have specificity for a particular confor-

Copyright © 1989 by Academic Press, Inc.
All rights of reproduction in any form reserved.

mation of a macromolecule or recognize a small molecule belonging to almost any chemical class. Antibody–antigen reactions occur at moderate ionic strengths, neutral pH, and mild temperatures; therefore, they can be performed under conditions that do not markedly alter the structure of the nucleosome and can be used to study both the *in situ* organization and the *in vivo* cellular function of defined chromosomal components.

The application of immunochemical techniques to the study of nucleosomes and their components is complicated by the fact that the nucleosome is a relatively complex and dynamic nucleoprotein structure. Purification of antigens necessary to obtain specific antisera requires procedures that not only disrupt the native structure of the nucleosome, but often also denature the antigen. Thus, the successful application of immunochemical approaches to the study of nucleosomes and chromatin is dependent on three major steps: (1) production and characterization of antisera specific for purified, defined, antigens; (2) evidence that the antibodies, even though elicited against purified and denatured chromosomal components, recognize and bind the antigens complexed in their native state; and (3) methods used to detect and quantify antigen–antibody reactions have to be adapted to the study of nucleosomes, chromatin, and chromosomes.

The purpose of this chapter is to summarize the methods used to elicit and characterize immunochemical probes for nucleosomes and some of their components and to describe the use of these probes in a variety of studies related to the specificity, structure, and function of nucleosomes. Most of the information is relevant to the specificity and organization of nucleosomal proteins, with particular emphasis on our experience with histones and HMG proteins. The main principles involved in the production and application of immunological probes are described. However, for all the methodological details it is best to consult the original literature.

Preparation and Characterization of Probes

Antisera Preparation

The best way to ensure a specific immune response is to use a pure and defined immunogen. In principle, the monoclonal approach may circumvent the necessity for absolute antigen purity. However, attempts to obtain large libraries of monoclonals against undefined mixtures of nuclear extracts failed to yield a large battery of useful antibodies. The underlying problem with the monoclonal approach is the difficulty in defining the specificity of the antibody, and, consequently, there is always a certain degree of uncertainty as to the meaning of the results.

Histone Antibodies. Histones are poor immunogens probably because they are conserved during evolution, they are ubiquitously distributed among nucleated cells, and they contain few aromatic amino acid residues. Therefore, even a slight contamination with other antigens may result in antiserum preparations that contain disproportionately large amounts of antibodies against these contaminants. Even if the histones have been purified by proven procedures, the purity of histones used as immunogens should be checked by several criteria, such as overloaded, two-dimensional polyacrylamide gels, amino-terminal determinations, and peptide mapping. The amino-terminal determinations are particularly useful since most histones have blocked amino-termini and detection of unexpected amino-terminal amino acids would indicate the presence of contaminants. Because histones are highly conserved during evolution, antibodies elicited against histones purified from one species can be used to study these proteins in most nucleated cells.

The most frequent contaminants in a histone preparation are other histones and nonhistone proteins, some of which elicit species- and even tissue-specific antibodies. Complications arising from the presence of antibodies against species-specific nonhistone contaminants can be minimized by using antisera elicited against histones derived from one source to study histones and nucleosomes from a heterologous source. Antibodies elicited by histone fractions purified from calf thymus interact with histones, chromatin, and nucleosomes from a variety of sources. Therefore, they can be used as "standard" reagents in a variety of experimental systems.

The following antigens have been used to elicit antihistone sera: (1) total histone, (2) purified histone classes, (3) histone complexes (i.e., dimer, octamer), (4) histone fragments, and (5) synthetic peptides. Antibodies specific for total histones or for a histone class can be elicited in rabbits by immunizing with insoluble histone–RNA complexes as follows:[1,2]

1. For each rabbit, mix 50–300 μg pure histone with 20–100 μg RNA in a total of 0.5–1.0 ml 0.15 M NaCl, 10 mM sodium phosphate buffer, pH 6.8. The appearance of a fine precipitate indicates the formation of insoluble complexes.

2. Add 2 volumes of Freund's complete adjuvant and emulsify thoroughly.

3. Inject the rabbit in at least 10 intradermal sites, 20–50 μl per site. The remaining material should be injected into 3–4 intramuscular sites and into footpads.

[1] B. D. Stollar and M. Ward, *J. Biol. Chem.* **245**, 1261 (1970).
[2] D. Goldblatt and M. Bustin, *Biochemistry* **14**, 1689 (1975).

4. Administer a booster injection, identical to the initial immunizing injection, 14 days after the first injection.

5. Administer three additional booster injections at biweekly intervals. These injections should be intramuscular, and the antigen should be prepared in Freund's incomplete adjuvant.

6. One week after the last boost, administer an intravenous injection of 50–100 μg histone dissolved in about 300 μl aqueous buffer, pH 6.8.

7. Starting 1 week after the intravenous boost, blood is collected weekly from the marginal ear vein. Each bleed can yield 30–50 ml blood. The first bleed should be tested by ELISA. If positive, collect about 10 bleeds, and mix the resulting sera.

Other protocols have also been used;[3] however, the protocol described above consistently yields antisera with high titer. The antisera elicited by this protocol have a relatively high content of antibodies against cytoskeletal components, most probably due to the tissue damage caused by repeated intradermal immunizations. Such complications can be eliminated by preadsorbing the sera with cytoplasmic extracts or by affinity purification of the histone-specific Ig molecules.

Histone H5, which is species-specific, is more immunogenic than the core histones. Therefore, antibodies can be elicited by a somewhat simpler protocol:[4]

1. Emulsify 100–200 μg H5 in 2.0 ml 66% Freund's complete adjuvant.

2. Inject at multiple intradermal sites and intramuscularly.

3. Administer an identical boost 14 days later.

4. Two weeks later, administer 2–3 intramuscular injections, totaling 100 μg H5 emulsified in 1.5 ml Freund's incomplete adjuvant.

5. Two weeks later, administer an intravenous boost of 25 μg H5 dissolved in 300 μl phosphate-buffered saline (PBS). (Higher amounts of antigen may precipitate an anaphylactic shock.)

6. Starting 1 week after the intravenous boost, the rabbit can be bled weekly from the marginal ear vein. The bleeds should be treated as described above for the histones.

Antibodies against large proteolytic fragments of the core histones can be elicited by the protocol described for core histones. Sera specific for the globular region of histone H5 can be elicited by the protocol described for H5. Small proteolytic fragments and synthetic peptides are usually coupled

[3] M. Bustin, in "The Cell Nucleus" (H. Busch, ed.), Vol. IV, p. 195. Academic Press, New York, 1978.
[4] J. Allan, B. J. Smith, B. Dunn, and M. Bustin, J. Biol. Chem. 257, 10533 (1982).

to a carrier, and the conjugates are used to immunize rabbits by the procedure described for H5 or any other procedure used to elicit antibodies against a good immunogen. The preparation of antibodies against synthetic histone peptides is described elsewhere in this volume.[5]

Description of the methodology for the production of monoclonal antibodies is beyond the scope of this chapter. So far, very few laboratories have succeeded in producing monoclonal antihistone antibodies. The production of a series of monoclonal antibodies against histone H5[6] and against nonhistone HMG1/2[7] has been described.

Antibodies to Nonhistone Proteins. The nonhistone proteins display species, and even tissue, specificity and are more immunogenic than histones. A variety of protocols have been used to elicit polyclonal and monoclonal antibodies against both pure proteins and mixtures of proteins belonging to this group. The choice of protocol depends on the characteristic of the protein and its expected immunogenicity. As pointed out earlier, antisera elicited against a poorly characterized antigen will be of limited use in studies on nucleosomes and chromatin. The protocol used to elicit antibodies against nonhistone chromosomal proteins, HMG 1, 2, and 17, can be used as an example for obtaining antibodies against a pure, evolutionarily conserved, nonhistone protein:[8]

1. Dissolve protein at 1.0 mg/ml in 10 mM Tris-Cl, pH 9, 0.1 M NaCl buffer.

2. To 0.4 ml of above, add 0.8 ml Freund's complete adjuvant, and emulsify thoroughly.

3. Inject 0.8 ml at multiple intradermal sites, and 0.4 ml intramuscularly.

4. Boost with an identical dose 14 days later.

5. Administer two additional boosts, in Freund's incomplete adjuvant, at 14-day intervals.

6. Administer an intravenous boost, consisting of 100–200 μg protein in about 0.3 ml PBS, 14 days after the previous boost.

7. Bleed rabbit at weekly intervals starting 1 week after the final boost.

In some cases, it may be advantageous to immunize directly with a pure protein, present as a band in preparative sodium dodecyl sulfate (SDS) gels:[9]

[5] S. Muller and M. H. V. Van Regenmortel, this volume [12].

[6] E. Mendelson and M. Bustin, *Biochemistry* **23**, 3459 (1984).

[7] J. N. Vanderbilt and J. N. Anderson, *J. Biol. Chem.* **260**, 9336 (1985).

[8] M. Bustin, R. B. Hopkins, and I. Isenberg, *J. Biol. Chem.* **253**, 1694 (1978).

[9] M. Silver and S. C. R. Elgin, *Chromosoma* **28**, 101 (1978).

1. Fractionate protein(s) by electrophoresis on SDS–polyacrylamide gels.
2. Stain gels and identify band of interest.
3. Excise band, lyophilize, and grind the dry material to fine powder.
4. Suspend in PBS, add an equal volume of Freund's complete adjuvant, and emulsify.
5. Immunize by administering four subdermal injections.
6. Administer an identical boost 4 weeks after immunization.

The rabbit can be bled 10 days after the second injection. However, if material is available, it is recommended to give at least one additional boost, prepared with Freund's incomplete adjuvant.

Antibodies to Nucleic Acid Components. A large repertoire of antibodies specific to nucleic acid components is available. For immunization, polynucleotides are usually complexed with methylated bovine serum albumin (BSA), while short oligonucleotides are coupled to carriers. A description of the available sera and the methods used to prepare them can be found in several recent reviews.[10,11]

Antibodies to Nucleosomes. Antisera against nucleosomes[12] can be elicited by techniques similar to those described for histones and nonhistone proteins:

1. Immunize rabbit with 1.0 mg of nucleosomes (determined as DNA) emulsified in 2.5 ml of 66% Freund's complete adjuvant, by injecting at multiple intradermal sites and at several intramuscular sites.
2. Administer an identical boost 14 days after the initial injection.
3. Administer two additional boosts at 14-day intervals with the nucleosomes emulsified in 66% Freund's incomplete adjuvant.
4. Administer an intravenous injection of 0.3 mg nucleosomes in 0.3 ml PBS 1 week after the last boost.
5. Bleed at weekly intervals and pool sera.

The antisera, which were elicited against mononucleosomes, did not react with DNA extracted from the immunizing nucleosomes, with core histones, with cross-linked histone octamer, or with HMG proteins. Digestion with DNase I resulted in about 50% loss of antigenicity, while digestion with proteases resulted in a total loss of antigenicity. The antisera reacted strongly with trimers, dimers, monomers, core particles, and even subnucleosomal fragments (<145 base pairs) prepared from the same tissue as the immunogen, but they displayed a significant degree of species

---

[10] B. D. Stollar, *CRC Crit. Rev. Biochem.* **20**, 1 (1986).
[11] B. D. Stollar and A. Rashtchiau, *Anal. Biochem.* **161**, 387 (1987).
[12] C. S. M. Tahourdin and M. Bustin, *Biochemistry* **19**, 4387 (1980).

specificity when tested with heterologous nucleosomes. It was concluded that most of the immunoglobulins present in the sera were directed against antigenic determinants composed of the amino-termini of the histones and DNA and that antibody binding is dependent on contact between the peptide and DNA segments and is independent of the integrity of the entire nucleosome.

To facilitate a better understanding of the processes associated with the binding of an IgG molecule to nucleosomes, the interaction has been simulated by the interactive computer surface graphics technique[13] (see Fig. 1). Steric considerations observed from the model indicate that the binding of antibodies to the nucleosomes is dependent primarily on short-range interactions. Antibodies will bind to fragments of nucleosomes as long as part of the histone–DNA contacts remain intact. Most of the histone antigenic sites are sterically hindered. Therefore, only a fraction of the antibodies elicited against free histones can bind to intact nucleosomes. The projections also reveal that, in most cases, immunoelectron microscopic examination will fail to resolve the contour of an antibody bound to the nucleosome (see below, Applications of Immunological Probes). The graphic display presented in Fig. 1 visualizes some of the limitations inherent to the serological approach.

Assay of the Probes

Chromosomal proteins tend to aggregate, and histones form nonspecific precipitates with several components present in sera. Therefore, immunological assays based on the formation of a specific precipitate between an antigen and an antibody often are not feasible for assays involving chromosomal components. The precipitin assays, however, are useful for certain pure nonhistones and nucleic acids. Of special interest is the microcomplement fixation technique. This technique, although cumbersome, is extremely sensitive to small differences between related antigens and can be used to study evolutionary relations between molecules. A detailed description of the technique can be found elsewhere.[3,14]

The methods of choice for antisera analysis are solid-phase enzyme immunoassay (ELISA), solid-phase radioimmunoassay, and immunoblotting. A typical protocol for ELISA involves the following steps:[15]

---

[13] M. Bustin and R. J. Feldman, *J. Theor. Biol.* **92,** 97 (1981).

[14] A. B. Champion, K. L. Soderberg, D. Wacher, and A. Wilson, *in* "Biochemical and Immunological Taxonomy of Animals" (C. A. Wright, ed.), p. 397. Academic Press, New York, 1974.

[15] E. Engvall and P. Perlman, *J. Immunol.* **109,** 129 (1972).

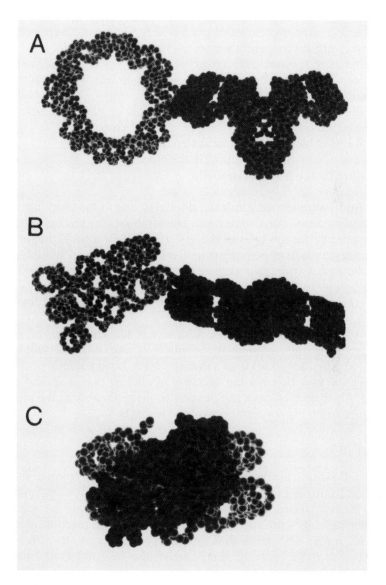

FIG. 1. Computer graphic projections of the surface topology of an IgG molecule bound to nucleosomal DNA. (A) Front view. (B) Side view (90° rotation of A out of the plane of the page, looking down the Fc axis). (C) The nucleosomal DNA fully obstructed by the IgG. Note that an IgG molecule recognizes only a small portion of the nucleosomal surface. (From Ref. 13.)

1. Coat microtiter plates with 100 μl solution containing serial dilutions of antigen. The antigen can be dissolved in practically any aqueous buffer, at ionic strength below 0.5, pH 5.5–9.5. The serial dilutions can be started with antigen concentrations of 1–10 μg/ml.

2. Incubate at room temperature for at least 4 hr or overnight at 4°.

3. Remove unadsorbed antigen by washing at least 4 times with PBS containing 0.02% Tween 20.

4. Block potential unsaturated binding sites by *filling* the wells with 1% horse serum or BSA in PBS for 1 hr at room temperature.

5. Remove blocking agent as in step 3.

6. Add 100 μl antisera, diluted in the blocking agent, for 2 hr at room temperature. Shake plate occasionally.

7. Remove unbound antisera as in step 3.

8. Add 100 μl second antibody, diluted in one of the blocking agents mentioned in step 4, for 1 hr at room temperature. The second antibody, which is specific for the IgGs of the first antibody, is a conjugate of IgGs with an enzyme such as peroxidase or alkaline phosphatase. It is advisable to use the purest conjugate available, such as a conjugate of the F(ab)$_2$ fraction.

9. Remove unbound second antibody by washing 8 times as described in step 3.

10. Add appropriate substrate. A variety of kits containing all necessary ingredients are commercially available. Color starts developing as early as 10 min after addition of substrate. More rapid color development is undesirable and indicates the necessity of diluting one or several of the reagents. Color development can be speeded up by incubating the plates at 37° or increasing the concentration of the reagents.

11. The color developed in the wells can be read with an automatic ELISA reader.

The following controls should be included, and the color developed subtracted from the experimental results: (a) All reactants containing the highest concentrations of reagents used in the test should be added to microtiter wells that do not contain antigens. (b) The reaction obtained with nonimmune or heterologous sera, at concentrations 10–20% higher than those used for immune sera, should also be subtracted from the experimental results.

The results of a typical ELISA assay are presented in Fig. 2, which illustrates the color obtained over an antigen concentration range from 1.0 to 0.008 μg/ml with antisera dilutions ranging from 1:1000 to 1:100,000. Note that the reaction is dependent on both antigen and antibody concentration.

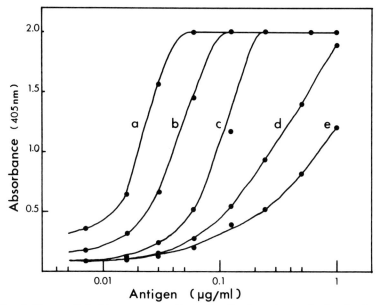

FIG. 2. Enzyme-linked immunoassay, illustrating the dependence of color development on the concentration of either the antigen or antiserum. Microtiter plates were coated with 100 μl antigen at the concentration indicated. The antigen was reacted with the antiserum at dilutions of 1 : 1000 (a), 1 : 5000 (b), 1 : 20,000 (c), 1 : 50,000 (d), and 1 : 100,000 (e). The absorbance values were read with a multiscan (Flow Laboratories, McLean, VA) 30 min after the addition of the substrate. (From Ref. 4.)

The protocols for solid-phase radioimmunoassays[16] are similar to those described for ELISA. The presence of the first antibody bound to antigens is detected with radioactively labeled second antibody or protein A. The amount of radioactivity bound can be determined either by autoradiography, by placing the microtiter plate directly on a film, or by cutting the plate and counting the wells in a counter.

Further evidence of specific antigen–antibody reactions in solid-phase immunoassays can be obtained by inhibition studies. The protocols are very similar to those described above; however, antiserum is preincubated with inhibitor prior to addition to the wells. For positive reactions in solid-phase immunoassays, it is sufficient that antibody binds to antigen via one Fab site, while complete inhibition of binding by free inhibitor requires saturation of both antibody binding sites. Therefore, the amount of inhibitor necessary to detect inhibition is significantly higher than the

[16] M. Romani, G. Vidali, C. S. M. Tahourdin, and M. Bustin, *J. Biol. Chem.* **255**, 468 (1980).

amount of antigen bound to the plates. The results of typical inhibition studies[17] demonstrating the presence of HMG 17 on nucleosomes are presented in Figs. 3 and 4.

A variety of protocols have been used to demonstrate antibody specificity by the immunoblotting (Western) technique. Depending on the type of solid support used, there are minor variations in the protocol, and it is best to follow the protocols recommended by the manufacturer. The following main steps are involved:

1. Antigens (proteins, modified nucleic acids, or nucleoproteins) are fractionated by electrophoresis on polyacrylamide or agarose gels.

2. The fractionated antigens are transferred to a solid support (nylon membranes, nitrocellulose, DBM paper, etc.) by the capillary or electroblotting procedure.

3. The binding sites on the paper are blocked with 3% BSA in PBS or with a 5% nonfat dry milk.

4. The membranes are incubated with antisera diluted in the blocking agents described in step 3.

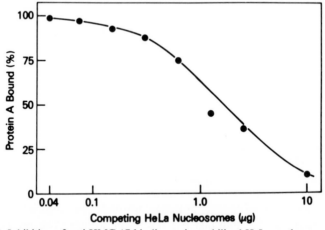

FIG. 3. Inhibition of anti-HMG 17 binding to immobilized HeLa nucleosomes by preincubation with free nucleosomes and detection by solid-phase radioimmunoassay. Antiserum at 1:400 dilution was incubated (18 hr at 4°) with an equal volume of 1% BSA in PBS containing increasing amounts of antigen. After centrifugation, the supernatant was added to microtiter plate wells with 50 ng of nucleosomes. Protein A bound is expressed as percentage of control (no inhibitor).

[17] C. S. M. Tahourdin, N. K. Neihart, I. Isenberg, and M. Bustin, *Biochemistry* **20,** 910 (1981).

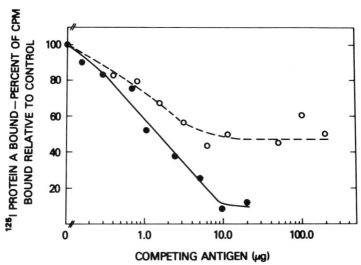

FIG. 4. Inhibition of anti-HMG 17 binding to immobilized HMG 17 by preincubation with HeLa nucleosomes (O) or HMG 17 (●) as competing antigen and detection by solid-phase radioimmunoassay. Antiserum at 1:100 dilution was incubated with increasing amounts of antigen; after centrifugation, the diluted supernatant was added to microtiter plates pretreated with 50 ng of HMG 17. Data are expressed as percent protein A bound relative to control. (From Ref. 17.)

5. Unbound antibodies are washed 3 times in PBS.

6. Membranes are incubated with enzyme-conjugated, affinity-pure IgG against immunoglobulins of the first antisera. The second antibody should also be diluted in blocking agent.

9. Unbound antibodies are removed with three washes of PBS.

10. Substrate is added, and the location of bound antibodies visualized.

The specific antibody–antigen complexes on the membrane can also be visualized by incubating the membranes in step 6 with radioactive second antibodies or radioactive protein A. Typical results, obtained by immunoblotting histones and HMG proteins to DBM paper, are presented in Fig. 5. The advantage of DBM paper is that proteins are bound covalently to the solid support. Nylon membranes, such as Zetabind or Nytran, are significantly more convenient to use, although some proteins are not retained quantitatively.

Purification of Probes

Polyclonal antisera contain, in addition to the immunoglobulin induced by the immunogen, the entire repertoire of antibodies present in the

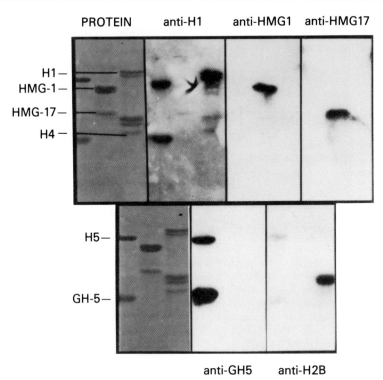

FIG. 5. Western blotting of nucleosomal proteins on DBM paper. Chromosomal proteins (left) were fractionated by electrophoresis in 18% polyacrylamide gels containing 0.1% SDS and transferred to DBM paper. The paper was treated with the antibody indicated in each panel. The location of the bound antibody was visualized by treatment with ^{125}I-labeled second antibody. All antisera, except anti-GH5, were diluted 1:300; anti-GH5 was diluted 1:1000. Note that anti-H1 cross-reacts with H5 and GH5, whereas anti-GH5 does not cross-react with any other chromosomal protein. GH5, globular region of histone H5.

immunized recipient. Some of these interact nonspecifically with cytoplasmic components and with chromatin and nucleosomes, giving rise to background reactions. Purification of the immunoglobulin by ammonium sulfate or DEAE chromatography does not alleviate this problem since the "nonspecific" background is due to *bona fide* immunoglobulin molecules that interact specifically with an antigen. In some cases, it is possible to reduce the background by adsorbing the sera with heterologous tissues or with purified cytoplasm. The method of choice for reducing background and enhancing the specificity of the polyclonal antiserum, however, is affinity chromatography on columns containing immobilized antigen:

1. Immobilize purified antigen on solid supports, such as Sepharose 4B, according to the manufacturer's instructions (for Sepharose 4B, use

CNBr-activated Sepharose from Pharmacia, Piscataway, NJ). Other supports are available.

2. Add antiserum to the immobilized antigen, make the slurry 0.3 M in NaCl and 0.2% in Triton X-100, and shake overnight. The following guidelines can be used to calculate the ratio of antigen to sera: assuming that the concentration of immunoglobulin in serum is 10.0 mg/ml and that 5% of the immunoglobulin (molecular weight 1.5×10^5) is specific, it can be calculated that 1.0 ml antiserum contains 5 μg of specific antibody. For an antigen with a molecular weight of 1.5×10^4 (like a core histone), 5 μg of antigen can adsorb all the antibodies present in 10.0 ml of antiserum.

3. Pour the slurry into a column. For 5–10 mg antigen immobilized on 1.0 ml Sepharose, the column can be constructed from a Pasteur pipette. Pass the serum through the column 2 times. Wash the column with PBS made 0.3 M in NaCl and 0.2% in Triton X-100 with 10 column volumes. Continue washing with the same solution, omitting the Triton X-100, until A_{230} of the eluate is below 0.05.

4. Elute the bound immunoglobulin with a chaotropic agent such as 6 M guanidinium-Cl, pH 7, 2 M KSCN, 0.1 M NH$_4$OH, or 0.1 M glycine, pH 2.5. Follow elution of the antibodies by measuring A_{280}. Most of the antibody elutes as a sharp peak in a small volume. Collect eluate into test tubes containing 0.2 ml 1.0 M sodium phosphate buffer, pH 6.8.

5. Dialyze overnight against PBS at 4°. Centrifuge to remove insoluble material, most probably denatured IgG.

When the quantity of antigen is very small and pure antigen is not available, it is possible to cut the band containing the antigen from an immunoblot (Western) and use it to purify antibodies according to the principles described above. Antibodies obtained by affinity chromatography are highly specific; however, there is some decrease in the affinity of the sera since the most specific immunoglobulin molecules have very high affinity constants and cannot be eluted from the column under conditions that do not cause irreversible denaturation.

Epitope-Specific Probes

Probes specific to defined regions of chromosomal proteins could be potentially useful in detailed studies on the nucleosomal organization of the proteins. The following approaches have been used for this purpose:

1. Large proteolytic fragments can be prepared by conventional techniques, and antisera elicited by one of the methods described above.

2. Synthetic peptides can be prepared, coupled to a carrier, and antibody elicited. The antisera can be tested against the intact protein. This approach is described in detail elsewhere in this volume.[5]

3. Monoclonal antibodies can be prepared, and binding of these to defined proteolytic fragments can be tested by immunoblotting. The specificity of monoclonals against various regions of the protein can thus be studied. Competitive ELISA (see Ref. 22) can be used to determine the proximity of epitopes, as was done for histone H1[0] (see Fig. 6).

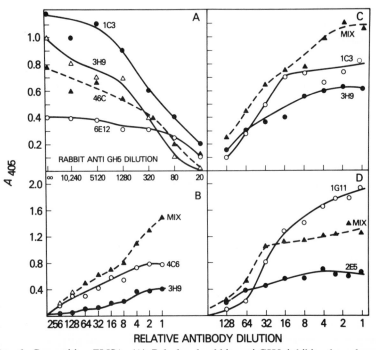

FIG. 6. Competitive ELISA. (A) Polyclonal rabbit anti-GH5 inhibits the subsequent binding of monoclonal antibodies. Monoclonal antibodies 1C3, 3H9, 4C6, and 6E12 were mixed with various dilutions of polyclonal rabbit anti-GH5 and added to microtiter plates containing H5. The relative amount of mouse monoclonal bound was determined by ELISA using alkaline phosphatase-labeled goat anti-mouse immunoglobulin. Note that increasing the rabbit anti-H5 concentration decreased the amount of mouse immunoglobulin bound. The experiment demonstrates that it is possible to use ELISA to detect competition between antibodies for the same antigenic site. Note also that the anti-rabbit sera contained antibodies which competed with all the monoclonals tested. (B–D) Competition experiments in which the monoclonal antibodies were added to the plates either as individual components or as a mixture. Note that in the absence of significant steric hindrance, the mix always gives higher A_{405} values than for each monoclonal. This is most obvious at low dilutions, when all sites in H5 are saturated. In contrast, when mutual interference to binding is observed (D), the mix has higher values than those for each individual antibody at high antibody dilution, that is, when the epitope is not fully saturated, and has an average of A_{405} at low dilution. 1G11 is IgM and therefore gives higher A_{405} than 2E5 which is IgG. When two IgG molecules bind to same site, the mix has the same value as each antibody added separately. (From Ref. 6.)

Applications of Immunological Probes

This section describes various applications of immunochemical probes in studies designed to understand the structure and function of chromatin, nucleosomes, and their components. The main pitfalls and limitations of such studies are pointed out. The probes have been applied to a variety of experimental systems. Therefore, rather than presenting detailed protocols of the various methods, the discussion emphasizes the principles and rationale of the procedures used in various experiments.

Specificity and Evolutionary Studies

Since nucleosomes and their major constituents have no assayable function, immunochemical techniques can be used as a means to identify similar proteins derived from different species and to express species differences between nucleosomes and their components in quantitative terms. For histones and HMG proteins, the reaction of proteins extracted from a new source is compared to that obtained with proteins extracted from calf or rat thymus, which can serve as standards. Reliable comparisons can be made by the ELISA technique, described above, using several antisera concentrations, or by the immunoblotting technique. In these tests, it is important to compare similar amounts of antigens. Both the ELISA test and the immunoblotting technique are relatively *insensitive* to small differences between antigens because the detection system measures the number of antibodies bound to the antigen. Even though a complex molecule has practically an unlimited number of antigenic determinants, the number of antibody molecules that can bind simultaneously to a molecule, such as lysozyme, is limited to three to five. It is possible to imagine a situation where two distinct but related antigens could share only a fraction of their antigenic determinants yet would bind the same number of antibody molecules and, therefore, give the same color intensity in ELISA.

Yet, the specificity of antibodies is sufficiently sharp to distinguish between two related proteins in which a single amino acid replacement occurred in an immunogenic region or to detect structural differences between similar nucleotides. The most suitable immunochemical technique to detect and quantitate differences between related antigens is the microcomplement fixation technique. Compared to other serological analyses, the microcomplement fixation technique is somewhat cumbersome. Briefly, the method involves the following steps:[3,14]

1. Various concentrations of antigens are reacted with several dilutions of antiserum. A series of complement fixation curves over a range of 20–90% maximal complement fixation are obtained.

2. For each antigen, the maximum complement fixed is plotted against the logarithm of the antiserum dilution.

3. The data obtained from the above plot are used to determine the index of dissimilarity, which is the ratio of the serum dilution that gives 50% maximum complement fixation with the homologous antigen to the dilution that gives 50% maximum complement fixation with the heterologous antigen.

4. The index of reciprocity (i.e., reaction of anti-A with antigen B and the reaction of anti-B with A) is established.

5. The index of dissimilarity can be converted to immunological distance:

$$\text{Immunological distance} = 100 \times \log (\text{index of dissimilarity})$$

6. Empirical observations indicated that for a number of protein families, the immunological distance (E) is linearly related to percent sequence difference (X) by $E = 5X$.

The microcomplement technique has been used to assess phylogenetic relationships between proteins. Obviously, the technique is not as rigorous as determination of the primary sequence. However, it is more reliable than peptide map analysis or any other physical measurement. Using microcomplement fixation, and more recently additional techniques, it was shown that chromatin, nonhistones, and nucleosome particles exhibited some degree of species specificity. Hnilica and co-workers noted that dehistonized chromatin elicits sera that can distinguish between dehistonized chromatin prepared from various tissues, between preparations of normal and transformed cells, and even between dehistonized chromatin prepared from cells at different stages of cell differentiation.[18]

Immunochemical Analysis of Nucleosome Composition and Organization

Specific Interaction of Antibodies with Nucleosomes Demonstrated by Gel Retardation. The interaction of antibodies with nucleosomes can be demonstrated by ELISA techniques performed as described above. However, in certain cases, it is desirable to use alternative methods that can provide additional information. Since the electrophoretic migration of macromolecules in various gels is dependent mainly on the molecular weight, it can be expected that a nucleosome–antibody complex will have a mobility that is distinguishable from that of the nucleosome. Indeed,

---

[18] W. F. Glass and R. C. Briggs, *in* "Chromosomal Nonhistone Protein" (L. S. Hnilica, ed.), Vol. II. CRC Press, Boca Raton, Florida, 1983.

nucleosome–antibody complexes can be detected in nondenaturing poly-acrylamide gels, as follows:

1. Purified nucleosomes or core particles are end-labeled with ^{32}P by accepted techniques.

2. The end-labeled nucleosomes are incubated with purified antibodies, in the presence of EDTA and protease inhibitors, in 50 mM Tris-Cl, pH 7, for 1 hr at room temperature and 4–18 hr at 4°. For antibody preparations purified by DEAE chromatography, the molar ratio of IgG to nucleosome should be at least 20.

3. The mixture is applied directly to 4% neutral polyacrylamide gels and electrophoresed under standard conditions.

4. The gels can be autoradiographed either after drying or while still wet, after wrapping them in Saran Wrap or similar material.

Typical results of such an experiment are shown in Fig. 7. By this

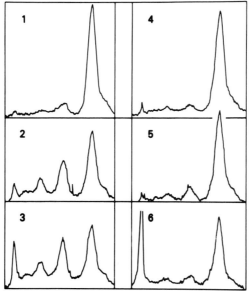

FIG. 7. Antibody binding to 5′-^{32}P-labeled nucleosomes detected by electrophoresis in neutral polyacrylamide gels. Ninety nanograms of nucleosomes was incubated with increasing quantities of IgG and then fractionated by using neutral polyacrylamide gels. The gels were exposed to X-ray film. Densitometer tracings of the autoradiograph are shown. Traces 1–3: 1, 12, and 25 μg of antinucleosome per incubation; traces 4–6: 1, 12, and 25 μg of normal rabbit IgG. Migration was from left to right. Note that in the presence of antibodies, peaks migrating slower than the free nucleosomes are observed. (From Ref. 12.)

approach, it is possible to distinguish free nucleosomes, nucleosomes with one antibody bound, and nucleosomes to which two IgG molecules are bound. In the control experiments, done with nonimmune IgG, nucleosome–antibody complexes were not observed. Under the conditions of the experiment, free IgG do not enter the gel and accumulate at the top. In some cases, high concentrations of IgG can block the entrance of the nucleosome into the gel (see Trace 6 in Fig. 7). This technique can be used for studies on subpopulations of nucleosomes.

Nucleosome Composition Studied by Immunosedimentation. The availability of antisera specific to histone subfractions, to nonhistone proteins, to modified nucleic acid components, and to other nucleosomal components allows examination of possible heterogeneity in the chromatin particle. The immunosedimentation approach is based on the same principle as the gel retardation assay, namely, that the antibody–nucleosome complex can be separated from antibodies and unreacted nucleosomes by sedimentation through sucrose gradients:

1. The nucleosomal DNA is radioactively labeled either by end-labeling or by growing cells in radioactive thymidine.

2. Affinity-purified antibodies are prepared, or, if these are not available, DEAE-purified IgG fractions can be used.

3. The nucleosomes are incubated with a 5- to 50-fold molar excess of the antibodies in 10 mM Tris-Cl, pH 8, 10 mM NaCl, 1 mM EDTA in the presence of protease inhibitors, for 1 hr at room temperature and 4–18 hr at 4°.

4. The mixture is sedimented through sucrose gradients in 0.25 mM EDTA, pH 7, with a meniscus concentration of 5% (w/w) sucrose. The gradients should be isokinetic for a particle with a density of 1.51 at 4° and sedimented in a Beckman SW40 rotor for 16 hr at 38,000 rpm and 4°.

5. Fractions are collected, and the distribution of the radioactivity along the gradients is determined.

6. Samples can be recovered from the sucrose gradients by dialysis and lyophilization, and the protein content of the particles examined by electrophoresis in polyacrylamide gels or another technique.

Typical results are shown in Fig. 8. Uncomplexed nucleosomes sediment as a sharp zone with a sedimentation value of 11 S. A gradual increase in antibody-to-nucleosome ratio results in a gradual shift of the nucleosome sedimentation band to higher value and also broadens the sedimentation zone. With histone antibodies, all the nucleosomes have sedimentation values higher than 11 S, indicating that each nucleosome contains at least one copy of each of the core histones. The broadening of

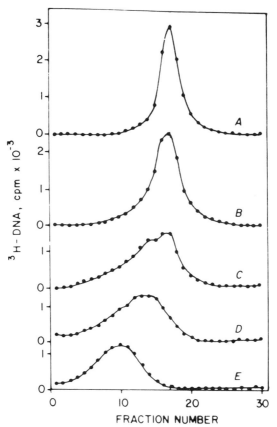

FIG. 8. Altered sedimentation behavior of nucleosomes incubated with anti-H2B antibody. Samples are (A) nucleosomes alone; (B–E) 0.8 g of nucleosomes (as DNA) plus (B) 5, (C) 15, (D) 30, and (E) 75 g of anti-H2B antibody. At the highest concentration of antibody, about 25% of the nucleosomes sediments to the bottom of the gradient. (From Ref. 19.)

the zone results from different amounts of antibodies bound. The nucleosomes in the front of the zone contain three IgG molecules per nucleosome, while those in the slowly sedimenting boundary have one IgG per nucleosome. The results seem to be dependent on the ionic strength at which the antibodies were added to the nucleosome. In the case of histone antibodies, low ionic strength increases the antibody binding, presumably by loosening the nucleosome structure and increasing the exposure of antibody determinants. With some antisera, low ionic strength increases

[19] R. T. Simpson and M. Bustin, *Biochemistry* **15,** 4305 (1976).

nonspecific binding of IgG to nucleosomes. Therefore, it is important to optimize conditions so as to maximize the specific binding.

Exposure of Antigenic Determinants in Nucleosomes Determined by Immunoadsorption. Conformational changes associated with nucleosome formation render many of the antigenic determinants, present in isolated histones and nonhistones, inaccessible to antibodies. By using chromatin and nucleosomes as an immunoadsorbent, it is possible to measure the availability of antigenic determinants in the native nucleoprotein structure and to purify the IgG molecules specific for either sterically hindered or exposed epitopes. The method is based on the fact that chromatin and H1-containing nucleosomes are highly insoluble at neutral pH and 0.15 M NaCl.

1. Chromatin or nucleosomes are prepared by accepted techniques and dispersed in 10 mM Tris, pH 7.5, 1 mM EDTA. Protease inhibitors are added.

2. Antisera at various dilutions in the same buffer are added, and the mixture is incubated with occasional shaking for 1 hr at 37° and 4–18 hr at 4°.

3. The mixture is made 0.15 M in NaCl by incremental addition of 1.5 M NaCl and centrifuged 15 min at 27,000 g.

4. The supernatant contains nonadsorbed IgG. The specificity of unadsorbed antibodies can be tested by one of the techniques described earlier.

5. The IgG bound to the chromatin can be recovered by exposing the pellet to a chaotropic agent, dialysis, and DEAE chromatography, or affinity chromatography on Sepharose–protein A columns.

At appropriate sera and antigen dilutions, this method is a very sensitive way to detect and to quantitate specific binding of antibodies to chromatin. As with other immunoassays, proper controls, such as the reaction with nonimmune sera or with heterologous sera specific against determinants not present in chromatin or nucleosomes, should also be performed.

Immunocytochemical Approaches

Cellular Localization Visualized by Immunofluorescence. The intracellular location and organization of chromosomal components, at the resolution afforded by the light microscope, can be visualized by several approaches. Details of the various protocols for immunofluorescence and related techniques can be found elsewhere.[20,21] The following steps are involved:

1. Sample preparation: The manner in which the tissue or cells are prepared can markedly affect the results. Components may be degraded, rearranged, migrate from the nucleus to the cytoplasm, or even leak out of the cells. Cells can be fixed onto microscope slides by immersing in methanol and acetone at $-20°$. This method both fixes the cells and permeabilizes their membranes to antibody penetration, but it does not preserve all the structural details. An alternative fixation method is treatment of cells with a cross-linking agent, such as formaldehyde, followed by a short treatment with 0.1% Triton X-100 to permeabilize the cell membranes. Cross-linking can result in loss of antigenic determinants, especially in nuclear antigens, since the antibodies may not be able to penetrate the nuclear membrane.

2. Antisera preparation: To avoid possible artifacts, it is important to use the best available antibody preparations. High-affinity monoclonals or IgG purified by affinity chromatography are the reagents of choice. To minimize nonspecific interactions, it is best to dilute the primary antibody in 1% horse or bovine serum.

3. Detection of the bound primary antibody: The binding of the primary antibody can be detected with an anti-antibody that is conjugated to a fluorochrome such as fluorescein, rhodamine or Texas Red. Peroxidase or avidin conjugates of the second antibody can also be used.

4. Controls: Because cytochemical techniques are especially prone to nonspecific effects, controls in which the primary antibody has been omitted or replaced by heterologous antisera have to be examined.

Typical immunofluorescent micrographs, demonstrating that histones are found in the cell nucleus, are presented in Fig. 9.

Visualization of Components in Metaphase Chromosomes. Specific antisera against chromosomal proteins or nucleic acid constituents can be useful tools for various studies on metaphase chromosomes. The fluorescent pattern obtained using antihistone sera with metaphase chromosomes, mildly treated with methanol–acetic acid, varies among chromosomes, suggesting that immunochemical techniques can be an additional

---

[20] M. C. Willingham and I. Pastan, "An Atlas of Immunofluorescence in Cultured Cells." Academic Press, New York, 1985.
[21] G. D. Johnson, E. J. Holbrow, and J. Darling, *in* "Handbook of Experimental Immunology" (D. M. Weir, ed.), p. 151. Blackwell, Oxford, 1979.

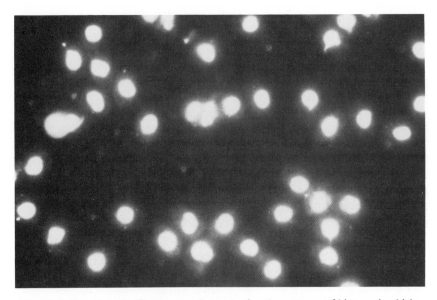

FIG. 9. Indirect immunofluorescence demonstrating the presence of histones in chicken erythrocyte nuclei, using monoclonal anti-H5. (From Ref. 6.)

tool for chromosome identification (see Fig. 10).[22] Miller and Erlanger[23] visualized the location of methylcytosine and other nucleic acid components in metaphase chromosomes. A key finding in their studies was that effective binding of various antibodies to nucleic acid in the chromosomes requires denaturation of the double-stranded DNA by treatment with NaOH, formamide, UV irradiation, photooxidation, or mild heat. Even a region containing only two unpaired nucleosides can be detected immunologically. The methodological approach for these studies is very similar to that described above. It is especially important to ensure that chromosomal proteins are not lost during chromosome preparation.

Polytene Chromosomes. Polytene chromosomes offer a uniquely advantageous system to study gene structure and function. Their large size offers the advantage of amplification in studying the location of a particular chromosomal protein. Furthermore, they can be experimentally manipulated so as to induce transcriptional activity in defined genetic loci. Transcriptional activity can be detected as "puffs" observable by phase-

[22] M. Bustin, D. Goldblatt, H. Yamasaki, E. Huberman, M. Shani, and L. Sachs, *Exp. Cell Res.* **97,** 440 (1976).
[23] O. J. Miller and B. F. Erlanger, *Pathobiol. Annu.* **5,** 71 (1975).

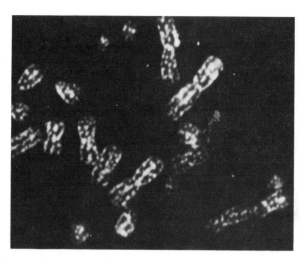

FIG. 10. Immunofluorescence of metaphase chromosomes. Chromosomes were exposed to methanol–acetic acid and stained with anti-H2B. (From Ref. 22.)

contrast microscopy or, more reliably, by autoradiographic techniques using labeled uridine. As mentioned in the previous sections, the results are highly dependent on the method of sample preparation and on the quality of the antibodies.

The polytene chromosome system has been used by many investigators to clarify various structural and functional aspects of the genome. For example, antibodies to histones and nonhistone proteins have been used to visualize the composition of various regions of the chromosome and to study structural rearrangements associated with induction of gene activity. Antibodies to RNA polymerase II and to DNA–RNA hybrids can be used to visualize transcriptionally active loci. Antibodies to chemically modified DNA have been used to demonstrate site-specific carcinogen binding to DNA in polytene chromosomes. The intensity of fluorescence and the resolution of the individual bands in these chromosomes are highly dependent on the concentration of the reactants.[24] At low serum dilutions, the chromosome often fluoresces along its entire length, and the resolution of individual bands is very poor. At very high serum dilutions, however, only the regions containing high concentrations of exposed antigenic sites can be detected (Fig. 11). Thus, several antiserum concentrations have to be tested, and a compromise between resolution and fluorescence intensity

[24] M. Bustin, D. D. Kurth, E. N. Moudranakis, D. Goldblatt, and W. B. Rizzo, *Cold Spring Harbor Symp. Quant. Biol.* **42,** 379 (1977).

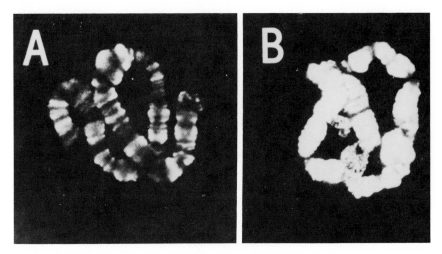

FIG. 11. Effect of antiserum concentration on banding pattern of polytene chromosomes. Chromosome 1 of *Chironomus thummi* was mixed with acetic acid and stained with anti-H2B by the indirect immunofluorescence technique. (A) Serum dilution 1:80; (B) 1:20. Note the differences in the resolution of the bands. (From Ref. 24.)

has to be reached. The immunofluorescent pattern reflects the *relative* availability of antigenic determinants rather than the absolute presence of the antigen. In certain cases, the presence of antigens in a region could be visualized only after components that sterically hindered antibody binding were removed.

Immunoelectron Microscopy. Immunoelectron microscopy techniques are suitable for studies on the organization of chromosomal components at the level of resolution of the single nucleosome. Antibodies to modified bases can be localized along the contour of the DNA fiber. It may be possible to define the domain occupied by a histone within the nucleosome, to locate a particular nonhistone protein on the particles, and to visualize the association of histones and other proteins with the transcriptionally active chromatin fiber. Initial studies used electron microscope grids coated with BSA so as to prevent nonspecific binding.[25] Chromatin or nucleosomes were spread on the grid, antiserum added, the grids incubated for 10 min in a moist atmosphere, the unbound antibodies washed off, and the grids negatively stained with aqueous solutions of uranyl acetate. To verify that antibodies were indeed bound to the nucleosomes, the grids were treated with ferritin-labeled anti-antibody. A subsequent modifica-

[25] M. Bustin, D. Goldblatt, and R. Sperling, *Cell* 7, 297 (1976).

tion of this approach was to preincubate the antiserum with the sample prior to the application of the sample to the grid by the Miler technique. This method does not require the use of albumin to precoat the grid, and therefore increases the resolution. Ferritin-labeled second antibody was added to visualize the location of the antibodies bound to the nucleosomes. Second antibodies, labeled with gold particles of defined size, seem to give higher resolution and lower background than the ferritin conjugates. Detailed descriptions of various methods used in the immunoelectron microscopy of nucleosomes and chromatin are described elsewhere in this book.[26]

Chromatin spread on electron microscope grids coated with BSA appears as a beaded structure. The beads are uniform in size and, depending on the degree of hydration, have a diameter of about 104 Å. Addition of nonimmune sera does not markedly change the shape of chromatin except that the average size of the particle increases by about 20%. Addition of antisera to chromatin or histones results in a preparation of heterogeneous particles with an average diameter of 300 Å. In some cases, it is possible to resolve the contour of an antibody at the periphery of the swollen particle. The interaction of antibodies with nucleosomes can be imagined as a situation where a nucleosome with a diameter of about 100 Å is encircled by several antibodies, each 110 Å in length. If only part of the antigenic sites are saturated or the antibodies do not bind to the periphery of the nucleosome, particles with a diameter of less than 300 Å will be observed. Structures with still larger diameters probably result from several adjacent, unresolved particles. The projections presented in Fig. 1 are a useful way to imagine the interaction of antibodies with nucleosomes. The appearance of chromatin spread on albumin-coated electron microscope grids is presented in Fig. 12.

Heterogeneity in Nucleosomes Demonstrated by Immunoelectron Microscopy. The immunoelectron microscopic techniques, described above and in the chapter dealing in detail with this topic,[26] allow examination of nucleosome composition and heterogeneity at the resolution level of the individual nucleosomes. In these experiments, nucleosomes are spread on electron microscope grids and reacted with several dilutions of antisera. Since unreacted nucleosomes have a diameter of about 100 Å and those reacting with nonimmune sera a diameter of about 120 Å, particles with a diameter of 150 Å and above are scored as antibody–nucleosome complexes. Histograms are constructed, and the percentage of nucleosomes reacting with an antibody of a known specificity is determined. Examination of the projections in Fig. 1 indicates that the projection allowing a full and clear resolution of an antibody bound to the nucleosome (i.e., Fig. 1A)

[26] C. L. F. Woodcock, this volume [9].

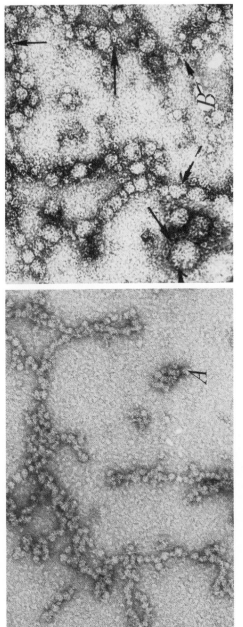

FIG. 12. Visualization of the interaction of chromatin particles with antibodies by immunoelectron microscopy. (A) Electron micrograph of chromatin spread on grids precoated with BSA. Note the uniformity in size of the particles. (B) Electron micrograph of chromatin reacted with antichromatin elicited in rabbits and with ferritin-labeled goat anti-rabbit IgG. Arrows point to ferritin molecules surrounding swollen chromatin particles reacted with antibodies. Note the heterogeneity in the size of the particles. (From Ref. 25.)

is only one of the possible views and that, in most cases, the photographs will fail to resolve the contour of the antibody attached to the nucleosome. The increase in size of the particle on antibody binding most probably reflects the attachment of several antibodies to the particle. Counterstaining with a labeled second antibody will increase the likelihood of detection of a single antibody molecule bound to the nucleosome. Using this approach, it was determined that nucleosomes are heterogeneous with respect to their content of nonhistone proteins and that they also differ with respect to the availability of histone antigenic determinants.[27]

Functional Studies on Chromatin and Nucleosomes

This section illustrates additional applications of serological approaches to questions relating to the structure and function of the chromatin fiber.

Composition of Transcriptionally Active Chromatin

Immunofluorescence on polytene chromosomes. Polytene chromosomes, prepared from the salivary glands of *Drosophila melanogaster* and *Chironomus thummi,* were used to study the distribution of chromosomal proteins in bands, interbands, and transcriptionally active puffs. The transcriptionally active regions can be identified with antisera specific to DNA–RNA hybrids. The fluorescence pattern obtained is highly dependent on the experimental conditions used, in particular to serum concentration. The relative fluorescence of the various regions along the chromosome axis indicates the *availability* of antigenic sites in particular regions. The possibility of steric hindrance to antibody binding must always be taken into account. Figure 13 demonstrates the use of indirect immunofluorescence to visualize the presence of nonhistone chromosomal protein HMG 14 in transcriptionally active puffs.[28]

Immunoelectron microscopy of transcriptionally active chromatin. Chromatin prepared from staged *Drosophila melanogaster* embryos or from amphibian oocytes can be spread on electron microscope grids under conditions that allow visualization of antibody binding to the chromatin regions between nascent ribonucleoprotein (RNP) fibers. The presence of histones and RNA polymerase II on transcriptionally active nonribosomal genes was demonstrated.

Microinjection of antibodies into amphibian oocytes.[19] The nuclei of amphibian oocytes contain lampbrush chromosomes consisting of several thousand highly active genes, which appear in the light microscope as lateral loops radiating from a central axis and in the electron microscope as

[27] D. Goldblatt, R. Sperling, and M. Bustin, *Exp. Cell Res.* **121,** 1 (1978).
[28] R. Westermann and U. Grossbach, *Chromosoma* **90,** 355 (1986).

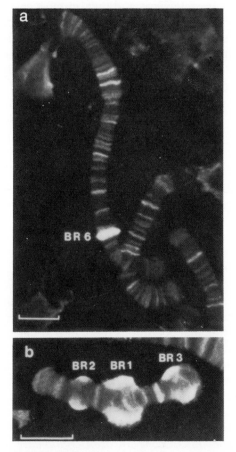

FIG. 13. Detection of chromosomal protein HMG 14 in transcriptionally active regions of polytene chromosomes, indirect immunofluorescence with anti-HMG 14 antibody of squash preparations of salivary gland chromosomes from *Chironomus pallidivittatus* larvae that were treated with 0.5% galactose in culture medium for several days. Note the intense fluorescence of the newly induced Balbiani ring 6 (BR6 in a) and the intensity of fluorescence in Balbiani ring 1 (BR1 in b). Bar, 25 μm. (From Ref. 28.)

typical transcriptional units densely covered by numerous transcriptional complexes. Antibodies against chromosomal proteins or RNA polymerase can be injected into the nuclei of the oocytes, and the appearance of the lampbrush chromosomes, prepared from the injected nuclei, can be used to judge the involvement of the antigen in transcriptional processes. It is important to note that, although the microscopic observation is done with fixed material, the actual antibody–antigen complex is formed under *in*

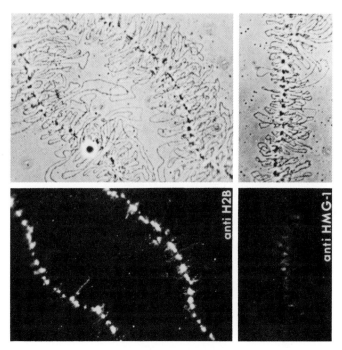

Fig. 14. Immunofluorescence of lampbrush chromosomes reacted with antisera to histone H2B and to calf thymus HMG 1. (Note that the central axis stains more intensely than the transcription loops radiating from the axis.) (Photograph by Dr. U. Scheer.)

vivo conditions since the antibody binds to the active, unfixed transcription complex. Details of the microinjection procedure can be found elsewhere.[29] Immunofluorescence revealed that antibodies to nucleosomal proteins, such as histones and HMG, bind specifically to lampbrush chromosomes (Fig. 14).

Chromosomes prepared from nuclei that have been injected with antibodies to histones[30] or HMG[31] proteins have a drastically changed morphology. Electron microscope spreads reveal a marked reduction in the density of the RNP lateral fibers. The kinetics of inhibition of transcription could be visualized by light microscopy (see Fig. 15). Within 30 min after

[29] U. Scheer, *J. Embryol. Exp. Morphol.* **97**, Suppl., 223 (1986).
[30] U. Scheer, J. Sommerville, and M. Bustin, *J. Cell Sci.* **40**, 1 (1979).
[31] T. A. Kleinschmidt, U. Scheer, M. C. Dabauvalle, M. Bustin, and W. W. Franke, *J. Cell Biol.* **97**, 838 (1983).

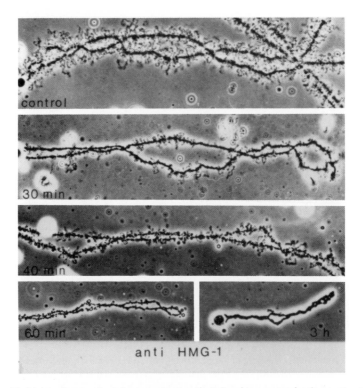

Fig. 15. Light micrographs (phase-contrast optics) showing progressive loop retraction of lampbrush chromosomes after injection of antibodies to HMG 1 protein into nuclei of *Pleurodeles waltlii* oocytes. All micrographs show, at the same magnification (×640), chromosome bivalent IV. (Control) Injection of nonimmune IgG (1.5–2.0 mg/ml) does not alter the structural appearance of the chromosomes. Partial retraction of lateral loops into the chromosome axis is seen 30 min after injection of antibodies to HMG-1. After prolonged exposure (1 hr, 3 hr) all lateral loops are almost completely retracted into the chromomeric axis. Note the drastic size reduction of the chromosomes after antibody-induced transcriptional inactivation. (From Ref. 31.)

antibody injection, loop retraction can be observed. The gradual retraction of transcription loops continues for several hours until the chromosome size is markedly decreased. In these experiments, it is imperative to show that nonimmune or heterologous serum does not affect the appearance of the chromosomes. The retraction of the transcription loops clearly shows that histones and HMG proteins are present in transcriptionally active regions of the chromosomes. Interestingly, transcription of ribosomal genes was not affected by microinjection, indicating differences between the composition and structure of transcriptionally active ribosomal genes and those transcribed by RNA polymerase II.

Microinjection of antibodies into somatic cells. Microinjection of antibodies, specific for chromosomal proteins, into living cells can be used to examine the role of proteins in transcription by observing changes in uridine incorporation subsequent to microinjection. The following protocol was used by Einck and Bustin:[32]

1. Fibroblasts, in logarithmic phase growth, were transferred to Titer-Tek microslide chambers.

2. Sixty minutes prior to antibody microinjection, the uridine pool was depleted by adding to the media glucosamine to a final concentration of 20 mM.

3. Cells were injected by the method of Graessmann using an inverted Leitz microscope and glass needles. Approximately 1×10^{-11} ml solution containing 1 – 10 mg/ml purified fluorescein or rhodamine-labeled IgG was injected. The efficiency of injection can be monitored by epifluorescent illumination immediately after injection.

4. After injection, the cells were incubated in [^{3}H]uridine in preconditioned medium for 60 min.

5. The cells were washed in Puck's saline, fixed in 10% formalin in saline for 30 min in the dark, and the slides processed for autoradiography using Kodak NTB-3 emulsion at room temperature for 7 – 10 days. After development in Dektol (Kodak) 1:4, they were examined by phase and fluorescence microscopy.

The relative amount of antibodies introduced into the cells could be judged from the fluorescence intensity. The effect on transcription can be evaluated from the relative number of autoradiographic dots. It is important to remember that intact IgG do not traverse the nuclear membrane. Therefore, the antibodies have to be injected directly into the nucleus. F(ab)$_2$ fragments, however, traverse the nuclear membrane and retain activity and can therefore be injected into either the nucleus or the cytoplasm. Antibodies to histones and HMG 17 inhibited uridine incorporation. Even monovalent Fab' fragments inhibited transcription, suggesting that the effect is not due to cross-linking of nucleosomes. The appearance of fibroblasts, injected with fluoresceinated antibodies to H2B and autoradiographed, is shown in Fig. 16.

Immunofractionation of Chromatin and Nucleosomes. The principles of immunoaffinity chromatography can be used to fractionate nucleosomes according to their content of antigenic determinants.[33] Purified

[32] L. Einck and M. Bustin, *Proc. Natl. Acad. Sci. U.S.A.* **80,** 6735 (1983).

[33] E. Mendelson, D. Landsman, S. Druckmann, and M. Bustin, *Eur. J. Biochem.* **160,** 283 (1986).

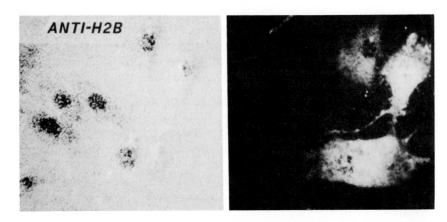

FIG. 16. *In vivo* inhibition of transcription by microinjection of antibodies to histone H2B into human fibroblasts. Human fibroblasts, microinjected with fluorescein-labeled, affinity-purified anti-H2B, were incubated with [³H]oridine and processed for autoradiography. The intensity of fluorescence is an indication of the amount of antibody microinjected, whereas the density of autoradiographic dots is an indication of RNA synthesis in the cells. Note the inverse correlation between the intensity of fluorescence (right) and the density of autoradiographic dots, observed under bright-field illumination (left). (From Ref. 32.)

antibodies of determined specificity are coupled to Sepharose according to the manufacturer's instructions. Nucleosomes are suspended in 5 mM Tris-Cl, pH 7.5, 1 mM EDTA, 30 mM NaCl and applied to the affinity column equilibrated with the same buffer. Unbound nucleosomes are washed with the buffer until A_{260} is less than 0.05. The nucleosomes can be eluted with a chaotropic agent, as described in Purification of Probes above. If the antigen is released from DNA at relatively low ionic strength, it is possible to recover the DNA associated with the protein by simply washing the column with 0.35–0.5 M NaCl. The DNA is recovered and analyzed. The antigen is subsequently removed by a chaotropic agent.

One of the uncertainties with this approach, which is inherent to all nucleosome preparations, is whether the nucleosome preparation indeed represents the entire population present in the nucleus. Preferential loss of a specific subset of nucleosomes can be minimized by preparing nucleosomes from chromatin rather than from intact nuclei, and by using relatively high micrococcal nuclease concentrations. During the entire procedure, the NaCl concentration must be kept below 40 mM so as to prevent rearrangement between chromosomal proteins during chromatin and nucleosome manipulations. In fact, the immunoaffinity technique described can be used as a stringent test for protein rearrangement by the following procedure:[34] Nucleosomes and oligonucleosomes prepared from one species (i.e., rat) are mixed with salt-stripped nucleosomes prepared from a second species (i.e., chicken). The mixture is applied to the column, and the DNA fraction recovered from the material bound to the column is analyzed with either chicken- or rat-specific DNA probes. The presence of chicken DNA in nucleosomes retained on affinity columns of antibodies specific for either H1^0 or HMG 17 indicates migration of the protein among the nucleosomes.

Another problem inherent to these methods is nonspecific binding of nucleosomes to the columns under the low ionic strength conditions necessary to prevent rearrangement of antigens. Using antibodies to H1^0 and HMG 17, it was found that a small fraction of the nucleosomes bound nonspecifically. It is important, therefore, to demonstrate that the nucleosomes indeed fractionated according to their content of antigenic determinants. This can be done by ELISA and immunoblotting of the original, the unbound, and the bound fraction.

In another immunofractionation approach, Dorbic and Wittig[35] used monoclonal anti-HMG 17 antibodies to immunoprecipitate oligonucleosomes, prepared from several different tissues, by the following procedure: 0.2 A_{260} units of chromatin were mixed with 10 μl containing 5 μg of antibodies and incubated in 50 mM NaCl, 50 mM Tris-Cl, pH 7.4, 5 mM EDTA containing 0.25% BSA. The antibody–antigen complexes were precipitated with formalin-fixed *Staphylococcus* A cells. The washed pellet was suspended in SDS–polyacrylamide gel sample solution, and the proteins were analyzed by electrophoresis. Analysis of the DNA from the same preparation indicates that in chromatin, transcribed sequences are enriched in HMG 17.

To avoid potential problems of protein rearrangement during chromatin fractionation, Blanco *et al.*[36] first cross-linked the nuclear proteins to DNA by UV irradiation. The irradiated nuclei were lysed in 6 M guanidinium hydrochloride, 20 mM Tris-Cl, pH 7.5, 2 mM EDTA, 14 mM EDTA, 0.1 mM phenylmethylsulfonyl fluoride (PMSF) overnight at 4°. The lysate was fractionated on CsCl gradients, and the DNA peak recovered and dialyzed. The dialyzed, cross-linked DNA–protein complexes were bound to anti-HMG affinity columns, essentially as described above. After washing the unbound fraction, the agarose–IgG–protein–DNA complex was digested with restriction enzymes. The DNA retained on the

[34] D. Landsman, E. Mendelson, S. Druckmann, and M. Bustin, *Exp. Cell Res.* **163**, 95 (1986).
[35] T. Dorbic and B. Wittig, *Nucleic Acids Res.* **14**, 3363 (1986).
[36] J. Blanco, J. C. States, and G. H. Dixon, *Biochemistry* **24**, 8021 (1985).

column by the anti-HMG antibodies was recovered by digestion with proteinase K and phenol extraction. Assuming that the cross-linking protocol did not cause protein rearrangement and that the cross-linked complexes formed randomly, this technique could become the method of choice for immunofractionation of chromatin and nucleosomes. Blanco *et al.* concluded that, in trout testis, HMG proteins are not randomly dispersed in the genome.[36]

Immunodetection of Specific DNA Structures and Modified DNA Bases. Most of the experiments described in the previous sections illustrate various applications of immunochemical approaches to studies on the protein components of nucleosomes. Clearly, the same approaches can be used for studies involving DNA. Aspects of various applications of immunochemical probes to studies of nucleic acid components can be found in recent reviews.[10,11] Three of these applications illustrating the relevance to nucleosome structure and function are briefly described below.

Occurrence of Z-DNA structures. Antibodies elicited against nucleic acid structures in the Z-DNA conformation have been used to determine whether this conformation occurs naturally in the genome. Anti-Z-DNA antibodies have been added to supercoiled plasmids, and the complex trapped on glass filters that bind protein–DNA complexes but do not retain free DNA. Immunofluorescence with polytene chromosomes, which have been treated by a variety of fixatives, has also been frequently used. In spite of these studies, it is not presently clear whether the Z-DNA structure occurs naturally or is induced in chromosomes by fixation procedures or other treatments designed to expose the antigenic determinants of the Z-DNA to antibody binding.

Localization of 5-methylcytosine in H1-containing nucleosomes.[37] Antibodies to 5-methylcytosine (m^5C) were elicited by injecting rabbits with m^5C-labeled BSA. Affinity-purified antibodies were prepared by passing the antiserum on columns prepared of cytidine and 5-methyluridine. The use of a heterologous affinity resin minimizes the loss of high-affinity antibodies, which frequently remain bound to the affinity column. The antibodies could distinguish between methylated and nonmethylated DNA. Nuclei, labeled with [^3H]thymidine, were digested with micrococcal nuclease, the soluble nucleosome fraction resolved on nucleoprotein gels, and the contents of the gel transferred to DBM paper. The position of the nucleosomes on the DBM paper was indicated by fluorography, while the content of m^5C was visualized with affinity-purified antibodies and ^{125}I-labeled antibodies. The results indicated that nucleosomes containing H1 and depleted of HMG are about 2-fold enriched in m^5C. Similar ap-

[37] D. J. Ball, D. S. Gross, and W. T. Garrard, *Proc. Natl. Acad. Sci. U.S.A.* **80**, 5490 (1983).

proaches can be used to study the nucleosomal distribution of other naturally occurring modified bases.

Mapping the binding of a chemical carcinogen to the genome. The structure of the chromatin fiber influences the manner in which DNA modifying agents, such as chemical carcinogens, X-rays, or UV radiation, interact with the genome. Furthermore, repair of DNA damage seems to involve reorganization of nucleosome structure. Studies on the binding of an active metabolite of the carcinogen benzo[a]pyrene (BP) to DNA are used as an example of the application of immunochemical probes to questions pertaining to the manner in which various agents modify DNA and nucleosomes.[38]

Antibodies to BP–DNA recognize modified DNA and RNA, but do not bind to unmodified bases. Specific binding of anti-BP–DNA to nuclei of cells modified by the carcinogen could be visualized by immunofluorescence. Repair of the damage, at the resolution of a single cell, can also be observed. At various times after removal of the carcinogen from the cell culture media, samples are removed and examined by immunofluorescence. The intensity of immunofluorescence is a measure of the presence of carcinogen. The rate of carcinogen removal varies among individual cells.

The distribution of carcinogen in cellular or viral chromatin and DNA can be determined by the following procedure: nuclei or viral chromatin, isolated from cells treated with BP, are digested with micrococcal nuclease, and the purified DNA is separated by gel electrophoresis. The DNA is transferred to nitrocellulose or DBM paper and treated with anti-BP antibodies. The paper is incubated with ^{125}I-labeled second antibody, and the location of the bound antibodies is visualized by autoradiography. The amount of antibodies bound is directly proportional to the extent of DNA modification. The DNA of the cells can be labeled with $^{32}PO_4$, and the specific activity of the various DNA fragments can be determined by cutting the paper and counting both isotopes. Using this technique, it could be demonstrated that the carcinogen binds preferentially to linker DNA.

The distribution of the carcinogen in a viral sequence can also be studied. Viral chromatin is isolated from ^{32}P-labeled, carcinogen-treated cells. The DNA is purified, restricted, and the fragments separated on gels. The DNA is transferred to a solid support which is treated with antibodies and ^{125}I-labeled second antibody, as described above. The resulting autoradiogram identifies the DNA fragments containing the carcinogen. Figure 17 shows results representative of this approach. Immunofluorescence with

[38] M. Seidman, H. Slor, and M. Bustin, *J. Biol. Chem.* **258**, 5215, 1983.

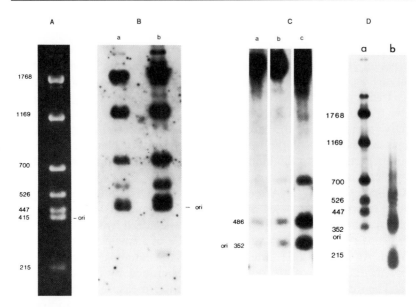

Fig. 17. Immunochemical analysis of BPDE binding to various regions of SV40 DNA. (A) Ethidium bromide stain of SV40 digested with *Hin*dIII and *Hpa*II. (B) Autoradiogram of the BPDE-modified SV40 DNA digested as in A: (a) 2 g of virus DNA modified *in vivo* to a level of 4 BPDE/DNA; (b) 0.5 g DNA modified *in vitro* to 20 BPDE/DNA. (C) The *Bgl*I, *Hpa*II, *Hae*II digest of DNA: (a) modified *in vivo* to 0.8 BPDE/DNA; (b) modified *in vivo* to 4 BPDE/DNA; and (c) *in vitro* 20 mmol BPDE/DNA. (D) (a) The *Bgl*I, *Hpa*II, and *Hin*dIII digest of SV40 DNA modified *in vivo* to about 2 BPDE/DNA; and (b) micrococcal nuclease digestion of same type of preparation. The DNA fragments were transferred from agarose to DBM paper, and the paper was incubated with antibodies specific to BPDE–DNA adducts. The location of the bound IgG was detected with ^{125}I-labeled antibody. (From Ref. 38.)

polytene chromosomes isolated from *Chironomus thummi* larvae, which have been exposed to BP, revealed that the carcinogen binds to the chromosomes in a site-specific manner.[39]

Concluding Remarks

The purpose of this chapter is to provide general guidelines for the preparation and use of immunochemical probes for various questions relating to nucleosome structure and function. The examples presented illustrate the wide applicability of this approach and its ability to yield information that cannot be obtained by other techniques. The newer

[39] P. Kurth and M. Bustin, *Proc. Natl. Acad. Sci. U.S.A.* **82**, 7076 (1985).

methods of antibody production and purification, novel, automated and highly sensitive methods of detecting antibody–antigen reactions, and the constantly expanding repertoire of specific antibodies indicate that the serological approach will continue to have widespread applications in studies on the structure and function of chromatin and chromosomes.

Acknowledgments

I wish to thank Drs. D. Landsman and F. Friedman for critical review of the manuscript.

[12] Use of Antihistone Antibodies with Nucleosomes

By Sylviane Muller and Marc H. V. Van Regenmortel

Introduction

Antibodies directed against defined regions of histone molecules represent one of the most specific probes for studying the surface topography of nucleosomes and chromatin and for monitoring posttranslational histone modifications. In recent years, about two dozen antigenic determinants or epitopes of the four core histones have been identified. Most of these epitopes, known as continuous epitopes, correspond to linear stretches of 6–20 amino acid residues. It is noteworthy that all terminal regions of the four histones, with the exception of the carboxy-terminal residues of H4, correspond to continuous epitopes. Antibodies directed against these epitopes can be used to determine which regions of histone molecules are exposed at the surface of nucleosomes and are useful probes for analyzing chromatin structure and function. Four approaches have been used to obtain antibodies suitable for these types of studies: (1) immunization with histone molecules; (2) immunization with synthetic peptide fragments of histones; (3) immunization with nucleosomes; and (4) preparation of hybridomas from nonimmunized, autoimmune mice.

Preparation of Immunogens

Histones

The validity of all immunochemical studies of chromatin ultimately depends on the specificity of the antibody probes that are used. In the case

Copyright © 1989 by Academic Press, Inc.
All rights of reproduction in any form reserved.

of histone antibodies, the main difficulty lies in the fact that histones tend to copolymerize, with the result that during antiserum production the animals are frequently immunized with more than one histone at a time. Particular care must therefore be given to the preparation of pure histone fractions free of mutual contamination.

Purified histones can be kept in lyophilized form for several months under dry conditions at room temperature. After six months, it may be advisable to control their antigenic reactivity to ensure absence of degradation. Lyophilized histones should be suspended in distilled water prior to dilution in suitable buffers. When kept in solution at 4°, histones should not be stored longer than about 2 weeks, since they form homoaggregates, especially at high ionic strength.

A variety of procedures have been described to isolate and purify the different histones (see von Holt et al., this volume [23]). Most commonly, the selective extraction method described by Johns[1] is used followed by the purification procedure of Michalski-Scrive et al.[2] A milder method, known as the saline method, was described by Van der Westhuyzen and von Holt.[3] In this procedure, extreme pH conditions are avoided, and protamine is used to displace histones from DNA. A series of runs on Sephadex G-100 columns, equilibrated in 50 mM sodium acetate successively at pH 5 and 4, are used to separate the five histone fractions, H1, H2A, H2B, H3, and H4. A method for extracting the erythrocyte-specific histone H5 from purified nuclei was described by Champagne et al.[4] The tertiary structure of histone molecules appears to be affected by the purification procedure since the antigenic properties of histones isolated by the salt and acid extraction procedures are not the same, as shown by Mihalakis et al.[5] for histone H1 and by Feldman and Stollar[6] for histones H3 and H4.

Since the different histones are very similar in molecular weight, charge, and secondary structure, they are difficult to separate. Recently, high-performance liquid chromatography (HPLC) was shown to represent an attractive alternative to the classic methods of histone purification.[7,8]

Acetylated histones have also been used both as immunogens and

[1] E. W. Johns, Biochem. J. 105, 611 (1967).
[2] C. Michalski-Scrive, J. P. Aubert, M. Couppez, G. Biserte, and M. H. Loucheux-Lefebvre, Biochimie 64, 347 (1982).
[3] D. R. van der Westhuyzen and C. von Holt, FEBS Lett. 14, 333 (1971).
[4] M. Champagne, A. Mazen, and X. Wilhelm, Bull. Soc. Chim. Biol. 50, 1261 (1968).
[5] N. Mihalakis, O. J. Miller, and B. F. Erlanger, Science 192, 469 (1976).
[6] L. Feldman and B. D. Stollar, Biochemistry 16, 2767 (1977).
[7] H. Lindner, W. Helliger, and B. Puschendorf, Anal. Biochem. 158, 424 (1986).
[8] M. C. McCroskey, V. E. Groppi, and J. D. Pearson, Anal. Biochem. 163, 427 (1987).

antigens.[9-11] The extraction and purification of H4 molecules presenting different degrees of acetylation were described by Couppez et al.[12] The di- and triacetylated forms of H4, which are naturally present in large amounts in cuttlefish testes, were obtained by ion-exchange chromatography on Bio-Rex 70 (Bio-Rad, Richmond, CA).

The purity of histone fractions can be evaluated by SDS–polyacrylamide gel electrophoresis according to the procedure described by Weintraub et al.[13] However, since the limit of detection in these gels is about 0.5 μg histone, it may be difficult to ascertain in a reliable manner the degree of purity that has been achieved. The sensitivity of immunochemical detection methods is much greater since histones can be detected by ELISA at a concentration of 25 ng/ml and by dot immunoassay at a level of 1 ng/spot. When such methods are used to assess the purity of histone preparations, it is not uncommon to find that fractions which appear to be pure by biochemical criteria are in fact contaminated by as much as 1% of other histone classes.[14] It should also be stressed that animals are usually immunized with quantities of immunogen as great as 100–1000 μg, i.e., with much more than the minimal immunizing dose. Since there is not necessarily a direct relationship between the dose of immunogen given to an animal and the level of immune response, the contaminating histone present in a preparation may sometimes induce a stronger response than the major histone component of interest.[15] One way to obviate these problems completely is to use synthetic peptides corresponding to a fragment of one histone for the production of antipeptide antibodies that will cross-react only with the parent histone.

Synthetic and Natural Peptides of Histones

Synthetic peptides of histones, including acetylated peptides,[10] have been prepared by the solid-phase method of Barany and Merrifield.[16] Additional terminal residues can be added to the peptide sequence to

[9] S. Muller, M. Erard, E. Burggraf, M. Couppez, P. Sautière, M. Champagne, and M. H. V. Van Regenmortel, *EMBO J.* **1,** 939 (1982).

[10] S. Muller, A. Isabey, M. Couppez, S. Plaué, G. Sommermeyer, and M. H. V. Van Regenmortel, *Mol. Immunol.* **24,** 779 (1987).

[11] T. R. Hebbes, A. W. Thorne, and C. Crane-Robinson, *EMBO J.* **7,** 1395 (1988).

[12] M. Couppez, A. Martin-Ponthieu, and P. Sautière, *J. Biol. Chem.* **262,** 2854 (1987).

[13] H. Weintraub, K. Palter, and F. Van Lente, *Cell* **6,** 85 (1975).

[14] S. Muller, E. Jockers-Wretou, C. E. Sekeris, M. H. V. Van Regenmortel, and F. A. Bautz, *FEBS Lett.* **182,** 459 (1985).

[15] C. S. M. Tahourdin and M. Bustin, *Biochemistry* **19,** 4387 (1980).

[16] G. Barany and R. B. Merrifield, in "The Peptides" (E. Gross and J. Meienhofer, eds.), Vol. 2, p. 3. Academic Press, New York, 1980.

facilitate subsequent coupling to carriers. Peptides are purified on a preparative MPLC (middle-pressure liquid chromatography) column, and the final purity is assessed by HPLC on a reversed-phase C_{18} column (Ultraspher ODS, 4.5 × 250 mm, Altex, San Ramon, CA). The amino acid composition of the peptides should be controlled.

Natural fragments of histones have also been used in immunological studies. However, because they can be obtained in only very small quantities and cross-contamination with other peptides can never be totally excluded, it is not advisable to use them for immunization. Fairly large fragments are produced by cleavage of whole histones with cyanogen bromide[17,18] or 0.25 M acetic acid at 105°[19] or with enzymes such as V8 staphylococcal protease,[20,21] trypsin,[12] and pepsin.[19]

The purity of histone fragments can be checked by polyacrylamide slab gel electrophoresis at pH 2.7 in 2.5 M urea[22] using a 17% (w/v) polyacrylamide concentration or by thin-layer chromatography on cellulose sheets (Polygram Cel 300, Macherey and Nagel, Düren, FRG) in 1-butanol : pyridine : acetic acid : water (7 : 5 : 2 : 6). The amino acid composition of fragments should be controlled.

Chemical acetylation of peptide 1–23 of H4 was described by Couppez et al.,[12] who succeeded in acetylating all four lysine residues, 5, 8, 12, and 16, present in this fragment. The acetylation of peptide 1–37 of H4 was described by Pfeffer et al.[23]

When short histone peptides are used as immobilized antigen in solid-phase immunoassays or as immunogen, it is necessary to use them in the form of peptide–carrier conjugates. The choice of conjugation procedure is important since the antigenic activity of a peptide may be drastically affected by different coupling procedures.[24] In the case of highly hydrophilic peptides such as those obtained from histones, the coupling reaction should be performed after chemical protection of any internal lysine, glutamic acid, and tyrosine residues present in the peptide. For this reason,

[17] D. Di Padua Mathieu and B. D. Stollar, *Biochemistry* **19**, 2246 (1980).
[18] G. Sapeika, D. Absolom, and M. H. V. Van Regenmortel, *Immunochemistry* **13**, 499 (1976).
[19] S. Muller, S. Plaué, M. Couppez, and M. H. V. Van Regenmortel, *Mol. Immunol.* **23**, 593 (1986).
[20] E. Tailllandier, L. Fort, J. Liguier, M. Couppez, and P. Sautière, *Biochemistry* **23**, 2644 (1984).
[21] B. A. Neary and B. D. Stollar, *Eur. J. Biochem.* **168**, 161 (1987).
[22] S. Panyim and R. Chalkley, *Arch. Biochem. Biophys.* **130**, 337 (1969).
[23] U. Pfeffer, N. Ferrari, and G. Vidali, *J. Biol. Chem.* **261**, 2496 (1986).
[24] J. P. Briand, S. Muller, and M. H. V. Van Regenmortel, *J. Immunol. Methods* **78**, 59 (1985).

coupling by means of reagents such as glutaraldehyde and carbodiimides should be used with caution. By introducing additional residues such as tyrosine or cysteine at the terminal end of peptides, it becomes possible to conjugate the peptide by means of other coupling reagents.[10,19,24,25] The yield of conjugation and the intactness of residues after coupling can be controlled by amino acid analysis as described by Briand et al.[24] It may be necessary to adjust coupling procedures to the particular properties of the peptide. For instance, when 1-ethyl-3-(3-dimethylaminopropyl)carbodiimide hydrochloride (EDC) was used to conjugate peptide 1–17 of H4 to BSA, the final conjugate was found to be insoluble. However, this was not the case when MCDI [N-cyclohexyl-N'-(2-morpholinoethyl)carbodiimide metho-p-toluol sulfonate] was used instead of EDC.[10]

Chromatin and Nucleosomes

Numerous methods can be used to prepare native and H1-depleted chromatin, nucleosomes, and core particles (see Kornberg et al., this volume [1]). We generally use chromatin obtained from chicken erythrocytes[26] or from rat liver nuclei prepared by the procedure of Hewish and Burgoyne[27] as described by Marion et al.[28] Long chains of chromatin comprising 35–50 nucleosomes, as well as fractions containing tri-, di-, and mononucleosomes, can be obtained by zonal centrifugation on 5–28% sucrose gradients in 10 mM Tris-HCl, 1 mM EDTA, 0.1 mM phenylmethylsulfonyl fluoride, pH 7.4, at 4°. Pooled fractions from different tubes are dialyzed against this buffer and concentrated by ultrafiltration through ultrathimbles (UH 100/25, Schleicher and Schüll, Dassel, FRG).

Chicken erythrocyte core particles are obtained by the method of Lutter[29] as described by de Murcia et al.[30] After two successive digestions by micrococcal nuclease, it is preferable to isolate core particles by zonal centrifugation instead of by chromatography on Sepharose 6B since the latter method has been shown by electron microscopy and circular dichroism measurements to induce a partial unfolding of core particle DNA.

[25] S. Muller, M. Couppez, J. P. Briand, J. Gordon, P. Sautière, and M. H. V. Van Regenmortel, Biochim. Biophys. Acta 827, 235 (1985).

[26] A. Mazen, G. de Murcia, S. Bernard, J. Pouyet, and M. Champagne, Eur. J. Biochem. 127, 169 (1982).

[27] D. R. Hewish and L. A. Burgoyne, Biochem. Biophys. Res. Commun. 52, 504 (1973).

[28] C. Marion, B. Roux, L. Pallota, and P. R. Coulet, Biochem. Biophys. Res. Commun. 114, 1169 (1983).

[29] L. C. Lutter, J. Mol. Biol. 124, 391 (1978).

[30] G. de Murcia, A. Mazen, M. Erard, J. Pouyet, and M. Champagne, Nucleic Acids Res. 8, 767 (1980).

The quality of nucleosome and core particle preparations should be controlled by electron microscopy and circular dichroism measurements. The DNA content can be monitored by electrophoresis on 2.5% (w/v) polyacrylamide–agarose slab gel[31] after extraction of the DNA as described by Mazen et al.[32] The histone content can be analyzed by electrophoresis on a 18% (w/v) polyacrylamide–sodium dodecyl sulfate (SDS) gel by loading total chromatin directly on the gel after treating the samples for 2 min at 100° in 1% (w/v) SDS, 2.5% (v/v) 2-mercaptoethanol, 0.25 M sucrose, 62.5 mM Tris-HCl, pH 6.8.

Preparation of Polyclonal Antibodies

Immunization Protocols

Isolated core histones are poor immunogens. This weak immunogenicity has been ascribed to the fact that their sequence has been highly conserved during evolution and contains very few aromatic amino acid residues.[33,34] However, when animals are immunized with histone–RNA (or histone–DNA) complexes instead of with free histones, there is no difficulty in inducing the formation of antibodies to the four core histones.[35] Antisera to histones have mostly been prepared in rabbits, although guinea pigs have also been used.[36]

In our laboratory, the following immunization protocol is used: equal volumes of yeast RNA (70 μg in 1 ml water) and core histone (200 μg in 1 ml water) are mixed, and the resulting suspension is emulsified in Freund's complete adjuvant (CFA) (1 : 1, v/v). This produces an emulsion corresponding to the dose needed to immunize one rabbit by the intramuscular route. The immunization is repeated several times at intervals of 2 weeks using Freund's incomplete adjuvant (IFA) instead of CFA. After two injections, the animals are bled regularly, 1 week after each injection, until an antiserum of sufficient titer is obtained.

Antisera to H1 and H5 have been obtained by immunizing animals either with free histones or with histone–RNA complexes.[37–40] Rabbits are given an immunizing dose ranging from 100 μg to 1 mg administered in

[31] A. C. Peacock and C. W. Dingman, *Biochemistry* **7**, 668 (1967).

[32] A. Mazen, M. Champagne, M. Wilhelm, and X. Wilhelm, *Exp. Cell Res.* **117**, 431 (1978).

[33] B. D. Stollar, *Methods Cell Biol.* **18**, 105 (1978).

[34] M. Bustin, *Curr. Top. Microbiol. Immunol.* **88**, 105 (1979).

[35] B. D. Stollar and M. Ward, *J. Biol. Chem.* **245**, 1261 (1970).

[36] C. V. Mura, A. Mazen, J. M. Neelin, G. Briand, P. Sautière, and M. Champagne, *Eur. J. Biochem.* **108**, 613 (1980).

[37] N. Sotirov and E. W. Johns, *J. Immunol.* **109**, 686 (1972).

[38] M. Bustin and B. D. Stollar, *J. Biol. Chem.* **247**, 5716 (1972).

the form of an emulsion in CFA for the first injection and emulsions in IFA for subsequent injections. Different routes of immunization can be used, for example, three intradermal injections followed by intravenous boosters (without adjuvant) at monthly intervals[41] or injections in the foot pads and at multiple intradermal sites.[42] There is no evidence that foot pad injections, which produce considerable discomfort to the animals, are required to obtain suitable antisera. Recently, it has been shown[40,43] that antibodies raised against H5 or H1 complexed with RNA did not recognize the same regions in the H5 and H1 molecules as antibodies induced by free H5 and H1. Since most analyses of the antigenic structure of core histones were carried out with antibodies raised against histone – RNA complexes, it cannot be excluded that the results of all these studies were biased by the type of immunogen used.

Antibodies directed against large fragments of H1 and H2B were obtained by immunizing rabbits with these fragments complexed with RNA.[17,44] Dimitrov et al.[45] obtained antibodies to the noncomplexed globular domain of H5 (GH5) by injecting rabbits repeatedly at multiple intradermal sites with 50 μg of GH5 emulsified with adjuvant.[46]

Peptides that are at least 15 – 20 residues long can be injected in the free form, whereas shorter peptides are usually injected in the form of carrier – peptide conjugates.[19] Generally, 10 to 12-week-old female rabbits are given 200 μg peptide emulsified in CFA (1 : 1 ratio, v/v). Multiple subcutaneous inoculations at 5 – 10 injection sites are used with unconjugated peptides, whereas intramuscular inoculations are mostly used with conjugated peptides. Further injections are performed in the same manner every two weeks except that, in the case of conjugated peptides, IFA is used instead of CFA. Bleedings are performed 8 days after each injection starting from the fifth week.

Antisera to nucleosomes have been prepared in several laboratories,[15,47,48] but there have been no reports describing the production of

[39] L.-L. Y. Frado, C. V. Mura, B. D. Stollar, and C. L. F. Woodcock, *J. Biol. Chem.* **258,** 11984 (1983).
[40] L. Srebreva and J. Zlatanova, *FEBS Lett.* **200,** 123 (1986).
[41] L. Srebreva, J. Zlatanova, G. Miloshev, and R. Tsanev, *Eur. J. Biochem.* **165,** 449 (1987).
[42] E. R. Markose and M. R. S. Rao, *J. Biol. Chem.* **260,** 16263 (1985).
[43] B. A. Neary, C. V. Mura, and B. D. Stollar, *J. Biol. Chem.* **260,** 15850 (1985).
[44] M. Bustin and B. D. Stollar, *Biochemistry* **12,** 1124 (1973).
[45] S. I. Dimitrov, V. R. Russanova, and I. G. Pashev, *EMBO J.* **6,** 2387 (1987).
[46] H. Yasuda, K. A. Logan, and E. M. Bradbury, *FEBS Lett.* **166,** 263 (1984).
[47] L. Einck, R. Dibble, L.-L. Y. Frado, and C. L. F. Woodcock, *Exp. Cell Res.* **139,** 101 (1982).
[48] S. Muller, D. Bonnier, M. Thiry, and M. H. V. Van Regenmortel, *Int. Arch. All. Appl. Immunol.* (in press) (1989).

antisera to core particles. In our laboratory, antibodies to mononucleosomes have been obtained by immunizing rabbits with freshly prepared mononucleosome preparations obtained as described above (50 μg/rabbit in terms of DNA) and emulsified with an equal volume of IFA. A series of monthly intramuscular injections are given over a period of 4 months, and rabbits are bled 1 week after each injection, from the third injection onward.

Immunological Assays

Enzyme-Linked Immunosorbent Assay (ELISA). The ELISA has been widely used for measuring the level of antihistone antibodies present in antisera and for studying the structure of nuclear antigens. Various ELISA formats have been used to locate epitopes on the surface of histones and nucleosome particles to follow changes in accessibility of these epitopes under different ionic strength and pH conditions.[49,50] The technique is highly versatile and allows the investigator to study chromatin components under a wide range of experimental conditions.

Indirect ELISA using immobilized histones or peptides. The test consists of the following steps:

1. Microtiter wells (polyvinyl plates, Falcon, Oxnard, CA) are coated by incubation overnight at 37° with 200 μl of 0.02–0.4 μg histone/ml or 0.2–2.5 μM peptide solution diluted in carbonate buffer, pH 9.6 (15 mM Na$_2$CO$_3$, 35 mM NaHCO$_3$, 0.2 g/liter NaN$_3$). Under these conditions, nearly 90% of the antigen becomes attached to the plastic as measured with ^{14}C-labeled H2B and H4. If the peptide does not become adsorbed when carbonate buffer is used, other buffers in the pH range 6–8 should be tried. When peptide conjugated to a carrier protein is used for coating the wells, the concentration of conjugate expressed as peptide should also be in the range of 0.2–2.5 μM (usually 1 μM).
2. Repeated washings (at least 3) are carried out with phosphate-buffered saline pH 7.4, containing 0.05% Tween 20 (PBS-T).
3. Blocking of remaining sites on the plastic is accomplished by incubation with 10 mg/ml bovine serum albumin in PBS-T, for 1 hr at 37°.
4. Repeat washings with PBS-T.
5. Incubate with the antiserum to be tested, diluted in PBS-T, for

[49] J. Romac, J. P. Bouley, and M. H. V. Van Regenmortel, *Anal. Biochem.* **113**, 366 (1981).
[50] S. Muller, E. Bertrand, M. Erard, and M. H. V. Van Regenmortel, *Int. J. Biol. Macromol.* **7**, 113 (1985).

1 – 2 hr at 37°. The range of two-fold dilutions appropriate for a peptide antiserum is usually from 1/500 to 1/4000 and for antihistone antisera from 1/1000 to 1/8000.

6. Repeat washings with PBS-T.
7. Incubate with a suitable antiimmunoglobulin – enzyme conjugate, diluted in PBS-T in the range of 1/500 to 1/8000 (generally 1/2000) for 2 hr at 37°. If the histone or peptide antiserum were raised in a rabbit, the enzyme conjugate could be goat anti-rabbit globulin conjugated to alkaline phosphatase.
8. Repeat washings with PBS-T.
9. Incubate with an appropriate enzyme substrate for 1 – 3 hr at 37°. If the enzyme is alkaline phosphatase, the substrate p-nitrophenyl phosphate at 1 mg/ml in 0.1 M diethanolamine buffer, pH 9.8, is used. This substrate is available as tablets (Sigma, St. Louis, MO) or in powder form (Boehringer-Mannheim, FRG). Several other enzyme – substrate systems are commercially available.[51] When enzyme conjugates prepared with horseradish peroxidase are used, the reaction can be revealed with the substrate ABTS (Boehringer-Mannheim, Sigma).
10. Read the optical density with ELISA readers such as the Titertek Multiskan MC (Flow Laboratories, Irvine, Scotland) or the Autoreader Micro ELISA (Dynatech, Billingshurst, UK).

Indirect ELISA using immobilized chromatin or its subunits. The same procedure described for histones is used except, since chromatin binds poorly to the plastic at 4°, the coating step is done at 37° in the presence of PBS-T (pH 7.4). The antigen concentration used to coat the plates is generally 25 – 200 ng/ml expressed as DNA.

Antibody double-sandwich ELISA. The antibody double-sandwich ELISA procedure has been rarely used with histones and chromatin mainly because of the tendency of histones to aggregate and to bind nonspecifically to various components as well as to the plastic solid phase. Assays in which histones are used to coat the solid support are therefore more convenient. The following double-sandwich procedure has been used successfully with H4 and specific monoclonal antibodies:[10]

1. Coat wells by incubation (2 hr at 37°) with 2 μg/ml anti-H4 rabbit immunoglobulins (Ig) diluted in coating buffer, pH 9.6.
2. Repeat washings with PBS-T.
3. Block with 1% BSA in PBS-T, followed by washings.

[51] P. Tijssen, *in* "Laboratory Techniques in Biochemistry and Molecular Biology" (R. H. Burdon and P. M. Van Knippenberg, eds.), p. 540. Elsevier/North-Holland, Amsterdam, 1985.

4. Incubate with H4 (25–200 ng/ml) in PBS-T for 2 hr at 37°.
5. Repeat washings with PBS-T.
6. Incubate with diluted ascitic fluids diluted 1/10,000–1/100,000 in PBS-T for 2 hr.
7. Repeat washings with PBS-T.
8. Incubate with sheep anti-mouse Ig coupled to alkaline phosphatase (Jackson, Avondale, PA) diluted 1/4000 in PBS-T for 2 hr at 37°. The final step of the assay is as described above.

ELISA inhibition test. The antigenic activity of histone peptides can be measured in an ELISA inhibition format in which the peptide is first incubated with histone antiserum overnight at 4° or for 2 hr at 37°. The mixture is then added to wells precoated with the histone antigen and incubated for 2 hr at 37°, as described above. A molar excess of peptide over histone in the range of $10^3 – 10^5$ is commonly used in this type of assay.

Solid-Phase Radioimmunoassay. The various steps of solid-phase radioimmunoassay (RIA) are similar to those of ELISA except that the enzyme conjugate is replaced by ^{125}I-labeled protein A. Following the incubation step with radiolabeled reagent (for instance, ^{125}I-protein A from Amersham Radiochemical Centre, Amersham, UK, 30–50 μCi/μg, using 4×10^4 cpm/ml incubated for 1 hr at 37° and repeated washings to remove the excess of radiolabeled reagent, the wells are cut from the plate and read in a gamma counter.[52] Other variations of solid-phase RIA have been used by Blankstein *et al.*[53] and Romani *et al.*[54]

Microcomplement Fixation Assay. During the 1970s, the serological technique of microcomplement fixation replaced the classic precipitation technique, which had frequently been found to lead to nonspecific reactions when applied to basic proteins such as histones. Microcomplement fixation tests require a concentration of histone antigen in the range of 0.25–2.0 μg/ml and a dilution of antiserum around 1/100 to 1/500; under these conditions, nonspecific reactions are minimized.

The main advantage of complement fixation tests is their ability to provide highly accurate quantitative data. This allows, for instance, the antigenic reactivity of different histone fragments to be compared on a quantitative basis.[18] The disadvantage of complement fixation tests is that they require a narrowly defined ionic environment which severely restricts the range of environmental conditions under which chromatin can be

---

[52] S. Muller, K. Himmelspach, and M. H. V. Van Regenmortel, *EMBO J.* **1,** 421 (1982).
[53] L. A. Blankstein, B. D. Stollar, and S. B. Levy, *Anal. Biochem.* **104,** 168 (1980).
[54] M. Romani, G. Vidali, C. S. M. Tahourdin, and M. Bustin, *J. Biol. Chem.* **255,** 468 (1980).

studied. In order to give meaningful data, complement fixation tests require careful standardization of all the reagents. The method has been described in detail by Levine and Van Vunakis[55] and Stollar.[33]

Dot Immunoassay. The dot immunoassay is particularly useful for testing large numbers of sera and culture supernatants of hybridomas. The procedure described by Hawkes *et al.*[56] is commonly used. Aliquots of 0.5 μl of purified histone preparation dissolved in PBS-T are applied to nitrocellulose sheets already printed with a 3×3 mm grid (Millipore, Bedford, MA). After successive washings of the filters for 5 min in PBS-T and blocking for 1 hr at room temperature with PBS-T containing 1% BSA, dilutions of the antibody are incubated for 2 hr at room temperature. Nitrocellulose sheets are then washed 3 times for 5 min with PBS-T and incubated for 2 hr with antiglobulin–peroxidase conjugates (peroxidase-linked goat anti-mouse Ig or peroxidase-linked goat anti-rabbit Ig). A final extensive washing is performed in PBS-T. A suitable substrate is *o*-dianisidine (2% [w/v]) – H_2O_2 (0.06% [v/v]) in 10 mM Tris-HCl buffer (pH 7.4). When the stain is clearly visible, the reaction is stopped by washing the sheets in distilled water.[14] In the case of histones, the sensitivity of the assay was not increased by using the avidin–biotin or the gold-staining detection methods.

Immunoblotting. Proteins separated by SDS–polyacrylamide gel electrophoresis are blotted electrophoretically on hydrophobic membrane (Immobilon, Millipore) at 250 mA for 75 min at 4°. After blocking remaining protein binding sites present on membrane by a 3-hr incubation with PBS-T containing 1% BSA, the filters are incubated overnight at room temperature with specific antisera diluted in PBS-T–BSA (generally 1/100 – 1/1000 for antipeptide antisera and 1/1000 – 1/10,000 for antihistone antisera). Following extensive rinsing with PBS-T, ^{125}I-labeled protein A (30 – 50 mCi/mg, Amersham) diluted in PBS-T to a final concentration of 350 – 700 nCi/ml is added. After a 2-hr incubation at room temperature, the sheets are extensively washed to remove unbound protein A. The blots are air-dried and exposed for autoradiography at −70° with an Illford (Aba-Geigy, Basildon, Essex, UK) intensifying screen.

Other immunoblotting procedures used with nuclear antigens have been described by Markose and Rao,[42] Srebreva *et al.,*[41] Pfeffer *et al.,*[23] and Whitfield *et al.*[57] Avramova *et al.*[58] described an indirect immunofluores-

[55] L. Levine and H. Van Vunakis, this series, Vol. 2, p. 928.
[56] R. Hawkes, E. Niday, and J. Gordon, *Anal. Biochem.* **119,** 142 (1982).
[57] W. G. F. Whitfield, G. Fellows, and B. M. Turner, *Eur. J. Biochem.* **157,** 513 (1986).
[58] Z. Avramova, A. Uschewa, E. Stephanova, and R. Tsanev, *Eur. J. Cell Biol.* **31,** 137 (1983).

cence test for the detection of antigen–antibody complexes in immuno-blotting.

Preparation of Monoclonal Antibodies for Immunized Mice

To date, the only histone against which monoclonal antibodies have been obtained by the classic approach, namely, fusing spleen cells from a mouse immunized with the antigen, is histone H5.[59–61] However, monoclonal antibodies reacting with histone H2B have been obtained from mice immunized with human chromatin[62] and with the total histone fraction from trout,[57] and one monoclonal antibody reacting with H3 was obtained using a rat liver nuclear extract as immunogen.[14] Ten monoclonal antibodies to histone H4 were obtained from mice immunized with cuttlefish triacetylated H4 complexed with RNA.[10] The fine specificity of these antibodies was studied by ELISA using various populations of acetylated H4, nonacetylated H3/H4 tetramers, and histone octamers as well as acetylated and nonacetylated peptides of H4. Surprisingly, none of these monoclonal antibodies was found to recognize the triacetylated H4 that had been used as immunogen. Only five of them bound to diacetylated, monoacetylated, and nonacetylated H4. One antibody was specific for H4 associated in the form of histone octamers but did not bind to any nonace-tylated or acetylated forms of H4 monomers. None of the monoclonal antibodies was completely specific for acetylated forms of H4. These results clearly demonstrate the advantage of using several antigens other than the immunogen (i.e., different acetylated forms, tetramers, and octamers) during the screening step of hybridoma production.

Preparation of Monoclonal Antibodies from Nonimmunized Autoimmune Mice

Monoclonal antibodies to histones have also been obtained from non-immunized autoimmune mice that develop a syndrome close to human systemic lupus erythematosus.[63] Spleen cells from autoimmune 9-month-old NZB/NZW female mice were fused with myeloma cells. Among a large

[59] A. F. M. Moorman, P. A. J. de Boer, M. T. Linders, and R. Charles, *Cell Differ.* **14,** 113 (1984).

[60] E. Mendelson and M. Bustin, *Biochemistry* **23,** 3459 (1984).

[61] M. Rozalski, L. Lafleur, and A. Ruiz-Carrillo, *J. Biol. Chem.* **260,** 14379 (1985).

[62] B. M. Turner, *Chromosoma* **87,** 345 (1982).

[63] R. Laskov, S. Muller, M. Hochberg, H. Giloh, M. H. V. Van Regenmortel, and D. Eilat, *Eur. J. Immunol.* **14,** 74 (1984).

series of clones, two hybridomas able to interact with histones were studied in detail. One of the monoclonal antibodies was specific for histone H2B and reacted with the histone free in solution or when present as an H2A–H2B complex. Its fine specificity was analyzed with peptides of H2B.[25] The second monoclonal antibody recognized a specific conformation in the H3–H4 complex.[63]

Other clones reacting with histone H2B were obtained using autoimmune MRL 1pr/1pr strain of mice, and their specificity was studied with overlapping histone peptides[25] and with native, H1-depleted, and phosphorylated chromatin.[64]

Conclusion

It should be stressed that antigenicity studies always suffer inevitably from one or another operational bias.[65] The method used for purifying histones and the immunization protocol, using histones complexed with RNA, influence the conformation of histone molecules and will thus affect their antigenic determinants. There is also evidence that polymerization of histones into chromatin components produces novel antigenic determinants that derive their specificity from the tertiary and quaternary structure. Changes in environmental conditions that alter the structure of chromatin or posttranslational modifications can thus be expected to modify its antigenic reactivity. Furthermore, the type of assay used for measuring the antigenic activity of histone peptides also clearly determines whether a peptide is identified as corresponding to a continuous epitope of a histone.[19] Since the type of immunological reagent (monoclonal or polyclonal), the assay format, and the chemical environment will all influence the reactivity of chromatin antigens, great care should be used in the interpretation of immunochemical data. However, the fact that immunological probes are very sensitive to the conformation of histones makes them valuable reagents for the study of chromatin activity and the concomitant structural changes occurring in nucleosomes.[11,23,64,66] Compared to biophysical measurements, which always reflect statistical averages, the use of specific antibodies makes it possible to obtain data that pertain to only a fraction of the population of chromatin subunits that is nevertheless functionally significant.[50]

[64] S. Muller, A. Mazen, A. Martinage, and M. H. V. Van Regenmortel, *EMBO J.* **3**, 2431 (1984).
[65] M. H. V. Van Regenmortel, *Trends Biochem. Sci.* **11**, 36 (1986).
[66] U. Pfeffer, N. Ferrari, F. Tosetti, and G. Vidali, *Exp. Cell Res.* **178**, 25 (1988).

[13] Digestion of Nucleosomes with Deoxyribonucleases I and II

By Leonard C. Lutter

Introduction

Nucleases have been used extensively as probes of chromatin and nucleosome structure. Noll's striking observation[1] that deoxyribonuclease I (DNase I) produces nicks spaced at multiples of about 10 bases in the DNA of chromatin indicated that the enzyme "sees" DNA bound to a surface and led to its extensive use in nucleosome core studies. Indeed, the original DNA "footprint" studies were carried out on the nucleosome core. Subsequently, other enzymes, among them deoxyribonuclease II (DNase II), were also found to produce nicks spaced at multiples of about 10 bases in DNA bound to a surface such as that found in chromatin. When it was demonstrated that DNase I showed considerable sequence specificity in its attack on naked DNA, it became clear that several nuclease probes should be employed in structural studies so that eccentric contributions of any single nuclease to the interpretation can be kept to a minimum. Thus, for example, digestion studies on the nucleosome core have employed not only DNase I and DNase II but micrococcal nuclease as well.

The results of DNase I and DNase II digestions have contributed substantially to the development of our current picture of the nucleosome core, but their use is still finding important application in detection of binding of the histone octamer to specific DNA sequences. However, sequence specificity of cutting requires more involved analysis of data derived from digestion of such defined sequences. Therefore, in this chapter both conditions for digestion as well as data analysis methods are described.

Digestion with DNase I

Methods of digestion of mixed-sequence nucleosome cores are described, but the digestion protocol applies equally well to the experiment in which histone octamers are reconstituted onto a unique sequence. Such

---

[1] M. Noll, *Nucleic Acids Res.* **8,** 1573 (1974).

Copyright © 1989 by Academic Press, Inc.
All rights of reproduction in any form reserved.

experiments are commonly carried out with the unique-sequence DNA being radioactively end-labeled and then reconstituted into nucleosomes by exchange in the presence of native, mixed-sequence, unlabeled nucleosomes (methods for such reconstitution are described elsewhere in this volume). Thus, as far as digestion conditions are concerned, the bulk of the sample is still mixed-sequence nucleosome cores.

A typical DNase I digestion[2] is carried out as follows for a 25 μl reaction:

1. Mix 12.5 μl end-labeled nucleosome core preparation (2 mg/ml) and 2.5 μl 10 × DNase I buffer (50 mM MgCl$_2$, 40 mM dithiothreitol, 500 mM Tris, pH 8) and add H$_2$O to 22.5 μl.
2. Incubate 2 min at 37°, then add 2.5 μl DNase I solution (100 Sigma units/ml).
3. Remove 5-μl samples at 20, 40, 60, 120, and 300 sec, adding each to 5 μl of 10 mM EDTA/2% sodium dodecyl sulfate (SDS) at room temperature.
4. Add 1 μl 5 mg/ml proteinase K and incubate 30 min at 37°.
5. Add 90 μl 50 mM Tris-Cl (pH 8) and extract twice with phenol/ chloroform.
6. Precipitate the DNA. Add 2 volumes ethanol, 0.1 volume 3 M sodium acetate (pH 5.2), and incubate in a dry ice–ethanol bath for 20 min. Centrifuge 20 min in an Eppendorf centrifuge. Wash the pellet once with 70% ethanol and dry.
7. Resuspend the pellet in 10μl gel-loading buffer [80% (v/v) deionized formamide, 50 mM Tris borate (pH 8.8), 1 mM EDTA, 0.1% (w/v) xylene cyanol, 0.1% (w/v) bromphenol blue].
8. Separate fragments by electrophoresis in a 16% polyacrylamide gel under denaturing conditions. Unique-sequence DNA can simply be separated on a sequencing gel,[3] while mixed-sequence DNA is best fractionated on a modified acrylamide gel containing a high methylenebisacrylamide : acrylamide ratio.[4]
9. Autoradiograph the gel.

Figure 1 shows an example of an experiment employing such a protocol on native, mixed-sequence nucleosome cores.

[2] L. C. Lutter, *J. Mol. Biol.* **124,** 391 (1978).
[3] T. Maniatis, E. F. Fritsch, and J. Sambrook, "Molecular Cloning." Cold Spring Harbor Lab., Cold Spring Harbor, New York, 1982.
[4] L. C. Lutter, *Nucleic Acids Res.* **6,** (1979).

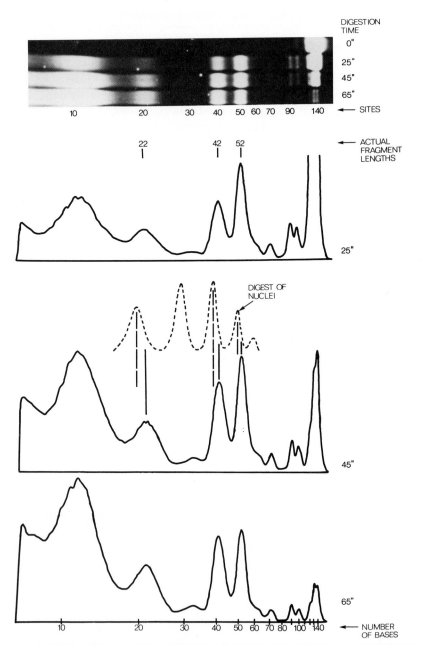

FIG. 1. DNase I digestion of 5′ end-labeled nucleosome cores. Nucleosome cores labeled with ³²P at the 5′ ends of their DNA were digested with DNase I for the times shown, after which the extracted DNA was fractionated by electrophoresis in a 16% acrylamide gel (36 cm long) under denaturing conditions. At the top is shown an autoradiograph of the gel, with the length of DNase I digestion in seconds shown at the right. Electrophoresis is from right to left.

DNase II Digestion

DNase II digestion[5] can be carried out in a manner similar to that described for DNase I by substituting 10 × DNase II digestion buffer [300 mM sodium cacodylate (pH 6.0), 1 mM EDTA] for the 10 × DNase I digestion buffer and DNase II solution (2000 Sigma units/ml) for the DNase I solution. These buffer conditions (sodium cacodylate, pH 6.0) can also be used for DNase I digestion by adding MgCl$_2$ to a final concentration of 5 mM and increasing the final DNase I concentration 10-fold. Unlike DNase I, DNase II does not require divalent cation, but it is active in the presence of MgCl$_2$, so digestion under identical ionic conditions[5] by the two nucleases can be performed using a 10 × buffer of 300 mM sodium cacodylate (pH 6.0) plus 50 mM MgCl$_2$. The range of time points suggested for the given enzyme concentrations should provide usable digestion data, that is, significant digestion, but not so much that sites far from the end-label are not represented. If usable data are not obtained, the amount of enzyme should be adjusted appropriately.

Quantitation of Digestion Data

The original autoradiograph of the polyacrylamide gel should provide immediate evidence for site-specific cutting by the characteristic gaps in the banding pattern. For example, the autoradiograph at the top of Fig. 1 shows immediately that sites which are 10, 20, 40, and 50 bases from the 5' end are cut well, but sites which are 30, 60, and 80 bases from the 5' end are not. Further information can be obtained from a quantitative representation of the degree of cutting. A pseudo-first-order rate constant for the cleavage of each site can be determined.[2]

The processing of the data for such a quantitative analysis involves two features: (1) a correction for cuts that occur between the site of interest and the radioactive end-label and (2) a "background" subtraction of the cutting that occurs in bare DNA when analysis of a nucleosome on a unique sequence is carried out. The first correction becomes very important for a sample that is substantially digested with a nuclease. When significant

[5] L. C. Lutter, *Nucleic Acids Res.* **9**, 4251 (1981).

Below are shown densitometer tracings of the channels of the three digested samples. At the bottom is indicated a scale specifying the number of bases from the 5' end. The number of cuts per strand for 25, 45, and 65 seconds of digestion are 1.76, 2.67, and 3.49, respectively. The dashed trace in the center indicates the relative mobilities of bands produced from a DNase I digestion of nuclei. (From Ref. 2.)

A   TIME

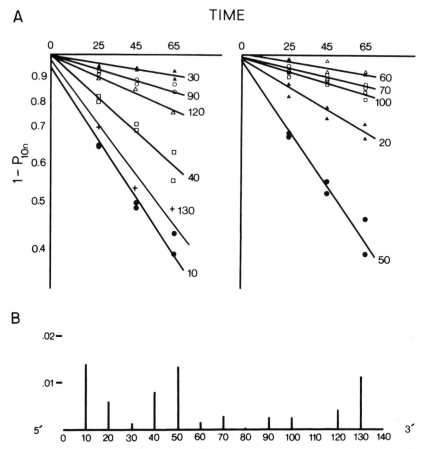

FIG. 2. Kinetic analysis of corrected DNase I digestion data. The fraction of strands left uncut at a given site $(1 - P_{10n})$ was determined by correcting the data in Fig. 1 for underrepresentation of cutting at sites far from the 5′ end. Corrected data were then plotted (A) for each time point for each site. Sites (cf. Fig. 1) are indicated to the right of each plot. (B) The pseudo-first-order rate constant for the attack at each site is represented as a bar graph. The height of a bar is proportional to the slope of the respective line in (A) and provides a quantitative representation of exposure of that site to DNase I.[2]

cutting has occurred, for example, greater than one cut per strand on the average, cutting at a site far from the end-label will not be scored on strands where a second cut has occurred between the site and the end-label. Correction for this underrepresentation involves an estimation of cutting at the site by dividing the radioactivity in the band corresponding to the site n by the sum of the radioactivity in the bands representing only fragments equal to and larger than that of site n, that is, not by dividing by the *total* radioactivity in the channel. This correction procedure has been described in detail.[2]

Once the data have been corrected, a pseudo-first-order rate constant for DNase I attack at each site can be determined from the appropriate graphic analysis of the data. Such an analysis for the results in Fig. 1 is shown in Fig. 2A. The rate constant for each of the various sites was determined from the slope of the respective line in Fig. 2A, and these constants are shown graphically in Fig. 2B. Figure 2B provides a quantitative representation of the degree of exposure of each site in the nucleosome to DNase I.

A second correction, which is essential for the qualitative treatment of digestion data on a nucleosome bound to a specific sequence, involves subtracting the "background" cutting of the bare DNA.[6] This is important because each nuclease shows some degree of sequence specificity, that is, all phosphodiester bonds in bare DNA are not cut with equal probability by a nuclease. Thus, a correction for the sequence preference of the nuclease should be made in the digestion data of the nucleosome-bound DNA. This has been accomplished by determining the probability of cutting of a given phosphodiester bond in bare DNA (from a separate nuclease digestion time course of bare DNA) and subtracting that from the probability of cutting of that bond in nucleosome-bound DNA. Details of the correction procedure are described elsewhere.[6]

[6] H. R. Drew, and A. A. Travers, *J. Mol. Biol.* **186**, 773 (1985).

[14] Analysis of Hypersensitive Sites in Chromatin

By CARL WU

Introduction

Although the precise cause and effect of the phenomenon of nuclease hypersensitivity in chromatin remains to be fully elucidated, the identification of nuclease hypersensitive sites in cellular chromatin is useful as a marker of nonnucleosomal organization that could also reflect an underlying sequence-specific protein–DNA interaction. In our experience with the *Drosophila* heat-shock gene system, initial crude mapping of DNase I hypersensitivity in noninduced cells localized some of these sites to the heat-shock promoter regions.[1] Higher resolution mapping next revealed a

[1] C. Wu, *Nature (London)* **286**, 854 (1980).

METHODS IN ENZYMOLOGY, VOL. 170

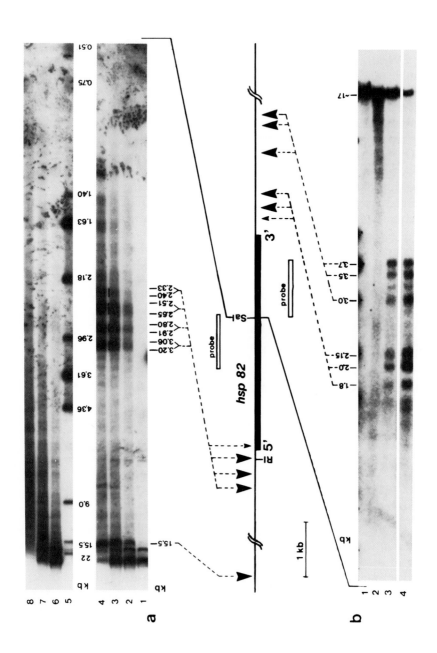

30-bp region over the TATA box sequence that was resistant to DNase I cleavage amid the several hundred base pair (bp) hypersensitive sites. The presence of specific protein binding at the TATA sequence was subsequently confirmed by an exonuclease (ExoIII) protection technique that also revealed the binding of a second factor, the heat-shock activator protein, in response to heat-shock stimulation of the cell.[2] The binding site of heat shock activator protein was located over the heat-shock regulatory element (HSE), about 20 bp upstream of the TATA box.

The above sequence of experiments can serve as a general strategy for surveying the chromatin organization of a eukaryotic gene or region of special interest, with the intention of further analyzing those regions in which the break in nucleosomal arrangement is reflected by a change in nuclease sensitivity. Such an *in vivo* analysis is an important study of gene regulation as it provides information on the protein–DNA interactions that are actually occurring in the cell nucleus.

Crude Survey

Isolated nuclei (or cell homogenates) are partially digested with DNase I, and the DNA is purified (there should be one cut on average within the selected region). The purified DNA is then digested completely with one (or two) restriction enzymes whose sites of cleavage bracket the region of interest. DNA fragments are separated by agarose gel electrophoresis and blotted to nitrocellulose or another suitable membrane. Choosing a short, P[32]-labeled hybridization probe that is homologous to sequences

[2] C. Wu, *Nature (London)* **309**, 229 (1984).

FIG. 1. (a) Autoradiogram showing the partial DNase I cuts in chromatin upstream from the *Sal*I site in the hsp82 transcription unit. *Sal*I was used for secondary cleavage; the DNA fragments (10 µg per lane) were sized on a 40-cm, 1% agarose gel. Lane 1, *Sal*I digest on purified Schneider Line 2 (SL2) DNA. (A minor polymorphism of the upstream *Sal*I site is observed.) Lanes 2–4, DNase I digests of SL2 cell nuclei in digestion buffer (1) [similar to nuclear buffer (see text) but contains 3 mM MgCl$_2$, 50 µM CaCl$_2$] with DNase I at 0.5, 1, and 2 units/ml, respectively, for 3 min at 25°. Prominent cleavages are noted by arrows. Lane 5, Marker DNA. Lanes 6–8, DNase I digests on naked DNA: DNase I at 0.025, 0.05, and 0.1 unit/ml was reacted with purified genomic DNA at 0.4 mg/ml for 1 min at 25°. (b) Partial DNase I cuts in chromatin downstream from the *Sal*I site. The DNA fragments were sized on a 0.8% agarose gel. Lane 1, *Sal*I digest of SL2 cell DNA. Lane 2, DNase I digest of naked DNA, as in (a), lane 6. Lanes 3 and 4, DNase I digests of SL2 cell nuclei, at a final DNase I concentration of 6 and 8 units/ml, respectively. This experiment was separate from that in (a), lanes 3 and 4; the overall extents of cleavage are comparable. Two other transcription units not shown lie near the hypersensitive sites on the 5' and 3' side of the hsp82 gene [D. O'Connor and J. T. Lis, *Nucleic Acids Res.* **9**, 5075 (1981); R. Blackman and M. Meselson, *J. Mol. Biol.* **188**, 499 (1986)]. (From Ref. 1.)

abutting or nearly abutting one end of the restriction fragment will cause the display of only one of the pair of specific subfragments resulting from a single DNase I cut—the subfragment that overlaps the hybridization probe. The length of the displayed subfragment locates the position of the DNase I cut relative to that restriction site, and the position of each DNase I cut in the specific restriction fragment population is likewise mapped by measuring the lengths of multiple subfragments displayed on the autoradiogram. This technique is very powerful as it allows a survey of the relative nuclease sensitivity of many kilobase pairs of chromatin in a single gel lane. Figure 1 shows DNase I hypersensitive site mapping over ˊ ~40 kbp region around the *Drosophila* hsp82 gene. Figure 2 shows tissue-specific hypersensitive sites around the rat preproinsulin II gene. By analogy to a similar principle, applied previously to mapping the cleavage positions on a P^{32}-end-labeled DNA fragment, the procedure has been referred to as indirect end-labeling.[1] Nedospasov and Georgiev[3] independently developed the same procedure to map the nuclease cleavage sites in the SV40 minichromosome (see Nedospasov *et al.,* [21], this volume).

Fine Mapping

When the sites of DNase I hypersensitivity have been localized at low resolution, a higher resolution analysis can be performed by selecting a new restriction site as point of reference (and a new hybridization probe) such that the cleaved subfragment is of a length (0.5–1.5 kbp) that is maximally resolvable on long agarose gels. (For resolution of shorter DNA fragments, denaturing acrylamide gels may be utilized; the electrophoresed DNA can be blotted to a nylon membrane, and the fragments displayed by hybridization to a single-stranded probe according to the genomic footprinting method of Church and Gilbert.[4])

Inspection of the DNase I hypersensitive site at this level of resolution for microregions of enhanced resistance (crude DNase I footprints) should suggest sites of specific protein binding that could be confirmed independently by the ExoIII protection assay. The positions of these microregions of resistance can be mapped to within 5–10 bp precision relative to known restriction sites in the DNA sequence by coelectrophoresis on adjacent lanes of genomic DNA cleaved with the restriction enzyme. The DNA sequence underlying the whole hypersensitive site is then studied for possible restriction sites from which ExoIII may gain entry. Figure 3 shows a

---

[3] S. A. Nedospasov and G. P. Georgiev, *Biochem. Biophys. Res. Commun.* **92,** 532 (1980).
[4] G. Church and W. Gilbert, *Proc. Natl. Acad. Sci. U.S.A.* **81,** 1991 (1984).

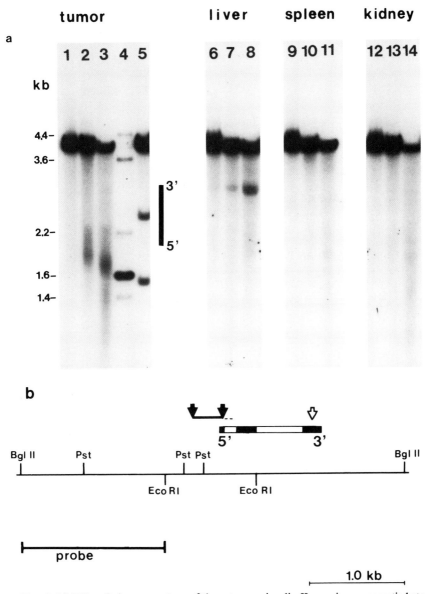

FIG. 2. (a) DNase I cleavage pattern of the rat preproinsulin II gene in a pancreatic beta cell tumor and other rat tissues. Purified nuclei (~ 500 μg DNA/ml) were digested in digestion buffer (1) with DNase I at the following concentrations: tumor, 0, 8, 16 units/ml; liver and kidney, 0, 16, 32 units/ml; spleen, 0, 32, 64 units/ml for 3 min at 25°. DNA was purified and digested to completion with *Bgl*II. Lane 5 is a partial *Eco*RI digest of *Bgl*II-cut (spleen) DNA. Thirty-five micrograms of DNA from each sample was electrophoresed on a 25-cm-long, 1% agarose gel. (b) Restriction map of the rat preproinsulin II gene and flanking sequences. Filled bars are exons, open bars are introns. The hypersensitive site in the beta cell tumor is located between the filled arrows; the open arrow indicates the hypersensitive site found only in liver. [From C. Wu and W. Gilbert, *Proc. Natl. Acad. Sci. U.S.A.* **78,** 1577 (1981).]

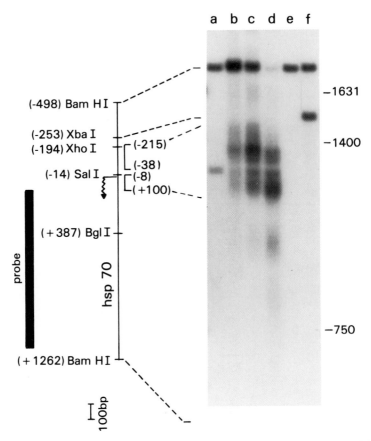

Fig. 3. DNase I hypersensitivity at the 5′ end of the noninduced hsp70 gene. Nuclei from SL2 cells were isolated and reacted in digestion buffer (1) with DNase I at 6, 8, and 4 units/ml (a separate reaction), respectively (lanes b–d). The DNA was purified, restricted with *Bam*HI, electrophoresed at 10 μg/lane on a 40-cm-long, 1.4% agarose gel, blotted, hybridized to probe, and autoradiographed. For markers, SL2 DNA was restricted partially with *Sal*I, *Bam*HI, and *Xba*I, respectively, followed by complete *Bam*HI restriction and processing as above (lanes a, e, and f). (From Ref. 2.)

fine mapping analysis of the DNase I hypersensitive site over the hsp70 gene promoter. The DNase I cleavages in chromatin span a broad region around the 5′ end of the hsp70 gene, and some suppression and enhancement is now clearly evident (lanes b–d). The hsp70 sequences hypersensitive to DNase I in chromatin extend from positions $+100$ to -8 and from -38 to -215, with a peak at -93. The 30-bp region is between, from -8

to -38, is relatively resistant to DNase I cleavage. This resistant site provided the first clue to *in situ* protein binding over the TATA box.

ExoIII Protection

A schematic diagram of the *in situ* ExoIII protection assay is shown in Fig. 4A. Since ExoIII requires a free end or nick in order to proceed with digestion, a cut has to be introduced in chromatin within the hypersensitive site, preferably with a single-cut restriction enzyme. The position of the initial cut should be located far enough away from the presumptive ExoIII block (about 40 bp or more for analysis by agarose gels) such that the resulting subfragment is electrophoretically separated from the parental subfragment. In case no appropriate restriction site is available, one can use general nucleases, such as DNase I and micrococcal nuclease, as the means to initiate ExoIII digestion. After the initial cleavage and ExoIII digestion, the DNA is purified, and the single-stranded tails are trimmed with S_1 nuclease. Upon further cleavage with flanking restriction enzymes, it is possible to determine the relative start and stop positions of ExoIII by indirect end-labeling of the subfragments produced from endo- and exonucleolytic cleavages. The other border of the protein binding site is determined by convergent ExoIII digestion from an entry point on the other side and the use of a new hybridization probe.

Figure 4B shows an application of the *in situ* ExoIII protection technique to the *Drosophila* hsp70 gene prior to heat-shock induction. The initial cut is made by each of four different endonucleases: DNase I, micrococcal nuclease, a Ca^{2+},Mg^{2+}-stimulated endogenous nuclease, and the restriction enzyme *Xho*I, which cleaves at position -194 inside the distal boundary of the DNase I hypersensitive site. ExoIII is added simultaneously in increasing amounts, except for micrococcal nuclease, which is applied first and inactivated with EGTA without affecting ExoIII added subsequently. The purified and S_1-trimmed DNA is restricted with *Bam*HI and *Xba*I, and the relative positions of the endo- and exonuclease cleavages are located by indirect end-labeling. Figure 4B shows that no matter which enzyme is used to make the initial cut, an ExoIII-resistant barrier is clearly observed at position -40 under limit ExoIII digestion conditions. To exclude the possibility than an artifactual cleavage is made by S_1 nuclease on the purified DNA or that an artifactual block to ExoIII is presented by the DNA sequence alone, the control experiment (Fig. 4C), using protein-free DNA as substrate, shows that no specific cuts are made by S_1 nuclease on free DNA at the position of blockage in chromatin, nor at any other position in the region analyzed (lane c); also, no impediments against ExoIII digestion caused by the DNA sequence are observed (lane d).

A

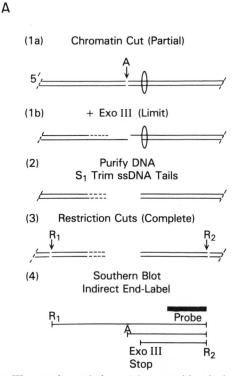

FIG. 4. (A) The ExoIII protection technique. A hypersensitive site in chromatin is drawn as a region of DNA with a hypothetical protein bound to it. (B) ExoIII resistance in chromatin in the noninduced hsp70 gene. Lane a, No enzyme control. SL2 nuclei were mock-incubated in nuclear buffer. Lanes b–e, SL2 nuclei were digested with DNase I, 10 units/ml, and ExoIII for 11 min at 30°. Lanes f–i, 0- to 18-hr embryonic nuclei were digested with micrococcal nuclease, 9 units/ml, 3 min, 25°. The enzyme was inactivated with EGTA to 0.5 mM, ExoIII was introduced, and digestion was continued at 30° for 7 min (lanes f–h) or for 30 min (lane i*). Lanes j–l, SL2 nuclei were digested with an endogenous nuclease stimulated by digestion in nuclear buffer plus 1 mM CaCl$_2$, at 37° for 10 min. CaCl$_2$ was chelated with EGTA to 1.5 mM, ExoIII was added, and digestion was continued at 37° for 15 min. Lanes m and n, SL2 nuclei were digested with *Xho*I at 2000 units/ml and ExoIII for 10 min, 30°. (C, p. 278) S$_1$ nuclease and ExoIII reactions on protein-free DNA. Lane a, *Xho*I plus ExoIII reaction on chromatin for comparison. Lanes b–d, 20 μg purified SL2 DNA was restricted with *Xho*I in 25 μl volume. After the reaction, 25 μl of nuclear buffer was added, and 5 μl of the mixture was removed for secondary *Bam*HI restriction (lane b). Lane c, DNA was purified from 22.5 μl of the mixture. The DNA was treated with 2500 units/ml S$_1$ nuclease, purified, and digested with *Bam*HI. Lane d, the last 22.5 μl of the mixture was digested with 4000 units/ml ExoIII at 30° for 10 min. DNA was purified, S$_1$-trimmed, repurified, and restricted with *Bam*HI. Samples were electrophoresed, blotted, and hybridized to hsp70 probe as in (B). (From Ref. 2.)

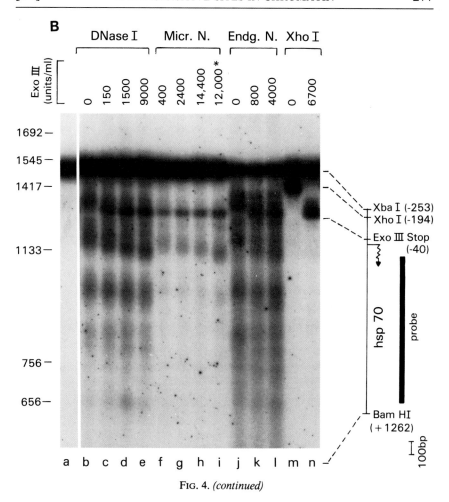

FIG. 4. *(continued)*

Hence, the resistance to ExoIII digestion is due to nucleoprotein structure in chromatin. The ExoIII protection technique has been successfully applied to mammalian systems.[5,6]

In addition to the demonstration of constitutive protein binding at the TATA box of the noninduced hsp70 gene, the ExoIII protection technique also reveals binding of *Drosophila* heat-shock activator protein in heat-shock-induced cells. This binding is most clearly visualized on the hsp82 gene promoter. Figure 5A shows that ExoIII, digesting from an *Eco*RI cut

[5] M. G. Cordingly, A. Tate Riegel, and G. L. Hager, *Cell* **48**, 261 (1987).
[6] I. Kovesdi, R. Reichel, and J. R. Nevins, *Science* **231**, 719 (1986).

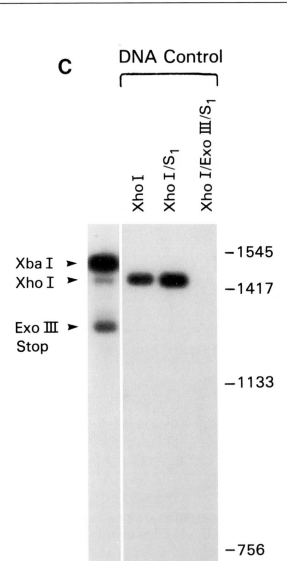

FIG. 4. *(continued)*

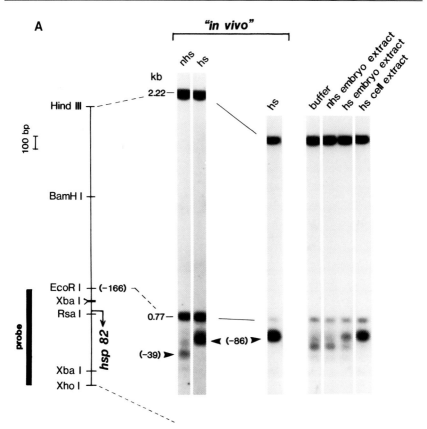

FIG. 5. (A) *In vivo* and *in vitro* binding of heat-shock activator protein. For *in vivo* experiments, purified nuclei from 2- to 7-hr non-heat-shocked and heat-shocked (36°, 15 min) *Drosophila melanogaster* embryos were digested with *Eco*RI and ExoIII; the DNA was purified, trimmed with S₁ nuclease, repurified, digested with *Hin*dIII and *Xho*I, electrophoresed, blotted to nitrocellulose, and hybridized with probe. For *in vitro* experiments, nuclei of non-heat-shocked embryos were incubated with the indicated extracts before digestion with *Eco*RI and ExoIII. Samples on the left were electrophoresed on a 40-cm-long gel, those on the right on a 25-cm-long, 1.4% agarose gel. (B, p. 280) *Xba*I cleavage of hsp82 gene chromatin. Isolated nuclei from normal and heat-shocked embryos were digested with 1400 units/ml *Xba*I at 35° for 15 min. The purified DNA was restricted with *Bam*HI and *Xho*I, electrophoresed, and hybridized to probe. (From Ref. 7.)

in chromatin at position − 166, encounters a new barrier at − 86, the distal border of the hsp82 HSE in heat-shock-induced *Drosophila* embryos. Furthermore, the binding can be reconstituted *in vitro* by incubating nuclei of nonshocked embryos with extracts of heat-shocked embryos or SL2 cells before performing the restriction enzyme and *Exo*III reactions.[7]

[7] C. Wu, *Nature (London)* **311**, 81 (1984).

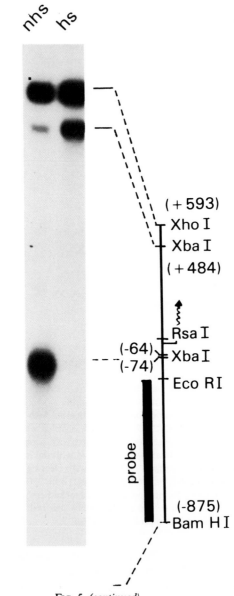

Fig. 5. *(continued)*

Restriction Endonuclease Protection

When a region that is protected from DNase I or ExoIII cleavage happens to include a site for a restriction endonuclease, it is also possible to employ accessibility to that endonuclease as a probe of protein binding. For example, Fig. 5B shows that two *Xba*I sites (positions -64 and -74) located within the heat-shock-activator binding site, normally hypersensitive to *Xba*I cleavage, are almost completely protected from *Xba*I cleavage in chromatin of heat-shock-induced *Drosophila* embryos.[2] In contrast, the *Xba*I site located within the hsp82 transcription unit at position $+484$ becomes more sensitive to *Xba*I digestion, consistent with the overall nuclease sensitivity of transcriptionally active sequences.

Limitations of the ExoIII Protection Technique

There are strong and weak blocks to ExoIII digestion in chromatin, and it is possible that some specific binding proteins may not present any impediment to ExoIII digestion. *In vitro* DNA-binding studies with heat-shock activator protein indicate that the binding site on the hsp82 gene, which contains three overlapping HSEs, is the highest affinity site among heat shock genes; when occupied *in vivo*, it is impenetrable to a limit ExoIII reaction. HSEs of other heat shock genes that have weaker binding *in vitro* show weaker resistance to ExoIII *in vivo*. An impenetrable ExoIII block in chromatin will mask the presence of other blocks downstream, in the direction of ExoIII digestion, from a restriction cut unless the positions of ExoIII entry are randomized, for example, by using DNase I.

Methods

Isolation of Nuclei

For convenience, a single "nuclear buffer" is used for isolating nuclei and for nuclease digestion reactions. The buffer, a deviant from the original Buffer A of Hewish and Burgoyne,[8] promotes nuclear and chromatin stability and also supports the activity of common endonucleases, exonucleases, and restriction enzymes. Because of the presence of Mg^{2+} in nuclear buffer, materials that have very high levels of endogenous nuclease may suffer autodigestion; if this cannot be tolerated, isolation of nuclei in Hewish and Burgoyne's Buffer A is recommended instead.

[8] D. R. Hewish and L. A. Burgoyne, *Biochem. Biophys. Res. Commun.* **52,** 504 (1973).

Reagents

Nuclear buffer

	Final concentration	Stock	Combine
KCl	60 mM	1.5 M	10 ml
NaCl	15 mM	0.375 M (56 g/500 ml)	10 ml
MgCl$_2$	5 mM	1 M (Fisher, Fair Lawn, NJ: 1 M solution)	1.25 ml
EGTA	0.1 mM	50 mM (3.8 g/ 200 ml final; titrate to pH 7.0 with NaOH)	0.5 ml
Tris-HCl, pH 7.4	15 mM	0.375 M (22.7 g Tris base/500 ml final; titrate with HCl)	10 ml
Dithiothreitol (DTT)	0.5 mM	1 M (3.08 g/ 20 ml)	125 μl
Phenylmethylsulfonyl fluoride (PMSF)	0.1 mM	25 mM [0.87 g/ 200 ml 50% (v/v) ethanol]	1 ml
Sucrose	300 mM	Solid, nuclease free	25.7 g
			H$_2$O to 250 ml

Nuclear buffer/1.7 M sucrose: change sucrose to 1.7 M
Nuclear buffer/5% glycerol: include glycerol to 5%
Procedure. All procedures are performed at 4°.

1. Thaw 5 g dechorionated and frozen *Drosophila* embryos into 50 ml nuclear buffer in a Dounce homogenizer. Stir with a spatula or rod to disperse large clumps.
2. Homogenize with around 10 complete strokes of the loose pestle (Dounce B); beware of squirting!

3. Homogenize with about 10 complete strokes of the tight pestle (Dounce A).
4. Filter homogenate through nylon mesh [Nitex: HD 3-63 (0.0025 inch mesh opening); manufacturer: Tetko, Elmsford, NY; supplier: Sargent Welch, Skokie, IL, special order].
5. Centrifuge filtrate in a Beckman JS13 or Sorvall HB4 rotor, 7000 rpm, 4 min (5000 g_{av}).
6. Aspirate milky supernatant and lipid skin; wipe inner wall of tube with a clean Kimwipe.
7. Resuspend pellet in 9 ml nuclear buffer by vortexing at moderate speed.
8. Layer suspension of nuclei over 25 ml nuclear buffer/1.7 M sucrose; stir interface, leaving buffer undisturbed about 2 cm from tube bottom. (During centrifugation, nuclei can often aggregate at the 1.7 M sucrose interface; to prevent this, layer nuclei over a 25 ml nuclear buffer 0.3 – 1.7 M sucrose gradient before centrifugation.)
9. Centrifuge in a Beckman JS13 rotor at 12,500 rpm, 15 min, or Sorvall HB4 rotor, 11,000 rpm, 20 min.
10. Resuspend nuclear pellet in 2 ml nuclear buffer/5% glycerol by vortexing and pipetting up and down through the constricted tip of a Pasteur pipette.
11. Check concentration by making a 1/100 dilution and counting (in a Petroff–Hausser counter); readjust volume to make 5 × 10^8 nuclei/ml *(Drosophila),* or 1 – 5 × 10^7/ml (mammals).
12. Dispense 100- to 500-μl aliquots into 1.5-ml Eppendorf centrifuge tubes, on ice.

Notes. (a) To obtain nuclei from tissue culture cells, frozen cell pellets are homogenized in 3 – 4 packed cell volumes of nuclear buffer using a Dounce homogenizer (tight pestle) until cell lysis is complete as gauged by microscopic inspection. Continue steps 8 – 11. (b) For mammalian tissues, frozen samples wrapped in aluminum foil are pulverized in a mortar and pestle, chilled on dry ice before homogenization (1 g tissue/25 ml nuclear buffer) in a glass homogenizer with rotating Teflon pestle. Filter the homogenate through a Nitex filter. Pellet the crude nuclei at 3000 – 5000 g for 3 min; resuspend the pellet in 9 ml nuclear buffer. Continue steps 8 – 11. (c) If we assume that the DNA content of a diploid *Drosophila* nucleus is approximately 0.4 pg, then a 100-μl aliquot of nuclei at 5 × 10^8/ml contains 20 μg DNA. If we assume 10 pg DNA/diploid mammalian nucleus, then 100 μl nuclei at 1 – 5 × 10^7/ml contains 100 – 500 μg DNA. (d) We usually use freshly prepared nuclei, although limited experience with *Dro-*

sophila embryo nuclei pelleted in a microfuge for 10 sec and snap frozen in liquid N_2 suggest that hypersensitive chromatin structures remain intact after several years of storage at $-70°$. Stability of nuclear chromatin in storage is best tested for each individual system. (e) In many instances, for example, limited amounts of cell material or difficulty in obtaining a clean nuclear preparation, one could forego the ritual of obtaining purified nuclei and perform nuclease digestion reactions directly on a cell or tissue lysate. Note that higher concentrations of DNase I may be required since crude lysates contain inhibitors of the enzyme. (f) Some bound proteins may have fallen off the DNA by the time nuclei have been purified. At least in one instance, the *in situ* detection of a protein factor on the β-interferon promoter was achieved only by DNase I digestion of a crude cell homogenate; washing nuclei, even in low salt buffers (<50 mM KCl), resulted in loss of the specific DNase I protection pattern.[9] Crude lysates may be used as an adjunct to, if not a substitute for, purified nuclei. Preparation of crude lysates requires only minutes of time, and there is no significant dilution of material—two factors that should preserve the integrity of protein–DNA complexes. A disadvantage of this procedure could be the retention of high endogenous nuclease activity, but the effects of endogenous nuclease can be suppressed by digesting on ice with high levels of DNase I.

Nuclease Digestion

Reagents

DNase I (Worthington, DPRF; other manufacturers probably equivalent): Dissolve in 100 mM NaCl, 10 mM Tris-HCl, pH 7.4, 100 μg/ml BSA at 20 units/μl. Freeze 25-μl aliquots in liquid N_2 and store at $-70°$; the enzyme remains stable for years. Thaw one aliquot and make fresh serial dilutions: 2, 1, $\frac{1}{2}$, $\frac{1}{4}$, $\frac{1}{8}$, $\frac{1}{16}$ units/μl in nuclear buffer plus 0.4 mM CaCl$_2$.

Micrococcal nuclease (Worthington): Dissolve in 10 mM NaCl, 10 mM Tris-HCl, pH 7.4, 100 μg/ml BSA at 16 units/μl. Freeze 50-μl aliquots in liquid N_2 and store at $-70°$; the preparation is stable for years. Thaw an aliquot and make fresh serial dilutions 4, 2, 1, $\frac{1}{2}$, $\frac{1}{4}$, $\frac{1}{8}$ units/μl in nuclear buffer plus 0.4 mM CaCl$_2$.

Restriction enzyme (All manufacturers): call for highest specific activity available.

Exonuclease III (Boehringer-Mannheim, Pharmacia, Bethesda Research Laboratories, New England Biolabs, and others): activity can

---

[9] K. Zinn and T. Maniatis, *Cell* **45**, 611 (1986).

fluctuate from one lot to the next. Use at 100–200 manufacturer's units/μl (note that different manufacturers have somewhat different unit definitions).

Endonuclease Digestions

	500 μl volume	100 μl volume
Nuclei	475 μl	95 μl
DNase I	25 μl	5 μl

React 3 min in a 25° bath.

Nuclei	475 μl	95 μl
Micrococcal nuclease	25 μl	5 μl
0.1 M CaCl$_2$	2 μl	0.4 μl

React 3 min in a 25° bath.

Restriction Endonuclease Digestion

Nuclei	100 μl
Restriction enzyme	5–10 μl
0.2 M MgCl$_2$	1 μl

React 10–30 min in a 30° or 37° bath.

Restriction Endonuclease/Exonuclease III Digestions

Nuclei	100 μl
Restriction enzyme	5–10 μl
ExoIII	1–5 μl
0.2 M MgCl$_2$	1 μl

React 10–30 min in a 30° or 37° bath.

Stop all nuclease digestions with one volume stop solution (20 mM EDTA, 1% SDS).

Notes. (a) The DNase I and micrococcal nuclease concentrations recommended should provide an adequate range (20–0.5 kbp) of cleaved fragments for analysis, although individual systems will differ with respect to extent of cleavage at a particular nuclease concentration. The overall size of cleaved fragments for a DNase I hypersensitive analysis on agarose gels should be 20–5 kbp; occasionally one may have to interpolate between the recommended nuclease concentrations in order to attain optimal fragmentation. For each system, the digestion conditions, once set, are quite reproducible, and it may suffice to perform 3 or fewer digestion points in subsequent experiments. (b) The time and temperature of digestion can be varied as desired. To suppress high levels of endogenous nuclease, use more exogenous enzyme at a lower temperature; even digest on ice. For restriction enzymes, one has to balance the specific activity of commercial preparations with cost and the effects of endogenous nucleases: 15 min at 30° is a compromise. Do not underestimate the good use to which endogenous endonuclease and exonuclease can be (and has been) put. (c) Owing to the relatively lower specific activities of commercial sources of *Bal*31 and λ exonuclease, we have not used these enzymes extensively for exonuclease protection; other workers have reported such studies.[10]

In Vitro Binding of Nuclear Factors to Isolated Nuclei

Procedure

1. Centrifuge 100 μl purified nuclei in suspension for 10 sec in an Eppendorf microfuge. Aspirate supernatant.
2. Add 100 μl crude nuclear extract (see note below) or column fraction in nuclear buffer.
3. Mix homogeneously, pipetting up and down in a 0–200 μl pipette tip (yellow tip).
4. Incubate at 25° for 10–20 min for the binding to occur, longer if necessary.
5. Repellet nuclei by centrifugation for 10 sec in microfuge.
6. To the nuclear pellet add 100 μl nuclear buffer, 5–10 μl restriction enzyme, 1–5 μl ExoIII, and 1 μl 0.2 M MgCl$_2$. Repeat step 3.
7. React 15 min at 30°.
8. Stop with 1 volume of stop solution.

Notes. (a) Prepare nuclear extract following the protocol of Dignam *et al.*[11] or Wu.[7] (b) Use DNase I, micrococcal nuclease, or restriction enzyme alone if specific protein binding can be revealed by protection from these nucleases.

DNA Purification

Procedure

1. To nuclear lysate add proteinase K (1 mg/ml stock) to 50 μg/ml, and digest overnight at 37°.
2. For volumes up to 500 μl, add 1 volume of an equal mixture of phenol and chloroform : isoamyl alcohol (24 : 1, v/v). Shake and

[10] W. A. Scott, C. F. Walter, and B. L. Cryer, *Mol. Cell Biol.* **4**, 604 (1984).
[11] A. D. Dignam, P. L. Martin, B. S. Shastry, and R. G. Roeder, this series, Vol. 101, p. 582.

vortex to form emulsion. Highly concentrated and viscous DNA solutions may require energetic shaking to emulsify.

3. Centrifuge 1–5 min in a microfuge to separate phases.
4. Remove lower organic phase by pipette.
5. Centrifuge 1–5 min to compact the protein interface.
6. Transfer aqueous phase to a fresh Eppendorf tube; viscous solutions can be transferred using a $0-200 \mu l$ (yellow) or 1 ml (blue) micropipette tip with the end cut off to create a wider opening.
7. At this stage, the overall extent of cleavage by DNase I or micrococcal nuclease can be gauged by electrophoresing a few microliters of the aqueous phase on an agarose gel and staining the DNA with ethidium bromide; only the tubes that have the desired levels of digestion are purified further.
8. Repeat the organic extraction twice.
9. Add 0.1 volume of 2 M sodium acetate.
10. Add 2.5 volumes absolute ethanol.
11. Chill in a dry ice–ethanol bath for 5 min.
12. Centrifuge the nucleic acid precipitate in a microfuge for 10 min at 4°. The pellet should be visible.
13. Aspirate the supernatant carefully.
14. Add 1.4 ml 70% ethanol; cap the tube and invert to mix.
15. Centrifuge briefly; pipette off the supernatant.
16. Centrifuge briefly; remove remaining supernatant with micropipette.
17. Air dry for approximately 15 min (do not over dry, especially under vacuum — high molecular weight DNA will take forever to dissolve).
18. Dissolve the pellet in 10 mM Tris, pH 7.4, 5 mM NaCl, 1 mM EDTA at room temperature. To assist dissolution, the solution can be vortexed, even vigorously, or kept at 37° for 1 hr.

S_1 Nuclease Digestion

The S_1 nuclease reaction is required for removal of single-stranded DNA resulting from ExoIII digestion. Reaction volumes can be scaled up or down as desired.

Reagents

S_1 nuclease reaction buffer
    50 mM sodium acetate, pH 4.5
    1 mM ZnSO$_4$
    250 mM NaCl

S_1 nuclease (Boehringer-Mannheim). Dissolve the lyophilized enzyme at 500 units/μl in:

40 mM sodium acetate, pH 4.5
1 mM $ZnSO_4$
50 mM NaCl

The S_1 solution is stable for many months at 4°.

Procedure

1. In an Eppendorf tube combine 20 μl DNA ($\sim$ 1 μg/μl), 200 μl S_1 nuclease reaction buffer, and 1 μl S_1 nuclease (500 units/μl).
2. Incubate 15 min, 30°.
3. Stop with 10 μl 1 M Tris-HCl, pH 8.0, 6 μl 0.2 M EDTA.
4. Immediately start organic extraction of the DNA as for DNA Purification, Steps 2–6 and 9–18.

Restriction Endonuclease Digestion of Purified DNA

The restriction enzyme digestions are done in the smallest convenient volume to facilitate loading on gels directly without further purification.

Reagents

10× low, medium, and high salt restriction buffers according to the CSH Molecular Cloning Manual, p. 453.[12]
RNase A (Sigma): 1 mg/ml in 10 mM Tris-HCl, pH 7.4
RNase T_1 (Sigma): 40 μg/ml in 10 mM Tris-HCl, pH 7.4
Mix RNase A and RNase T_1 1:1, heat to 80° for 30 min to inactivate DNase, and freeze aliquots at −20°
Restriction enzyme stop solution
20% glycerol
10% Ficoll-400
0.1 M EDTA
0.1% bromphenol blue
0.1% xylene cyanol

Procedure

1. For a typical restriction enzyme digestion combine 10 μl DNA ($\sim$ 1 μg/μl), 1.4 μl 10× restriction buffer, 1 μl 1 μg/μl BSA, nuclease free, 1–2 μl restriction enzyme, and 0.1 μl RNase A + T_1. Add H_2O to make 14 μl volume.
2. Mix well by pipetting up and down about 10 times in a 0–200 μl micropipette tip.

[12] T. Maniatis, E. F. Fritsch, and J. Sambrook, "Molecular Cloning." Cold Spring Harbor Lab., New York, 1982.

3. Incubate at the appropriate temperature, usually 37°, for 4 hr; use more or less time for digestion, depending on activity of the enzyme.
4. Stop digestion by adding 4 μl of restriction enzyme stop solution. Vortex to mix.
5. Load sample directly on an agarose gel.

Agarose Gel Electrophoresis/Southern Blotting/Labeling Probe/Filter Hybridization

Consult the molecular cloning manual by Maniatis et al.[12] for these procedures. For agarose gel electrophoresis, we have used agarose concentrations up to 1.4% and both Tris–acetate and Tris–borate buffers. We prefer Tris–acetate for better resolution of DNA fragments of 1–1.5 kbp and Tris–borate for fragments of 0.3–0.8 kbp. (FMC Corporation, Rockland, ME, has a NuSieve agarose product that can resolve DNA fragments of 10 to several hundred base pairs.) We run normal 25-cm-long gels that can be cast in commercial molds. For clear separation of fragments differing in size by 10–20 bp, we use 40-cm-long gels cast on a thick glass plate taped all around the sides and electrophoresed in a homemade submarine apparatus or connected by paper wicks to individual buffer tanks. The relevant section of the gel is cut out after electrophoresis and blotted. We use nitrocellulose for blotting by the conventional Southern procedure, but newer membranes and procedures can be more advantageous.[13] We have previously used the conventional nick-translation method to label the hybridization probe but have now switched to the oligolabeling technique using random primers that results in extremely high specific activity, above 10^9 dpm (disintegrations per minute)/μg.[14]

Alternative Technologies

High resolution analysis of protein–DNA interactions in vivo and in nuclei can also be obtained by mapping of nuclease or chemical cleavage sites using genomic footprinting,[4,15] primer extension,[16–18] or S_1 mapping procedures.[19,20]

[13] K. C. Reed and D. A. Mann, Nucleic Acids Res. 13, 7207 (1986).
[14] A. P. Feinberg and B. Vogelstein, Anal. Biochem. 132, 6 (1983).
[15] P. B. Becker and G. Schutz, Genetic Engineering 10, 1 (1988).
[16] J. D. Gralla, Proc. Natl. Acad. Sci. U.S.A. 82, 3078 (1985).
[17] J. M. Huibregtse and D. R. Engelke, Gene 44, 151 (1986).
[18] R. L. Buchanan and J. D. Gralla, Mol. Cell Biol. 7, 1554 (1987).
[19] P. D. Jackson and G. Felsenfeld, Proc. Natl. Acad. Sci. U.S.A. 82, 2296 (1985).
[20] H. Weintraub, Mol. Cell Biol. 5, 1538 (1985).

[15] Application of Nucleases to Visualizing Chromatin Organization at Replication Forks

By MICHAEL E. CUSICK, PAUL M. WASSARMAN,
and MELVIN L. DEPAMPHILIS

Introduction

Semiconservative replication of double-stranded DNA produces forks in the parental template strands that generally contain newly replicated DNA on both arms, although the proximity of nascent DNA to the fork is difficult to assess, and DNA synthesis may be discontinuous (i.e., repeated initiation of Okazaki fragments) on one or both arms.[1] These types of replication forks have been visualized by electron microscopy of simian virus 40 and polyoma virus replicating DNA,[2] and of chromosomal DNA from budding yeast,[3,4] slime mold,[5] fruit fly embryos,[6-10] sea urchin embryos,[10-12] and mammalian cells.[13] Not all replication forks contain nascent DNA on both sides. Adenovirus and mitochondrial DNA replication proceed by displacing one template strand while replicating the other.[1] An even more extreme version of this mechanism may occur in frog embryos where mostly unbranched, single-stranded DNA is found, suggesting that strand separation can sometimes be uncoupled completely from DNA synthesis.[14]

However, although electron microscopy has confirmed that newly rep-

---

[1] A. Kornberg, *in* "DNA Replication." Freeman, San Francisco, California, 1980.

[2] M. L. DePamphilis and P. M. Wassarman, *in* "Organization and Replication of Viral DNA" (A. S. Kaplan, ed.), p. 37–114. CRC Press, Boca Raton, Florida, 1982.

[3] C. S. Newlon, T. D. Petes, L. H. Hereford, and W. L. Fangman, *Nature (London)* **247,** 32 (1974).

[4] T. D. Petes and C. S. Newlon, *Nature (London)* **251,** 637 (1974).

[5] S. Funderud, R. Andreassen, and F. Haugli, *Nucleic Acids Res.* **6,** 1417 (1979).

[6] H. J. Kriegstein and D. S. Hogness, *Proc. Natl. Acad. Sci. U.S.A.* **71,** 135 (1974).

[7] A. B. Blumenthal, H. J. Kriegstein, and D. S. Hogness, *Cold Spring Harbor Symp. Quant. Biol.* **38,** 205 (1974).

[8] C. S. Lee and C. Pavan, *Chromosoma* **47,** 429 (1974).

[9] V. A. Zakian, *J. Mol. Biol.* **108,** 305 (1976).

[10] C. T. Baldari, F. Amaldi, and M. Buongiorno-Nardelli, *Cell* **15,** 1095 (1978).

[11] M. P. Kurek, D. Billig, and P. J. Stambrook, *J. Cell Biol.* **81,** 698 (1979).

[12] M. Shioda and T. Shiroya, *Zool. Sci. (Tokyo)* **3,** 899 (1986).

[13] M. S. Valenzuela, G. C. Mueller, and S. Dasgupta, *Nucleic Acids Res.* **11,** 2155 (1983).

[14] M. F. Gaudette and R. M. Benbow, *Proc. Natl. Acad. Sci. U.S.A.* **83,** 5953 (1986).

Copyright © 1989 by Academic Press, Inc.
All rights of reproduction in any form reserved.

licated eukaryotic DNA is organized into chromatin,[15-19] microscopy has not been able to: (1) determine whether DNA synthesis occurs before or after the parental DNA template is assembled into chromatin; (2) determine whether Okazaki fragments are contained within nucleosomes; (3) measure the distance from the 5' to 3' ends of newly synthesized DNA to the first nucleosome on each side of the fork that contains newly replicated DNA; or (4) distinguish the structure of newly replicated chromatin from the structure of nonreplicated chromatin.

Therefore, these aspects have been investigated by evaluating the accessibility of nascent DNA in newly replicated chromatin to digestion by various nucleases (nuclease specificities are summarized in Table I). This approach has been applied most extensively to purified simian virus 40 (SV40) replicating chromosomes, and therefore results with SV40 are presented in detail to illustrate the various techniques. Only nonspecific endonucleases have been used to analyze replicating cellular chromatin, but these results are consistent with the model proposed for SV40. Blotting–hybridization techniques using sequence-specific probes to detect digestion products should allow all of these techniques to be applied *in situ* to any chromosome.

Structure of Chromatin at SV40 Replication Forks

SV40 replicates in nuclei of monkey or human cells as a small (5.2 kilobase pairs), circular, double-stranded DNA chromosome whose histone composition and nucleosome structure are indistinguishable from host chromatin (reviewed in Refs. 2, 20, and 21). The only protein provided by the virus is large tumor antigen that is required for initiation of viral DNA replication and may continue to act as a helicase at replication forks; all remaining parts and labor are provided by the host cell. Replication proceeds bidirectionally from a unique DNA site to create a replication bubble. RNA-primed nascent DNA chains 40 to 290 (typically 135)

[15] C. Cremisi, A. Chestier, and M. Yaniv, *Cold Spring Harbor Symp. Quant. Biol.* **42,** 409 (1978).

[16] M. M. Seidman, C. F. Caron, and N. P. Salzman, *Nucleic Acids Res.* **5,** 2877 (1978).

[17] S. L. McKnight and O. L. Miller, *Cell* **12,** 795 (1979).

[18] S. Busby and A. H. Bakken, *Chromosoma* **79,** 85 (1980).

[19] J. M. Sogo, H. Stahl, T. Koller, and R. Knippers, *J. Mol. Biol.* **189,** 189 (1986).

[20] M. L. DePamphilis and M. K. Bradley, *in* "The Papovaviridae" (N. P. Salzman, ed.), p. 99. Plenum, New York, 1986.

[21] M. L. DePamphilis, *in* "Molecular Aspects of the Papovaviruses" (Y. Aloni, ed.), pp. 1–40. Nijhoff, The Hague, 1987.

TABLE I

NUCLEASE SPECIFICITY ON PURIFIED DNA AND CHROMATIN SUBSTRATES[a]

Nuclease	DNA substrate	Chromatin substrate	Refs.
Micrococcal (staphylococcal) nuclease (MNase)	Prefers ssDNA to dsDNA, 10 : 1; prefers AT-rich sequences; cuts dsDNA leaving 5′ end extended by 2 nt	Rapidly degrades internucleosomal and nonnucleosomal DNA into nucleotides leaving 145-bp nucleosomal core; core DNA is eventually fragmented	b–d
Pancreatic deoxyribonuclease (DNase I)	Prefers dsDNA to ssDNA, 3 : 1; no apparent sequence preference	Rapidly degrades internucleosomal and nonnucleosomal DNA to oligonucleotides; nicks nucleosomal DNA at 10.4-bp intervals resulting in dsDNA fragments with 3′ end extended 2 nt	e–i
Spleen deoxyribonuclease (DNase II)	Prefers dsDNA to ssDNA, 2 : 1; no apparent sequence preference	Rapidly degrades internucleosomal and nonnucleosomal DNA; fragments nucleosomal DNA into multiples of 10 nt with 3′ end extended 4 nt	j–m
Escherichia coli exonuclease III (ExoIII)	Degrades dsDNA only; nonprocessive release of mononucleotides beginning at 3′-OH or 3′-PO₄ ends; prefers recessed 3′ end; works equally well at internal nicks or gaps; exhibits RNase H (degrades RNA from RNA–DNA duplex) and 3′-phosphatase activities; pauses at G-rich sequences	Degrades nonnucleosomal DNA from a nick or gap but does not penetrate nucleosomal DNA in intact chromatin; slowly degrades DNA in isolated nucleosome monomers, pausing at 10-nt intervals	n–s
E. coli phage T7 exonuclease (T7 Exo)	Degrades dsDNA only; releases mononucleotides from 5′-OH or 5′-PO₄ termini at ends of linear DNA or at internal nicks or gaps; prefers recessed 5′ ends; exhibits RNase H activity	Degrades nonnucleosomal DNA from nicks or gaps, but does not penetrate nucleosomal DNA in intact chromatin	t–v, q, r
Aspergillus oryzae S₁ nuclease (S₁ nuclease)	Degrades ssDNA only; attacks dsDNA at gaps greater than 5 nt or if negative, superhelical tension promotes "breathing" of easily denatured sequences or noncomplementary base pairs	Degrades ssDNA regions in nonnucleosomal DNA, but not within nucleosomes	w–z, aa
Neurospora crassa endonuclease	Same as S₁	Same as S₁	z, bb

Restriction endonucleases	Cuts dsDNA only at a unique sequence, generally 4–6 bp	Cuts dsDNA in nonnucleosomal or internucleosomal DNA, but not within nucleosomes[cc, dd, ee]

[a] dsDNA, Double-stranded DNA; ssDNA, single-stranded DNA; nick, single phosphodiester bond interruption; gap, small ssDNA region in dsDNA; nt, nucleotides; bp, base pairs.

[b] C. Dingwall, G. P. Lomonosoff, and R. A. Laskey, Nucleic Acids Res. 9, 2643 (1981); B. R. Shaw, T. M. Herman, R. T. Kovacic, G. S. Beaudreau, and K. Van Holde, Proc. Natl. Acad. Sci. U.S.A. 73, 505 (1976).

[c] R. D. Camerini-Otero, B. Sollner-Webb, and G. Felsenfeld, Cell 8, 333 (1976).

[d] E. R. Shelton, P. M. Wassarman, and M. L. DePamphilis, J. Biol. Chem. 255, 771 (1980).

[e] M. Laskowski, in "The Enzymes" (P. D. Boyer, ed.), 3rd Ed., Vol. 4, p. 289. Academic Press, New York, 1971.

[f] M. Noll, Nucleic Acids Res. 1, 1573 (1974).

[g] L. Lutter, J. Mol. Biol. 124, 391 (1978).

[h] L. Lutter, Nucleic Acids Res. 6, 41 (1979).

[i] A. Prunell, R. D. Kornberg, L. Lutter, A. Klug, M. Levitt, and F. H. C. Crick, Science 204, 855 (1979).

[j] G. Bernardi, in "The Enzymes" (P. D. Boyer, ed.), 3rd Ed., Vol. 4, p. 271. Academic Press, New York, 1971.

[k] W. Hörz, F. Miller, G. Klobeck, and H. G. Zachau, J. Mol. Biol. 144, 329 (1980).

[l] B. Sollner-Webb, W. Melchior, Jr., and G. Felsenfeld, Cell 14, 611 (1978).

[m] W. Altenburger, W. Hörz, and H. G. Zachau, Nature (London) 264, 517 (1976).

[n] S. G. Rogers and B. Weiss, this series, Vol. 65, p. 201.

[o] A. Prunell, Biochemistry 22, 4887 (1983).

[p] A. Prunell and R. D. Kornberg, Philos. Trans. R. Soc. London 283, 269 (1978).

[q] T. M. Herman, M. L. DePamphilis, and P. M. Wassarman, Biochemistry 20, 621 (1981).

[r] M. E. Cusick, T. M. Herman, M. L. DePamphilis, and P. M. Wassarman, Biochemistry 20, 6648 (1981).

[s] W. L. Linxweiler and W. Hörz, Nucleic Acids Res. 10, 4845 (1982).

[t] C. Kerr and P. D. Sadowski, J. Biol. Chem. 247, 311 (1972).

[u] K. Shinozaki and T. Okazaki, Nucleic Acids Res. 5, 4245 (1978).

[v] M. Engler and C. C. Richardson, J. Biol. Chem. 258, 11197 (1983).

[w] V. M. Vogt, Eur. J. Biochem. 33, 192 (1973).

[x] T. E. Shenk, C. Rhodes, P. W. J. Rigby, and P. Berg, Proc. Natl. Acad. Sci. U.S.A. 72, 989 (1975).

[y] R. C. Wiegand, G. N. Godson, and C. M. Radding, J. Biol. Chem. 250, 8848 (1975).

[z] T. M. Herman, M. L. DePamphilis, and P. M. Wassarman, Biochemistry 18, 4563 (1979).

[aa] D. E. Pulleybank, D. B. Haniford, and A. R. Morgan, Cell 42, 271 (1985).

[bb] S. Linn and I. R. Lehman, J. Biol. Chem. 240, 1287 (1965).

[cc] R. Fuchs and R. Blakesley, this series, Vol. 100, p. 3.

[dd] L. C. Tack, P. M. Wassarman, and M. L. DePamphilis, J. Biol. Chem. 256, 8821 (1981).

[ee] T. Igo-Kemenes, W. Hörz, and H. G. Zachau, Annu. Rev. Biochem. 51, 89 (1982).

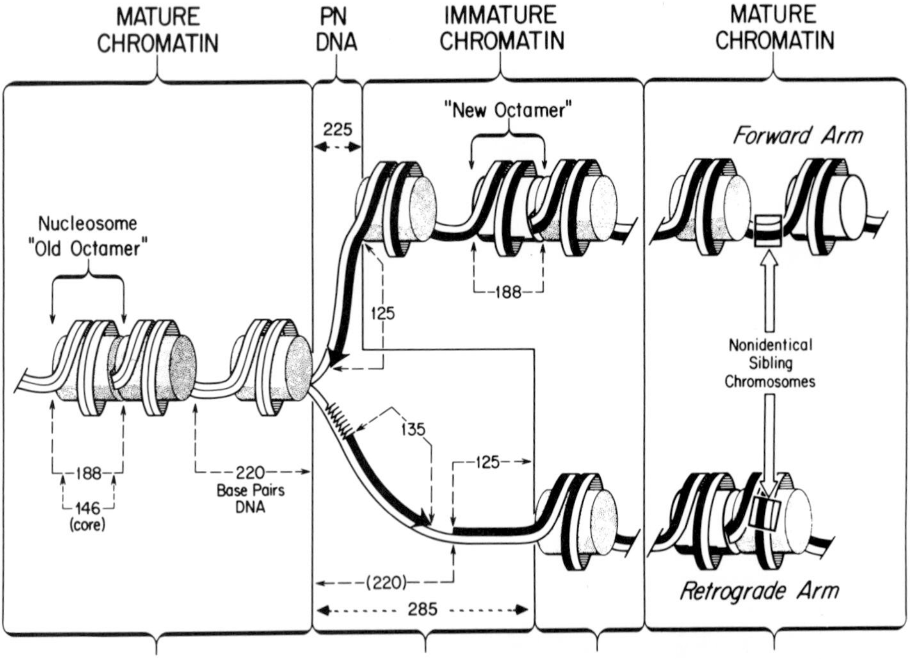

FIG. 1. Nucleosome organization at SV40 replication forks. Nucleosomes are represented as histone octamers (cylinders) around which is coiled double-stranded DNA (white ribbon). Nascent DNA is represented as a black ribbon with one RNA-primed Okazaki fragment on the retrograde arm. Numbers indicate the average distance in nucleotides based on experiments described in the text, except for the figure 220 in parentheses which is the expected average length of single-stranded DNA template that will be generated if replication forks pause at each nucleosome core.

nucleotides in length are synthesized predominantly, perhaps exclusively, on the retrograde arm of the two forks, that side of the fork where the direction of DNA synthesis must be opposite to the direction of fork movement (Fig. 1). Therefore, DNA synthesis on the forward arm is a relatively continuous process. Each replicating chromosome contains an average of 0.25 to 1 Okazaki fragment per fork, and 50% of the Okazaki fragments contain RNA covalently attached to their 5' end. Approximately $\frac{3}{4}$ of this RNA is 5'-(p)ppA/G(pN)$_{6-9}$. Polyoma virus DNA replication is essentially the same as in SV40.[22] Replication forks in eukaryotic chromosomes are similar to those in SV40 and polyoma virus, although, in some

[22] M. L. DePamphilis, E. Martinez-Salas, D. Y. Cupo, E. A. Hendrickson, C. E. Fritze, W. R. Folk, and U. Heine, in "Eukaryotic DNA Replication" (B. Stillman and T. Kelly, eds.), Vol. 6, pp. 165–175. Cold Spring Harbor Lab., Cold Spring Harbor, New York, 1988.

instances, the ratio of Okazaki fragments to long nascent DNA strands indicates either that several Okazaki fragments are present on retrograde arms or that Okazaki fragments are present on forward as well as retrograde arms.[23]

The structure of chromatin at SV40 DNA replication forks can be divided into at least four domains that appear equally applicable to polyoma virus and cellular replication forks as well (Fig. 1; reviewed in Refs. 2, 20, and 23). Comparison of replicating and mature SV40 chromosomes by analysis with micrococcal nuclease (MNase),[24,25] restriction endonucleases,[26] and electron microscopy[15,16,19] revealed that the size and arrangement of nucleosomes on newly replicated chromatin were essentially the same as those on mature chromatin (i.e., nonreplicating chromatin). SV40 chromosomes contain 24 ± 1 nucleosomes per genome,[27,28] each with a DNA "core" of 146 ± 3 base pairs (bp)[30] and an average repeat distance of 187 ± 11 bp *in situ*,[29] 198 ± 5 bp when isolated under various conditions,[30,51] and 177 ± 4 bp when isolated in the absence of the major capsid protein, VP1.[51] This is consistent with the average repeat distance for host cell chromatin of 182 ± 6 bp[29] to 188 ± 5 bp.[30,51] Both electron microscopy and MNase digestion[29] reveal that SV40 nucleosomes are separated by highly variable (0 to 60 bp) regions of internucleosomal DNA. A value of 27 ± 2 nucleosomes per genome has been calculated from electron microscopic analysis of DNA isolated from psoralen-cross-linked SV40 chromosomes.[19]

The actual sites of DNA synthesis are free of nucleosomes and therefore referred to as prenucleosomal (PN) DNA.[24,31] Fragments of newly replicated DNA in this region are rapidly released by MNase from both sides of forks in isolated SV40 chromosomes.[24] Okazaki fragments are released by single-strand-specific endonucleases.[31] Nascent DNA released by either nuclease is not contained within nucleosomes. Nascent DNA on forward arms along and in Okazaki fragments is excised by ExoIII ($3' \rightarrow 5'$), while nascent DNA on retrograde arms and in Okazaki fragments is excised by T7 Exo ($5' \rightarrow 3'$). These exonucleases do not excise nascent nucleosomal

[23] M. L. DePamphilis and P. M. Wassarman, *Annu. Rev. Biochem.* **49**, 627 (1980).

[24] M. E. Cusick, T. M. Herman, M. L. DePamphilis, and P. M. Wassarman, *Biochemistry* **20**, 6648 (1981).

[25] K.-H. Klempnauer, E. Fanning, B. Otto, and R. Knippers, *J. Mol. Biol.* **136**, 359 (1980).

[26] L. Tack, P. M. Wassarman, and M. L. DePamphilis, *J. Biol. Chem.* **256**, 8821 (1981).

[27] S. Saragosti, G. Moyne, and M. Yaniv, *Cell* **20**, 65 (1980).

[28] G. Moyne, K. Freeman, S. Saragosti, and M. Yaniv, *J. Mol. Biol.* **149**, 735 (1981).

[29] E. R. Shelton, P. M. Wassarman, and M. L. DePamphilis, *J. Mol. Biol.* **125**, 491 (1978).

[30] E. R. Shelton, P. M. Wassarman, and M. L. DePamphilis, *J. Biol. Chem.* **255**, 771 (1980).

[31] T. M. Herman, M. L. DePamphilis, and P. M. Wassarman, *Biochemistry* **18**, 4563 (1979).

DNA in replicating chromosomes.[32] Thus, the enzymes responsible for DNA synthesis utilize nonnucleosomal DNA templates. Based on exonuclease digestion of nascent DNA, PN DNA consists of an average of 123 ± 20 nucleotides of newly replicated DNA on the forward arm and up to one Okazaki fragment plus an average of 126 ± 20 nucleotides of the newly replicated daughter duplex DNA on the retrograde arm.[32] Electron microscopy of psoralen-cross-linked DNA from SV40 replicating chromosomes revealed that the distance from the branch point in replication forks to the first nucleosome is 225 ± 145 nucleotides on the forward arm and 285 ± 120 nucleotides on the retrograde arm.[19] Thus, newly replicated DNA is organized into nucleosomes as rapidly as sufficient double-stranded DNA becomes available. Nucleosome assembly during *in vitro* DNA replication occurs preferentially at active replication forks, suggesting that single-stranded regions at replication forks allow the release of torsional strain that forms during nucleosome assembly.[34] "Old" prefork histone octamers are distributed in an apparently random fashion to both arms of the fork, and "new" histone octamers are assembled on both arms.[19,33] Although the fate of "old" histones in SV40 was studied only in the presence of cycloheximide to block synthesis of new histones, the results are in agreement with studies on cellular chromatin, some of which were done in the absence of protein synthesis inhibitors (reviewed in Refs. 19 and 33).

The initial structure of newly replicated chromatin is immature chromatin, because it is hypersensitive to nonspecific endonucleases such as MNase, DNase I, and DNase II.[24,25,35] It is digested about 5-fold faster and about 25% more extensively than DNA in mature viral chromosomes. Immature chromatin can be distinguished from PN DNA in two ways. Replicating chromatin predigested with ExoIII and T7 Exo is still hypersensitive to nonspecific endonucleases, and nascent nucleosomal oligomers excised from replicating SV40 chromosomes are more susceptible to MNase digestion than oligomers from mature chromosomes. Both PN DNA and newly assembled immature chromatin contribute to the commonly observed hypersensitivity of newly replicated chromatin to endonucleases. Hybridization analysis with released DNA fragments reveals that both PN DNA and immature chromatin exist on both sides of replication forks.[24,35]

[32] T. M. Herman, M. L. DePamphilis, and P. M. Wassarman, *Biochemistry* **20**, 621 (1981).

[33] M. E. Cusick, M. L. DePamphilis, and P. M. Wassarman, *J. Mol. Biol.* **178**, 249 (1984).

[34] B. Stillman, *Cell* **45**, 555 (1986).

[35] M. E. Cusick, K.-S. Lee, M. L. DePamphilis, and P. M. Wassarman, *Biochemistry* **22**, 3873 (1983).

Postreplicative, mature chromatin consists of newly assembled nucleosomes that cannot be distinguished from mature chromatin. No differences in the nucleoprotein products from MNase digestion of replicating and mature SV40 chromosomes have been observed that correspond to chromatin maturation.[24,35] Digestion of replicating and nonreplicating chromosomes with various restriction endonucleases revealed no differences in the accessibility of different sequences to endonuclease attack.[26] Thus, the structure, spacing, and phasing with respect to DNA sequence of nucleosomes on replicating viral chromosomes is indistinguishable from mature viral chromosomes. Furthermore, the accessibility of a restriction endonuclease cleavage site on one arm of a replication fork is unrelated to the accessibility of the same site on the other arm. This is consistent with a nearly random phasing of chromatin structure on both arms and suggests that chromatin assembly occurs independently on the two sibling molecules of a single replicating chromosome.

Materials and Methods

Enzyme Digestion of Replicating Chromosomes

Micrococcal Nuclease (MNase). Micrococcal nuclease is from Worthington Biochemicals (now a subsidiary of Flow General Corp). Stock solutions are prepared by dissolving the enzyme at a concentration of 15,000 units/ml in distilled water containing 1 mg/ml bovine serum albumin. The enzyme solution is divided into 200-μl aliquots, and stored at $-20°$ with no detectable loss of activity after 2 years.

For digestion, samples of SV40 chromosomes, adjusted to $5-10$ μg of DNA/ml, are incubated with MNase (0.1 unit/μg of DNA) at 37° in a reaction volume of 0.65 ml containing 10 mM Tris-HCl (pH 7.5), 1 mM CaCl$_2$, and 1 mM MgCl$_2$. To determine the acid-insoluble radioactive DNA remaining, at various times in the digestion a 25- to 50-μl aliquot of the reaction is taken and added to 3 ml of 1 N HCl, 0.5% sodium pyrophosphate. After leaving the tubes for 10 min on ice, the DNA precipitates are collected on Whatman GF/C filters under vacuum. The filters are washed with $2-3$ ml of ice-cold 1 N HCl, then with $2-3$ ml ethanol. The filters are dried under a heat lamp, placed in a standard toluene-based liquid scintillation cocktail, and then radioactivity is measured in any liquid scintillation system.

Pancreatic Deoxyribonuclease I (DNase I). DNase I is purchased from Sigma (St. Louis, MO) or Worthington Biochemicals. Aliquots of DNase I, dissolved in 50 mM Tris-HCl (pH 7.6) at 2000 units/ml, are stored at $-20°$. For digestion, SV40 chromosomes, adjusted to $5-10$ μg of DNA/

ml, are incubated with DNase I (0.2 unit/μg of DNA) at 37° in a 0.65-ml reaction volume containing 10 mM Tris-HCl (pH 7.5) and 3 mM MgCl$_2$. Acid-insoluble radioactivity remaining is determined at various times during the digestion as described for MNase.

Spleen Deoxyribonuclease II (DNase II). DNase II is purchased from Sigma (Type DN-II-HP). Aliquots are prepared and stored as for DNase I. For digestion, samples of SV40 chromatin, adjusted to 5 – 10 μg DNA/ml, are incubated in a 0.65-ml reaction volume containing 10 mM Tris-HCl (pH 7.5), 10 μM EDTA, and DNase II at 100 units/μg DNA. Alternatively, the reaction contains 10 mM Tris-HCl (pH 7.5), 1 mM CaCl$_2$, and DNase II at 1000 units/μg DNA.[36] Both conditions give identical results. Acid-insoluble radioactivity is determined at various times during the digestion as described for MNase.

S$_1$ Nuclease. S$_1$ nuclease is obtained from Sigma. For optimal digestion of SV40 chromatin, isolated SV40 chromosomes are adjusted to 10 – 25 μg DNA/ml and incubated at 37° in a reaction volume of 0.1 ml containing 30 mM sodium acetate (pH 5.3), 5 μM zinc acetate, 75 mM NaCl, and 2.5 – 10,000 units/ml of S$_1$ nuclease. The reactions are stopped by addition of EDTA and Tris-HCl (pH 7.5) to final concentrations of 10 and 50 mM, respectively.

Neurospora crassa Endonuclease. The optimal conditions for *N. crassa* nuclease digestion of SV40 chromosomes are 100 mM Tris-HCl (pH 7.5), 10 mM MgCl$_2$, 0.1 mM cobalt acetate, and SV40 chromosomes at 5 – 10 μg/ml, in a 0.1-ml reaction volume with 0.5 – 32 units/ml of nuclease incubated at 37°. Following the digestion the reaction is stopped by the addition of EDTA to a final concentration of 10 mM.

Escherichia coli Exonuclease III (ExoIII). ExoIII is obtained from New England Biolabs (Beverly, MA). For digestion, SV40 chromosomes are adjusted to 5 – 10 μg DNA/ml and incubated at 20° with 40 units of ExoIII/μg DNA, in a 0.30-ml reaction volume containing 50 mM Tris-HCl (pH 8.0), 0.26 mM MgCl$_2$, 60 μM EDTA, and 1 mM dithiothreitol (DTT). Aliquots of 25 – 30 μl are removed at various times during the digestion, and the acid-insoluble radioactivity remaining is determined as described above for MNase.

Bacteriophage T7 Exonuclease (T7 Exo). T7 Exo purified by the method of Engler and Richardson[37,45] is now available from the United States Biochemical Corp. (Cleveland, OH). For digestion, SV40 chromosomes are adjusted to 5 – 10 μg DNA/ml and incubated at 20° with 16 units of T7 Exo/μg DNA, in a 0.30-ml reaction volume containing 50 mM

[36] W. Hörz, F. Miller, G. Klobeck, and H. G. Zachau, *J. Mol. Biol.* **144**, 329 (1980).
[37] M. Engler and C. C. Richardson, *J. Biol. Chem.* **258**, 11197 (1983).

Tris-HCl (pH 8.0), 20 mM KCl, 0.15 mM MnCl$_2$, 50 μM EDTA, and 1 mM DTT. Aliquots are removed during digestion as described above for ExoIII. When replicating SV40 chromosomes are digested concurrently with ExoIII and T6 Exo the conditions for T7 Exo are used. ExoIII retains full activity under T7 Exo conditions.

Restriction Endonucleases

All restriction endonucleases are purchased from New England Biolabs or Bethesda Research Laboratories (BRL, Bethesda, MD). Generally, the conditions used for chromatin digestion are the same as those provided by the supplier for use with naked DNA. For chromatin digestions, 10–100 times more enzyme should be used than needed to digest an equivalent amount of DNA. Incubations proceed for 15–30 min at 37°. Reactions are halted by placing the tubes in ice water and adding EDTA to 20 mM.

Preparation and Radiolabeling of Replicating SV40 Chromosomes

The preparation of chromatin is covered in other chapters in this volume (see Kornberg et al. [1] and Oudet et al. [2]), so it is not our purpose here to review preparation methods in detail. We review only those techniques used to obtain replicating SV40 chromosomes. Replicating chromosomes are specifically radiolabeled by addition of radioactive nucleotide precursors for very short time periods (generally known as pulse-labeling). Radiolabeling is then stopped, and the chromosomes, radiolabeled for only a short stretch of DNA at the replication fork, are isolated and subjected to enzymatic analysis. SV40 chromosomes can be radiolabeled in vivo in intact cells or in vitro as isolated chromosomes in nuclear extracts. Both methods give identical results with regard to the structure of chromatin revealed by enzymatic analysis. We prefer the in vitro methods, for they allow incorporation of radiolabeled precursors at higher specific activities and permit a rapid halt of further radiolabeling by addition of unlabeled precursors during a chase period, in which further replication proceeds.

In Vivo Radiolabeling

Replicating SV40 chromosomes are pulse-labeled in infected cells at 36–38 hr postinfection, when the rate of viral DNA synthesis is maximal. First, the dTTP pools are equilibrated with radioactive dTTP by preincubation with [³H]thymidine for 10 min at 0°. Equilibration is necessary to maximize the subsequent incorporation of [³H]dTTP.[31,38] DNA synthesis

---

[38] D. Perlman and J. Huberman, *Cell* **12**, 1029 (1977).

does not occur during the preincubation, for the temperature is too low. After preincubation, the plates of infected cells are shifted to a higher temperature for several minutes to allow DNA synthesis to proceed. The pulse of DNA synthesis is then stopped by returning the plates to 0°.

Plates (100 mm) of SV40-infected cells are washed with 5 ml of TS buffer [20 mM Tris-HCl (pH 7.4), 5 mM KCl, 0.5 mM MgCl$_2$, 1.0 mM CaCl$_2$, 137 mM NaCl], drained, and floated on an ice-water bath. Thirty seconds later 100 μCi of [^{3}H]thymidine (purchased from New England Nuclear, Boston, MA; 55 Ci/mmol) in 0.6 ml of TD buffer [20 mM Tris-HCl (pH 7.4), 5 mM KCl, 1 mM Na$_2$HPO$_4$, 137 mM NaCl] containing 5% fetal calf serum is added to each dish. The plates must be shaken briefly to evenly distribute the [^{3}H]thymidine. After a 10-min incubation on ice, each plate is floated onto water at 20° or 30° for 1–2 min. The temperature of 30°, used in more recent experiments in our lab, allows greater incorporation of [^{3}H]thymidine with no detectable changes in chromatin structure as judged by subsequent enzymatic analysis. The plates of cells are then refloated on ice water, 5 ml of ice-cold buffer A [15 mM Tris-HCl (pH 7.6), 2 mM MgCl$_2$, 25 mM NaCl] is added, then quickly drained into a radioactive waste bottle, and 5 ml more buffer A is added.

Extraction of Chromosomes

The cells are then processed for SV40 chromosome extraction. Two methods have been used: isotonic extraction from isolated nuclei[31,39] and hypotonic extraction from nuclei.[40,41] The results of enzymatic analysis of chromatin structure with the two methods are identical. With isotonic extraction, DNA synthesis cannot continue *in vitro,* presumably because the higher NaCl concentration used in this method strips chromosomes of factors essential for DNA replication. Hypotonically extracted chromosomes are capable of further DNA synthesis for up to 1–2 hr following extraction. Thus, the fate of pulse-labeled ^{3}H-DNA, labeled *in vivo,* can be followed. No further incorporation of residual [^{3}H]thymidine is observed *in vitro.*

Isotonic Conditions. All procedures following the washing of the plates are done in the cold room. All equipment (centrifuge tubes, pipettes, homogenizer, and rubber policemen) is precooled to 4°. The plates of cells are drained of buffer A, and the cells are scraped from the dishes with a

M. H. Green, H. I. Miller, and S. Hendler, *Proc. Natl. Acad. Sci. U.S.A.* **68,** 1032 (1971).
[40] R. T. Su and M. L. DePamphilis, *Proc. Natl. Acad. Sci. U.S.A.* **73,** 3466 (1976).
[41] R. T. Su and M. L. DePamphilis, *J. Virol.* **28,** 53 (1978).
[42] J. D. McGhee and G. Felsenfeld, *Annu. Rev. Biochem.* **49,** 1115 (1980).
[43] T. Igo-Kemenes, W. Hörz, and H. G. Zachau, *Annu. Rev. Biochem.* **51,** 89 (1982).

rubber policeman. The cells are transformed into a Dounce homogenizer with a prewetted pipette, and lysed with 3 strokes of the tight-fitting pestle B. The homogenate is transferred to a centrifuge tube, and nuclei are pelleted by centrifugation at 1500 g for 5 min at 4°. The supernatant is discarded, and the nuclear pellet is resuspended in 2.5 ml of buffer B [10 mM Tris-HCl (pH 7.6), 2 mM EDTA, 200 mM NaCl] per original 10 dishes (100 cm) of cells. One-tenth volume of 10% Triton X-100 is added. The suspension of nuclei is incubated at 30° for 30 min, with vortexing every 5–10 min to ensure that the nuclei remain suspended. The nuclei are then pelleted by centrifugation at 2000 g for 5 min at 4°.

The supernatant is kept and the pellet discarded. The supernatant containing the replicating SV40 chromosomes is layered onto a 5–30% neutral sucrose gradient in buffer B and centrifuged for 5 hr at 25,000 rpm, 4°, in a Beckman SW27 or SW28 rotor. Gradients are fractionated, from the top to avoid contamination with virions in any pellet, into about 30 fractions. Sometimes chromosomes were sedimented through sucrose gradients in hypotonic solution [10 mM HEPES (pH 7.8), 5 mM KCl, 0.5 mM MgCl$_2$, 0.5 mM DTT] plus 200 mM NaCl, but no difference is observed in the experimental results. A 25-μl aliquot of each fraction is spotted onto a piece of Whatman GF/C paper, dried, and counted, to find the fractions containing the peak of replicating SV40 chromosomes (~65 S). The peak fractions are pooled, then concentrated into TE [10 mM Tris-HCl (pH 7.5), 0.1 mM EDTA] by vacuum dialysis or by pressure dialysis under N$_2$ gas. This procedure usually yields about 3–5 μg of SV40 DNA per dish of cells (~6 × 10^6 cells). Chromosomes can be stored at 4° for 1–2 weeks. After 2 weeks they seem to fall apart, so chromosome preparations greater than 10 days old should not be used for any experiments.

Hypotonic Conditions. Nuclear extracts containing SV40 chromosomes are prepared for SV40-infected cells at 36–38 hr postinfection by a slight modification of the hypotonic extraction method of Su and DePamphilis.[35,40,41] Plates of infected cells are drained of media and taken into a cold room. Again, all equipment that touches the cells must be precooled to 4°. The plates are washed twice with 5 ml of ice-cold hypotonic solution [10 mM HEPES (pH 7.8), 5 mM KCl, 0.5 mM MgCl$_2$, 0.5 mM DTT], then drained of excess fluid by aspiration. Cells are scraped off the dishes with a rubber policeman and then transferred to a Dounce homogenizer where they are disrupted by three strokes with a tight-fitting pestle B. The homogenate is transferred to a centrifuge tube, and nuclei are pelleted by centrifugation at 3000 g for 5 min at 4°. The supernatant is removed and saved for preparation of cytosol. The nuclear pellet is resuspended by vortexing in hypotonic solution, 0.8 ml of solution per original 10 dishes

(100 cm) of cells. The nuclear suspension is incubated on ice for 1–2 hr, with occasional vortexing to maintain a uniform suspension. The suspension of nuclei is then centrifuged at 8000 g for 10 min at 4°. The supernatant is the nuclear extract, containing replicating SV40 chromosomes. The SV40 chromosomes are then purified by sedimentation through neutral sucrose gradients, collected, and concentrated, all as described earlier for isotonic extraction. The hypotonic extraction method routinely yields 1–2 μg of SV40 DNA per dish (~6 × 10^6 cells). Nuclear extract must be used immediately; it is not stable in storage.

In Vitro Radiolabeling

Replicating SV40 chromosomes in nuclear extracts are pulse-labeled with [α-^{32}P]dCTP (purchased from New England Nuclear, either 600 or 3000 Ci/mol) for 1–5 min at 30° in the presence of added cytosol. The standard replication sample contains 0.2 units of nuclease extract, 0.7 units of cytosol, and 0.1 units of an assay mix consisting of a buffered salt solution with deoxyribo- and ribonucleotide triphosphates and an ATP-regenerating system. The final concentration of each component is as follows: 39 mM HEPES (pH 7.9), 24.5 mM KCl, 4.5 mM MgCl$_2$, 1 mM EDTA, 0.45 mM DTT, 2 mM rATP, 5 mM phosphoenolpyruvate (PEP), 30 μg/ml pyruvate kinase in glycerol, 100 μM each of rGTP, rUTP, and rCTP, 20 μM each of dATP, dGTP, and dTTP, and 2 μM of dCTP. Cytosol and assay mix are first mixed together at 0°, then DNA replication is started by adding nuclear extract and transferring the mixture to 30°. To stop further incorporation of radiolabel (end the pulse), a 100-fold excess of unlabeled dCTP is added. The replication reaction can then be continued at 30° for up to 1 hr (the chase) without further incorporation of radiolabel. To stop replication at any time the samples are placed on ice, and EDTA is added to a final concentration of 10 mM. The radiolabeled replicating SV40 chromosomes are purified by sedimentation through neutral sucrose gradients, collected, and concentrated, as described above under isotonic conditions.

The cytosol fraction used in the *in vitro* radiolabeling of replicating SV40 chromosomes is made from the supernatant left over from the preparation of nuclei for the nuclear extract. The supernatant is centrifuged in a Beckman 50Ti rotor at 100,000 g for 60 min if obtained from infected cells, or at 150,000 g for 90 min if obtained from infected cells. Any pellet is discarded, and potassium acetate is added to the supernatant (the cytosol) to a final concentration of 50 mM. Cytosol can be used immediately, or stored at −70° for up to 6 months with no apparent loss of activity. Although a variety of methods exist for radiolabeling and preparing replicating SV40 chromosomes, there is no difference between chromosomes prepared by different methods as shown by subsequent enzymatic analysis.

Nuclease Analysis of Replicating Chromatin

Identification of Immature Chromatin and Chromatin Maturation

It is commonly observed that newly replicated cellular or viral chromatin is more rapidly and extensively degraded by nonspecific endonucleases such as MNase, DNase I, and DNase II (Table I) than is nonreplicating chromatin.[23-25,35] This is thought to reflect differences in nucleosome structure and spacing in newly assembled, "immature" chromatin compared to nonreplicating, "mature" chromatin. The presence of immature chromatin and the process of chromatin maturation can be quantitated by digesting a mixture of replicating chromatin containing pulse-labeled nascent DNA and mature chromatin at various times after the newly replicated DNA is labeled.

SV40 chromosomes, pulse-labeled for a short time to radiolabel only a short stretch of nascent DNA right at the replication fork, or pulse-labeled and then chased for various times *in vitro* so as to examine the stretch of radiolabeled nascent DNA as the replication fork moves beyond it, are mixed with mature SV40 chromosomes and digested together with MNase, DNase I, or DNase II, as described in Materials and Methods. The inclusion of mature, uniformly labeled SV40 chromosomes is important for it provides a control or reference digestion to compare to the digestion of replicating chromosomes. The mature SV40 chromosomes are radiolabeled by adding [^{3}H]- or [^{14}C]thymidine to the culture media of SV40-infected cells for 12 hr before those cells are taken for chromosome extraction. If replicating chromosomes are to be pulse-labeled with [^{3}H]thymidine, then [^{14}C]thymidine is used (purchased from Schwarz-Mann, 50 mCi/ml; apply 0.5 μCi/ml of media). If pulse-labeling is done with [α-^{32}P]dCTP then [^{3}H]thymidine is used (purchased from New England Nuclear, 85 Ci/mmol; apply 2.5 μCi/ml of media). Mature chromosomes can be labeled in the same cells used to extract replicating chromosomes. If so, the two types of chromosomes are copurified by the methods described earlier. Alternatively, mature chromosomes can be purified from a separate batch of infected cells, then mixed with replicating chromosomes just before the endonuclease digestion is initiated. Both approaches produce equivalent results.

It is important to keep the final chromatin concentration in the reaction mixture above 5 μg/ml. If the concentration falls below $1-2$ μg/ml then nucleosomes seem to unravel, releasing their DNA,[44] and the chromatin digests with kinetics akin to the digestion of naked DNA.[45] At low concentrations the addition of cellular chromatin is necessary to stabilize viral chromatin. Cellular chromatin is prepared from uninfected cells by methods described elsewhere in this volume (Kornberg et al. [1]).

The time course for the endonuclease digestion is plotted as percent chromosomal DNA digested versus time of endonuclease digestion (Fig. 2A). Regardless of the endonuclease used, replicating SV40 chromatin is digested faster ($\sim$2- to 5-fold) and more extensively ($\sim$20–25% more) than mature SV40 chromatin is digested in the same tube. The sensitivity of replicating SV40 chromosomes to MNase, DNase I, or DNase II gradually becomes comparable to that of mature chromatin during a chase of radiolabel from replication forks. This process, known as maturation, has been observed for both cellular and viral chromatin.[23]

Since many times the digestions do not plateau at a limit value at late times of digestion, but continue upward slightly, the extent of digestion cannot be used as a valid measure of chromatin maturation. Instead, we have found that a convenient method is to measure the areas between the curve describing replicating chromatin and the curve describing mature chromatin for various chase times (Fig. 2A). These areas are then normalized so that the difference between replicating and mature chromatin seen at the beginning of the chase is defined as 0% chromatin maturation and the difference seen at the end of the chase is defined as as 100% chromatin maturation. This method makes one experiment comparable to another. The half-time ($t_{1/2}$) for maturation is then the chase time at 50% chromatin maturation (Fig. 2B).

Nucleosome Structure and Spacing in Immature Chromatin

The average repeat length between contiguous nucleosomes (nucleosome spacing), the average size of DNA in nucleosome monomers and core particles, and the characteristic folding of DNA around histone octamers is generally determined from the DNA products released by endonuclease digestion and analyzed by polyacrylamide gel electrophoresis. The size and shape of nucleosomes is generally determined from nucleoprotein products analyzed either by velocity sedimentation in neutral sucrose gradients[24,32,35] or by electrophoresis in polyacrylamide gels (cf. Garrard[35]). The average repeat length is calculated by dividing the nucleotide length for the center of each oligomeric DNA band by the number of

[44] R. W. Cotton and B. A. Hamkalo, *Nucleic Acids Res.* **9**, 445 (1981).
[45] M. E. Cusick, Ph.D. Thesis, Harvard Medical School, 1983.

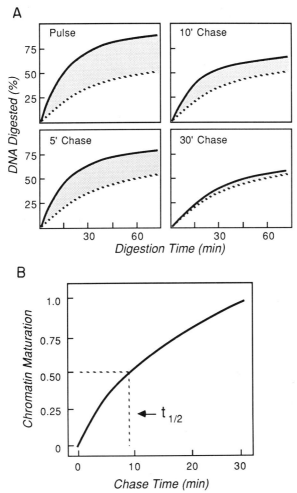

FIG. 2. Use of endonucleases to assess the maturation of chromatin as accessibility of internucleosomal, nucleosomal, and prenucleosomal DNA with time after DNA replication. (A) Typical time courses for micrococcal nuclease (MNase) digestion of replicating SV40 chromatin are shown. Replicating SV40 chromosomes (——) containing DNA that was radiolabeled for a short period of time (pulse-labeled, see Materials and Methods) were allowed to continue replication in the absence of radiolabeled deoxyribonucleotides for the indicated times (chase period). These replicating chromosomes were then mixed wih mature SV40 chromosomes (– – –) containing DNA that was uniformly labeled with a different radioisotope and then digested with MNase. The percentage of acid-soluble radioactivity remaining is measured as a function of the time of digestion. (B) The rate of chromatin maturation can be measured using the time courses determined in (A). The area between the curve describing replicating chromatin and mature chromatin is measured (shaded area), and the area measured at 0 min chase time is defined as 0% chromatin maturation while the area measured at 30 min chase time is defined as 100% chromatin maturation. The remaining areas are then normalized to these values. We found that the difference between replicating and mature chromatin at chase times longer than 30 min was essentially the same as at 30 min. The half-time ($t_{1/2}$) for maturation is the time at which 50% maturation occurs.

nucleosomes per oligomer. DNA length is obtained by comparison with a series of DNA restriction fragments run in parallel on the same gel. To correct for "nibbling" of the ends of each oligonucleosome as it is released by the nuclease, nucleosome monomers are not included in the calculation, and the average repeat length obtained at several different times during digestion is plotted against the fraction of acid-soluble DNA present in order to determine the repeat length at 0% digestion.[30] An alternative method is to plot the length of each nucleosome oligomer against the number of nucleosomes per oligomer.[53] Sufficient oligomers should be present in the digest so that a straight line is obtained. One could plot the reciprocals of these numbers in order to extrapolate to "infinitely" long oligomers, thus, negating any nibbling of oligomer ends.

Samples of SV40 chromosomes are digested with MNase or DNase I (see section on Materials and Methods). Digestion is terminated by placing the samples on ice and adding EDTA to a final concentration of 10 mM. DNA products are extensively purified before fractionation by electrophoresis by incubating samples in 0.25% sodium dodecyl sulfate (SDS) and 10 μg/ml of proteinase K for 1 hr at 37°. Samples are extracted once with 2 volumes of $CHCl_3$–isoamyl alcohol (24:1, v/v). After centrifugation to clear the extraction, the upper aqueous phase is digested with RNase A at 50–100 μg/ml for 1 hr a 37°. The stock of RNase A is boiled for 5 min to inactivate any DNase present. Samples are then extracted twice with $CHCl_3$–isoamyl alcohol. NaCl is added to the aqueous phase to a final concentration of 0.3 M followed by 2 volumes of ethanol to precipitate DNA. After at least 2 hours on dry ice, the pellet is collected by centrifugation. The pellet is washed once with 70% ethanol to remove residual salt, dried *in vacuo,* and then dissolved in 10–20 μl of TE buffer before loading onto a gel.

The DNA samples are mixed with an equal volume of gel dye solution [9 mM Tris–borate (pH 8.3), 0.25 mM EDTA, 50% sucrose (or glycerol), 0.05% bromphenol blue, and 0.05% xylene cyanol FF]. For analysis of duplex DNA fragments from MNase digestions, the DNA is electrophoresed on either 2% agarose or 5% polyacrylamide in TAE buffer [40 mM Tris–acetate (pH 7.9), 20 mM sodium acetate, 2 mM EDTA].[30,46] For analysis of denatured DNA fragments from DNase I digestions, the DNA is electrophoresed on 12% polyacrylamide gels in TBE buffer [10 mM Tris–borate (pH 8.3), 0.2 mM EDTA] containintg 7 M urea.[30]

For detection of radiolabeled DNA on gels, several methods are available. If the DNA is from replicating SV40 chromosomes pulse-labeled with

---

[46] E. R. Shelton, J. Kang, P. M. Wassarman, and M. L. DePamphilis, *Nucleic Acids Res.* **5,** 349 (1978).

[32]P, then simple autoradiography will suffice. The gel is dried down under vacuum then exposed to Kodak XR-5 X-ray film at − 70° against a regular intensifying screen. Fluorography is used to detect DNA from replicating chromosomes pulse-labeled with [3]H.[47] For digestions of replicating and mature chromosomes mixed together, where two different radiolabels are present (either ^{3}H and ^{14}C, or ^{32}P and ^{3}H, for replicating and mature chromosomes, respectively), counting of individual gel slices is necessary. The gels are sliced into 2-mm fractions, and each slice is incubated overnight with shaking at 37° in 10 ml of Liquifluor (New England Nuclear) containing 3% NCS (Amersham Radiochemical Centre, Amersham, UK) before measuring the amount of radioactivity in a liquid scintillation counter.[32]

To analyze the nucleoprotein products formed by MNase digestion or replicating SV40 chromosomes, the digestions are stopped by adding EDTA to 10 mM final concentration and placing the tubes on ice. For analysis by velocity sedimentation the samples are layered onto 5–20% neutral sucrose gradients in TEN$_{50}$ [10 mM Tris-HCl (pH 7.5), 0.1 mM EDTA, 50 mM NaCl]. Higher NaCl concentrations (100 or 200 mM) can be used; at least 50 mM salt is needed, or else DNA binds to soluble protein, making results difficult to interpret.[35,48] Sedimentation is routinely for 14 hr at 35,000 rpm in a Beckman SW41 rotor at 4°. About 32–36 equal-volume fractions are collected from the bottom of the tube. The contents of each fraction are precipitated by the addition of 3 ml of ice-cold 1 N HCl, and precipitates are collected under vacuum on Whatman GF/C filters. Filters are washed with ethanol and dried under a heat lamp. The radioactivity is measured in a standard toluene-based liquid scintillation cocktail. Nucleosome monomers (11 S) usually sediment to the middle of the gradient under these conditions. Larger sucrose gradients can be used to isolate oligonucleosomes on a preparative level.[35]

Identification of Prenucleosomal DNA

ExoIII begins degrading the nascent DNA strand at replication forks from its 3′ end and continues until it encounters a nucleosome core.[32] This stop signal is much stronger than sequence-specific pause sites observed in purified DNA,[54] and it can easily be distinguished from these sites by increasing either the amount of enzyme or the time of incubation. T7 Exo behaves similarly to ExoIII except that it degrades the nascent DNA strand

[47] R. A. Laskey, this series, Vol. 65, p. 363.
[48] E.-J. Schlager, *Biochemistry* **21**, 3167 (1982).
[49] M. Yaniv and S. Cereghini, *CRC Crit. Rev. Biochem.* **21**, 1 (1986).
[50] D. A. Prentice and L. R. Gurley, *Biochim. Biophys. Acta* **740**, 134 (1983).

from its 5' end. Thus, ExoIII degrades PN DNA from forward arms of forks plus Okazaki fragments from retrograde arms, and T7 Exo degrades all of the PN DNA (including Okazaki fragments) from retrograde arms of forks. Therefore, these enzymes can be used to measure the amount of PN DNA at replication forks.

Replicating SV40 chromosomes, pulse-labeled for a short time to radiolabel only a short stretch of nascent DNA right at the replication fork, or pulse-labeled and then chased for various times *in vitro* so as to examine the stretch of radiolabeled nascent DNA as the replication fork moves beyond it, are mixed with mature SV40 chromosomes and digested together with ExoIII, T7 Exo, or both, as described in Materials and Methods. Mature, uniformly labeled SV40 chromosomes (prepared as described earlier) are not digested by either exonuclease. They do not have free, accessible 5' and 3' ends. They are included to check that a nonspecific endonuclease is not active during the exonuclease digestion. The exonuclease conditions that are used (see Materials and Methods) suppress the activity of an endogenous endonuclease that sometimes copurifies with SV40 chromosomes.[32]

The results of the exonuclease digestions are plotted as percent chromosomal DNA digested versus the time of exonuclease digestion (Fig. 3A). Initially a large portion (40–60%) of the pulse-labeled nascent DNA is susceptible to exonuclease. This figure represents the fraction of labeled nascent DNA that is present as prenucleosomal DNA at replication forms.[32,35] With increased radiolabeling time,[32] or during a chase,[35] the amount of DNA susceptible to exonuclease digestion declines. The time course for loss of exonuclease susceptibility is easily determined (Fig. 3B). Unlike endonuclease digestions, digestions with exonucleases plateau at a particular percent digestion. This plateau, or limit digestion, indicates that exonucleases do not penetrate and degrade nucleosomal DNA at replication forks.[32] Increasing the concentration of either exonuclease will not increase the extent of digestion of replicating chromosomes but increases only the rate at which the plateau is reached. Thus, to obtain a maturation curve the difference between the plateau of the digestion curve for replicating chromatin and that for mature chromosomes is plotted as a function of the chase period (Fig. 3B).

Determining Average Distance from 3' and 5' Ends of Nascent DNA to First Nucleosome on Each Arm of Replication Fork

The fraction of radioactivity released from nascent DNA by either ExoIII or T7 Exo is proportional to the average distance in nucleotides traveled by the exonuclease. The fraction of ^{3}H-DNA released from the 3' end by ExoIII or 5' ends of long nascent DNA chains (Fig. 1) is converted

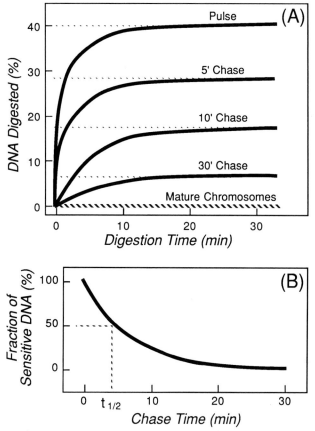

FIG. 3. Use of exonucleases to assess the maturation of chromatin as a decrease in the fraction of radiolabel in prenucleosomal DNA with time after DNA replication. (A) Typical time courses for ExoIII, T7 Exo, or ExoIII plus T7 Exo digestion of replicating (- - -) and mature (——) SV40 chromatin are shown. The experimental strategy is the same as with endonuclease digestion (Fig. 2A). (B) The rate of chromatin maturation can be measured using the time courses determined in (A). The difference between the plateau (dotted lines) of the digestion curve for replicating chromatin at 0 and 30 min chase is measured for each chase time. These values are normalized so that the value at 0 min chase is defined as 100% sensitivity to exonuclease and the value at 30 min chase is defined as 0% sensitivity. The half-time ($t_{1/2}$) for loss of radiolabel in prenucleosomal DNA is the chase time at which 50% sensitivity remains.

to nucleotides by measuring the total number of nucleotides incorporated into nascent DNA during the pulse-labeling period[32] and then multiplying that number by the fraction of radioactivity released from either the 3′ or 5′ ends (the plateau value for the pulse in Fig. 3A). The number of nucleotides incorporated into replicating DNA is determined by ExoIII digestion of purified replicating ^{3}H-DNA in the presence of a uniformly labeled unique ^{32}P-DNA restriction fragment (200–400 bp) containing 3'-OH recessed ends as an internal standard that will be digested at the same rate as nascent DNA at replication forks. A 10-fold excess of unlabeled linear DNA with 3'-OH recessed ends is included to ensure that small variations in the amounts of labeled DNA will not affect their rate of digestion by ExoIII. A maximum of 50% of the ^{32}P-DNA is digested because ExoIII degrades only duplex DNA. The time required to digest 10% of a DNA fragment is proportional to its length.[32] Therefore, (number of ^{3}H-labeled nucleotides released from replicating DNA)(% ^{3}H cpm released) = (number of ^{32}P-labeled nucleotides released from restriction fragments)(% ^{32}P cpm released). From these data, the total number of nucleotides incorporated during the pulse-label (N) is calculated and shown to be proportional to the radiolabeling period.

The shorter the labeling period, the greater the fraction of radioactivity in Okazaki fragments that must be subtracted from the total amount of labeled DNA before calculating the average number of nucleotides released when replicating chromosomes are digested. This problem can theoretically be circumvented by increasing the labeling time until the fraction of label in Okazaki fragments is insignificant. However, this also results in less exonuclease-sensitive nascent DNA. Another approach is to first excise Okazaki fragments with the alternative exonuclease. If ExoIII is used to measure the length of PN DNA on the forward arm, then chromosomes can be digested first with T7 Exo to remove Okazaki fragments. All three approaches gave the same results with SV40.[32] Very short labeling periods lead to additional corrections because preequilibration of cells with labeled nucleosides at on ice (see *In Vivo Radiolabeling*) results in a rapid burst of radioactivity incorporated into nascent DNA. This means that DNA synthesized during the first minute will not be uniformly labeled. The specific radioactivity will be greater initially, and therefore the fraction of label in nucleosomal DNA will be disproportionately higher than in PN DNA. Furthermore, labeling of 5' ends of long nascent DNA strands on retrograde arms will be delayed because label first enters Okazaki fragments which are then joined to the 5' ends of growing daughter strands before the 5' ends can become labeled.

Nucleosome Structure and Spacing in the Absence of Prenucleosomal DNA

To determine the nucleosome structure on replicating chromosomes after exonuclease digestion, a second digestion with MNase is done. To do this the reaction conditions for chromosome digestion are changed by gel filtration. The first exonuclease reaction is stopped by adding EDTA to

5 mM and placing the samples on ice. The reaction mixture is filtered through a Sephadex G-50 (Pharmacia, Piscataway, NJ) column equilibrated in TE [10 mM Tris (pH 7.5), 0.1 mM EDTA]. SV40 chromosomes elute in the void volume. Normally a 3- to 4-ml column, placed in a cut-off 5-ml pipette stoppered with sterile glass wool, is used. Five-drop fractions are collected, and a small aliquot of each fraction is counted to find the fractions containing radiolabeled SV40 chromosomes. The reaction conditions are then adjusted for MNase digestion (see Materials and Methods). The digestion products can be analyzed by methods already described.

Release of Okazaki Fragments from Replication Forks

Replicating forks in replicating chromosomes are highly sensitive to single-strand DNA (ssDNA)-specific endonucleases, while DNA in non-replicating chromatin is relatively resistant. At least 50% of Okazaki fragments can be released as 4–6 S fragments of double-stranded DNA (dsDNA) by digestion of replicating SV40 chromosomes with a single-strand-specific endonuclease.[31] The remaining Okazaki fragments cannot be released because at least one of the ssDNA regions flanking them (Fig. 1) is too small to be recognized by the nuclease under the incubation conditions used. At least 90% of the Okazaki fragments released are not contained in nucleosomes, even when chromosomes are pretreated with formaldehyde to cross-link proteins and DNA. Therefore, most Okazaki fragments are joined to the growing daughter strand prior to nucleosome assembly.

S_1 nuclease is the enzyme of choice for this type of experiment. It has very high specificity for ssDNA and can recognize small mismatched regions of ssDNA within dsDNA (Table I). *Neurospora crassa* endonuclease (Table I) can also be used, with results being similar to those obtained with S_1.[31] However, *N. crassa* nuclease preparations contained contaminating dsDNA nuclease activity and were less efficient than S_1 at recognizing small ssDNA regions in dsDNA. S_1 has maximal activity at pH 4.6, but the solubility of chromatin decreases dramatically below pH 5.5. Therefore, digestion conditions were developed in which chromatin remained soluble yet the enzyme retained significant activity.[31] While isolated SV40 is completely soluble at pH 5.5, all of it is precipitated at pH 4.6. On the other hand, the rate of digestion of ssDNA by S_1 decreases approximately 10-fold as the pH is increased from 4.6 to 5.3. At pH 7.0, S_1 activity is not detectable. Under conditions optimal for chromatin digestion (pH 5.3, see Materials and Methods), S_1 nuclease has 10% of the activity seen at pH 4.6, and 80% of the chromatin is soluble.[31] Chromatin solubility also depends on the concentration of Zn^{2+}, a required cofactor for S_1. At pH 5.3, chromatin solubility increases steadily as Zn^{2+} concentration is lowered from 0.14 to 0.05 mM. Chromatin has its maximal solubility under conditions optimal for chromatin digestion (0.05 mM Zn^{2+}; see Materials and Methods), and S_1 activity is not reduced significantly.[31]

After S_1 digestion, the DNA is purified as previously described. DNA products can be analyzed by agarose gel electrophoresis or velocity sedimentation in neutral sucrose gradients to examine dsDNA fragments as previously described. The size distribution of ssDNA Okazaki fragments can be determined by velocity sedimentation in alkaline sucrose gradients[31] or electrophoresis in gels under denaturing conditions.

Relationship of Chromatin Structure to DNA Sequence at Replication Forks

The relationship between chromatin structure and DNA sequence (commonly referred to as nucleosome phasing) can be assessed by measuring the accessibility of unique DNA sequences to cleavage by an appropriate restriction endonuclease.[2,23] The accessibility of a particular DNA site depends both on chromatin structure at that site and on the restriction endonuclease chosen.[26] Although nucleosomes are probably the dominant factor in protecting a DNA sequence from cleavage by a restriction endonuclease, the presence or absence of other chromosomal proteins cannot be discounted.[26] If nucleosome phasing is unique, then a particular DNA sequence will be either protected or accessible in every chromosome in the population. If nucleosome phasing is random, then a particular site will be protected in only a fraction of the chromosomes, and that fraction will be proportional to the number of nucleosomes per chromosome. If nucleosome phasing has preferred sequences and/or if sequence-specific DNA binding proteins are present, then intermediate accessibilities will be observed.

The effect of replication on nucleosome phasing can be viewed in two ways. First, the accessibility of a particular DNA sequence in replicating chromatin can be compared with its accessibility in nonreplicating chromatin, or the accessibility of a site in replicating chromosomes can be compared before and after replication forks have passed through. If no change is evident, then nucleosome phasing is not altered by chromatin replication. If changes are observed, they can be related to their proximity to replication forks. Second, the accessibility of the same DNA sequence in two sibling chromosomes (i.e., the two arms of a single replication fork) reveals whether nucleosome phasing on one arm of the fork is identical to nucleosome phasing on the other arm. This measurement is independent of whether phasing in the total chromosome population is unique, random, or preferred.

For simplicity in data analysis, a restriction enzyme should be chosen that cuts the chromosome at a single site. Accessibility is then measured as the fraction of chromosomes cleaved. This is accomplished either of two ways. The preferred method is to incubate SV40 chromosomes with increasing concentrations of restriction endonuclease for a short period of time (e.g., 15 min at 37°) until the maximum number of chromosomes are cleaved. An alternative method is to incubate chromosomes for increasing times with a single amount of enzyme that is sufficient to cleave rapidly all the sites in bare DNA when present at the same concentration as the chromatin sample. In either method, results will appear as in Fig. 3A, and the fraction of accessible chromosomes can be determined by extrapolation of the plateau to zero enzyme concentration or zero digestion time (dotted lines in Fig. 3A). If unbranched chromosomes such as mature SV40 chromosomes are analyzed, then the fraction of linear DNA molecules (nuclease-sensitive chromosomes) can be assessed by fractionating the purified DNA products by gel electrophoresis to separate linear and circular forms of viral DNA.[26] If branched chromosomes such as replicating SV40 chromosomes are analyzed, then the fraction of chromosomes cut by a restriction endonuclease can be analyzed by electron microscopy of the purified DNA products.[26]

Assuming that nucleosomes limit the accessibility of DNA to restriction endonucleases, three models for nucleosome phasing can be distinguished (Fig. 4). If nucleosomes are uniquely phased with respect to DNA sequence, then a given restriction site will be exposed in either all or none of the chromosomes; therefore, the probability of cutting (X) at a particular DNA sequence is either 0 or 100%. If nucleosomes are randomly phased, then a given restriction site will be exposed in a fraction of the chromosomes; thus, the fraction of mature chromosomes cut is X and the fraction of mature chromosomes not cut is $1 - X$. For replicating chromosomes that have already duplicated the restriction site, the fraction of uncut molecules is $(1 - X)^2$, the fraction with two cuts is X^2, and the fraction cut once is $1 - (1 - X)^2 - (X)^2$ or $(2X - 2X)^2$. Sequential cutting of mature chromosomes with two different restriction endonucleases revealed that the extent of cleavage at the second site is not altered by cleavage at the first site.[26] If nucleosomes are randomly phased, but the two sibling molecules in a single replicating chromosome have identical chromatin structures, then either both sibling chromosomes will be resistant or both will be cut. The data for SV40 are consistent with this model.

Interpretation of the actual extent of cleavage by each restriction enzyme is more complex.[26] Accessibility of SV40 chromosomes to six different restriction endonucleases varied from 13 to 49%, but no significant differences were detected between mature and replicating chromatin. If only the 146-bp nucleosomal core DNA is protected from cleavage, then a

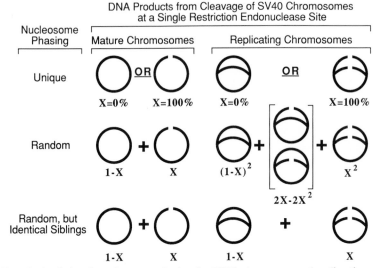

FIG. 4. Analysis of nucleosome phasing in SV40 chromosomes by digestion with a restriction endonuclease that cuts at a single site in nonreplicated or in newly replicated DNA. X is the fraction of restriction sites that are cleaved.

random distribution of 22–24 nucleosomes over 5243 bp of SV40 DNA would predict 33–39% cleavage at any restriction site; protection of the 165 bp of chromosomal DNA would predict 25–31% cleavage; and protection of the 187 bp of nucleosomal core plus linker DNA would predict 15–22% cleavage. However, the extent of cleavage at each site varied with the method of chromosome preparation, and two isoschizomers did not cut chromosomes to the same extent, particularly when chromatin was isolated under hypotonic conditions where many nonhistone chromosomal proteins are present. For example, in the absence of VP1, the major SV40 capsid protein, the fraction of viral chromosomes with accessible restriction sites in the origin–promoter–enhancer region increases as much as 2-fold[52] and the average nucleosome repeat length decreases 10%,[51] revealing that VP1 can protect nonnucleosomal DNA, perhaps by altering nucleosome spacing. Therefore, unambiguous interpretation of the extents of chromosomal cleavage with different restriction endonucleases cannot be made.

[51] V. Blasquez, A. Stein, C. Ambrose, and M. Bina, *J. Mol. Biol.* **191,** 97 (1986).
[52] C. Ambrose, V. Blasquez, and M. Bina, *Proc. Natl. Acad. Sci. U.S.A.* **83,** 3287 (1986).
[53] J. O. Thomas and R. J. Thompson, *Cell* **10,** 633 (1977).
[54] W. Linxweiler and W. Horz, *Nucleic Acids Res.* **10,** 4845 (1982).

TABLE II
ANALYSES OF REPLICATING CELLULAR CHROMATIN

Cell system	Radiolabeling	Chromatin	Enzymes	Analytical methods	Refs.[a]
Tissue culture					
HeLa	[³H]TdR pulse *in vivo*	Isolated nuclei	MNase, DNase I	Kinetics, DNA gel electrophoresis, nucleoprotein (NP) gel electrophoresis	1–5
HTC (Hepatoma tissue culture)	[³H]TdR pulse *in vivo*	Isolated nuclei	MNase	DNA gels, NP gels	6–8
CHO (Chinese hamster ovary)	[³H]TdR pulse *in vivo*	Isolated nuclei	MNase	Kinetics, DNA gels, neutral sucrose gradients (NSGs)	9–11
EAT (Ehrlich ascites tumor cells)	[³H]TdR pulse *in vivo*	Isolated nuclei	MNase	Kinetics, NSGs	12
Friend leukemia cells	[³H]TdR pulse *in vivo*	Isolated nuclei	MNase	Kinetics, DNA gels	13
Isolated cells					
Activated bovine lymphocytes	Isolated nuclei labeled *in vitro* with [α-³²P]dATP	Isolated nuclei	MNase	DNA gels, NSGs	14–16
Chicken erythrocytes	[³H]TdR pulse *in vivo*	Isolated nuclei	DNase I	Kinetics, DNA gels	17
Mouse sarcoma cells	Permeable cells pulse labeled with [³H]dTTP	Whole permeable cells	DNase I	Kinetics	18
Organisms					
Sea urchin blastulas	[³H]TdR pulse *in vivo*	Isolated nuclei	MNase, DNase I	Kinetics, DNA gels, NP gels	19–21
Plant embryos	[³H]TdR pulse *in vivo*	Isolated chromatin	MNase	Kinetics, DNA gels, NSGs	22
Regenerating liver	[³H]TdR pulse perfused into whole animal	Isolated nuclei	DNase I	Kinetics	23

[a] Key to References: (1) R. L. Seale, *Nature (London)* **255**, 247 (1975); (2) R. L. Seale, *Proc. Natl. Acad. Sci. U.S.A.* **75**, 2717 (1978); (3) A. T. Annunziato, R. K. Schindler, C. A. Thomas, and R. L. Seale, *J. Biol. Chem.* **256**, 11880 (1981); (4) A. T. Annunziato and R. L. Seale, *Biochemistry* **21**, 5431 (1982); (5) A. T. Annunziato and R. L. Seale, *Nucleic Acids Res.* **12**, 6179 (1984); (6) V. Jackson and R. Chalkley, *Cell* **23**, 121 (1984); (7) V. Jackson, S. Marshall, and R. Chalkley, *Nucleic Acids Res.* **9**, 4563 (1981); (8) P. A. Smith, V. Jackson, and R. Chalkley, *Biochemistry* **23**, 1576 (1984); (9) C. E. Hildebrand and R. A. Walters, *Biochem. Biophys. Res. Commun.* **73**, 157 (1976); (10) C. Vaury, C. Gilly, D. Alix, and J. J. Lawrence, *Biochim. Biophys. Res. Commun.* **110**, 811 (1983); (11) D. J. Roufa, *Cell* **13**, 129 (1978); (12) V. Pospelov, B. Anachkova, and G. Russev, *Biochim. Biophys. Acta* **699**, 241 (1982); (13) R. F. Murphy, R. B. Wallace, and J. Bonner, *Proc. Natl. Acad. Sci. U.S.A.* **75**, 5903 (1978); (14) E.-J. Schlaeger and K.-H. Klempnauer, *Eur. J. Biochem.* **89**, 567 (1978); (15) E.-J. Schlaeger and R. Knippers, *Nucleic Acids Res.* **6**, 645 (1979); (16) E.-J. Schlaeger, *Biochemistry* **21**, 3167 (1982); (17) D. Hewish, *Nucleic Acids Res.* **4**, 1881 (1977); (18) S. Seki, S. Mori, and T. Oda, *Acta Med. Okayama* **40**, 183 (1986); (19) A. Levy and K. M. Jakob, *Cell* **14**, 259 (1978); (20) G. Galili, A. Levy, and K. M. Jakob, *Nucleic Acids Res.* **9**, 3991 (1981); (21) K. M. Jakob, S. B. Yosef, and I. Tal, *Nucleic Acids Res.* **12**, 5015 (1984); (22) K. Yakura and S. Tanifuji, *Biochim. Biophys. Acta* **609**, 448 (1980); (23) L. A. Burgoyne, J. D. Mobbs, and A. J. Marshall, *Nucleic Acids Res.* **3**, 3293 (1976).

Enzymatic Analysis of Replicating Cellular Chromatin: General Comments

The techniques that are used to analyze replicating viral chromatin can also be adapted to the analysis of replicating cellular chromatin. Table II lists the cellular systems that have been examined and the techniques that have been used. The basic approach remains the same. Replicating cells are radiolabeled with deoxyribonucleotide precursors for a short time to label only nascent DNA at the replication fork. Then, the chromatin structure of the labeled nascent DNA is examined by nuclease digestion. The radiolabeling is usually done with [³H]thymidine *in vivo,* in the same manner that replicating SV40 chromosomes are radiolabeled *in vivo* in SV40-infected cells (see Materials and Methods). Newly replicated cellular chromatin can also be radiolabeled in isolated nuclei.[46]

Digestions of cellular chromatin are done in isolated nuclei. There is great variability in the methods used to produce isolated nuclei. Each cell line requires different methods to break open the cells and isolate nuclei, depending on factors such as the strength of the cell membrane, the size of the cell and its nucleus, the adherence of cytosol contents (e.g., ribosomes) to the nuclei, and the stability of the nuclei. The researcher must find and use those methods for his chosen cell type which yield the best nuclei. Nuclease accessibility to the chromatin may vary. Nuclei are usually permeable to most enzymatic probes. But perhaps condensation of chromatin, or possible chromatin linkage to subnuclear structures such as the nuclear matrix or nuclear lamina, will restrict accessibility of an enzyme to the replication fork. Seemingly small changes in nuclear preparation procedures may produce changes in nuclease accessibility.[50] Great care should be taken to assure reproducibility of the nuclear preparations. SV40 chromosomes have been digested in nuclei, in nuclear extracts, and as isolated chromosomes, but no changes in chromatin structure have ever been detected.[29,35] The studies of replicating cellular chromatin have so far used only MNase and DNase I. There seems no reason why the other nucleases described in this chapter could not be applied to replicating cellular chromatin.

[16] Nuclease Digestion of Transcriptionally Active Chromatin

By Maria Bellard, Guy Dretzen, Angela Giangrande, and Philippe Ramain

Introduction

We can define active chromatin as the structure within which a specific gene and its flanking sequences (up to 100 kilobases DNA[1]) are organized in the phases prior to, during, and postexpression in the cell type(s) in which the gene is expressed. In general, no more than 20% of the genome is organized as active chromatin in a given cell type. Active chromatin appears to differ from inactive or bulk chromatin at all levels of organization, from the histone content of nucleosomes to its higher order structure. These differences involve protein modifications as well as structural changes in the DNA filament.[2-5] Active chromatin is less compact than bulk chromatin, and this is reflected in its increased accessibility to enzymes. This can be seen at several levels. First, we can compare the structure surrounding a specific gene in terms of its nuclease sensitivity in a cell type where there can be no expression to a cell type where there has been, is, or will be expression of that gene. In general, the structure will be more sensitive in the latter case, even when gene expression is not evident. Second, further changes are associated with gene activity and expression during cell development that are due to modifications in chromatin structure of flanking sequences and increased sensitivity in coding sequences, probably reflecting changes in nucleosomal structure during the passage of polymerase. A particular feature in flanking regions is the appearance of hypersensitive sites (or regions) some of which have been correlated with the presence of regulatory proteins.[4,5] Because of this, we should consider active chromatin as a highly labile structure.

[1] G. M. Lawson, B. J. Knoll, C. J. March, S. L. C. Woo, M.-J. Tsai, and B. W. O'Malley, *J. Biol. Chem.* **257**, 1501 (1982).

[2] R. Reeves, *Biochim. Biophys. Acta* **782**, 343 (1984).

[3] J. C. Eissenberg, I. L. Cartwright, G. H. Thomas, and S. C. R. Elgin, *Annu. Rev. Genet.* **19**, 485 (1985).

[4] D. S. Pederson, F. Thoma, and R. T. Simpson, *Annu. Rev. Cell. Biol.* **2**, 117 (1986).

[5] G. H. Thomas, E. Siegfried, and S. C. R. Elgin, *in* "Chromosomal Proteins and Gene Expression" (G. R. Reeck, G. H. Goodwin, and P. Puigdomènech, eds.), p. 77. Plenum, New York, 1985.

Copyright © 1989 by Academic Press, Inc.
All rights of reproduction in any form reserved.

Nuclease digestions may also be used to investigate changes in nucleosome organization and positioning around a given gene in different cell types and stages. Several nucleases, or chemical treatments, are currently used to study chromatin structure. These include DNase I, DNase II, micrococcal nuclease (MNase), S_1 nuclease, and copper/phenanthroline, which have differences in their specificity for DNA sequences or structures.[6] In addition, restriction enzymes may be employed,[7] as well as more recent reagents for the nonenzymatic cleavage of chromatin.[8] In this chapter, we focus on the use of DNase I and MNase to study chromatin structure in intact nuclei directly.

Preparation of Nuclei and Homogenates

As active chromatin is a labile structure, the aim of all procedures is to treat isolated nuclei as rapidly as possible while conserving their physiological structure. It is essential that the nuclei are uniformly accessible to nucleases following the removal of cytoplasm, but there is no particular advantage in purifying nuclei beyond this point. Extended treatments may lead to changes in nuclear structure, which introduce artifacts into the experiment.

Examples of Preparation of Nuclei from Different Sources

All steps should be carried out at 4° wherever possible. (See General Hints and Troubleshooting for notes on improving techniques.)

Chicken Erythrocyte Nuclei.[9] Blood is collected either following decapitation of the chicken or by puncturing the wing vein with a disposable syringe. To avoid coagulation and to evaluate the quantity rapidly, the blood is added at room temperature to a known volume of buffer E.1 (75 mM NaCl, 24 mM EDTA) chosen so that the blood to buffer ratio is between 1 : 1 and 1 : 5. A laying hen will typically yield 50–70 ml of blood. After filtration through nylon gauze (to remove feathers and clots), the sample is centrifuged for 5 min (650 g at 4°). The supernatant is removed with a Pasteur pipette connected to a water vacuum pump to avoid dislodging the pellet. If necessary as an aid to purification, the upper layer of white cells may be partially removed. The erythrocytes are resuspended in a volume of buffer E.1 that is twice the initial volume of blood collected and recentrifuged. The supernatant is again removed, and the pellet is

---

[6] H. R. Drew, *J. Mol. Biol.* **176**, 535 (1984).
[7] J. D. McGhee, W. I. Wood, M. Dolan, J. D. Engel, and G. Felsenfeld, *Cell* **27**, 45 (1981).
[8] I. L. Cartwright and S. C. R. Elgin, this volume [18].
[9] X. Wilhelm and M. Champagne, *Eur. J. Biochem.* **10**, 102 (1969).

resuspended in buffer E.1 (4 times the initial volume of blood) in a Potter homogenizer with a Teflon piston. Small agglomerates are dispersed by gentle strokes of the piston.

The sample is poured into a beaker on ice, and $\frac{1}{40}$ volume of saponin stock solution (10% in water, stored at $-20°$) is added drop by drop with gentle swirling. The sample is left for 20 min with occasional swirling. The lysis of erythrocytes is evident by clarification of the solution. The solution is centrifuged for 5 min (650 g at 4°) in a transparent tube. The supernatant is removed with a Pasteur pipette, and the nuclei are resuspended in 4 volumes of buffer E.2 (75 mM NaCl, 1 mM EDTA). Centrifugation and resuspension are repeated 5–6 times until the pellet of nuclei is no longer pink. The final pellet is rinsed by resuspending in 4 volumes of the buffer to be used for nuclease digestion (see Typical Digestion Conditions, below), centrifuged 5 min, and the pellet resuspended in a standard volume of digestion buffer (some 2–3 times the volume of blood collected). An aliquot is taken for a rapid assessment of DNA content (see Evaluation of DNA Content of Preparations, below), and the remainder recentrifuged. From the results of the DNA content determination, the pellet is resuspended at a concentration of 1 mg DNA equivalent/ml of digestion buffer.

Nuclei from Chicken Organs. This procedure[10] is a one-step modification of Hewish and Burgoyne[11] technique. Dissected tissues (e.g., spleen, liver, or kidney) are placed in a beaker on ice and rinsed with chilled buffer A.1 [15 mM Tris-HCl, pH 7.5, 15 mM NaCl, 60 mM KCl, 15 mM 2-mercaptoethanol, 2 mM EDTA, 0.5 mM EGTA, 0.15 mM spermine, 0.5 mM spermidine, 1.9 M sucrose, 0.1% Triton X-100, 0.1 mM phenylmethylsulfonyl fluoride (PMSF)—the PMSF should be added just before use from a 100 mM stock solution in 80% 1-propanol, stored at room temperature]. The tissue is then cut into small pieces with dissecting scissors and rinsed again with buffer A.1. It is then placed in a Potter homogenizer with a Teflon piston in 10 volumes of buffer A.1. The piston is mechanically driven in a drill chuck, and the cylinder, held in ice, is passed approximately 10 times in the space of 3–4 min. In the case of laying hen oviduct, the pieces of the magnum portion are treated in a tissue press to remove the membranes and fibers. The tissue is collected in 10 volumes of ice-cold buffer A.1 and homogenized by 20 strokes in a Dounce homogenizer (pestle A). In both cases, the homogenates are filtered through a nylon gauze (200 μm) to remove debris and then diluted by the addition of an equal volume of buffer A.2 (= buffer A.1 without Triton X-100). After mixing, an aliquot of the sample is examined in a refractometer, and the

[10] M. Bellard, G. Dretzen, F. Bellard, P. Oudet, and P. Chambon, *EMBO J.* 1, 223 (1982).
[11] D. R. Hewish and L. A. Burgoyne, *Biochem. Biophys. Res. Commun.* 52, 504 (1973).

index of refraction is adjusted to 1.40–1.42 by the gradual addition of buffer A.3 (= buffer A.2 without sucrose).

The sample is centrifuged for 90 min (90,000 g in the middle of the tube) at 5°. The supernatant is poured off, the walls of the tube are immediately wiped with a tissue, and 2 volumes of buffer A.4 (15 mM Tris-HCl, pH 7.5, 15 mM NaCl, 60 mM KCl, 15 mM 2-mercaptoethanol, 0.15 mM spermine, 0.5 mM spermidine, 0.34 M sucrose, 0.1 mM PMSF) is added. The pellet is dislodged with a glass rod, and the sample (pooled from several tubes, if appropriate) is placed in a Potter homogenizer. It is then resuspended by several strokes of the piston, centrifuged 5 min at 3000 g, and the pellet resuspended in a volume equivalent to that of the starting tissue. An aliquot is then taken for a rapid assessment of DNA content (see below), the remainder is recentrifuged, and the pellet is resuspended at a concentration of 1 mg/ml of buffer A.4. A variant of this technique[10] uses buffers without EDTA and EGTA, in which spermine and spermidine are replaced by divalent cations (e.g., 3–5 mM MgCl$_2$).

Nuclei from Cultured Cells.[12] Cells are harvested from suspension cultures (or by scraping, in the case of cells growing in monolayers), rinsed in phosphate-buffered saline (PBS), pH 7.4 (8.1 g NaCl, 0.22 g KCl, 1.14 g Na$_2$HPO$_4$, 0.27 g KH$_2$PO$_4$ per liter), and pelleted by centrifugation at 160 g for 3 min. The pellet is resuspended in buffer C (10 mM Tris-HCl, pH 7.5, 1 mM CaCl$_2$) at a concentration of approximately 2 × 10^7 cells/ml. The cells are incubated on ice 10 min during which time they swell, as can be seen by light microscopy. The cells are then lysed mechanically in a Dounce homogenizer (pestle B). The lysate is then layered over 3 ml of 1.7 M sucrose in buffer C and centrifuged 1 hr at 650 g. This purification using a sucrose cushion is optional and may be omitted in the interest of rapidity; in such a case, the nuclei should be pelleted by centrifugation for 5 min at 160 g, resuspended in buffer C, repelleted, and then resuspended in digestion buffer (~ 100 μl for 1 ml of cells) (see Typical Digestion Conditions).

An alternative procedure uses a detergent-based lysis.[13,14] Cells are harvested as above and resuspended in buffer D (5 mM PIPES, pH 8, 85 mM KCl, 1 mM CaCl$_2$, 5% sucrose) with 0.5% Nonidet P-40 (NP-40) added. The cells are left on ice 10 min during which lysis occurs, and the nuclei are pelleted by centrifugation at 1000 g for 3 min. The nuclei are rinsed 3 times by resuspension and centrifugation in buffer D and, finally,

[12] J. L. Compton, M. Bellard, and P. Chambon, *Proc. Natl. Acad. Sci. U.S.A.* **73**, 4382 (1976).
[13] J. Lengyel, A. Spradling, and S. Penman, *Methods Cell Biol.* **10**, 195 (1975).
[14] P. Ramain, M. Bourouis, G. Dretzen, G. Richards, A. Sobkowiak, and M. Bellard, *Cell* **45**, 545 (1986).

are resuspended at 0.5 mg/ml in digestion buffer. Note that buffer D may be used for DNase I digestion (see below).

Nuclei from Drosophila Embryos.[15] Overnight or staged embryos are collected from a population cage by washing onto a sieve made from nylon gauze (150 μm) and rinsing thoroughly to eliminate gross particles of food. Collection is facilitated by the use of a firm agar laying medium to prevent females from inserting their eggs into the medium. Embryos can be used immediately or frozen in liquid nitrogen and stored at $-80°$. A good population cage will yield at least 1 g of embryos in an overnight (0–18 hr) collection. We routinely prepare nuclei from 0.5–3 g of stored embryos (Fig. 1).

Embryos are dechorionated in 50% Clorox/0.02% Triton X-100 for 2 min at room temperature. After several washes with distilled water, the dechorionated embryos are rinsed with 70% ethanol and air-dried by placing the sieve on a cushion of paper tissues over a sheet of aluminum foil in an ice bucket. All subsequent steps are performed on ice. The dried embryos stick together and are transferred with a spatula to a Dounce

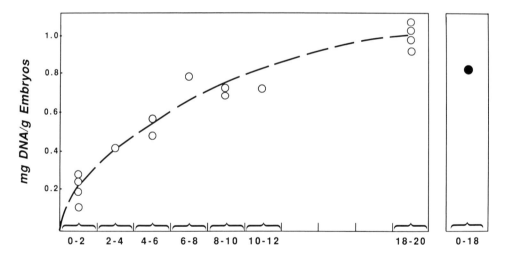

Age of Embryos in hours

FIG. 1. The yield of DNA from nuclei prepared from *Drosophila* embryos of different ages. Eggs are collected during a 2-hr laying period and either treated immediately (0–2 hr sample) or incubated at 25° for varying periods. The yield from an overnight collection (0–18 hr) is shown at right.

[15] C. Wu, *Nature (London)* **309**, 229 (1984).

homogenizer and disrupted in 3 volumes of buffer D (see above) by 10–15 strokes with a B pestle. The homogenate may be passed through filter paper, but this step does not greatly improve the purity of nuclei, in our experience, and may reduce yields. The homogenate is centrifuged at 160 g for 1 min at 4° in a swinging rotor to eliminate the cuticles. The supernatant is centrifuged at 2000 g for 5 min at 4° to pellet crude nuclei. The pellet is resuspended in 2–5 volumes of buffer D and centrifuged through an equal volume of buffer D with 1.7 M sucrose added at 7000 g for 15 min at 4°. The upper layer (lipids) is carefully removed by pipetting, and all the supernatant is discarded. The walls of the tube are wiped with a paper tissue to eliminate any sucrose residue. Purified embryonic nuclei are resuspended for digestion in 2 volumes in buffer D. The yield of nuclei depends on the stage of the embryos. The graph (Fig. 1) shows the yield (mg DNA/g embryos) at different embryonic stages (0–2 hr, 2–4 hr, etc.). The overnight collection will yield approximately 0.8 mg DNA/g embryos.

Drosophila Salivary Gland Homogenates. In some cases, the preparation of nuclei from limited quantities of starting material is impractical. In these cases, studies of chromatin structure can be undertaken in homogenates containing released nuclei. We have successfully studied polytene salivary gland nuclei of staged *Drosophila melanogaster* larvae using this approach.[14]

Hand-dissected salivary glands from different developmental stages are accumulated in aliquots of 50 pairs of glands in 150 μl of buffer D (see above and General Hints and Troubleshooting) in 1.5-ml Eppendorf tubes on ice and then stored at −20°. We routinely use 200 pairs of glands (which yield around 2 μg of DNA) for each gel slot. Several aliquots are placed on ice to thaw and then homogenized by hand with a resin-based pestle preformed to the shape of the tube. The release of the nuclei is monitored by microscopy (see below), and the homogenates are pooled. The homogenization tubes are rinsed with 50 μl of buffer D, which is added to the pooled material. The material is then realiquoted for immediate digestion. The pooling ensures an even redistribution of nuclei for digestion.

Evaluation of DNA Content of Preparations

In order to obtain reproducible results with chromatin digestions, it is essential to be able to monitor both the yield and quality of each sample of nuclei prior to digestion. Once an extraction technique is mastered, there should be little variation from one experiment to another. However, each source of nuclei will have its own characteristics, which must be determined in a series of preliminary experiments. For this, an aliquot of the

TABLE I
CHARACTERISTICS OF NUCLEI PREPARATIONS[a]

Source of nuclei	UV absorption ratios		DNA (μg/ml) corresponding to 1 OD unit	Protein/ DNA ratio
	230/260	280/260		
Laying hen oviduct, 0.5 mg DNA/g tissue, yield ~ 50%	1.24 [1.4–1.29]	0.65 [0.63–0.67]	28 [33–25]	3.3 [2.2–4.2]
Mature hen erythrocytes, 3 mg of DNA/ml of blood, yield ~ 80%	0.94 [0.90–1.04]	0.61 [0.59–0.66]	35 [40–21]	2.0 [1.6–2.8]

[a] In each case the mean for 10 experiments is given, with extreme values shown in square brackets. The DNA equivalence (in μg/ml) per OD unit measured when using the rapid determination technique is derived from an accurate determination of the DNA content using the Burton technique[16] (see text). Note the difference in mean values between the two types of nuclei for each column.

starting material is set aside, and the total DNA content is determined following extraction by a standard technique, such as that described by Burton[16] or the DABA technique.[17] A similar determination is made from an aliquot of purified nuclei. These are used to calculate the yield of nuclei (DNA from isolated nuclei/total DNA). In a reliable purification procedure, the yield should not be lower than 50%. Below 50%, there is a danger that the purification is no longer random and that certain nuclei are preferentially selected. The quality or purity of the sample is characterized by its UV absorption at 230, 260, and 280 nm. The ratios 230/260 and 280/260 reflect the protein contained both within the nuclei and from cytoplasmic contamination and should remain constant for a given source of nuclei extracted by a given protocol. The different values obtained from these ratios indicate the contribution to absorption at 260 nm of molecules other than DNA. If the protein content is determined independently,[18,19] the protein/DNA content may also be calculated and will characterise the purification procedure. Examples of these ratios are shown in Table I.

The rapid evaluation of DNA content, which must be performed for each sample, is based on knowledge of the equivalence of DNA in μg/ml of 1 OD (optical density) unit (260 nm) and should be applied for a given

[16] K. Burton, *Biochem. J.* **62**, 315 (1956).
[17] B. Fiszer-Szafarz, D. Szafarz, and A. G. de Murillo, *Anal. Biochem.* **110**, 165 (1981).
[18] O. H. Lowry, N. G. Rosebrough, A. L. Farr, and R. J. Randall, *J. Biol. Chem.* **193**, 265 (1951).
[19] M. M. Bradford, *Anal. Biochem.* **72**, 248 (1976).

source of nuclei. For example, 1 OD unit of chick oviduct (prepared as above) corresponds to approximately 28 μg/ml (see Table I) as opposed to 50 μg/ml for purified DNA. For rapid evaluation, an aliquot of nuclei is lysed in 10–20 volumes of 2 M NaCl, and the DNA is sheared by vigorous vortexing. The UV absorption of the solution at 230, 260, and 280 nm is read against a 2 M NaCl standard, and the DNA content is estimated.

General Hints and Troubleshooting

For each new source of nuclei, it will be necessary to undertake preliminary experiments so as optimize the experimental conditions. Wherever possible, the nuclei should be examined by phase-contrast light microscopy at each stage of the extraction. This investment of time and effort should avoid unfortunate artifacts. Differences in the size or density of nuclei may cause them to behave differently in centrifugation steps, and it may be necessary to examine both the pellet and supernatant. The examination by microscope will also provide information on both the release of the nuclei from the cell debris (i.e., quality of homogenization) and the physiological state of the nuclei (choice of buffers).

The basic components of a nuclear extraction buffer are the following: (1) buffer (Tris-HCl, pH 7.5, PIPES, etc.); (2) KCl and NaCl in a combined amount of approximately 80 mM to ensure with the other components the physiological ionic strength of the buffer; (3) dithiothreitol (DTT) or 2-mercaptoethanol to stabilize sulfhydryl groups in nuclear proteins; (4) divalent cations (Mg^{2+}, Ca^{2+}) or polyvalent cations (spermine, spermidine) to maintain nuclear membrane integrity and chromatin structure;[20] (5) protease inhibitors (e.g., PMSF); and (6) sucrose at 0.3–2.0 M. In preliminary experiments, it is advisable to vary the composition of a given extraction buffer and to verify the state of the nuclei, especially when starting work with a new tissue or species. An example is shown in Fig. 2 where *Drosophila* polytene nuclei are compared in extracts using a classic vertebrate buffer[11] and a buffer chosen experimentally to preserve maximally the internal structure. In other cases, for example, in hypotonic buffers, a distinct swelling of nuclei may occur so that internal structures are no longer apparent; in extreme cases, there may be lysis of the nuclei seen as distorted and broken spheres. Lysis will also be evidenced by a dramatic increase in the viscosity of the solution due to released DNA.

If regions of cytoplasm remain attached to the nuclei following homogenization, they will prevent reproducible nuclease digestions. It is advisable to change the homogenizer or increase the rate or duration of homo-

---

[20] P. R. Walker, M. Sikorska, and J. F. Whitfield, *J. Biol. Chem.* **261**, 7044 (1986).

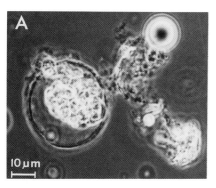
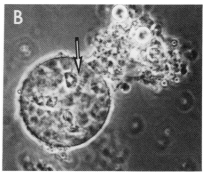

FIG. 2. (A) Polytene nuclei from third instar *Drosophila melanogaster* salivary glands homogenized in buffer A.2 (see text). (B) The same nuclei homogenized in buffer D (see text). In (A) the chromatin is compacted, and a large part of the nucleus appears empty. In (B) the internal structure is preserved, and chromosome banding patterns (arrow) are apparent.

genization and control by examination with the microscope. If homogenization is poor, many nuclei will remain trapped in the tissue debris, and this will result in a low overall yield. Small amounts of detergents (e.g., Triton X-100 or NP-40) may help the homogenization and improve yield, but they should be used sparingly and diluted as soon as possible following homogenization (see Nuclei from Chicken Organs). If the nuclei are isolated, but the nuclear pellet contains granules of cytoplasmic debris, this will be reflected in the UV measurements (see Evaluation of DNA Content of Preparation). This contamination may be removed by changing either the sucrose concentration and/or centrifugation conditions. Isolated nuclei should not stick together. The formation of agglomerates may indicate either an excessive use of detergents, which may modify the external layer of the nuclear membrane,[21] or centrifugation speed that is too high, especially in the final steps in low sucrose concentrations. If, despite these precautions, a few agglomerates are found in the preparation that cannot be resuspended, they may be removed by passing the sample through a nylon filter (chosen so as to permit the passage of individual nuclei) held in a stainless steel support (e.g., Millipore XX30 025 00). It will be obvious that agglomerates are a serious problem if clumps are seen falling rapidly to the bottom of the tube leaving a clear supernatant following resuspension, which normally results in an opaque solution.

If it is necessary to work with frozen material, it is preferable to freeze

---

[21] R. P. Aaronson and G. Blobel, *J. Cell Biol.* **62**, 746 (1974).

tissues rather than isolated nuclei. If a procedure has been established with freshly prepared nuclei, it will not necessarily give the same results with frozen tissues. This is a consequence of the high lability of chromatin structures, which may be differentially sensitive to the freezing procedure.

Digestion of Nuclei by Nucleases

Although it is possible to study the activity of endogenous nucleases on chromatin structure exclusively,[22,23] the majority of experiments use exogenous nucleases. Nonetheless, the potential of an important contribution from endogenous nucleases, in certain types of nuclei (e.g., liver or pancreatic cells), should not be overlooked in the experimental design. In most experiments, we are interested in the rate of disappearance of DNA fragments (or the appearance of their digestion products). Because of the many variables involved, in practice it is not possible to reproduce exactly the same digestion conditions for each experiment. For this reason, all protocols include a series of progressive digestion points, ensuring an overlap between experiments while permitting a comparison within experiments of the behavior of individual fragments.

DNase I

Principle of Digestion. DNase I (bovine pancreas, MW 31,000; EC 3.1.21.1) digests double-stranded DNA by cutting one of the two strands by a nucleophilic attack on the $O-3'$-P bond. Although it has been suggested that the enzyme shows a specificity for AT-rich regions,[24] more recently it has been proposed that preferred sites are regions showing structural modifications of the DNA helix.[6] Digestion within single-stranded regions is two orders of magnitude lower than that of double-stranded DNA. Digestion is optimal at pH 7.8 and requires the presence of divalent cations (Mg^{2+}, Ca^{2+}, or Mn^{2+}). In the presence of Mg^{2+} alone, the enzyme cuts only one of the two strands, and it has been suggested that addition of Ca^{2+} both stabilizes the enzyme and facilitates a second cut in double-stranded DNA in close proximity to the first.[25] Current protocols use 3 or 5 mM Mg^{2+} with 0.1 mM Ca^{2+}. Under these conditions, digestion is rapid and difficult to control. We prefer to use 1 mM Ca^{2+}, in the absence of Mg^{2+}, which results in a digestion some 15 times slower, in combination with a higher concentration of enzyme (see below).

[22] H. P. Fritton, A. E. Sippel, and T. Igo-Kemenes, *Nucleic Acids Res.* **11**, 3467 (1983).
[23] C. A. Pinkert, D. M. Ornitz, R. L. Brinster, and R. D. Palmiter, *Genes Dev.* **1**, 268 (1987).
[24] P. P. Nelson, S. C. Albright, and W. T. Garrard, *J. Biol. Chem.* **254**, 9194 (1979).
[25] P. A. Price, *J. Biol. Chem.* **247**, 2895 (1972).

Preparation of DNase I Solutions. DNase I is supplied as a lypohilized powder. We recommend that the entire contents of a vial be dissolved by the addition of 1 ml/mg of DNase I stock buffer (20 mM Tris-HCl, pH 7.5, 50 mM NaCl, 1 mM DTT, 100 μg/ml BSA, 50% glycerol), aliquoted, and stored at $-20°$. Under these conditions, the enzyme is stable for several months. Working dilutions are in 15 mM NaCl or in the chosen digestion buffer. Using a DNase I preparation at 1600 U/mg protein, mild digestion conditions require 0–30 μg/ml (i.e., 0–50 U/ml), extended digests 0–150 μg/ml (see below). Because the enzyme is sensitive to physical denaturation, vortexing and shaking should be avoided. Mix by inverting gently. A given preparation of DNase I must be characterized for its activity. A formal analysis is given by Kunitz,[26] but in practice, it is sufficient to rely on the supplier's analysis for initial experiments. However, it is important to test new batches in progressive digests by comparison with previous batches, preferably well before the current stock is exhausted.

Effect of Divalent Cations. Because DNase I activity is dependent on divalent cations, this should be taken into account in the preparation of nuclei. If protocols include EDTA, EGTA, spermine, and spermidine (the last two replacing divalent cations), it will be necessary to remove the EDTA and EGTA by a series of washes in buffers containing spermine and spermidine or divalent cations and to add divalent cations prior to digestion. A similar procedure is necessary if the concentration of cations in which the nuclei have been prepared is to be adjusted.

Choice of Protocol. In all cases, it is necessary to examine samples from a progressive digestion of chromatin. This may be achieved either by a time course or by altering enzyme concentrations for a fixed time of digestion. The latter is preferable in that each sample is exposed to endogenous nucleases having similar activity. In addition, time sampling from a single digest is less homogeneous. Two zero points should be prepared. The first, without incubation, serves to monitor all steps of extraction, electrophoresis, transfer, etc. The second reveals the importance of endogenous nucleases during the time of incubation.

A further control is digestion of naked DNA, using either purified genomic DNA[27] or a fragment from a recombinant plasmid.[28] This permits comparison of the digestion of chromatin with digestion due to sequence specificity[29,30] or structural peculiarity of the naked DNA from the region

[26] M. Kunitz, *J. Gen. Physiol.* **33**, 349 (1950).
[27] M. Bellard, F. Gannon, and P. Chambon, *Cold Spring Harbor Symp. Quant. Biol.* **42**, 779 (1978).
[28] M. Bellard, G. Dretzen, F. Bellard, J. S. Kaye, S. Pratt-Kaye, and P. Chambon, *EMBO J.* **5**, 567 (1986).
[29] C. Dingwall, G. P. Lomonosoff, and R. A. Laskey, *Nucleic Acids Res.* **9**, 2659 (1981).
[30] W. Hörz and W. Altenburger, *Nuclei Acids Res.* **9**, 2643 (1981).

under study. If using genomic DNA, it should be prepared as high molecular weight DNA.[31] In the case of plasmid fragments, they should be mixed with exogenous DNA so as to maintain overall DNA–enzyme concentrations.

Because of the lability of chromatin structures, the temperature of digest is important. Digests are normally performed between 20° and 37°, but a control should also be set up at lower temperatures (see MNase protocol for further discussion).

Typical Digestion Conditions. We can distinguish two general categories of digestion with DNase I. For studies of DNase I hypersensitive regions,[32-34] mild conditions of digestion are employed. In contrast, where the extended digestion of short regions is under study (e.g., for liquid hybridization, see below), the concentration of enzyme, the time of digestion, and the temperature of digestion are increased.

For mild digestion, nuclei are suspended at 1 mg/ml in 85 mM KCl, 1 mM CaCl$_2$, 5 mM PIPES buffer, pH 7.5, 5% sucrose and incubated for 20 min at 20°. Digest between 150 and 1000 μg of DNA for each sample. DNase I is added at a concentration of 0–30 μg/ml for chromatin surrounding an active gene and 0–150 μg/ml for the same region in a tissue where the gene is inactive.

For extended digests, samples are incubated in the same buffer modified to contain 3 mM MgCl$_2$ and 0.1 mM CaCl$_2$, from 0–40 min at 37° with about 10 μg/ml DNase I.[27] The extent of digestion is characterized by the acid solubility of the resulting material. For this, NaCl (2 M final concentration) and perchloric acid (1 M final concentration) are added to an aliquot ($\sim$50 μg) at 0°, which is then centrifuged (600 g, 10 min, 4°). The soluble DNA in the supernatant is measured by UV absorption at 260 nm and the insoluble DNA in the pellet by the Burton[16] assay. It is normally more convenient to measure only the DNA in the supernatant. In this case, the total DNA content (i.e., 100%) is measured in the supernatant of a nondigested aliquot at the zero time point, which is treated with perchloric acid at 80° for 30 min before centrifugation. The remaining values for the digestion points are expressed as a percentage of total DNA.

In both cases, digests are stopped by the addition of an equal volume of stop buffer [50 mM Tris-HCl, pH 7.5, 150 mM NaCl, 15 mM EDTA, 0.3% sodium dodecyl sulfate (SDS)]. Following mild treatments, the extent

[31] M. Gross-Bellard, P. Oudet, and P. Chambon, *Eur. J. Biochem.* **36,** 32 (1973).
[32] J. S. Kaye, M. Bellard, G. Dretzen, F. Bellard, and P. Chambon, *EMBO J.* **3,** 1137 (1984).
[33] J. S. Kaye, S. Pratt-Kaye, M. Bellard, G. Dretzen, F. Bellard, and P. Chambon, *EMBO J.* **5,** 277 (1986).
[34] C. Wu, this volume [14].

of nuclease digestion is evaluated in the samples by gel electrophoresis after purification of the DNA (see below).

There are special cases when digestion time should be shortened to 2–5 min, such as a tissue containing an important endonuclease activity. Another example is *Drosophila* salivary gland material where each sample contains a small quantity of DNA in total gland homogenates (see above). In these cases, it is important to stabilize the temperature of tubes, solutions, etc., prior to the addition of enzyme so as to obtain reproducible digests.

Micrococcal Nuclease (MNase)

Principle of Digestion. Micrococcal nuclease (or staphylococcal nuclease, MW 16,800; EC 3.1.31.1) cuts both strands of a DNA molecule simultaneously via the O–5′–P bond and has both an endonuclease and exonuclease activity, the former being more important. Single-stranded DNA is cut about 100 times faster than double-stranded DNA.[35] The enzyme cuts pA and pT 30 times faster than pC and pG,[6] but surrounding sequences play a role in its specificity. Its activity has an absolute requirement for Ca^{2+}. Optimal activity is at pH 9.2. In chromatin, MNase cuts preferentially in the spacer regions between nucleosomes.[24]

Preparation of Solutions. Micrococcal nuclease is best used batchwise. The entire contents of a vial of MNase are dissolved in MNase buffer (50 mM Tris-HCl, pH 8, 0.05 mM $CaCl_2$; 20% glycerol) at 100 units/μl and are then aliquoted and stored at $-20°$. Final dilutions are made in the digestion buffer immediately prior to use. New batches should be tested simultaneously with the previous batch by performing a progressive digestion of nuclei or purified DNA, followed by electrophoresis on 1.7–2% agarose gels (Fig. 6A).

Choice of Protocol. The same general considerations apply as in the case of DNase I (see above). In view of the specificity of MNase for certain DNA sequences, the digests of naked DNA are especially important. Digestions are generally carried out between 20° and 37°. However, it has been shown that nucleosomes may "slide" along the DNA fiber. This effect is thought to be particularly important at high temperatures and for H1-depleted chromatin.[36,37] It has been suggested that active chromatin has alterations in its H1 content which are reflected in a destabilized nucleosomal structure.[38] For these reasons, it is important to undertake control digestions at

---

[35] M. L. Dirksen and C. A. Dekker, *Biochem. Biophys. Res. Commun.* **2,** 147 (1960).
[36] C. Spadafora, P. Oudet, and P. Chambon, *Eur. J. Biochem.* **100,** 225 (1979).
[37] F. Thoma, L. W. Bergman, and R. T. Simpson, *J. Mol. Biol.* **177,** 715 (1984).
[38] H. Weintraub, *Cell* **38,** 17 (1984).

0° where nucleosomal sliding is minimal. The following may serve as a guide to estimate the conditions necessary for digestion at 0°. By reference to digestion at 20° under our typical conditions (see below), digestions are some 10–15 times faster at 37° and 60–80 times slower at 0°. Enzyme concentrations should be adjusted accordingly.

Typical Conditions. MNase digestions may be used to study general sensitivity,[39] nucleosomal positioning,[40–42] hypersensitive sites,[32] nucleosomal ladders, and other structures[43,44] (see below). With the exception of the last application,[44] where digestion at 37° is often employed, material for all other techniques may be obtained by a progressive digestion at 20°, controlled by a digestion at 0° (see above). The extent of digestion is estimated by running aliquots of purified DNA on agarose gels (see Gel Analysis and Fig. 6A). In the case of extended digests, acid-soluble material is estimated as for DNase I above, except that in this case the Burton reaction[16] must be used with the pellet material as the supernatants will contain variable amounts of RNA, digested by the MNase.[45]

In a typical MNase digest, nuclei equivalent to 150–1000 μg of DNA suspended at concentrations of 1 mg/ml in digestion buffer (buffer A.4, see above, plus 1 mM CaCl$_2$) are digested for 20 min at 20° with 5–600 units of MNase/ml. For naked DNA, use 0.5–10 units/ml (for plasmid fragments add heterologous DNA as for DNase I). In the case of nuclei prepared in a buffer in which spermine and spermidine have been replaced by Mg^{2+}, they may be treated either following the addition of 1 mM CaCl$_2$ or by exchanging the Mg^{2+}-based buffer for one containing 1 mM CaCl$_2$. The stop buffer is as for DNase I (see above).

DNA Purification and Redigestion by Restriction Enzymes

For gel analysis, the DNA from partial nuclease digestion is purified. For some purposes, it is necessary to redigest the purified DNA to completion with restriction enzymes (see below). In all extractions, phenol is saturated with 500 mM Tris-HCl, pH 8,[31] with 8-hydroxyquinoline added (1 g/liter). Chloroform is a mixture of chloroform with 3-methyl-1-butanol (isoamyl alcohol) (24:1, v/v).

Standard Procedures. The nuclease digestion is terminated by addition

[39] M. Bellard, M. T. Kuo, G. Dretzen, and P. Chambon, *Nucleic Acids Res.* **8**, 2737 (1980).
[40] R. T. Simpson, *Bio Essays* **4**, 172 (1986).
[41] S. A. Nedospasov, A. N. Shakhov, and G. P. Georgiev, this volum [21].
[42] B. Neubauer and W. Hörz, this volume [31].
[43] Y. L. Sun, Y. Z. Xu, M. Bellard, and P. Chambon, *EMBO J.* **5**, 293 (1986).
[44] R. T. Simpson and L. W. Bergman, *J. Biol. Chem.* **255**, 10702 (1980).
[45] K. K. Reddi, this series, Vol. 12, p. 257.

of an equal volume of stop buffer (see above) which causes the lysis of the nuclei. The sample is treated for 1 hr at 37° with 50 μg/ml of RNase A (a 100× stock solution is prepared as 5 mg/ml in 150 mM NaCl, heated 30 min at 80°, and then stored at −20°), and then overnight at 37° with 50 μg/ml of proteinase K (Merck, Darmstadt, FRG). The DNA is phenol–chloroform extracted, reextracted with chloroform, and then precipitated overnight at −20° with 2 volumes of ethanol in the presence of 200 mM NaCl. The DNA is recovered by centrifugation (20,000 g, 20–45 min, depending on DNA concentration), washed with 70% ethanol, air-dried, resuspended in sterilized distilled water or TE (10 mM Tris-HCl, pH 7.5, 1 mM EDTA) and stored at −20°. The DNA content is estimated by reading the UV absorption of an aliquot diluted in distilled water at 230, 260, and 280 nm. The ratio 230/260 should be from 0.42 to 0.45, and 280/260 should be from 0.52 to 0.56. If these ratios are respected, 1 OD unit at 260 nm correspond to 50 μg of DNA/ml. At this point the extent of nuclease digestion may be examined by running samples of 1–5 μg on a 0.7% agarose gel (see Evaluation of DNA Content of Preparations, above).

Special Cases. In certain cases, purification by standard techniques results in a DNA sample that is refractory to redigestion by restriction enzymes. In the case of *Drosophila* salivary gland homogenates (see above), the digestion of nuclei from 200 salivary glands in a volume of 800 μl necessitates the purification of approximately 2 μg of DNA. To obtain material that is readily redigested by restriction enzymes, the reaction is stopped by the addition of an equal volume of stop buffer F (200 mM Tris-HCl, pH 9, 30 mM EDTA, 2% SDS). The sample is then treated for 1 hr at 37° with 100 μg/ml proteinase K and extracted with phenol–chloroform. The organic phase, together with the interphase, is reextracted with a small volume of TE (see above) which is added to the first supernatant. This is reextracted with chloroform, and the DNA is precipitated with ethanol overnight at −20°. The DNA is recovered by centrifugation (45 min at 20,000 g), washed with 70% ethanol and resuspended in TE. It is then treated with 50 μg/ml RNase A (see above) for 15 min at 37°, and an aliquot is taken to assess the DNase I digestion pattern. After addition of 10 μg of sonicated salmon sperm DNA to the remainder, the samples are precipitated with ethanol. The exogenous DNA increases the DNA concentration and apparently redistributes the contaminants that prevent redigestion of the *Drosophila* DNA.

Monitoring the Redigestion. If appropriate, the purified DNA is redigested with restriction enzymes, following suppliers' recommendations regarding digestion buffer. We routinely redigest 20–50 μg of DNA for each sample in a volume of 200–500 μl. The amount of enzyme necessary is determined by digesting aliquots of 400 ng of the DNA sample, mixed with 100 ng of viral or plasmid DNA, with a known digestion pattern for the enzyme (e.g., for *Eco*RI and *Bam*HI–adenovirus (Ad2); *Hin*dIII and *Pst*I–SV40; *Hpa*II–pSP). The aliquots are digested for 4 hr with increasing quantities of enzyme, and the digest is monitored by the appearance of viral or plasmid fragments. In this way, the minimum quantity of enzyme necessary is determined, and the reaction is scaled up to digest the sample (without viral DNA), respecting the ratio of DNA to enzyme to digestion volume of the test. Once the conditions have been established for a given combination of DNA and enzyme, they should, nonetheless, be routinely checked by taking a 400-ng aliquot from each sample after 30 min of incubation and adding this to 100 ng of control DNA. Both sets of digestions are continued for 4 hr, and the digestion of the viral DNA is analyzed by gel electrophoresis. If the test digestion is complete, the samples are precipitated with ethanol for gel electrophoresis. Although indirect, this test is usually a reliable guide for redigestion. If the test digestion is incomplete, it may be possible to add enzyme to the sample or to repurify the DNA and redigest it. We strongly recommend this procedure to avoid continuing the experiment with incomplete digests, which will give uninterpretable autoradiographs.

Gel Analysis

The fragments of DNA produced by the digestions are analyzed by separation on agarose gels, blotting to a filter support, and hybridization with radioactive probes. Both vertical and horizontal gels may be used; a simple horizontal gel electrophoresis apparatus, as used for the analysis of DNA restriction fragments, is probably the most convenient.

Choice of Gel Conditions. The concentration of agarose in the gel, usually between 0.7 and 2.5% is chosen according to the size of fragments expected. This can be utilized to expand regions of particular interest for the analysis (see Fig. 3). The thickness of the gel, as well as the width of the loading well, is chosen according to the amount of DNA to be deposited. We typically load 20–50 μg of DNA at a concentration of 1.5–2 μg/μl in loading solution (0.17% SDS, 0.017% bromphenol blue, 5% Ficoll 400 in 0.5 mM NaOH: prepare as a 10× stock solution and store at room temperature). We use 6 mm × 1–2 mm slots in a 3.5–4.5 mm thick gel to study single-copy genes from vertebrates. In the case of *Drosophila* with a 10-fold less complex genome, 2–5 μg is sufficient. Electrophoresis is typically for 1 hr at 3 V/cm followed by 6–8 hr at 4 V/cm in TAE buffer (40 mM Tris–acetate, pH 8, 2 mM EDTA) with circulation.

Internal Markers. Despite precautions in gel preparation and electro-

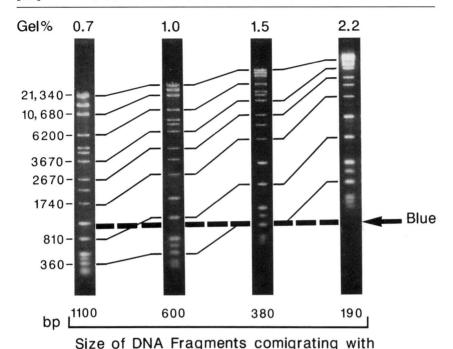

FIG. 3. Migration of DNA fragments and blue dye marker in agarose gels. DNA fragments are electrophoresed in gels of differing agarose content (%) in a horizontal apparatus as described in the text. The dye in the blue loading solution was run to a constant position (arrow), after which the gels were stained with ethidium bromide, rinsed, and photographed under UV illumination.

phoresis, there will always be differences in the mobility of molecules run in adjacent lanes. Apart from mechanical distortion during the blotting procedures, this is a consequence of intrinsic factors, such as local concentrations of DNA molecules resulting from the size distribution following digestion. For this reason, we recommend internal size markers in each slot. These are prepared as restriction-enzyme digests of viral DNA. Plasmid DNA should be avoided because of cross-reaction with contamination in probes (see Hybridization, below) and possibly with bacteria contaminating the original sample (e.g., gut content of *Drosophila*). We use approximately 2 ng per slot of a mixture of *Hin*dII-plus *Hin*dIII-cut SV40 and *Bam*HI- and *Eco*RI-cut adenovirus (Ad2), which produces markers from 200 to 20,000 bp (see Fig. 3). The markers are revealed by a second hybridization of the filter.

Blotting to DBM Paper

Although most commercially available filters may be used for blotting gels, we prefer to use DBM (diazobenzyloxymethyl)-treated paper.[46] DNA is fixed covalently to the filters, which have the particular advantage of strength and may be used for at least 20 cycles of hybridization. This enables a single filter to be hybridized with a variety of probes.

Before treatment, we cut Whatman cellulose 540 filters to standard sizes for our gels and then wash them 2 times for 10 min in water and 2 times for 10 min in acetone, which is allowed to evaporate. The filters are dried at 60° and stored in sealed plastic bags at room temperature. Filters are then treated to produce (nitrobenzyloxymethyl) (NBM) papers in batches, as described,[46] except that we use twice the amount of (1-[(m-nitrobenzyloxy)methyl]pyridinium chloride) (NBPC). They may be stored at 4° for at least 1 year at this step. The reduction of NBM to (aminobenzyloxymethyl) (ABM) filters is done as in Ref. 46, and papers may be stored at this stage for at least 1 week at 4° in water. The final activation, which takes approximately 45 min, should be started 5 min before the end of the NaOH treatment of the gel (see below).

After electrophoresis, the gel may be stained with ethidium bromide to reveal the distribution of the bulk DNA, but it is preferable to avoid this additional treatment. The DNA is denatured by treating the gel in 0.4 N NaOH, 1 M NaCl for 10 min (1% gel) or 15 min (2% gel) at room temperature. The gel is neutralized by three treatments with 1 M sodium acetate, pH 5.5 (prepared by mixing 500 ml of 1 M sodium acetate with 175 ml of 1 M acetic acid at 20°), for 10, 15, and 20 min, respectively, and then blotted overnight, essentially after Southern,[47] in the same buffer at room temperature. The paper is then washed 2 times for 20 min in 0.4 N NaOH and then 4 times for 5 min in distilled water with gentle agitation. The alkali treatment removes DNA that is not covalently bound and inactivates the remaining reactive groups on the DBM paper. At all steps, excess liquid may be removed by placing the filter on absorbent paper, but the filter should never be allowed to dry. The filter is transferred to prehybridization buffer (see Hybridization of Filters) and sealed in a plastic bag. It may be stored at 4° or used immediately by placing at 42°.

Notes on the Use of DBM Paper. (1) The maximum reactivity of DBM paper is short-lived, hence the need to prepare papers just before the transfer of gel. It is preferable that the gel waits for the paper rather than vice versa. (2) DBM paper reactivity is better maintained at low temperature, but there is a reduced rate of transfer DNA from the gel. Conversely,

---

[46] J. C. Alwine, D. J. Kemp, B. A. Parker, J. Reiser, J. Renart, G. R. Stark, and G. M. Wahl, this series, Vol. 68, p. 220.

[47] E. M. Southern, *J. Mol. Biol.* **98,** 503 (1975).

at 37°–42° transfer is faster, but the reactivity is rapidly lost. Our choice of room temperature for transfer is a compromise between these two factors. (3) Although DNA fragments of 150–250 bp appear to be fixed correctly to these papers, their hybridization under standard conditions is markedly reduced in comparison to longer fragments.

Hybridization

Labeling Techniques. Probes may be prepared using any of the current labeling techniques (e.g., nick-translation,[48] M13 single-stranded primer extension,[49] single-stranded RNA probe[50]). We currently use nick-translation of purified fragments (see below) or, if material to be hybridized is limited, a modified protocol for the synthesis of M13-derived, single-stranded probes (see Preparation of Probes).

Preparation of Fragments.[51] In our experience, probes should be chosen so as to be between 300 and 1000 bp in length. We prefer to label purified DNA fragments to avoid background resulting from labeled vector sequences. Fragments may be purified by any of the techniques described by Smith.[52] We routinely recover restriction fragments from acrylamide gels by electroelution but have encountered problems in labeling fragments recovered from agarose gels by this procedure. In this case, we use a modification of our previously published technique (Ref. 51 and Fig. 4). Plasmid DNA (15–25 μg) is digested, phenol–chloroform extracted, and ethanol precipitated. The DNA is resuspended in 400–500 μl of loading solution (see Choice of Gel Conditions, above) and applied to either a vertical or horizontal agarose gel in a single 12-cm-wide slot. Under these relatively diluted conditions, cross-contamination of bands is minimized. Electrophoresis is in TAE buffer (see above) at about 4 V/cm.

After electrophoresis, the gel is stained in ethidium bromide and the relevant band(s) is excised with a scalpel on a clear acrylic, ultraviolet-translucent (UVT), plastic plate placed on a UV source filtered for 300–360 nm (to avoid damage to DNA by shorter UV emissions). The gel fragment is cut into a series of fragments increasing in size, which are stacked on a strip of DEAE-cellulose Whatman DE-81 paper on a second gel, as shown in Fig. 4, for a vertical gel apparatus. If using a horizontal gel apparatus, an additional full-width gel fragment should be placed behind

[48] P. W. J. Rigby, M. Dieckmann, C. Rhodes, and P. Berg, *J. Mol. Biol.* **113**, 237 (1977).
[49] F. Sanger, S. Nicklen, and A. R. Coulson, *Proc. Natl. Acad. Sci. U.S.A.* **74**, 5463 (1977).
[50] D. A. Melton, P. A. Kreig, M. R. Rebagliati, T. Maniatis, K. Zinn, and M. R. Green, *Nucleic Acids Res.* **12**, 7035 (1984).
[51] G. Dretzen, M. Bellard, P. Sassone-Corsi, and P. Chambon, *Anal. Biochem.* **112**, 295 (1981).
[52] H. O. Smith, this series, Vol. 65, Part I, p. 371.

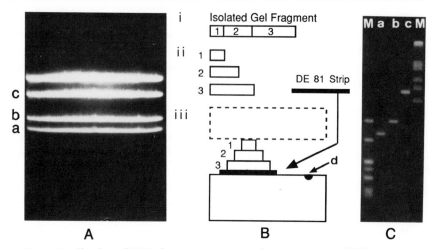

Fig. 4. Purification of DNA fragments on preparative agarose gels. (A) The restriction fragments (a, b, c, etc.) are separated using a slot of broad width. (B) The excised band (i) is then cut to give a series of fragments (ii) which are stacked on the DE-81 strip (iii) on top of a second gel (note that an unused portion of the first gel may be salvaged for this purpose). In the case of a horizontal gel, a second gel piece (dashed lines) is used to maintain contact between gel pieces and the DE-81 paper; d indicates a slot containing blue dye to monitor the second electrophoresis. (C) The purified fragments (a, b, c) are controlled with size markers (M).

the stack so as to maintain contact between the gel pieces and the paper. Strips of DEAE paper should be precut for the gel section and soaked 2 times for 2 hr in 2.5 M NaCl, washed 5 times for 10 min in distilled water, and stored in 1 mM EDTA at 4°. The second electrophoresis is monitored by the progress of the blue dye (Fig. 4), and the absorption of the DNA onto the DEAE paper is verified by UV illumination of the gel. The fluorescence should move from the gel fragments to the paper.

The strip of paper is then placed in a hemolysis tube in 600 μl of elution buffer (1.5 M NaCl, 20 mM Tris-HCl, pH 7.5, 1 mM EDTA) per 100 mm² of paper. It is dispersed by rapid vortexing and then incubated 1 hr at 37° with occasional vortexing. The slurry is pelleted by centrifugation (5 min, 5000 g). The supernatant is recovered and poured into a pierced 1.5-ml Eppendorf tube containing a siliconized glass wool plug placed in the mouth of a larger tube. The assembly is then centrifuged (5 min, 5000 g). The original pellet is reextracted with one-half volume of elution buffer for 15 min, the whole added to the pierced Eppendorf tube and centrifuged. The combined filtrates are extracted once with water-saturated 1-butanol to remove ethidium bromide, and the aqueous phase is precipitated overnight by adding 2 volumes of ethanol. The DNA is recovered by centrifugation for 1 hr at 20,000 g; the pellet is washed twice with 70% ethanol, dried, and resuspended in distilled water at a concentration of about 50 ng/μl. It is then freeze–thawed 3 times ($-20°$) and centrifuged for 15 min at 10,000 g, 4°, to remove any remaining traces of agarose. The concentration of DNA is checked by passing an aliquot on a gel, and the remainder is stored at $-20°$.

Preparation of Probes

Nick-translated probes. The purified fragments are nick-translated following the method of Rigby *et al.*[48] We routinely obtain a specific activity of $1-3 \times 10^8$ cpm/μg using three labeled nucleotides (800 Ci/mmol). We prepare an excess of approximately 10-fold labeled probe relative to the sequences to be hybridized. In the case of an 18-slot gel carrying 20–50 μg DNA per slot, we label 25–50 μg of a 0.5- to 1-kb fragment for a single-copy gene from vertebrates. The probe is purified from free nucleotides, using Sephadex G-50 with either a fixed or spun column procedure, in 25 mM Tris-HCl, pH 7.5, 2 mM EDTA, 0.1% SDS buffer and recovered in 500–700 μl. It is then denatured in the presence of 500 μg of sonicated salmon sperm DNA by boiling for 5 min, followed by rapid chilling on ice. The probe is then added to a final volume of 10 ml of hybridization buffer consisting of prehybridization buffer (see Hybridization of Filters) supplemented so as to contain 12% dextran sulfate 500,000.[53] The dextran sulfate should be prepared as a 50% stock solution in 20 mM EDTA (stored at 4°). The probe in hybridization buffer is incubated 4–6 hr at 42° with a sheet of Whatman cellulose 540 (prewetted in prehybridization buffer) to absorb impurities. The probe is now ready for hybridization (see below).

M13 probes. Suitable DNA fragments are cloned into the polylinker of M13, and single-stranded templates are obtained and purified by standard techniques [polyethylene glycol (PEG) precipitation followed by CsCl-gradient purification[54]]. The preparation is analyzed by passing an aliquot on an agarose gel, and the amount estimated by UV absorption (1 OD unit at 260 nm corresponds to 40 μg/ml). The probe is synthesized by hybridizing the universal primer (17-mer) to sequences flanking the polylinker, and the primer is then elongated with Klenow polymerase in the presence of four dNTPs. While maximum specific activity will be obtained by using all four nucleotides radiolabeled, such probes are subject to rapid radiolysis, and we prefer to use a single, marked nucleotide (dCTP, 800 Ci/mmol), which results in a probe that may be used and reused for at least 7–10 days. The labeled, single-stranded probe is purified on an agarose–formamide denaturing gel and recovered by electroelution. Such probes result in hybridization signals some 10 times stronger than for the same fragment labeled by nick-translation.

[53] G. M. Wahl, M. Stern, and G. R. Stark, *Proc. Natl. Acad. Sci. U.S.A.* **76**, 3683 (1979).
[54] G. Heidecker, J. Messing, and B. Gronenborn, *Gene* **10**, 69 (1980).

Regarding establishing conditions for probe synthesis, this technique is simplified in that a single elongation step is used and there is no redigestion by restriction enzymes (see Ref. 55 for standard technique). We commonly use probes of 500–1000 bases. As the rate of elongation on individual templates differs, for a probe of 500 bases, we optimize the yield of molecules between 300 and 800 bases long. This is done by limiting the amount of labeled dCTP in the reaction, the other nucleotides being present in excess (about 80×). Conditions are established in preliminary experiments in which the proportion of template to dCTP is varied. The rationale is as follows: 1 μg of single-stranded M13 contains 66.7 ng of a 500-base insert sequence, corresponding on average to 16.7 ng of each nucleotide. As 1 mol of nucleotides equals 300 g, 16.7 ng is equivalent to 55 pmol of each nucleotide. (In practice, the optimal size distribution of elongation products is obtained with 25 pmol of dCTP as further increases favor the longer molecules which continue into the M13 sequences beyond the polylinker.) For preliminary experiments, 24 pmol of dCTP plus 1 pmol of labeled dCTP (as tracer) are used in a set of reactions using 0.5–5 μg of template. The reaction products are analyzed on a gel with known size markers and detected by an autoradiographic exposure of the gel (~4 hr). From the autoradiograph, reaction conditions are chosen so as to give the maximum labeling in the 300–800 base range. In our experience, 2 μg of template should be labeled with 25 pmol dCTP. Conditions should be established for each size of probe insert.

The universal primer (6 ng) is hybridized in equimolar amounts to 2 μg of template (see above) in 17 mM Tris-HCl, pH 8, 17 mM MgCl$_2$ in a total volume of 6 μl. This is boiled in a water bath for 5 min and then maintained for 45 min at 65°. The heating element is switched off, and the sample is left to cool to room temperature over several hours. (The efficiency of hybridization may be checked by using a kinased primer and passing the hybrid on a nondenaturing gel to separate free primer from that hybridized to the template.) To this 6 μl is added 1 μl of cold dNTPs (2 mM each of dATP, dGTP, and dTTP), 2 μl of dCTP (12.5 μM, 800 Ci/mmol), and 1 μl of Klenow polymerase (5 units). The reaction is incubated 2 hr at 25° and stopped by the addition of 10 μl of formamide and 2 μl of 10× blue loading solution (see Gel Analysis). The sample is denatured in boiling water for 5 min and then cooled on ice. It is then loaded on a 1% agarose gel in TAE buffer (see Choice of Gel Conditions) containing 50% formamide. (The gel should be poured in a cold room.) The running buffer is also TAE, 50% formamide. After electrophoresis, the gel is covered with Saran Wrap and autoradiographed (15 min). The zone containing molecules in the desired size range is cut out, melted in a water bath, and then

55 N. Hu and J. Messing, *Gene* 17, 271 (1982).

poured into the well of an electroelution apparatus. Once solidified, the sample is subjected to routine electroelution[44] and is recovered in 2–3 ml of electroelution buffer. The probe is then purified as described above for nick-translated probes by adding it to 10 ml of hybridization buffer and incubating with a Whatman cellulose 540 filter for 4–6 hr at 42°.

Hybridization of Filters. The filters are prehybridized at 42° for 4–6 hr with gentle agitation in 10 ml (for a filter of 350 mm²) of prehybridization solution (47.4% deionized formamide, 695 mM NaCl, 40 mM NaH$_2$PO$_4$/Na$_2$HPO$_4$ buffer, pH 6.8 at 20°, 1.6× Denhardt's solution,[56] 1.7 mM EDTA, 0.8% SDS, 150 µg/ml denatured, sonicated salmon sperm DNA) in sealed plastic bags. For hybridization, a corner of the bag is cut and the prehybridization buffer carefully drained. The pretreated probe in hybridization buffer is added, the bag resealed, and the filter hybridized overnight (16 hr) at 42° with gentle agitation. The filter is then washed 3 times for 10 min at room temperature in washing solution (15 mM NaCl, 1.5 mM sodium citrate, 0.1% SDS), followed by 3 times for 15 min in the same solution at 58°–62°, depending on stringency desired, with gentle agitation throughout.

The filter is then placed on a piece of Whatman paper to remove excess liquid and mounted on a rigid (5 mm) plastic plate, covered with Saran Wrap and exposed for autoradiography with Kodak X O-MAT AR5 film and intensifying screens at −80°. After an overnight exposure to determine signal strength, the filter is reexposed for an appropriate time (2–10 days). The autoradiograms are analyzed following a second hybridization to reveal the internal standards (see Internal Markers). The position of bands can be ascertained by comparison with the reference curve obtained from the standards. An estimation of the relative importance of nuclease digestion can be derived from scanning the bands of the progressive digestion points.

Dehybridization and Storage of Paper. To remove a probe and prepare the filter for further hybridizations, the paper is treated with 0.4 N NaOH, and all subsequent steps are performed in the same way as following blotting (see above).

Different Uses of Nuclease Digestion in the Analysis of Chromatin Structure

Nuclease digestions are used to probe chromatin structures *in vivo* so as to reveal the accessibility of the DNA fiber. The assumption is that the accessibility of the DNA to nucleases will reflect its accessibility to regulatory molecules. Since its introduction, the technique has been modified

---

[56] D. T. Denhardt, *Biochem. Biophys. Res. Commun.* **23**, 641 (1966).

and adapted to several different problems, including nucleosome positioning (see Refs. 41 and 42) and hypersensitive regions (see Refs. 5 and 34). In this section, we describe three other general applications which are used for delineating the borders of regions of different sensitivity associated with active chromatin. All experimental procedures are found above; only protocols for new elements are given below.

Characterization of Regions Showing a General Sensitivity to Nucleases

It is often necessary to characterize the regions of active chromatin containing expressed genes and their flanking sequences. While this can most accurately be done using the liquid hybridization technique (see below), such an approach requires a detailed knowledge of the region and works most efficiently with fragments of 300–1000 bp. Larger fragments may be compared either with similar-size fragments in the same cell type or with the same fragment in a different cell type.[39] Fragments in the range of 1–15 kb are chosen from the restriction map of the region, prepared from recombinants, and labeled by standard techniques. When comparing two or more fragments in the same tissue, they should be chosen so as to have a similar size, thus presenting a similar target size for nuclease digestion. Nuclei are digested with increasing amounts of nuclease, the DNA is purified, redigested with restriction enzymes, separated on agarose gels, blotted, and the filters probed with the chosen fragments.

If the original fragments are similar in size but sufficiently different to be separated in the gel electrophoresis, they may be probed with a mixture of probes, providing that a similar specific activity is obtained for each probe. The corresponding bands are scanned for each digestion point, and the percentage of the starting material remaining is estimated. If smaller bands disappear markedly faster than large bands, they are no doubt more sensitive to digestion than the latter. However, if the larger bands disappear only slightly faster than the smaller bands, this is probably only a consequence of target size. By comparing fragments progressively further from the transcribed sequence, it should be possible to find the limits of the region of general nuclease sensitivity. If comparing the same fragment in two different cell types, it will be necessary to find a standard reference fragment so as to avoid artifacts resulting from differences such as the accessibility of different nuclei to nucleases.[39]

Precise Determination of Sensitive Regions by Liquid Hybridization

Originally designed to detect viral genomes in transformed cells,[57] the liquid hybridization technique is used to compare the nuclease sensitivity

---

[57] S. J. Flint, J. Sambrook, J. F. Williams, and P. A. Sharp, *Virology* 72, 456 (1976).

of genomic fragments in the range of 300–1000 bp. To define the borders of a nuclease-sensitive domain, a number of fragments chosen from the restriction map of the region may be compared. The sensitivity of each fragment is determined in turn. To do this, the cloned fragment is purified, denatured, and its rate of renaturation is established by standard Cot procedures. The renaturation rate is dependent on temperature, salt conditions, and specific DNA concentrations. In practice, all are held constant, except the specific DNA concentrations, which thus determine the rate of renaturation. If the quantity of specific DNA is doubled, its renaturation will be 2 times faster. This may be achieved by adding an amount of genomic DNA containing the same quantity of the specific DNA fragment as in the labeled probe fragment (P). The principle of the method thus lies in adding DNA from nuclei exposed to increasing nuclease digestion to the labeled (cloned) fragment. As nuclease digestion proceeds, the genomic DNA sequence homologous to the probe will be reduced to fragments that are too short to hybridize with the probe and are thus unable to alter the reannealing rate of the latter. This will be reflected by changes in the reannealing rate of the probe when incubated with fractions coming from samples that have been digested to different extents. The disappearance of the genomic fragment may thus be plotted as the percentage of initial quantity remaining at different points of digestion (expressed as time points, nuclease concentration, or percentage of acid-soluble material). If different fragments are tested with aliquots coming from the same digestion points, their rates of disappearance and hence their nuclease sensitivity may be compared.

Experimental Conditions. The fragments of interest are excised and purified using techniques such as in Preparation of Fragments. The chosen fragment is then labeled by nick-translation (see Ref. 48 and Nick-translated probes, above) to give a hybridization probe. For a probe of 500 bp, 40 ng of fragment is labeled to a specific activity of 1×10^8 cpm/μg in 20 μl of buffer (50 mM Tris-HCl, pH 7.5, 5 mM MgCl$_2$, 10 mM 2-mercaptoethanol, 50 μg/ml BSA) using labeled dATP, dCTP, and dTTP (800 Ci/mmol, 1.5 μM) and dGTP (20 μM) at 10° for 1 hr with DNA polymerase and DNase I. The amount of enzymes should be determined in preliminary experiments to maximize both the specific activity and the length of the labeled fragments. The relatively low temperature (10°) is chosen so as to reduce DNase I activity. The mixture is then heated 5 min at 65° to inactivate the polymerase and DNase I activity. ATP is then added to 10 mM, DTT to 5 mM, and the MgCl$_2$ concentration is adjusted to 5 mM final concentration. Ligase (2 units) is added, and the sample is then incubated 30 min at 37°. The reaction is stopped by the addition of 40 μl of 10 mM EDTA and 0.1% SDS, and the probe is purified on a

Sephadex G-50 column (see above). The peak fractions are pooled and contain about 70% of the initial DNA (~ 28 ng).

Aliquots of the probe are denatured in 30 mM NaOH and analyzed on a 1.5% NaOH denaturing agarose gel to verify the length of the single-stranded fragments. The gel is prepared by boiling the agarose in water equivalent to nine-tenths of the final volume, cooling to 50°, and then adding one-tenth volume of 300 mM NaOH, 20 mM EDTA. The gel is run at 30–40 V in 30 mM NaOH, 2 mM EDTA. If the single-stranded fragments are near to full length, the probe should renature to 90–95%. If degradation has occurred, the extent of renaturation will be markedly reduced. The probe is denatured by adding NaOH to 0.3 M and incubating 30 min at 37°. It is then neutralized to 90% with HCl and stored at −20°. For initial experiments, a quantity of unlabeled fragment is treated with NaOH, neutralized, and stored at −20° in the same way.

The homologous genomic DNA samples from the digestion time course and the heterologous DNA are prepared as follows. Nuclei are prepared and digested as above, being sure to include a zero digestion point. Suitable digestion points are 5%, 10–15%, 20–25%, 35–40% acid solubility (see Typical Digestion Conditions for DNase I). It may be necessary to choose these from an extended range of time points or enzyme concentrations. The DNA is purified (see above) and the amount of DNA estimated by the Burton reaction.[16] The DNA is then treated by heating or boiling with 0.3 M NaOH to reduce the DNA to fragments of 300–1000 bp. Examples of treatment are as follows: nondigested or 6–7% acid-soluble material, 20 min at 100°; 40%, 30 min at 37°. The NaOH is then neutralized to 90% by HCl (resulting in ~ 300 mM NaCl), and the sample is then diluted with water to 160 mM NaCl and stored frozen at −20°. Heterologous DNA is used to maintain the total DNA concentration during characterization of the reannealing of the probe. For this, salmon sperm DNA is treated with NaOH as above and stored at −20° in aliquots.

For each test point, 7 pg of probe (~ 700 cpm) is added with genomic DNA (heterologous or homologous, see below) to a final volume of 20 μl, which is 160 mM in sodium phosphate buffer, pH 6.8, and 200 mM in NaCl (final concentration, including that derived from DNA samples). To characterize the reannealing of the probe itself, a volume equivalent to 12 samples (240 μl) is prepared. The quantity of heterologous DNA to be added per sample equals 7 pg times the genome size (in bp) of the animal studied, divided by the length of the probe (e.g., for the chicken, with a genome size of 2.1×10^9 bp,[58] $7 \times 2.1 \times 10^9/500 = 2.94 \times 10^7$ pg =

[58] B. J. McCarthy, *Prog. Nucleic Acid Res. Mol. Biol.* **4**, 129 (1965).

29.4 μg). Eight 20-μl aliquots are taken up in siliconized glass capillaries (nominal volume 50 or 100 μl), which are immediately heat-sealed. These are used to prepare duplicates of four time points. Two are taken immediately for the zero point, and the rest are incubated in a water bath at 68° for 60, 120, and 180 min, respectively. To the remaining 80 μl is added 400 ng of the denatured, unlabeled probe fragment (see above), and the NaCl concentration is adjusted to 1 M. Two capillaries of 20 μl of this solution are incubated at 68° for 5 hr so as to ascertain the maximum reannealing capacity of the probe.

Following incubation, all capillaries are treated immediately by breaking the glass, expelling the contents into a siliconized hemolysis tube, and rinsing the walls of the capillary with 1 ml of 125 mM sodium phosphate buffer, pH 6.8, 0.05% SDS. The samples are then applied to hydroxyapatite columns (2 cm in a siliconized Pasteur pipette, prepared in the same buffer) jacketed at 62°. The sample is absorbed onto the column, which is then washed with 3 ml of the same buffer, and the effluent is collected directly in scintillation counting vials. This fraction contains the single-stranded DNA. The column is then washed with 4 ml of 400 mM sodium phosphate buffer, pH 6.8, which releases the reannealed double-stranded DNA. The two samples are then counted using Cerenkov conditions.[59] It is necessary to count each sample several times as the cpm/sample is low (30–700 cpm). Background may be estimated by counting vials containing 4 ml of sodium phosphate buffer. The same protocol is now used for each of the digestion points except that the heterologous DNA is replaced by the same amount of homologous DNA from the digestion and only the set of eight capillaries is used, assuming the maximum reannealing of the probe is constant.

The results are calculated as follows (theoretical considerations are found in Ref. 57). The counts per minute for each sample is corrected by subtracting background, and the single-stranded (ss) and double-stranded (ds) values are obtained for each incubation sample. The total counts for a sample X, TX, equals ss$X$ plus ds$X$ and the fraction of single-stranded DNA in sample X, fss$X$, equals ssX/TX. Calculate fss(0), which gives the fraction of single-stranded probe present initially, and fss(max), which gives the amount of residual nonreannealing, single-stranded probe. In a similar manner, fss(60), fss(120), and fss(180) are calculated and converted to FSS(60), etc., as follows: FSS(60) = [1 − fss(max) − fss(0) + fss(60)]/[1 − fss(max)]. These values are then used in a reciprocal plot as shown in Fig. 5A. In this way, a straight line is obtained for each digestion sample.

The experiment is repeated using either other fragments from the

[59] K. Asada, M.-A. Takahashi, and M. Urano, *Anal. Biochem.* **48**, 311 (1972).

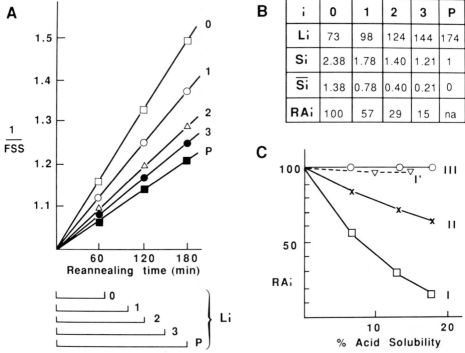

FIG. 5. Analysis of nuclease digestions by liquid hybridization. (A) Graphic representation of $1/fss$ versus time of reannealing for samples from different digestion points. P, Probe with heterologous DNA; 0, probe plus nondigested homologous DNA; 1, 2, and 3, probe with homologous DNA coming from samples subjected to increasing nuclease digestion. Below are shown the Li values obtained for each sample at the arbitrarily chosen value of 1.2 for $1/fss$. The ratio LP/Li is used for further calculations and is independent of the $1/fss$ value, which is chosen for convenience of measurement. (B) Calculation of RAi for the samples (0, 1, 2, 3, and P) shown in A. $Si = LP/Li$; $\overline{Si} = Si - 1$; $RAi = \overline{Si}/\overline{S0} \times 100$. na, Not applicable. (C) Plot of percentage of a given fragment that remains undigested versus degree of total digestion expressed as the percentage of acid-soluble material. The data for the fragment I (nuclease-sensitive chromatin) of A and B are shown together with data for fragments II (moderately sensitive) and III (insensitive chromatin) in the same tissue, and I′ for fragment I in a tissue in which it is insensitive to nucleases.

region or the same fragment in other cell types. To compare the sensitivity of the different fragments, the reannealing is calculated from the respective graphs such as in Fig. 5A. The calculation takes into account errors in the estimation of the labeled probe. The maximum increase in the rate of reannealing ($\overline{S0}$) of the probe is given by $\overline{S0} = S0 - 1 = (LP/L0) - 1$ (see Fig. 5A). Ideally $S0 = LP/L0 = 2$, that is, $\overline{S0} = 1$. In order to compare between probes, experimental values $\overline{Si}$ are corrected and expressed as the percentage of nondigested fragments (RAi) by converting the $\overline{S0}$ values arbitrarily to 100 (see Fig. 5B for examples). These values may then be represented graphically as in Fig. 5C. The percentage degradation of a given fragment ($\overline{RA}$) may also be calculated for each of the digestion points ($\overline{RA} = 100 - RA$).

Nucleosome Repeats and Ladders

As MNase digests DNA preferentially in the linker regions between nucleosomes, its effect on bulk chromatin is to free oligomers, correspond-

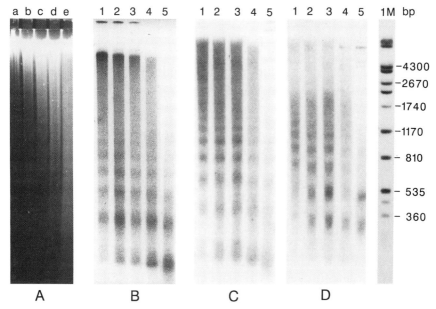

FIG. 6. (A) Progressive micrococcal nuclease digestion of purified (naked) laying hen erythrocyte DNA. DNA, 4 μg/slot, was digested with (lane a), 0.3, (b) 0.7, (c) 1.3, (d) 2.0, or (e) 2.7 units MNase/mg DNA/ml for 20 min at 20°. The 1.8% agarose gel is stained with ethidium bromide and photographed with UV illumination. (B) Nuclei from erythrocytes of 16-day-old chicken embryos were digested with (lane 1) 5, (2) 10, (3) 20, (4) 50, and (5) 400 units MNase/mg DNA equivalents/ml for 20 min at 20°. Purified DNA (4 μg/slot) was loaded on a 1.8% agarose gel. The gel was stained with ethidium bromide as in A. (C and D) Samples (45 μg) from the same digests as in B were separated on a 1.8% agarose gel and transferred to DBM paper. The filter was probed (C) with an intronic β-globin probe (probe d, Ref. 43) and then reprobed (D) with an exonic ovalbumin probe (probe from the 3' region of the seventh exon, Ref. 28). Note that the digestion of globin gene sequences is more rapid when compared to bulk DNA and the ovalbumin gene, which is not expressed in this tissue. Note also the differences in the ovalbumin and globin ladders on the same filter (see Ref. 43 for discussion). In lane M1 are the internal markers run in slot 1, probed with a mixture of labeled Ad2 and SV40 DNAs.

ing to 1, 2, 3 nucleosomes, etc. This is reflected in gel analyses by a ladder of bands whose sizes differ by the size of the DNA contained in one nucleosomal repeat unit. Although this value is constant within a tissue, it may differ between different cell types in the same organism (e.g., 196 bp, chick oviduct; 207 bp, chick erythrocytes; see Ref. 12) and probably reflects differences in linker length. At the bottom of the ladder ($n = 1$ to 4), the sizes clearly deviate from integrated multiples of this value owing to exonuclease trimming of the exposed ends of the oligomer. In mild digests of bulk chromatin this trimming is slight,[12] of the order of 20 bp, and is thus negligible when $n > 5$. As digestion progresses, two further intermediates are recognized. The first corresponds to a protected fragment of 166 bp, the chromatosome, and the second at 146, the core particle.[60] These two subnucleosomal particles are independent of the source of nuclei. These patterns may be seen in bulk chromatin either by UV illumination following ethidium bromide staining (Fig. 6B), or by probing DNA transferred to a filter with a radioactive probe from a region of inactive chromatin.

A number of different situations have been described when filters are probed with DNA from regions of active chromatin (e.g., Fig. 6C and 6D). In general there is an important increase in the general sensitivity to micrococcal nuclease. In certain cases the regular ladder is disrupted, and bands of novel size are detected (e.g., Ref. 11). In other cases a repeat structure is found but displaced when compared with that of the bulk chromatin.[43] In this case it appears that, even under conditions of mild digestion, the entire linker region and an important part of the core particle of the nucleosomes at the ends of the oligomer are digested. It has been suggested that this increased sensitivity reflects changes in both H1 and H2A–H2B content in regions of active chromatin (see Refs. 38 and 43 for discussion).

Acknowledgments

This work was supported by grants from the CNRS, the INSERM and the Fondation pour la Recherche Médicale, The Ministère de la Recherche et de l'Enseignement Supérieur, and the Association pour la Recherche sur le Cancer. A.G. is funded by the Consiglio Nazionale delle Ricerche Progetto Finalizzato "Ingegneria Genetica" U.O.47 No. 8686/51.

---

[60] M. Noll and R. D. Kornberg, *J. Mol. Biol.* **109**, 393 (1977).

[17] Analysis of RNA Polymerase III Transcription *in Vitro* Using Chromatin and Cloned Gene Templates

By JOEL M. GOTTESFELD

In the eukaryotes, transcription of the genetic material is carried out by three distinct RNA polymerases. The large ribosomal RNAs are transcribed by RNA polymerase I, messenger RNA by polymerase II, and the low molecular weight RNAs (including 5 S, tRNA, viral-encoded RNAs and the transcripts of repetitive DNA families) by RNA polymerase III. These enzymes have been highly purified from a variety of biological sources; however, these enzyme preparations are unable by themselves to initiate accurate transcription from deproteinized DNA templates (for reviews, see Refs. 1 and 2). In the case of RNA polymerase III, the enzyme is able to initiate transcription from chromatin templates.[3,4] In contrast to these findings, cell-free extracts prepared from frog oocytes or cells in culture are able to support accurate transcription initiation from cloned DNA templates, thus implying the existence of other factors, in addition to the RNA polymerases, that are required for transcription by each of the polymerases.[1,2,5-8] These results are true not only for the genes transcribed by RNA polymerase III, but also for those genes transcribed by polymerases II[9,10] and polymerase I.[11,12] Presumably, these factors remain bound to the DNA template in isolated nuclei or chromatin, thereby permitting transcription with exogenously supplied polymerase.

Progress toward the identification and purification of the transacting protein factors, which allow accurate transcription by each of the polymer-

[1] N. Heintz and R. Roeder, *Genet. Eng.* **4,** 57 (1982).

[2] C. H. von Beroldingen, W. F. Reynolds, L. Millstein, D. Bazett-Jones, and J. M. Gottesfeld, *Mol. Cell. Biochem.* **62,** 97 (1984).

[3] J. A. Jaehning, P. S. Woods, and R. G. Roeder, *J. Biol. Chem.* **252,** 8762 (1977).

[4] C. S. Parker and R. G. Roeder, *Proc. Natl. Acad. Sci. U.S.A.* **74,** 44 (1977).

[5] S.-Y. Ng, C. S. Parker, and R. G. Roeder, *Proc. Natl. Acad. Sci. U.S.A.* **76,** 136 (1979).

[6] P. A. Weil, J. A. Segall, B. Harris, S.-Y. Ng, and R. G. Roeder, *J. Biol. Chem.* **254,** 6163 (1979).

[7] G. J. Wu, *Proc. Natl. Acad. Sci. U.S.A.* **75,** 2175 (1978).

[8] E. H. Birkenmeier, D. D. Brown, and E. J. Jordan, *Cell* **15,** 1077 (1978).

[9] J. L. Manley, A. Fire, A. Cano, P. A. Sharp, and M. L. Gefter, *Proc. Natl. Acad. Sci. U.S.A.* **77,** 3855 (1980).

[10] P. A. Weil, D. S. Luse, J. A. Segall, and R. G. Roeder, *Cell* **18,** 469 (1979).

[11] I. Grummt, *Proc. Natl. Acad. Sci. U.S.A.* **78,** 727 (1981).

[12] K. G. Miller and B. Sollner-Webb, *Cell* **27,** 165 (1981).

Copyright © 1989 by Academic Press, Inc.
All rights of reproduction in any form reserved.

ases has been quite rapid over the past few years.[13] In this chapter some of the procedures used in the analysis of RNA polymerase III transcription from both cloned gene templates and from chromatin templates are presented. In addition, procedures used for the reconstitution of chromatin templates using *Xenopus* oocyte extracts are discussed.

Preparation of Nuclei and Chromatin

For *in vitro* transcription, nuclei have been prepared from both cells in culture and cells from tissues; here we discuss the former. Nuclei from HeLa and *Xenopus* kidney cells are prepared as follows. Cells are pelleted by centrifugation at 1200 rpm at $0°-4°$ in a clinical centrifuge and washed once in phosphate-buffered saline, once in buffer C (25 mM KCl, 1 mM Tris-HCl, pH 7.6, 0.9 mM MgCl$_2$, 0.9 mM CaCl$_2$, 0.14 mM spermidine[14]) and resuspended in buffer C containing 0.5 mM PMSF (phenylmethylsulfonyl fluoride, as a protease inhibitor). All subsequent steps are carried out at $0°-4°$. Cells are allowed to swell in buffer C containing 0.5 mM PMSF for 10–20 min and are lysed by homogenization with 8–12 strokes of a Dounce B (glass–glass) homogenizer. Nuclei are pelleted at 1500 rpm, washed once in buffer C, and finally resuspended in transcription buffer (see below). Soluble chromatin is prepared by digestion of nuclei with micrococcal nuclease. Generally, nuclei at 1 mg DNA/ml ($A_{260\ nm} = 20$) are digested on ice for 5–15 min with 1–10 units/ml of nuclease. Nuclei are pelleted at 2000 rpm for 5 min and lysed by resuspension in 0.2 mM EDTA, pH 8.0. Insoluble debris is removed by centrifugation at 12,000 rpm in a Brinkmann microfuge, and the supernatant is used as template for *in vitro* transcription.

An alternative method for preparation of nuclei and soluble chromatin from *Xenopus* cultured cells has been described by Schlissel and Brown.[15] In their method, nuclei are prepared from the kidney-derived cell line by first swelling cells in 1 mM Tris, pH 8.0, 0.1 mM EDTA, and 0.4 mM PMSF for 30 min at room temperature. The nonionic detergent Triton X-100 (Beckman) is added to 0.2% (v/v) and MgCl$_2$ to 3 mM. The swollen cells are lysed by 30 strokes with a Dounce B homogenizer, and the nuclei are pelleted by centrifugation for 5 min at 850 g at $0°-4°$. All subsequent steps are carried out either on ice or at $4°$. The nuclei are resuspended at a

[13] W. S. Dynan and R. Tjian, *Nature London* **316**, 774 (1985).
[14] M. M. Sanders, *J. Cell Biol.* **79**, 97 (1978).
[15] M. S. Schlissel and D. D. Brown, *Cell* **37**, 903 (1984).

concentration of about 2×10^8/ml in nuclei buffer [NB, 50 mM Tris, pH 7.5, 70 mM KCl, 7 mM MgCl$_2$, 0.1 mM EDTA, 2 mM dithiothreitol (DTT), 0.4 mM PMSF, and 25% glycerol]. Soluble chromatin is prepared by digestion of nuclei at 1×10^8/ml in NB containing 1 mM CaCl$_2$ with 2 units/ml of micrococcal nuclease at 37° for 15 min. EGTA is added to 2.5 mM, and the nuclei are pelleted at 850 g for 10 min. The nuclei are resuspended in 0.25 mM EDTA, pH 7.0, 1 mM DTT, and 0.4 mM PMSF, dialyzed against this buffer for 12–24 hr, and then dialyzed for a further 3 hr against chromatin storage buffer (10 mM Tris, pH 7.5, 0.1 mM EDTA, 1 mM DTT, 0.4 mM PMSF, and 10% glycerol). Insoluble material is removed by centrifugation at 1000 g for 5 min, and the soluble chromatin supernatant is stored at $-70°$ until use.

Methods for the preparation of chromatin from *Xenopus* embryos have been developed by Wormington and Brown.[16] Batches of staged embryos are dejellied in 2% cysteine–NaOH (w/v), pH 8.0, and rinsed extensively in water. After allowing the embryos to settle by gravity, excess liquid is removed, and the embryos are resuspended in 20 packed embryo volumes of 10 mM Tris, pH 7.5, 70 mM KCl, 7 mM MgCl$_2$, 0.1 mM EDTA, 2.5 mM DTT, and 0.25% (v/v) Triton X-100. Embryos are homogenized with 10 strokes of a loose-fitting Dounce homogenizer, and the homogenate is centrifuged at 650 g for 10 min at 0°–4°. The crude nuclear pellet is resuspended by Dounce homogenization as above in 20 packed embryo volumes of the same buffer without detergent but containing 0.25 M sucrose. The homogenate is layered over an equal volume of 0.88 M sucrose in the buffer above, and the nuclei are pelleted at 650 g for 20 min. This step is repeated until the nuclear pellet is freed of yolk and pigment granule contamination. The nuclei are resuspended in 10 mM Tris, pH 7.5, 10 mM EDTA, 2.5 mM DTT, and 0.34 M sucrose and layered over an equal volume of 2.2 M sucrose in this buffer. The chromatin is pelleted at 45,000 rpm for 60 min in a Beckman SW50.1 rotor. The chromatin pellet is resuspended in the last buffer and stored at $-70°$.

Preparation of Cell-Free Transcription Extracts

A variety of cell-free extracts have been used for the transcription of cloned DNA and chromatin templates. For the class III genes, these include extracts from mammalian[6,9] and amphibian cells in culture[17] and

[16] W. M. Wormington and D. D. Brown, *Dev. Biol.* **99**, 248 (1983).
[17] H. R. B. Pelham, W. M. Wormington, and D. D. Brown, *Proc. Natl. Acad. Sci. U.S.A.* **78**, 1760 (1981).

extracts from *Xenopus* oocytes, eggs, and embryos.[5,8,18,19] Methods for the preparation of each of these extracts are described below.

HeLa S-100

This procedure is adapted from Weil *et al.*[6] and has been used successfully for the transcription of cloned class III genes; it also supplies adequate levels of RNA polymerase III for transcription from nuclei and chromatin templates.[20] HeLa cells are grown in RPMI media to a density of approximately $5-10 \times 10^5$ cells/ml. Generally, 2 liters of cell culture are processed at one time. The cells are pelleted by centrifugation in an IEC preparative centrifuge at 1200 rpm (800 g) for 5 min at $0°-4°$; the supernatant is decanted, and the cell pellet is washed twice with sterile phosphate-buffered saline. The volume of the cell pellet is estimated, and the cells are washed once with 10 volumes of sterile hypotonic buffer (10 mM HEPES, pH 7.9, 1.5 mM MgCl$_2$, 10 mM KCl, 0.5 mM DTT, 0.5 mM PMSF). The cells are resuspended in two cell volumes of hypotonic buffer and allowed to swell on ice for 10–20 min. The cells are lysed by homogenization with 8–12 strokes of the Dounce B homogenizer. One-ninth total volume of a solution containing 0.3 M HEPES, pH 7.9, 1.4 M KCl, and 30 mM MgCl$_2$ is added, and the cell lysate is centrifuged at 28,000 rpm in the Beckman SW40 rotor (100,000 g) for 60 min at 4°. The supernatant is removed, and aliquots (250 μl) are stored frozen at $-70°$. The extract remains active for periods of greater than 6 months; however, aliquots should be used once and not refrozen.

Assay Conditions. For each preparation of extract, the optimum volume of crude extract per assay volume should be determined. Generally, 20 μl of S-100 per 50 μl reaction provides the optimum level of transcription with both cloned DNA templates and chromatin. Using *Xenopus* tRNA and 5 S RNA gene-containing plasmids, the reaction mixtures (50 μl) contain 100 ng plasmid DNA, 400–900 ng pBR322 carrier DNA, 12 mM HEPES, pH 7.9, 8% (v/v) glycerol, 66 mM KCl, 5 mM MgCl$_2$, 0.5 mM DTT, 600 μM each ATP, UTP, and CTP, and 40 μM GTP. These reaction conditions include the salt, divalent cation, and buffer contributions from the extract, chromatin, or nuclei and any other additions to the reaction. Generally, 1 μl of 10 mCi/ml [α-^{32}P]GTP (New England Nuclear, Boston, MA) is added per reaction. Incubations are at 22° for 2 hr. For chromatin transcriptions, approximately 1–10 μg DNA equivalents of nu-

[18] G. C. Glikin, I. Ruberti, and A. Worcel, *Cell* **37**, 33 (1984).
[19] D. D. Brown and M. S. Sclisssel, *Cell* **42**, 759 (1985).
[20] W. F. Reynolds, L. S. Bloomer, and J. M. Gottesfeld, *Nucleic Acids Res.* **11**, 57 (1983).

clei or soluble chromatin are used per reaction. To obtain the highest levels of incorporation of nucleoside triphosphates into RNA, the optimum amount of nuclei or chromatin should be determined. RNA is purified and analyzed by gel electrophoresis, as described below.

Xenopus Cultured Cell Extract

Pelham et al.[17] have utilized whole-cell extracts from Xenopus kidney-derived tissue culture cells for monitoring 5 S RNA and tRNA transcription from cloned gene templates. The extract was prepared by the method of Manley et al.[9] with the following modification. The final $(NH_4)_2SO_4$ precipitate is redissolved in one packed cell volume of buffer J [7 mM $MgCl_2$, 70 mM NH_4Cl, 0.1 mM EDTA, 10 mM HEPES, pH 7.5, 2.5 mM DTT, 6% (v/v) glycerol], dialyzed for 12 hr against 100 volumes of buffer J, and centrifuged to remove insoluble material; aliquots are stored at $-70°$. For transcription reactions, plasmids containing pol III genes were used at 4–5 μg/ml, and the extract comprised 50% of the final reaction volume. Unlabeled nucleoside triphosphates are added at 0.2 mM, and [α-^{32}P]GTP (10 μCi per reaction) is at 0.02 mM final concentration.

Oocyte S-150 Extracts

Xenopus oocyte S-150 extracts are prepared similarly to the method described by Glikin et al.[18] but with several important modifications. Mature Xenopus frogs are obtained from Nasco (Modesto, CA) or Xenopus I (Fort Atkinson, WI). Techniques for the care and handling of Xenopus have been described by Gurdon.[21] For the preparation of the extract, mature Xenopus females are anesthetized by hypothermia, and whole ovaries are removed surgically. After washing in calcium-free OR-2 medium (82 mM NaCl, 2.5 mM KCl, 1 mM $MgCl_2$, 1 mM H_3PO_4, 5 mM HEPES, pH 7.6), the ovaries are cut into several pieces with scissors and gently agitated for 4–8 hr at 22° in 0.75% collagenase (Sigma type III) in the same medium. Undigested tissue is removed, and the dispersed, defolliculated oocytes are allowed to settle under normal gravity. The oocytes are washed several times more in OR-2 medium. This serves to remove all traces of collagenase and to separate the mature stage VI oocytes from the smaller immature oocytes. At this point, 90% or more of the oocytes are visually identical to fresh oocytes. In the few cases where a noticeable fraction of the mature oocytes reveals polar discoloration, the preparation is discarded.

---

[21] J. B. Gurdon, Methods Cell Biol. 16, 125 (1977).

The oocytes are then washed twice in ice-cold buffer J (10 mM HEPES, pH 7.4, 70 mM NH$_4$Cl, 7 mM Mg$_2$Cl, 2.5 mM DTT, 10% glycerol, 1 mM EDTA[8]), followed by two washes in ice-cold extraction buffer (30 mM Tris-HCl, pH 7.9, 90 mM KCl, 2 mM EGTA, 1 mM DTT[18]). Oocytes are then transferred into SW50.1 tubes, and extraction buffer is added so that the total volume is twice that of the settled oocytes. Following centrifugation at 40,000 rpm (150,000 g) for 30 min at 4° in the SW50.1 rotor, the clear portion of the gradient is collected into a sterile syringe by puncturing the polyallomer tube at the bottom of the region to be collected. Care must be taken to avoid the lipid plug near the surface and the pellet. The supernatant is frozen in aliquots at −70° and can be stored for several months without loss of activity. These extracts have been used for the assembly of plasmids containing class III genes into chromatin and the subsequent analysis of the transcriptional activity of the chromatin reconstitutes.[22-24] These extracts also support transcription from templates that do not contain full complements of nucleosomes[25,26] (see below).

For chromatin assembly, the optimal ratio of extract to input DNA must be carefully titrated. The most common assay employed to monitor nucleosome assembly is change in linking number. Plasmid DNA is first treated with DNA topoisomerase I [Bethesda Research Laboratories (BRL), Bethesda, MD] under the conditions specified by the supplier and then deproteinized by proteinase K digestion in the presence of sodium dodecyl sulfate (SDS) and repeated phenol–chloroform and chloroform extractions. The DNA is freed of SDS and concentrated by ethanol precipitations. In our experience, one oocyte (3 μl) equivalent of extract can maximally supercoil approximately 30–50 ng of plasmid DNA in a 10 μl reaction during a 3–4 hr incubation at 22°. This reaction can be scaled up 10-fold. Worcel and colleagues[18] have described the divalent cation and ATP requirements of the chromatin assembly reactions. Mg^{2+} and ATP concentrations of 3 and 6 mM, respectively, were found to be optimum; however, these parameters must be established for each extract. Glycerol is also included in the reaction at 8–10%. We find that chromatin assembled in this manner can be recovered by centrifugation at 12,000 g for 15 min in a microfuge. This chromatin can be resuspended in 5–10 μl of 10 mM Tris-HCl, pH 7.5, 1 mM EDTA, and assayed by *in vitro* transcription in either the S-100 or S-150 systems or by oocyte microinjection.[21]

[22] E. B. Kmiec and A. Worcel, *Cell* **41**, 945 (1985).
[23] E. B. Kmiec, F. Razvi, and A. Worcel, *Cell* **45**, 209 (1986).
[24] E. B. Kmiec, M. Ryoji, and A. Worcel, *Proc. Natl. Acad. Sci. U.S.A.* **83**, 1305 (1986).
[25] L. J. Peck, P. Eversole-Cire, L. Millstein, J. M. Gottesfeld, and A. Varshavsky, *Mol. Cell. Biol.* **7**, 3503 (1987).
[26] L. Millstein, P. Eversole-Cire, J. Blanco, and J. M. Gottesfeld, *J. Biol. Chem.* **262**, 17100 (1987).

Conditions for *in vitro* transcription with the S-150 are given below. In these experiments, plasmid DNA, soluble chromatin, or chromatin reconstitutes are incubated in a 50 μl reaction volume containing, typically, 100 ng template plasmid DNA or reconstitute along with 400–900 ng pBR322 carrier DNA or, for cellular chromatin, approximately 1–10 μg DNA equivalents of chromatin. The final buffer conditions are 20 mM Tris-HCl, pH 7.9, 66 mM KCl, 8 mM MgCl$_2$, 8% glycerol, and nucleoside triphosphates, as described above for the HeLa S-100. The mono- and divalent cation contributions from the extract, template, and any other additions to the reactions should be considered in the final buffer composition. Generally 30–40 μl of extract are used per 50 μl reaction.

Under these conditions, supercoiled DNA is rapidly relaxed by the extract (10 min) and during a subsequent 2-hr incubation (the time we have used for *in vitro* transcription) does not acquire a significant level of supercoils; however, transcription begins after a lag of 20–30 min and is linear for at least 90 min to 2 hr. The lag time has been interpreted as the time required to assemble active transcription complexes, and the order of factor interactions with 5 S DNA has been established.[27] Transcription from class III templates in this system does not appear to require nucleosome assembly, as judged by the acquisition of negative supercoils.[28] Nonetheless, transcription from chromatin templates assembled *in vitro* is also highly efficient in the S-150 system.[18,22-24]

Xenopus Oocyte S-100 Homogenate

Xenopus oocyte S-100 homogenate has been used for transcription of DNA and chromatin templates and also serves as the starting point for the preparation of the 5 S gene-specific transcription factor TFIIIA.[29] The procedure is modified from that of Ng, Parker, and Roeder.[5] The ovaries of one or two large female *Xenopus* frogs are removed surgically and rinsed in cold 0.15 M NaCl, 15 mM trisodium citrate, pH 7.0 (1 × SSC). The ovaries are washed several times in a buffer containing 50 mM Tris-HCl, pH 7.9, 50 mM KCl, 25% (v/v) glycerol, 0.1 mM EDTA, 2 mM DDT and then minced with scissors in approximately 20–25 ml of this buffer. The oocytes are homogenized with a Kontes homogenizer with a Teflon pestle (8–12 strokes), and the homogenate is centrifuged in a Sorvall SS-34 rotor at 3000 g for 20–30 min at 4°. At this point, the yellow yolk forms a skin at the top and insoluble material forms a large pellet. The supernatant above the pellet and below the yolk is collected with a Pasteur pipette and

[27] J. L. Bieker, P. L. Martin, and R. G. Roeder, *Cell* **40**, 119 (1985).
[28] A. P. Wolffe, M. T. Andrews, E. Crawford, R. Losa, and D. D. Brown, *Cell* **49**, 301 (1987).
[29] D. R. Engelke, S.-Y. Ng, B. S. Shastry, and R. G. Roeder, *Cell* **19**, 717 (1980).

transferred to Beckman SW40 tubes and centrifuged in this rotor at 40,000 rpm (100,000 g) for 2 hr at 4°. The supernatant is aliquoted and stored at −70°. Assay conditions are similar to those for the oocyte S-150 extracts (described above) except that 20 μl of extract are used per 50 μl reaction.

Oocyte Nuclear (GV) Extract

Brown and colleagues have utilized oocyte nuclear extract extensively in the study of the transcription of cloned gene and chromatin templates. The following procedure is modified from that of Birkenmeier et al.[8] Ovaries are removed from adult female frogs and rinsed briefly in ice-cold distilled water. Fragments of ovary containing approximately 20–50 mature (stage VI) oocytes are primed by incubation on ice for 2–3 hr in 5 mM Tris, pH 7.8, and 10 mM MgCl$_2$. This serves to gel the nuclear contents for ease of subsequent manipulation. Fragments of ovary are next transferred to ice-cold buffer J (above) containing 2% (w/v) polyvinyl pyrrolidone. The oocytes are broken open using two watchmaker's forceps, and the oocyte nuclei, which remain on the bottom of the vessel (usually a petri dish), are taken up in approximately 2 μl of buffer J per nucleus with a plastic pipetteman tip. Nuclei are stored in an Eppendorf tube on ice until sufficient numbers of nuclei are collected. Excess buffer J is removed such that the final extract contains 2 μl of buffer per nucleus. To prepare the nuclear extract, the germinal vesicles are disrupted by brief vortexing, and the resulting lysate is centrifuged at 10,000 g for 1 min at 4°. Aliquots are stored frozen at −70°.

A 15–25 μl final volume transcription reaction with the GV extract contains 30–40 μg/ml plasmid DNA or 0.2–2 μg chromatin, 0.02–0.1 mM GTP (10 μCi[α^{32}P]GTP) and 0.5 mM unlabeled nucleoside triphosphates in buffer J plus 8–10 μl of extract. Incubations are at 22° for varying periods of time.

Under the conditions described above, supercoiled plasmid DNAs are rapidly relaxed by the extract and are not extensively supercoiled even on prolonged incubations with the nuclear extract. It is thus unlikely that this extract assembles nucleosomes on deproteinized DNA templates. Nonetheless, this extract supports very high levels of transcription from both plasmid DNA and chromatin templates (see below).

Unfertilized Egg S-100 Extract

Unfertilized egg S-100 extract has been used to monitor TFIIIA-dependent 5 S gene transcription since unfertilized eggs contain 20- to 50-fold lower levels than do mature oocytes of this transcription factor.[29] The extract is highly active in tRNA gene transcription and presumably contains adequate levels of RNA polymerase III and the general class III gene transcription factors IIIB and IIIC.[5,29] Unfertilized egg extracts have also been used for chromatin assembly (see Rhodes and Laskey, this volume [27]).

Unfertilized eggs are obtained from adult *Xenopus* females after injection with 1000 units of human chorionic gonadotropin into the dorsal lymph sac.[21] Eggs are collected overnight in Barth's modified saline solution [10 mM HEPES – NaOH, pH 7.6, 88 mM NaCl, 1 mM KCl, 2.4 mM NaHCO$_3$, 0.82 mM MgSO$_4$, 0.33 mM Ca(NO$_3$)$_2$, 0.4 mM CaCl$_2$[21]]. The eggs are washed once in Barth's modified saline, dejellied by incubation in 2% (w/v) cysteine-HCl, pH 7.8, for 10 min, and then washed extensively in Barth's modified solution, followed by several washes in 50 mM Tris-HCl, pH 7.9, 50 mM KCl, 25% glycerol, 0.1 mM EDTA, 1 mM DTT. The volume is adjusted to twice that of the settled eggs, and the eggs are homogenized with a glass – Teflon homogenizer (8 strokes) with a loose-fitting pestle. The homogenate is centrifuged at 2000 g for 20 min at 4° and then centrifuged for 2 hr at 100,000 g in the SW40 rotor at 4°. The supernatant is removed and aliquots frozen as for the S-150 oocyte extract. Transcription conditions are also similar to those for the S-150 extract, except 20 μl of extract is used per 50 μl reaction.

Transcription Efficiencies

The extracts described above support very high levels of transcription from cloned class III gene templates and from cellular chromatin templates. Both 5 S and tRNA genes of *Xenopus* give rise to approximately 5 – 15 transcripts per gene per hour of incubation in both the HeLa and *Xenopus* oocyte S-150 systems.[20,26] Under conditions employing lower levels of gene-containing plasmids and higher carrier DNA concentrations to maintain the optimum total DNA concentration, significantly higher levels of 5 S transcription have been achieved with the oocyte nuclear extract (rate-enhanced conditions[30]). These levels of transcription mimic those calculated *in vivo* in growing oocytes (approximately 350 transcripts/gene/hr). It should be noted that the oocyte S-150 extract displays the preference for somatic-type 5 S transcription over oocyte-type 5 S transcription observed *in vivo* in midblastula-stage embryos (~50-fold higher levels of somatic- over oocyte-type 5 S transcription[25,26]). Calculations of transcription efficiency entail slicing bands from polyacrylamide gels containing the RNA transcripts and counting in a liquid scintillation counter. After correction for counting efficiency, the number of disintegrations per

[30] A. P. Wolffe, E. Jordan, and D. D. Brown, *Cell* **44**, 381 (1986).

minute incorporated into RNA is determined and converted to number of transcripts, knowing the specific activity of the radiolabeled nucleoside triphosphate, the number of those nucleotide residues per transcript, and the number of genes per reaction. As a guide, 60,000 cpm incorporated into somatic-type 5 S RNA under the standard reaction conditions with the oocyte S-150 extract (100 ng gene plasmid, 3×10^6 MW, 41 guanine residues per 5 S RNA, and GTP specific activity of 10 Ci/mol, 2-hr incubation) corresponds to approximately 10 transcripts per gene per hour.

RNA Purification

After transcription incubations are complete, the reactions are stopped by the addition of an equal volume of 0.5% SDS, 50 mM EDTA, pH 8.0, containing 500 μg/ml proteinase K. The proteinase K is maintained as a concentrated stock solution (10–50 mg/ml) at −20°, and appropriate aliquots are added to the SDS–EDTA solution prior to use. The samples are vortexed and incubated at 42° for a minimum of 30 min. An equal volume of phenol–chloroform [1:1 (v/v), saturated with 1 M Tris-HCl, pH 7.5] is added, and the mixture is vortexed vigorously and then centrifuged in a microfuge for 5 min. The aqueous (upper) phase is removed and transferred to a fresh tube. One-ninth volume of 3 M sodium acetate and 2.5 volumes of 95% ethanol are added, and the mixture is vortexed vigorously and then centrifuged in a microfuge for 15 min or longer. The supernatant is carefully removed, and the pellet is washed with 70% ethanol. After removal of the supernatant, the pellet is briefly dried in a vacuum desiccator. Samples are resuspended in 90% formamide (redistilled, obtainable from BRL) containing 0.1% bromphenol blue and 0.05% xylene cyanol and heated in a boiling water bath for 5 min. Care should be taken that the samples are completely dissolved prior to analysis by gel electrophoresis. In most instances, samples are loaded onto 8% polyacrylamide sequencing gels[31] and subjected to electrophoresis at 1500 V until the bromphenol blue dye reaches the bottom of the gel (about 2.5 hr).

Partially Denaturing Gel System for the Resolution of Identically Sized RNAs of Different Sequence

Korn and Gurdon[32] and Wakefield and Gurdon[33] have utilized polyacrylamide gels to separate the identically sized oocyte- and somatic-type 5 S RNAs of *Xenopus* that differ in sequence by only six bases out of 120

[31] F. Sanger and A. Coulson, *FEBS Lett.* **87,** 107 (1978).
[32] L. J. Korn and J. B. Gurdon, *Nature (London)* **289,** 461 (1981).
[33] L. Wakefield and J. B. Gurdon, *EMBO J.* **2,** 1613 (1983).

nucleotides. Separation is presumably effected by differences in RNA secondary structure under partially denaturing conditions (25% urea). A modification of the method of Wakefield and Gurdon is given here.

Long 35- to 40-cm (0.35-mm-thick) gels are cast in two steps: a resolving gel and a stacking gel, which differ in both polyacrylamide concentration and in buffer composition. The running gel consists of 14.5% (w/v) acrylamide, 0.55% (w/v) bisacrylamide, 25% (w/v) urea, and $1 \times$ TBE buffer (88 mM Tris–borate, 2.5 mM EDTA, pH 8.3). To 50 ml of running gel solution, add 200 μl of freshly prepared ammonium persulfate (10%, w/v). The solution is filtered and degassed, and polymerization is initiated by addition of 15 μl of TEMED. The running gel is carefully overlaid with $1 \times$ TBE and allowed to polymerize for 1 hr. After polymerization, the TBE is removed, the stacking gel is poured, and the comb inserted. The stacking gel consists of 6% (w/v) acrylamide, 0.22% (w/v) bisacrylamide, 25% (w/v) urea, and $1 \times$ TBE, pH 6.8. TBE adjusted to pH 6.8 is prepared by titrating $10 \times$ TBE with HCl. The stacking gel solution is filtered and degassed, and polymerization is initiated by the addition of ammonium persulfate (100 μl/20 ml) and TEMED (15 μl/20 ml). After the gel has completely polymerized, it is subjected to preelectrophoresis at 5 mA, constant current, for 1–2 hr. The buffer for electrophoresis is $1 \times$ TBE, pH 8.3. The sample wells are washed thoroughly with a Pasteur pipette, and the samples are loaded and subjected to electrophoresis at 5 mA for 20–22 hr. During this time, the gel plates must not heat above room temperature, otherwise further denaturation of the RNAs will occur, and separation will not take place. An example of a separating gel is shown in Fig. 1.

Purification of TFIIIA from Immature Oocytes

The 5 S gene-specific transcription factor TFIIIA was first isolated by conventional biochemical fractionation and column chromatographic procedures.[29] Subsequently, this protein was shown to be identical to the protein bound to 5 S RNA in 7 S ribonucleoprotein particles, both by physical characterization and by transcriptional activity using the egg complementation assay (see above).[34,35] The following method is adapted from Pelham and Brown[34] with several significant modifications[36] (Reynolds, personal communication, 1986).

Ovaries are dissected from young *Xenopus* frogs (3–5 cm) and washed first in $1 \times$ SSC and then in buffer H (50 mM HEPES, pH 7.5, 5 mM

[34] H. R. B. Pelham and D. D. Brown, *Proc. Natl. Acad. Sci. U.S.A.* **77**, 4170 (1980).
[35] B. M. Honda and R. G. Roeder, *Cell* **22**, 119 (1980).
[36] D. J. Hazuda and C.-W. Wu, *J. Biol. Chem.* **261**, 12202 (1986).

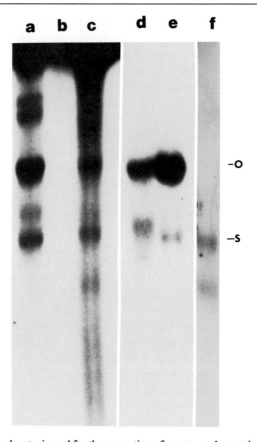

FIG. 1. Partially denaturing gel for the separation of oocyte- and somatic-type 5 S RNAs of equal length. The positions of the oocyte and somatic RNAs are indicated. Lane a, *Xenopus* cultured cell DNA template in the HeLa S-100 extract; lane b, cultured cell chromatin; lane c, cultured cell chromatin template plus oocyte S-100 extract; lane d, purified oocyte-type 5 S RNA; lane e, mixed oocyte and somatic 5 S RNA purified from a denaturing polyacrylamide gel; lane f, purified somatic-type 5 S RNA.

MgCl$_2$, 50 μM DTT, 100 μM ZnCl$_2$, 25 μM PMSF, 25 mM KCl) and homogenized in a Dounce B glass–glass homogenizer in a minimum volume (sufficient buffer to cover the ovaries). The homogenate is centrifuged in a microfuge for 5 min at 4°, and the supernatant is layered onto glycerol gradients. The gradients, prepared in SW41 tubes, consist of equal volume steps of 10, 12, 15, and 30% (v/v) glycerol in buffer H. Two hundred microliters of homogenate is loaded per gradient, and the gradients are centrifuged at 40,000 rpm for 20 hr in the SW41 rotor at 4°. Fractions of 0.5 ml are collected, and 5-μl aliquots of each are analyzed by

SDS–PAGE using the buffer system of Laemmli.[37] Fractions containing 7 S particles are identified by the presence of 5 S RNA, detected by ethidium bromide staining of the gel. Those fractions are pooled and subjected to chromatography on DEAE-cellulose (DE-52, Whatman) using 0.5 ml of resin equilibrated in buffer A (20 mM HEPES, pH 7.5, 100 μM ZnCl$_2$, 50 μM DTT) containing 160 mM KCl. The 7 S particles are eluted from the column with buffer A containing 320 mM KCl, and 100-μl fractions are collected. Fractions are assayed as above, but using Coomassie blue staining to locate the position of the 38.5-kDa TFIIIA protein.

TF fractions are pooled, and RNase is added to 50 μg/ml and incubated at 22° for 5–10 min. Then 2.2 volumes of buffer A (without KCl) is added to adjust the KCl concentration to 0.1 M. This solution is applied to a 0.5-ml column containing Bio-Rex 70 (Bio-Rad, Richmond, CA) equilibrated in buffer A plus 0.1 M KCl. The column is eluted first with 3 ml of this buffer, then with 3 ml of buffer A plus 0.5 M KCl, and, finally, TFIIIA is eluted with 3 ml of buffer A containing 1 M KCl. Fractions of 100 μl are collected, and 2-μl aliquots are assayed by SDS–PAGE and by *in vitro* transcription using the egg extract. All chromatographic steps are performed at 4°. Fractions containing TFIIIA are pooled, and, after adding an equal volume of glycerol, the TFIIIA is frozen in aliquots of 5–20 μl at −70°. The protein isolated by this procedure remains fully active for 3–6 weeks but, thereafter, loses transcriptional activity. DNA-binding activity as assayed by DNase I protection (footprinting) is more stable, and this activity remains on storage for at least 6 months.

Acknowledgments

I wish to thank Drs. J. Blanco, L. Millstein, and W. Reynolds for providing detailed methodologies. This work was supported by grants from the National Institutes of Health (GM 26453) and the American Cancer Society (FRA 292). This is publication number MB 4981 from the Research Institute of Scripps Clinic.

[37] U. K. Laemmli, *Nature (London)* **277**, 680 (1970).

[18] Nonenzymatic Cleavage of Chromatin

By Iain L. Cartwright and Sarah C. R. Elgin

The most common method currently in use for releasing free mono- and oligonucleosomes involves digestion of chromatin (either within isolated whole nuclei or as a purified or reconstituted preparation) with the

Copyright © 1989 by Academic Press, Inc.
All rights of reproduction in any form reserved.

enzyme micrococcal nuclease (also known as staphylococcal nuclease). While it is abundantly clear that this enzyme does digest a majority of the eukaryotic chromosome into discrete nucleosomal particles by virtue of preferential cleavage in the linker DNA between cores, it has become apparent from a number of studies that in some regions of eukaryotic chromatin, the enzyme cleaves primarily at DNA sequences for which it has a high degree of preference.[1,2] Thus, in studies designed to determine whether nucleosomes occupied distinct, fixed positions with respect to a given DNA sequence near certain *Drosophila* genes, micrococcal nuclease digestion of chromatin produced a pattern of cleavages identical to those observed on protein-free DNA.[3-5] Such a situation obviously drastically diminishes the extent to which one can draw conclusions regarding nucleosomal organization at such loci. This type of result might, in fact, be expected in a situation where the nucleosomal array was randomly positioned with respect to a given sequence. On the other hand, the possibility also exists that the structure of the nucleosomes spanning the particular DNA sequence under investigation is such that sites of preferential DNA sequence cleavage within specifically positioned cores are relatively exposed.

As a means of counteracting these problems and addressing questions of nucleosome positioning, it appeared that a more sequence-neutral probe of chromatin structure would be beneficial. Since most reputedly nonspecific endonucleases do, in fact, possess significant degrees of DNA sequence preference in their modes of action, we decided to explore the possibility that a chemical method of cleaving chromatin at nucleosomal linker regions might provide a less ambiguous set of data regarding nucleosomal organization in those regions that had proved refractory to the standard enzymatic analysis.

Principle

Utilizing the phenomenon of intercalation of a planar aromatic molecule into double-stranded DNA represents a logical initial approach to the problem of locating a potential DNA cleaving capability specifically at the chromatin fiber in a nuclear milieu. Previous investigations of photoinduced DNA cross-linking in chromatin with the intercalator 4,5',8-tri-

[1] W. Hörz and W. Altenburger, *Nucleic Acids Res.* **9,** 2643 (1981).
[2] C. Dingwall, G. P. Lomonosoff, and R. A. Laskey, *Nucleic Acids Res.* **9,** 2659 (1981).
[3] M. A. Keene and S. C. R. Elgin, *Cell* **27,** 57 (1981).
[4] I. L. Cartwright and S. C. R. Elgin, *EMBO J.* **3,** 3101 (1984).
[5] I. L. Cartwright and S. C. R. Elgin, *Mol. Cell. Biol.* **6,** 779 (1986).

methylpsoralen showed that such a molecule preferentially bound in the linker region between nucleosome cores.[6,7] It is assumed that the location of a preferred binding site in a chromatin fiber by such an intercalator would depend to a large extent on any inherent sequence preference of the molecule, the accessibility of the site to the solvent, and also on the ability of the DNA at the site to untwist by a small amount.

This last consideration is important because intercalative events are accompanied by a decrease in the relative twist angle of the adjacent base pairs between which the intercalator binds. It seemed possible that nucleosomal linker regions would be less constrained in their motion compared to core-located DNA. Based on such reasoning, we investigated the ability of the 1,10-phenanthroline–cuprous complex to recognize the nucleosomal structure of chromatin. This intercalating reagent was reported to be an efficient DNA cleavage agent.[8] We confirmed its ability to degrade DNA rapidly and found, moreover, that oligonucleosomal ladders, substantially similar to those produced by micrococcal nuclease, were rapidly released as a result of its action on isolated nuclei.[9] Unfortunately, the reagent was found to possess substantial sequence specificity in its cleavage of DNA[9,10] and was not considered further as a useful reagent for this type of chromatin analysis.

Methidiumpropyl-EDTA (MPE) is a synthetic molecule, which, in the form of a 1 : 1 complex with the ferrous ion [MPE · Fe(II)] cleaves double-stranded DNA in the presence of molecular oxygen, producing both nicks and double-stranded cuts.[11] It was designed and synthesized by P. Dervan and colleagues, who originally used it in experiments that defined the binding sites on DNA of various small molecules. MPE consists of the intercalating moiety, methidium, covalently linked by a 3-carbon tether to the metal-chelating moiety, ethylenediaminetetraacetic acid. The methidium group provides the essential binding preference for double-stranded DNA, and, in the presence of the ferrous ion and oxygenated electron donors (e.g., molecular oxygen or peroxide ion), reactive species (probably hydroxyl radicals) are produced very efficiently. These participate in oxidative degradation of the deoxyribose ring, after which base loss and β-eliminative cleavage of the phosphodiester backbone apparently

[6] C. V. Hanson, C.-K. J. Shen, and J. E. Hearst, *Science* **193**, 62 (1976).

[7] T. Cech and M. L. Pardue, *Cell* **11**, 631 (1977).

[8] L. E. Marshall, D. R. Graham, K. A. Reich, and D. S. Sigman, *Biochemistry* **20**, 244 (1981).

[9] I. L. Cartwright and S. C. R. Elgin, *Nucleic Acids Res.* **10**, 5835 (1982).

[10] B. Jessee, G. Gargiulo, F. Razvi, and A. Worcel, *Nucleic Acids Res.* **10**, 5823 (1982).

[11] R. P. Hertzberg and P. B. Dervan, *J. Am. Chem. Soc.* **104**, 313 (1982).

occur.[11,12] This mechanism leads predominantly to the production of nicks in double-stranded DNA, but double-stranded cuts are produced with some frequency,[11,12] presumably because the locally high concentration of hydroxyl radicals at a given binding site leads to secondary cutting opposite the primary nick. Most importantly, from the point of view of its potential use in chromatin cleavage studies, MPE·Fe(II) shows relatively little sequence specificity in its cleavage reaction on DNA from a variety of sources.[12-14]

Initial tests of the reagent demonstrated that oligonucleosome ladders could be readily released from nuclei under mild conditions and that the bulk population of nucleosomes released from *Drosophila* nuclei appeared very similar in size to those produced by the action of micrococcal nuclease.[15] Further studies have demonstrated the effectiveness of MPE·Fe(II) for the delineation of nucleoprotein organization in chromatin in a variety of experimental systems.[4,5,16,17] Not only is MPE·Fe(II) very effective in reporting on nucleosomal positions in specific regions, but, by virtue of its extremely small size, it readily detects other regions of chromatin accessibility, for example, hypersensitive sites, with considerable detail. Such a facility makes MPE·Fe(II) an extremely useful tool for studies of mapping DNA–protein interactions, both *in vitro* and *in vivo*, to single base pair resolution.

Procedure

Source and Storage of Methidiumpropyl-EDTA

The synthesis of MPE has been described.[11,12] As of this writing, there is no commercial supplier of MPE. However, samples are available for experimental purposes from Dr. Peter Dervan, Division of Chemistry and Chemical Engineering, California Institute of Technology, Pasadena, California 91125. MPE is a maroon solid, which we store in the dark at −20°. Aqueous solutions are also stable at −20°. We routinely store a concentrated aqueous stock solution (5 mM) for use as required. The concentration of MPE in solution is readily calculated from the intensity of its visible

[12] R. P. Hertzberg and P. B. Dervan, *Biochemistry* **23**, 3934 (1984).

[13] M. W. Van Dyke, R. P. Hertzberg, and P. B. Dervan, *Proc. Natl. Acad. Sci. U.S.A.* **79**, 5470 (1982).

[14] M. W. Van Dyke and P. B. Dervan, *Biochemistry* **22**, 2373 (1983).

[15] I. L. Cartwright, R. P. Hertzberg, P. B. Dervan, and S. C. R. Elgin, *Proc. Natl. Acad. Sci. U.S.A.* **80**, 3213 (1983).

[16] R. Benezra, C. R. Cantor, and R. Axel, *Cell* **44**, 697 (1986).

[17] I. L. Cartwright, *EMBO J.* **6**, 3097 (1987).

absorption peak at 488 nm ($\epsilon = 5994\ M^{-1}\ cm^{-1}$, see Ref. 12). Since the ferrous complex of MPE is reactive at very low concentrations, 1 mg of solid MPE provides enough material for numerous analyses.

Preparation of Methidiumpropyl-EDTA · Iron(II) Complex

We routinely prepare a 1 : 1 equimolar complex of MPE and ferrous ion at a concentration 10-fold higher than that desired for use in the nuclear digestion reaction. This entails mixing suitable amounts of MPE (from the concentrated aqueous stock) and ferrous ammonium sulfate (always freshly prepared as a 5 mM aqueous stock) in nuclear digestion buffer (60 mM potassium chloride, 15 mM sodium chloride, 15 mM Tris-HCl, pH 7.4, 0.25 M sucrose). Generally we have found that a suitable concentration of MPE · Fe(II) complex is $2.5 \times 10^{-4}\ M$ in digestion buffer. The complex is prepared literally a few moments before its use in the nuclear digestion reaction and kept at room temperature shielded from incident light until then. Immediately before use, a suitable reducing agent such as dithiothreitol (DTT) is added directly to the solution of MPE · Fe(II) to achieve a final concentration of 10 mM (from a freshly prepared 1 M stock) in order to activate the complex.

Digestion of Nuclei

Nuclei are prepared from cells by standard procedures and resuspended in nuclear digestion buffer. EDTA and/or EGTA can be added to final concentrations of 1 and 0.1 mM, respectively, in order to sequester free divalent cations and to inhibit any endogenous nuclease action. Nuclei are kept on ice until shortly before use, at which point they are equilibrated to 25°. Frequently we elevate the level of hydroxyl radical progenitor (i.e., over and above the level of dissolved molecular oxygen) by addition of hydrogen peroxide (using a 100 mM aqueous stock) to the nuclear suspension at a final concentration of 0.5 or 1 mM. This is done immediately before mixing with the MPE · Fe(II) complex. For *Drosophila* nuclei, we have routinely resuspended at approximately 5×10^8 nuclei/ml (haploid content of DNA per nucleus is ~0.2 pg). When digesting nuclei from other cell types it would probably be wise to make adjustments in the concentration of nuclei and/or MPE · Fe(II) in order to obtain the desired extent of digestion.

Immediately after the addition of DTT and hydrogen peroxide to the tubes, as described above, 1 volume of the MPE · Fe(II) complex is added to 9 volumes of the nuclear suspension, briefly vortexed, and incubated at 25°. It is undoubtedly also possible to incubate the digestion reaction at 37°, although we have not conducted our own investigations at this temperature. At suitable time intervals (e.g., every 5 min) an aliquot of reaction mixture is withdrawn, and the reaction is quenched by its addition to 0.1 volume of a 50 mM aqueous solution of bathophenanthroline disulfonate (Sigma, St. Louis, MO). A strong red coloration develops quickly in solution as the iron is strongly chelated by the bathophenanthroline, leading to cessation of the reaction. Standard procedures can then be adopted for the isolation of partially digested chromatin or DNA. The red bathophenanthroline–iron complex is freely soluble in the aqueous phase and is not extracted during subsequent DNA purification steps that involve phenol–chloroform extraction. However, most of it can be eliminated during the ethanol precipitation of DNA as it remains highly soluble under these conditions.

An example of a time course of MPE·Fe(II) digestion of *Drosophila* nuclei is shown in Fig. 1 where the resulting purified DNA fragments have been fractionated by electrophoresis and compared with the similar products of micrococcal nuclease digestion. Clearly, the bulk nucleosomal ladders appear virtually identical, displaying a characteristic repeating unit of approximately 190 base pairs.

Comments on the Procedure

1. We have routinely adopted a procedure that utilizes a fixed ratio of complex to target nuclear chromatin and performed a time course series of digestions. There is no reason why one could not perform a similar series with varying concentrations of complex relative to nuclei for a fixed digestion period. This might be advantageous in cases where there are concerns regarding extraneous perturbation of chromatin structure over the time course of the digestion series.

2. A very useful feature of the procedure is that it can be performed on solutions of nuclei or chromatin that contain EDTA and/or EGTA, thus minimizing the potential degradative action of divalent metal ion-activated endogenous nucleases. It is important that the MPE·Fe(II) complex itself be formed in a solution free of these agents since they apparently compete with MPE for the ferrous ion. Once formed, however, the complex seems relatively stable. It can be added to solutions that contain these chelators and still perform very effectively.

3. Addition of a reducing agent such as DTT to the complex is important since it ensures that the metal ion is kept in the reactive iron(II) (ferrous) oxidation state. Generation of hydroxyl radicals by the iron(II)– EDTA moiety is most likely a redox reaction that leaves the complex in the

FIG. 1. DNA fragment pattern derived from MPE·Fe(II) (lanes 1–3) or micrococcal nuclease (lanes 4–6) digestion of *Drosophila* embryo nuclei. Nuclei at 5×10^8/ml were digested at 25° with MPE·Fe(II) at 2.5×10^{-5} M, as described in the Procedure section, for 10 min (lane 1), 20 min (lane 2), or 30 min (lane 3). Nuclei at 10^9/ml were digested for 3 min at 25° with micrococcal nuclease at 47 U/ml (lane 4), 23.5 U/ml (lane 5), or 11.75 U/ml (lane 6). DNA was purified and size separated on 1.2% agarose, and the gel was stained with ethidium bromide (1 μg/ml). [Reproduced, with permission, from I. L. Cartwright and S. C. R. Elgin, *EMBO J.* **3**, 3101 (1984).]

iron(III) oxidation state. The presence of a reducing agent, such as DTT or ascorbate, ensures that the complex recycles to the iron(II) oxidation state and enhances the cleavage efficiency by up to nearly two orders of magnitude.[12]

4. The reaction is dependent on the presence of dissolved oxygen in solution,[12] and the rate can be enhanced significantly by addition of hydrogen peroxide up to millimolar concentrations. Under these conditions, we have not detected any adverse effect on the chromatin structure within nuclei in control reactions performed in the absence of MPE·Fe(II).

5. Recognizing that MPE·Fe(II) cleaves DNA by creating single-stranded nicks, we reasoned that in a chromatin sample digested to a low extent so that only very few double-stranded cuts are introduced, there should be a significant proportion of molecules present in the population containing nicks within the linker regions. Treatment of DNA isolated from such a preparation with S_1 nuclease results in a dramatic redistribution of fragment lengths to smaller sizes which, however, when resolved on an agarose gel, show the characteristic repeating nucleosomal ladder.[15] S_1 nuclease is apparently cutting DNA opposite the nicks created by MPE·Fe(II). Consequently, these nicks must have been sustained mainly in nucleosomal linkers; otherwise the S_1 treatment would be expected to lead to a smear of randomly sized DNA fragments. This observation with S_1 nuclease holds practical value since it allows the subsequent generation of nucleosomal ladders, after performing digestions with very low ratios of MPE·Fe(II) to DNA base pairs, conditions that should preclude any significant distortion of the native chromatin.[15] In performing nucleosome mapping experiments at the *Drosophila* histone locus, we detected no differences in the patterns of cleavage generated either with MPE·Fe(II) alone or after subsequent treatment with S_1 nuclease of DNA samples generated with lower concentrations of MPE·Fe(II) and/or shorter periods of digestion.[15]

Applications

Figure 2 shows the large amount of information concerning nucleoprotein organization that can be revealed by limited digestion of isolated nuclei with MPE·Fe(II). The autoradiogram shows the pattern of cleavages introduced by different reagents into the nuclear chromatin of a particular region of the *Drosophila* genome, encoding the genes for the

---

FIG. 2. Autoradiogram displaying chromatin fine structure in the vicinity of the *Drosophila hsp26* gene as revealed by an indirect end-labeling analysis. Nuclei derived from 6- to 18-hr embryos or protein-free genomic DNA were exposed to limited digestion with micrococcal nuclease (lanes 1–4), DNase I (lanes 5–7), or MPE·Fe(II) (lanes 8–13) at 25°. The DNA was purified, cut to completion with *Bam*HI, and size-fractionated on a 40-cm-long 1.2% agarose gel. After blotting to nitrocellulose, the filter was hybridized with a radioactive probe directly abutting the downstream *Bam*HI site. Samples were protein-free DNA (lanes 1, 7, and 13), non-heat-shocked nuclei (lanes 2, 5, and 8), heat-shocked nuclei (lanes 3, 4, 6, 9, and 10), and 0.35 M KCl (lane 11) or 0.5 M KCl (lane 12) extracted non-heat-shocked nuclei. Lane M shows size markers run in parallel. The symbols marked on the figure are explained in the text. [Adapted, with permission, from I. L. Cartwright and S. C. R. Elgin, *Mol. Cell. Biol.* **6,** 779 (1986).]

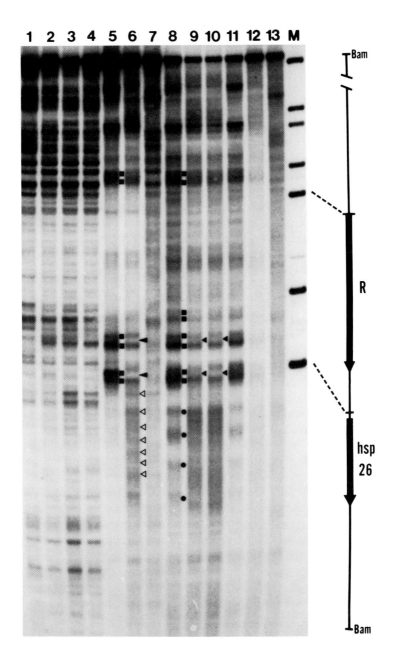

small heat-shock proteins, as analyzed by the indirect end-labeling technique.[18,19] As can be seen in lanes 1–4, the pattern of cleavages produced by micrococcal nuclease at this locus was extraordinarily similar on both protein-free DNA, nuclear chromatin derived from non-heat-shocked embryos (where *hsp26* is transcriptionally inactive), and nuclear chromatin derived from heat-shocked embryos (where *hsp26* is highly transcriptionally active). In contrast, MPE·Fe(II) digestion showed little strong specific cutting on protein-free DNA (lane 13) while revealing much detail regarding accessible regions in nuclear chromatin and the alterations that occur on transcriptional activation (lanes 8–10). This can be compared with the information provided by DNase I digestion of non-heat-shocked (lane 5) and heat-shocked (lane 6) nuclei.

The solid circles marked along the coding region of *hsp26* in lane 8 represent linker cleavage in an array of specifically positioned nucleosomes. This information is absent in both the corresponding DNase I (lane 5) and micrococcal nuclease (lane 2) digests. Interestingly, the pattern smears significantly through this region when the *hsp26* gene is activated (lanes 9 and 10). Immediately upstream of *hsp26,* MPE·Fe(II) reveals strongly hypersensitive regions (denoted by solid squares in lane 8) that appear to be the same as the sites of DNase I hypersensitivity (lane 5). In fact, all sites of DNase I hypersensitivity observed by us appear also to be sites of MPE·Fe(II) hypersensitivity. Additional sites of cleavage are seen throughout the neighboring *R* gene (a developmentally regulated transcription unit) that are not detected by DNase I. This demonstrates particularly well how the amount of information regarding nucleoprotein organization is increased substantially with MPE·Fe(II) as a probe when compared to DNase I.

Finally, it can be seen that both DNase I and MPE·Fe(II) display a markedly reduced ability to cleave in two specific regions upstream of *hsp26* when the gene is activated by heat shock (denoted by black arrowheads in lanes 6, 9 and 10). These regions of induced hyposensitivity (or "footprints") are most likely caused by the association of specific heat-shock gene transcription factor(s) with consensus transcriptional regulatory sequences that precisely coincide with the apparent footprint.[15] Clearly, at high resolution, MPE·Fe(II) possesses the ability to reveal sites of strong DNA–protein interaction to the precise base pair.

Along these lines, MPE·Fe(II) has recently found practical use in both *in vitro* and *in vivo* footprinting studies where its small size has been found, in a number of cases, to allow a more precise delineation of the boundaries

[18] S. A. Nedospasov and G. P. Georgiev, *Biochem. Biophys. Res. Commun.* **92,** 532 (1980).
[19] C. Wu, *Nature (London)* **286,** 854 (1980).

of some specific DNA–protein associations compared to DNase I. Among *in vitro* studies, the interaction of TFIIIA with 5 S genes,[20] the interaction of transcription factor Sp1 with its binding site,[21] and the interaction of adenovirus VA RNA coding sequences with various specific protein binding factors[22] have been reported. The *in vivo* interaction of the heat-shock transcription factor and other nuclear proteins with sequences upstream of the *Drosophila hsp26* gene has recently been mapped with single base pair precision using MPE·Fe(II).[23]

Acknowledgments

Work from the authors' laboratories was supported by grants from the National Institutes of Health. We thank Peter Dervan for multiple gifts of MPE.

[20] M. Sawadogo and R. G. Roeder, *Cell* **43**, 165 (1985).
[21] M. R. Briggs, J. T. Kadonaga, S. P. Bell, and R. Tjian, *Science* **234**, 47 (1986).
[22] M. W. Van Dyke and R. G. Roeder, *Mol. Cell. Biol.* **7**, 1021 (1987).
[23] G. H. Thomas and S. C. R. Elgin, unpublished observations, 1987.

[19] Chemical Radiolabeling of Lysines That Interact Strongly with DNA in Chromatin

By Jean O. Thomas

Introduction

The basic side chains of lysine and arginine residues in the histones are major contributors to DNA binding in chromatin. They are probably involved primarily in electrostatic interactions with DNA phosphates, as well as in hydrogen bonding to phosphate oxygens. A preference for certain sequence types at particular positions in the DNA wrapped around the histone octamer in the nucleosome core particle, which seems to be related to the bendability of the DNA,[1] raises the possibility of additional interactions of amino acid side chains with bases within the grooves of the DNA double helix.

Lysine ϵ-amino groups ($pK_a \sim 10.5$), in contrast to the guanidino side chains of arginine ($pK_a \sim 13$), are amenable to chemical modification under mild conditions and act as nucleophiles at pH values not far above

[1] S. C. Satchwell, H. R. Drew, and A. A. Travers, *J. Mol. Biol.* **191**, 659 (1986).

Copyright © 1989 by Academic Press, Inc.
All rights of reproduction in any form reserved.

neutrality. Reagents that react with lysine ϵ-amino groups are, therefore, useful probes of protein topography. Under conditions where reagent concentrations are not limiting and reaction times are relatively long, fine kinetic differences, arising from the different microenvironments of lysine residues whose general environment is broadly similar, are eliminated. (Contrast the competitive labeling procedure[2] that aims to distinguish fine differences in reactivity.)

In nucleosome core particles or chromatin, three classes of lysine residues may be distinguished on the basis of the availability of their side chains to chemical reagents: (1) Those that are accessible in the native state include lysines that are not engaged in interactions with DNA (e.g., lysines in the amino-terminal basic regions of some of the core histones in nucleosome core particles; there are unlikely to be any completely "free" lysines in chromatin) and lysines involved in electrostatic interactions which can nonetheless react with small chemical reagents. Such lysines occur in the very basic regions of the core histones and H1 (and its variants such as H5 in nucleated erythrocytes). These regions may act as delocalized polycations,[3] neutralizing the polyanionic DNA. However, despite the relative ease with which they can be chemically modified, they contribute substantially to the binding energy of histones and DNA. (2) Those lysines presumed to be involved in specific (hydrogen-bonded or ionic) interactions with DNA (e.g., at particular points on the surface of the core histone octamer, or at the nucleosome binding site on H5 or H1) react completely only when the octamer, or H5 and H1, respectively, are dissociated from DNA at high ionic strength. (3) Those that are internal, that is, located at subunit interfaces in the octamer or within the globular regions of particular histones, react only when the histones are denatured. There are at most a few lysines in this class, and their side chains are likely to exist in internal ion pairs with side-chain carboxylates of aspartic and glutamic acids.

If chemical modification of amino groups is carried out in three successive stages and under conditions appropriate for exposure of the three classes of lysine residues, these classes may be distinguished by using radiolabeled reagent (of high specific activity) at one stage only. This chapter is concerned particularly with the selective radiolabeling and identification of amino groups at "strong" DNA binding sites (class 2, above).[4,5]

[2] B. Malchy and H. Kaplan, *J. Mol. Biol.* **82**, 537 (1974).
[3] G. S. Manning, *Q. Rev. Biophys.* **11**, 179 (1978).
[4] S. F. Lambert and J. O. Thomas, *Eur. J. Biochem.* **160**, 191 (1986).
[5] J. O. Thomas and C. M. Wilson, *EMBO J.* **5**, 3531 (1986).

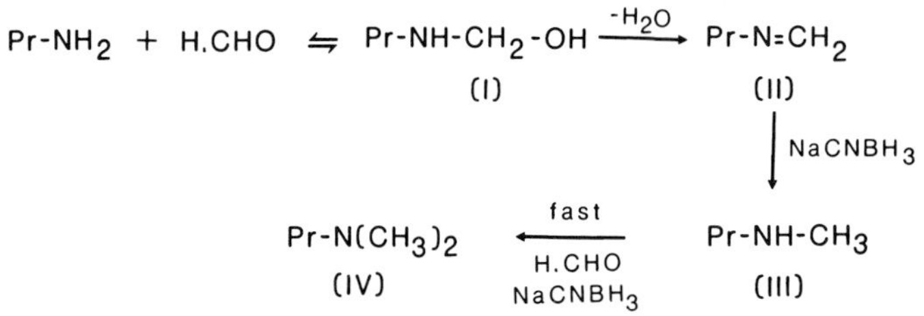

$$Pr-NH_2 \ + \ H.CHO \ \rightleftharpoons \ Pr-NH-CH_2-OH \ \xrightarrow{-H_2O} \ Pr-N=CH_2$$

$$(I) \qquad\qquad\qquad (II)$$

$$\Big\downarrow NaCNBH_3$$

$$Pr-N(CH_3)_2 \ \xleftarrow[\substack{H.CHO \\ NaCNBH_3}]{fast} \ Pr-NH-CH_3$$

$$(IV) \qquad\qquad\qquad (III)$$

FIG. 1. Reductive methylation of lysine side chains in proteins. Addition of the ϵ-amino group of lysine to the carbonyl group of formaldehyde results in a hydroxymethyl adduct (**I**); this is dehydrated to the imine (Schiff base) (**II**), which is then reduced with sodium cyanoborohydride to the ϵ-N-monomethyl derivative (**III**). Rapid reaction with a further equivalent of formaldehyde, followed by reduction, results in the ϵ-N,N-dimethyllysine derivative (**IV**). The pK_a values of the unmethylated and dimethylated ϵ-amino groups are essentially the same (~ 10.2) [T. A. Gerken, J. E. Jentoft, N. Jentoft, and D. G. Dearborn, *J. Biol. Chem.* **257**, 2894 (1982)].

Choice of Reagent

In principle, both amidation with monofunctional imidoesters[6,7] and reductive methylation[8-10] using formaldehyde and a reducing agent (Fig. 1) may be used to probe lysine accessibility in proteins and nucleoprotein assemblies. For example, in contrast to acetylation using acetic anhydride, both procedures ensure preservation of the positive charge on the lysine side chain (albeit, in the case of amidation, one atom further removed from the peptide backbone). However, the side reactions of imidoesters[7] that give rise to cross-linking between lysine residues at the pH values (near neutrality) that are necessary to avoid distortion of nucleosomes make amidation less suitable than reductive methylation for studies of chromatin topography.

Extensively methylated nucleosome core particles retain their structural integrity,[4] possibly because "nonspecific" electrostatic neutralization of DNA phosphates is not substantially affected by lysine methylation, and methyl groups are relatively small and innocuous. Both α- and ϵ-amino groups in proteins react with formaldehyde and are converted rapidly in the presence of reducing agent through the monomethyl to the dimethyl form (Fig. 1). The pK_a of the side chain of ϵ-N,N-dimethyllysine is close to

[6] M. J. Hunter and M. L. Ludwig, *J. Am. Chem. Soc.* **84**, 3491 (1962).

[7] J. K. Inman, R. N. Perham, G. C. DuBois, and E. Appella, this series, Vol. 91, p. 559.

[8] G. E. Means and R. E. Feeney, *Biochemistry* **7**, 2192 (1968).

[9] N. Jentoft and D. G. Dearborn, this series, Vol. 91, p. 570.

[10] G. E. Means, *J. Protein Chem.* **3**, 121 (1984).

that of the unmodified group (~0.4–0.6 units less), and ion-pair interactions involving lysine residues in proteins are not necessarily altered by methylation, as shown by a study of lysozyme[11] (e.g., ion-pair interactions of two amino groups with carboxyl groups are not affected by dimethylation), although methylation reduces the multiple hydrogen-bonding capacity of an amino group. Many enzymes retain substantial activity when methylated.[9,10]

The Approach

The strategy of "cold–hot–cold" modification involves three stages: (1) reductive methylation of all "accessible" amino groups in nucleosome core particles or chromatin with unlabeled "cold" formaldehyde and reducing agent; (2) release of the intact octamer (and H1,H5 in the case of chromatin) from the DNA in 2 M NaCl and selective radiolabeling of the newly exposed amino groups on the protein surfaces; and (3) extraction of the histones, free of DNA, and "cold" reductive methylation under denaturing conditions to ensure complete modification of any hitherto buried lysines. Because the differences in reactivity between lysines in classes 1 and 2 (p. 370) are kinetic rather than absolute, several times modification may be necessary at each of the first two stages in which "readily accessible" and "DNA-protected" lysines, respectively, are labeled, in order to gain a clear picture. Moreover, small chemical reagents (such as formaldehyde) may penetrate the "breathing" structure and so blur the distinction further; greater discrimination may be possible with larger reagents, provided the possibility of structural distortion is kept in mind. Despite these limitations, useful insights into histone topography in chromatin have emerged from application of the differential radiolabeling approach, in particular, concerning the location of the lysine residues whose side chains are engaged in the "tightest" interactions with DNA,[4,5] as described below.

To illustrate the approach, typical procedures are described for radiolabeling by reductive methylation of (1) residues on the surface of the histone octamer in nucleosome core particles that are (partially) protected by, and therefore presumably interacting directly with, DNA,[4] and (2) lysine residues in H5 that are (partially) protected when H5 is bound in chromatin and presumably present at the nucleosome binding site.[5] The strategy is essentially the same in the two cases. Methods are also described for the separation of reductively methylated histones by high-performance liquid chromatography (HPLC), for peptide mapping and fluorography, and for

---

[11] T. A. Gerken, J. E. Jentoft, N. Jentoft, and D. G. Dearborn, *J. Biol. Chem.* **257**, 2894 (1982).

the purification and characterization of radiolabeled tryptic peptides of ^{3}H-methylated histones by HPLC. The specific radioactivities of these peptides indicate the degree of protection of the constituent lysines by association with DNA in the native structure.

Reagents and Conditions

The general considerations in reductive methylation of proteins have been fully described.[9,10] Modification of chromatin is carried out at pH 7 – 7.5 to minimize the risk of structural distortion. The buffer must be free of amino groups; phosphate or HEPES (N-2-hydroxyethylpiperazine-N-2-ethanesulfonic acid) buffers are both suitable, but triethanolamine hydrochloride inhibits the reaction in a concentration-dependent manner (about 50% inhibition in 140 mM buffer relative to 20 mM).[12]

The reducing agent of choice is sodium cyanoborohydride,[7] which is effective at neutral pH, milder than sodium borohydride,[6] and reduces Schiff bases without reducing formaldehyde; dimethylamine borane, which was tested as an alternative reducing agent[13] under the same conditions, was found to be less effective.[12] For greatest effect, sodium cyanoborohydride (Sigma, St. Louis, MO) should be purified by precipitation from acetonitrile with dichloromethane[7] and stored in an evacuated desiccator. Aqueous solutions (1 M in buffer) should be prepared immediately before use.

Although radiolabel can be introduced from ^{14}C-labeled formaldehyde or ^{3}H-labeled NaCNBH$_3$, the former suffers from a relatively low specific activity and the latter from wasteful exchange of the labeled hydrogens with water, coupled with a need for a relatively high concentration to achieve reasonable reaction rates. Labeling is conveniently achieved with high specific activity [^{3}H]formaldehyde (New England Nuclear, Boston, MA), supplied as a 0.33 M solution at 70–80 Ci/mol. We use it diluted 2-fold with "cold" formaldehyde (AR grade) as a 0.5 M solution nominally at 35–40 Ci/mol, and we determine its effective specific activity in the modification reaction as described below (final section of this chapter). It is used in at least a 10-fold excess over amino groups[9] since the aim is to react amino groups in any particular reactivity class essentially to completion. Sodium cyanoborohydride is also used in at least 10-fold excess over amino groups and is added to the chromatin before the formaldehyde to minimize the risk of lysine–lysine cross-linking (by formaldehyde); reductive methylation of chromatin results in virtually no histone–histone

---

[12] S. F. Lambert, Ph.D. Thesis, Univ. of Cambridge, 1985.
[13] J. C. Cabacungan, A. I. Ahmed, and R. E. Feeney, *Anal. Biochem.* **124,** 272 (1982).

cross-linking, as assessed by polyacrylamide gel electrophoresis in the presence of sodium dodecyl sulfate (SDS).

Reductive methylation carried out under denaturing conditions, when the aim is to modify *all* lysines to completion, is best achieved in 8 M urea; 6 M guanidinium chloride appears to inhibit the reaction.[12] The urea should be recrystallized (from ethanol) or otherwise treated (e.g., with ion-exchange resin) to remove cyanates, which would react with amino groups.[14] (Note that cyanate is present in 8 M urea at a concentration of 20 mM at equilibrium![14])

Preliminary time courses are necessary to determine the conditions for the first ("cold") modification and the second ("hot") modification. In both situations, described in detail below, the first modification of accessible groups is fairly exhaustive, such that the only surface groups that remain incompletely reacted are those on the octamer surface protected by interaction with DNA and those on H5 protected by association with the nucleosome. These are then radiolabeled.

Experimental Procedures

Radiolabeling of Lysine-Containing DNA-Binding Regions on the Surface of the Histone Octamer in the Nucleosome Core Particle[4]

1. To modify all the completely accessible amino groups, treat nucleosome core particles (A_{260} ~25, histone concentration ~1.25 mg/ml) in 10 mM HEPES, 1 mM Na$_2$EDTA, 0.25 mM phenylmethylsulfonyl fluoride (PMSF), pH 7.5, with "cold" NaCNBH$_3$ (15 mM final concentration, added from a 1 M stock in the same buffer) and H·CHO (20 mM, from a 0.5 M stock in the buffer) for 4 hr at 20°. Quench the reaction with 1/20 volume of 1 M glycine, pH 8. (The reaction time is determined by preliminary experiments; the time chosen here leaves only the most tenacious DNA-binding lysines not completely modified.[4])

2. Dialyze the modified core particles into 10 mM HEPES, 0.25 mM PMSF, pH 7.5, containing 2 M NaCl to dissociate the octamer from the DNA, and radiolabel the newly exposed lysines by adding [^{3}H]H·CHO (0.5 M, nominally at 35–40 Ci/mol; see below) to a final concentration of 20 mM and NaCNBH$_3$ (0.1 M) to 15 mM. After 4 hr at 20° quench the reaction as above.

3. Extract the histones with 0.2 M H$_2$SO$_4$ at 4° and precipitate them with an equal volume of 50% (w/v) trichloroacetic acid (TCA). Wash the precipitate (acetone/0.1 M HCl, and then acetone), dissolve it in 8 M urea

[14] G. R. Stark, W. H. Stein, and S. Moore, *J. Biol. Chem.* **235**, 3177 (1960).

(recrystallized from ethanol) in 10 mM HEPES, pH 7.5, and methylate all remaining amino groups by two treatments of 2 hr each with "cold" H·CHO (20 mM at each addition) and NaCNBH$_3$ (15 mM at each addition). Dialyze the fully modified histones exhaustively against 0.5% (w/v) NH$_4$HCO$_3$, pH 8.2, and lyophilize the solution, or dilute the 8 M urea solution 3- to 5-fold with water and precipitate the histones with TCA, as described above. Dissolve the histones in water or 5 mM HCl, and estimate their concentration using $A_{230} = 4.2$ for 1 mg/ml.[15]

4. Check the quality of the material by SDS–18% polyacrylamide gel electrophoresis;[16] the four core histones should be present in the same (equimolar) ratios as in the initial chromatin, as judged by staining with Coomassie Brilliant Blue R. Complete reductive methylation causes a decrease in the electrophoretic mobility of the histones roughly in proportion to their lysine content, the effect decreasing in the order H2B > H2A = H3 > H4 (for which the effect is barely perceptible). To obtain a semiquantitative estimate of the relative radiolabeling of the core histones, soak the gel in Amplify (Amersham International plc, Amersham, UK) dry it onto Whatman 3 MM filter paper, and expose it to preflashed[17] Fuji RX X-ray film at −70°.

5. Separate the reductively methylated histones by reversed-phase HPLC using a large-pore (330 Å) C$_4$ column (Hi-Pore RP-304; 5 μm; 25 × 0.46 cm; Bio Rad, Richmond, CA) with a guard column attached, following the method of Lindner *et al.*[18] for unmodified histones. We load the sample (up to ~1 mg) onto the column in about 150 μl 0.3% (v/v) aqueous trifluoroacetic acid (TFA) (Pierce, Rockford, IL; Sequanal grade), and elute the histones with a gradient of acetonitrile (Koch-Light, Haverhill, UK; HPLC grade), 30–65% (v/v) in 0.3% TFA, applied over about 80 min, with a flow rate of 0.7 ml/min. Monitor the eluate spectroscopically at 220–230 nm and by electrophoresis in SDS–18% polyacrylamide gels;[16] reductively methylated chicken erythrocyte core histones are eluted in the order H2B, H2A, H4, H3. Lyophilize the fractions, keeping a close eye on them in the initial stages while the acetonitrile evaporates.

Adequate separation of reductively methylated core histones is also achieved on a C$_{18}$ column (Spherisorb 3 ODS-2, fully capped; 3 μm; 25 × 0.46 cm; HPLC Technology, Macclesfield, UK) following the

[15] J. O. Thomas and P. J. G. Butler, *J. Mol. Biol.* **116**, 769 (1977).
[16] J. O. Thomas and R. D. Kornberg, *Methods Cell Biol.* **18**, 429 (1978).
[17] R. A. Laskey and A. D. Mills, *FEBS Lett.* **82**, 314 (1977).
[18] H. Lindner, W. Helliger, and B. Puschendorf, *Anal. Biochem.* **158**, 424 (1986).
[19] L. R. Gurley, J. G. Valdez, D. A. Prentice, and W. D. Spall, *Anal. Biochem.* **129**, 132 (1983).

method of Gurley et al.[19] for unmodified histones. We load the sample in 0.3% (v/v) aqueous TFA and elute with a gradient of acetonitrile (35–70%, v/v) in 0.3% TFA over 2 hr, at a flow rate of 0.7 ml/min. The core histones elute in the same order as from the C_4 column, but H4 contaminates both H2A and H3.

6. Digest the individual histones with trypsin, visualize radiolabeled peptides directly by peptide mapping and fluorography, purify the radioactive peptides by HPLC, and determine the specific radioactivities of the tryptic peptides, as described below. Representative results for H4 are shown in Fig. 2 and Table I.

Radiolabeling of a Strong Nucleosome Binding Site on H5[5]

The strategy ("cold–hot–cold" modification) is as described above. After radiolabeling of the surface amino groups exposed when the histones are released from the DNA in 2 M NaCl (in the case of H1 and H5, the amino groups at the nucleosome binding site), H1 and H5 are extracted with perchloric acid, reductively methylated to completion under denaturing conditions with nonradioactive reagent, and separated for analysis.

1. The first "cold" modification of accessible amino groups is essentially as described above for nucleosome core particles, except that the concentration of histones is 5-fold lower; otherwise, partial precipitation of the chromatin occurs during the modification. Dilute chromatin[20] at A_{260} of about 50 in 0.2 mM Na$_2$EDTA, 5 mM HEPES, pH 7.5, to a concentration of ~0.5 mg/ml (A_{260} ~5) with 5 mM HEPES, 0.2 mM Na$_2$EDTA, 0.25 mM PMSF, pH 7.5. Add NaCNBH$_3$ (1 M), typically to 25 ml chromatin, to a final concentration of 15 mM and unlabeled ("cold") H·CHO (0.5 M) to 20 mM (both stock solutions in the 5 mM HEPES buffer). After 3 hr (based on preliminary experiments to establish conditions) at 23° stop the reaction with 1/20 volume 1 M glycine, pH 7.5, and dialyze the reaction mixture at 4° against 5 mM HEPES, 2 M NaCl, 0.25 mM PMSF, pH 7.5. To reduce the volume before radiolabeling, concentrate the solution in the dialysis bag about 5-fold using sucrose and redialyze against 5 mM HEPES, 2 M NaCl, 0.25 mM PMSF, pH 7.5. (Note that if the octamer is to be analyzed in the same experiment, which is not the case here, concentration serves to reverse any dissociation of the octamer that occurred in dilute solution.)

2. For the second ("hot") modification step, add NaCNBH$_3$ to 15 mM as above, followed by [^{3}H]H·CHO (0.5 M) to 10 mM. After 4 hr at 23°, stop the reaction with 50 mM glycine and dialyze the radiolabeled sample against 5 mM HEPES, 0.25 mM PMSF, pH 7.5, at 4°.

[20] M. Noll, J. O. Thomas, and R. D. Kornberg, Science 187, 1203 (1975).

A   Stain   B   Fluorogram

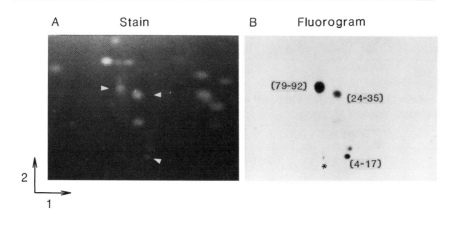

C

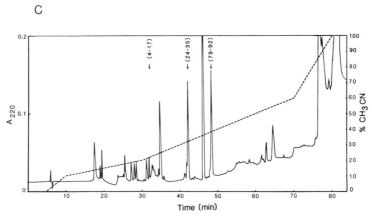

FIG. 2. The tryptic peptides from H4 isolated from nucleosome core particles subjected to "cold–hot–cold" reductive methylation to radiolabel lysines strongly protected by DNA. (A) Peptide map stained with fluorescamine. First dimension (1), electrophoresis at pH 6.5; second dimension (2), chromatography. (B) Fluorogram (2 days) of the map in A (origin indicated by asterisk). (C) Reversed-phase HPLC on a 3-μm C_{18} column eluted with a gradient of acetonitrile in 0.1% (v/v) TFA. The sample contained 70 μg of digested, methylated H2A in 60 μl 0.1% (v/v) TFA. For details, see text. Peaks from material eluting at above 60% acetonitrile appear on the absorbance profile of a "blank run" without sample and do not contain peptides. [From S. F. Lambert and J. O. Thomas, *Eur. J. Biochem.* **160**, 191 (1986).]

3. For the final ("cold") modification step, extract H1 and H5 from the radiolabeled chromatin with 5% (v/v) perchloric acid,[21] precipitate with an equal volume of 50% (w/v) TCA, and wash the precipitate as above. Dissolve the pellet in 2 ml 8 M urea, 20 mM HEPES, 0.25 mM PMSF,

[21] E. W. Johns, *Biochem. J.* **92**, 55 (1964).

TABLE I

"COLD–HOT–COLD" MODIFICATION: QUANTIFICATION OF RADIOLABELING OF LYSINE-
CONTAINING PEPTIDES IN H4[a]

Peptide position in sequence[b] (residues)	No. of lysine residues	Specific radioactivity (dpm/nmol peptide)[c]	Average specific radioactivity of N,N-dimethyllysine[d] (dpm/nmol)	Average degree protection of lysine[e] (%)	Amino-terminal sequence analysis[f]
4–17	4	11,000	2,750	2.2	G K G G . . .
24–35	1	8,083	8,083	6.5	D N I L . . .
79–92	2	38,800	19,400	15.7	K* T V . . .

[a] From S. F. Lambert and J. O. Thomas, *Eur. J. Biochem.* **160**, 191 (1986).
[b] R. J. Sugarman, J. B. Dodgson, and J. D. Engel, *J. Biol. Chem.* **258**, 9005 (1983).
[c] Peptide concentration determined by amino acid analysis, with norleucine as internal standard.
[d] Assuming the radioactivity is uniformly distributed between all the lysines in the peptide.
[e] Based on the known specific radioactivity of dimethyllysine in a reductively methylated synthetic peptide (123,959 dpm/nmol).
[f] K* indicates radiolabeled DABTH-Lys(Me$_2$) detected on polyamide sheets. Weakly radiolabeled lysines escape detection. _ indicates residue identified by sequence analysis.

pH 7.5, and methylate with 15 mM NaCNBH$_3$ and 20 mM unlabeled H·CHO at 23°; repeat the reagent additions after 2 hr. After a further 2 hr stop the reaction with 0.1 volume 1 M glycine. Dilute 5-fold with water, precipitate the completely methylated H1 and H5 with an equal volume of 50% (w/v) TCA, and wash the precipitate as above. The recovery is about 1 mg, roughly 80% that expected, based on A_{230} = 1.85 for 1 mg/ml.[22] Check the purity and integrity of the material by SDS–18% polyacrylamide gel electrophoresis.[16]

4. Dissolve the methylated H1 and H5 in 2 ml of 10 mM sodium phosphate, 0.25 mM PMSF, pH 7.0, and separate them on carboxymethylcellulose (Whatman CM-52) as described for unmodified H1 and H5,[23] using a gradient of 0 to 0.7 M NaCl in 10 mM sodium phosphate, pH 7.0. Collect the fractionated H1 and H5 by precipitation with 25% (w/v) TCA, and wash the precipitate as above. Alternatively, separate the methylated H1 and H5 by reversed-phase HPLC on a C$_4$ Hi-Pore column exactly as described above for the core histones (they elute in the order H5, H1 before the core histones) and lyophilize them.

5. Digest the individual histones with trypsin, and visualize the radioactive peptides by peptide mapping and fluorography. Purify the tryptic

[22] R. D. Camerini-Otero, B. Sollner-Webb, and G. Felsenfeld, *Cell* **8**, 333 (1976).
[23] D. J. Clark and J. O. Thomas, *J. Mol. Biol.* **187**, 569 (1986).

peptides by HPLC, and determine their specific radioactivities as described below. The results for H5 are summarized in Fig. 3 and Table II.

Complete Tryptic Digestion and Peptide Mapping of ³H-Reductively Methylated Histones

Dissolve 100–200 μg of pure, reductively methylated core histones or H5 at approximately 1 mg/ml in ammonium hydrogen carbonate (0.1%, w/v), pH 8.2, and digest with trypsin [EC 3.4.21.4; Worthington, Freehold, NJ; TPCK-treated to prevent any adventitious chymotryptic cleavages; 1:50 (w/w) enzyme:histone] for 4 hr at 37°. Take out about 1 nmol for peptide mapping (~12–16 μg for the core histones, ~20 μg for H5) and lyophilize both this and the bulk to be used for HPLC. The mapping procedure is based on that of Bates *et al.*[24] and is carried out on 0.25 mm silica-coated plates (20 × 20 cm, Polygram Sil G; Machery-Nagel and Co., Duren, FRG; purchased from Camlab, Cambridge, UK). The 20 × 20 cm plate will accommodate two samples for the first electrophoresis step and is marked gently in pencil as shown (Fig. 4). The points at which samples and markers should be applied are as indicated. (We normally set the origin halfway along the plate, although, since there are very few acidic tryptic peptides in the histones, it would be safe to set the origin closer to the anode.)

Dissolve the 1 nmol of sample in 3 μl acetic acid (5%, v/v) and, using a drawn-out capillary, apply the solution to the plate with cold air drying from a hair dryer, allowing the spot to become no larger than about 1 mm in diameter. Apply α-dinitrophenylaspartic acid as a visible marker similarly. Then apply electrophoresis buffer (see below) to the plate through two pieces of Whatman 3 MM paper placed gently onto the silica on either side of the origin and about 2 cm away from it. Allow the buffer from the two sides to meet by capillary action at the origin, thus focusing the sample into a sharp line. Carry out electrophoresis at pH 6.5 [10% pyridine, 0.5% acetic acid (v/v) in water] at 1000 V for about 25 min, until the yellow α-dinitrophenylaspartic acid marker has migrated toward the anode almost to the end of the plate; we use an electrophoresis tank (Model TLE 20) made by Savant Instruments, with white spirit containing 8% pyridine (v/v) as coolant. Dry the plate thoroughly (for at least 1 hr) in a fume cupboard in a stream of cold air. If two samples have been accommodated on one 20 × 20 cm plate for electrophoresis, cut the sheet in half carefully with a guillotine or with a pair of scissors. Carry out ascending chromatography at right angles to the direction of electrophoresis in butan-l-ol/acetic acid/water/pyridine (15:3:12:10, v/v) for about 2 hr, until the solvent

[24] D. L. Bates, R. N. Perham, and J. R. Coggins, *Anal. Biochem.* **68**, 175 (1975).

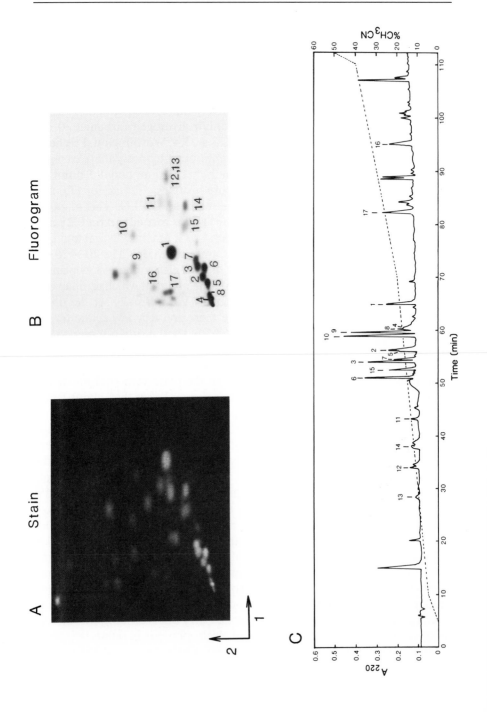

front reaches the top of the plate. Stain the maps by dipping them in a solution of fluorescamine (25 mg/liter) in acetone containing 0.5% (w/v) triethylamine, and photograph through a red filter under 365 nm UV light, using Kodak Tri-X film. To locate the radiolabeled peptides by fluorography,[25] dip the plates in a solution of 0.5% (w/w) 2,5-diphenyloxazole (PPO) in molten naphthalene (Sigma), drain off the excess liquid, and then expose the plate to preflashed Fuji RX X-ray film at −70°. Comparison of the stained maps (Figs. 2A and 3A) and their fluorograms (Figs. 2B and 3B) gives a *qualitative* indication of the relative radiolabeling of the various peptides (qualitative only, since the yields of the peptides may be different, as may their intrinsic staining with fluorescamine).

Fractionation of Tryptic Digests of Reductively Methylated Histones by Reversed-Phase HPLC and Peptide Purification

Tryptic digests of modified histones (~ 100–200 μg) may be fractionated by reversed-phase HPLC on a C_{18} column (Spherisorb 3 ODS-2, fully capped; 3 μm; 25 × 0.46 cm; HPLC Technology) with a guard column (Waters, Milford, MA).[5,24] Load the sample onto the column in 0.1 or 0.3% (v/v) aqueous TFA (50–150 μl), and apply a gradient of acetonitrile from 10 to about 60% in 0.1 or 0.3% TFA over 1–2 hr, with a flow rate of 0.7 ml/min (Figs. 2C and 3C).[4,5] Collect fractions of approximately 300 μl, and locate radiolabeled peptides by counting 15-μl samples for radioactivity in Triton X-100/toluene (3:7, v/v) containing PPO (5 g/liter). Assess the purity of the labeled peptides by peptide mapping and fluorography, and, where necessary, purify peptides further by HPLC, either by flattening the gradient through the elution range of the peptides or by changing the solvent.

Peptides that are difficult to purify by reversed-phase HPLC may be purified for identification by elution from silica-coated thin-layer plates after two-dimensional electrophoresis and chromatography as above. Locate fluorescamine-stained spots under 365 nm UV light, and mark their

---

[25] W. F. Bonner and J. O. Stedman, *Anal. Biochem.* **89**, 247 (1978).

FIG. 3. The tryptic peptides from H5 radiolabeled at the nucleosome binding site by the "cold–hot–cold" reductive methylation procedure. (A) Fluorescamine-stained map. (B) Fluorogram of the map in A. (C) Reversed-phase HPLC on a 3-μm C_{18} column eluted with a gradient of acetonitrile in 0.3% (v/v) TFA. H5 tryptic digest (355 μg) was loaded in 100 μl 0.3% (v/v) TFA. For details, see text. [From J. O. Thomas and C. M. Wilson, *EMBO J.* **5**, 3531 (1986).]

TABLE II

"COLD–HOT–COLD" MODIFICATION: QUANTIFICATION OF THE RADIOLABELING OF LYSINE-CONTAINING PEPTIDES IN H5[a]

Peptide no.	Peptide position in sequence (residues)[b]	No. of lysine residues	Specific radioactivity[c] (dpm/nmol dimethyllysine)	Average degree protection of lysine[d] (%)	Partial or complete amino-terminal sequence analysis[e]
1	82–94	2	11,298	9.1	K*QTK***G...
2	104–113	4	1454	1.2	SPGK*K*K*K*...
3	129–139	3	1622	1.3	SPAK*K*K*...
4	147–163	7	1244	1.0	ASPK*K*AK**K*PK*...
5	164–173	5	1460	1.2	K*ASK*AK*K*...
6	184–189	4	1045	0.8	K*SPK*K*K*
7	95–103	3	926	0.7	LAK...
8 (f)	147–159	6	1095	0.9	n.d.
9	48–53	1	3256	2.6	Q?SIQK**Y^g
10	54–58	1	4098	3.3	IK**SHY^g
11	16–20	1	2298	1.8	VK*A...
12 (f)	178–183	1	2612	2.0	n.d.
13 (f)	140–142/126–128	1	1329	1.1	n.d.
14 (f)	143–146	2	2350	1.9	n.d.
15 (f)	115–123	2	1448	1.2	n.d.
16 (f)	59–70	2	3782	3.1	n.d.^{g}
17 (f)	1–15	2	1123	0.9	n.d.

[a] Peptide number refers to the map in Fig. 4. From J. O. Thomas and C. M. Wilson, *EMBO J.* **5**, 3531 (1986).

[b] From the amino acid composition and, where indicated, amino-terminal sequence analysis of the peptide, based on the complete sequence of H5 [G. Briand, D. Kmiécik, P. Sautière, D. Wouters, O. Borie-Loy, A. Mazen, and M. Champagne, *FEBS Lett.* **112**, 147 (1980)].

[c] Determined by scintillation counting of the ε-N,N-dimethyllysine eluted from the amino acid analyzer column.

[d] From the average specific radioactivity of dimethyllysine (previous column) and the known specific radioactivity of ε-N,N-dimethyllysine in a reductively methylated synthetic peptide (see text).

[e] ¯ indicates residues identified by sequence analyses. K*, etc., indicates radiolabeled DABTH-Lys(Me₂) on the polyamide sheets; the number of asterisks gives a very rough indication of the relative labeling. Some very weakly labeled lysines might not be detected by this method. n.d., Not determined.

[f] Peptides purified or repurified by elution from silica-coated thin-layer plates; amino acid composition determined but sequence not determined.

[g] Peptides arising from chymotryptic cleavages; the trypsin used in this experiment was not TPCK-treated.

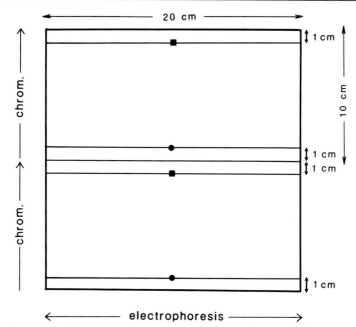

FIG. 4. Layout of silica gel-coated thin-layer plate for peptide mapping by electrophoresis at pH 6.5 followed by chromatography at right angles. The lines are drawn on the plate lightly in pencil. Sample is applied at ● and the marker at ■ (see text).

outline in pencil. Rinse the plate with acetone to remove excess fluorescamine, and extract the silica containing the required peptide in siliconized microfuge tubes, first with 250 μl 0.1 M HCl and then with 150 μl of 5% NH_3 solution. Evaporate the two extracts to dryness successively in an acid-washed Pyrex tube, and hydrolyze the peptide for amino acid analysis (see below). Peptides purified in this way can also be sequenced, provided a 10-fold lower fluorescamine concentration is used for staining.

Identification of Radiolabeled Peptides and Determination of Specific Radioactivities

Amino acid compositions alone are often sufficient to permit the identification of the tryptic peptides from reductively methylated histones, since the amino acid sequences of the histones from several species are known. Arginine is expected at the carboxy terminus of all the peptides (except those that arise from secondary cleavages and the carboxy-terminal peptide of the whole protein, unless this is arginine) because trypsin does not cleave after ϵ-N,N-dimethyllysine;[9,26] slow cleavage will occur after any residual monomethyllysine,[9,26] of which, even in the lysine-rich histones, there is very little under the conditions described. For amino acid analysis, hydrolyze about 0.5–1 nmol of the peptide in 6 M HCl containing 2-mercaptoethanol (0.5 ml/liter) and phenol (0.5 g/liter) *in vacuo* in chromic acid-washed Pyrex tubes at 105° for 24 hr. Include a known amount of norleucine as internal standard if the analysis is to be used to determine peptide *concentration* (see below). In our laboratory, analyses are carried out on an LKB 4400 automatic analyzer. ϵ-N,N-Dimethyllysine elutes as a well-resolved peak after histidine and just before lysine. The optical absorption factor at 570 nm for the ϵ-N,N-dimethyllysine adduct with ninhydrin is 0.822 of the value for the lysine–ninhydrin adduct.[8]

The specific radioactivities of radiolabeled peptides are proportional to the degree of protection of the lysine(s) owing to association with DNA in the native structure, enabling the relative degree of protection of lysine residues in different peptides to be easily deduced. Specific activities may be determined in two ways: first, from the concentrations of solutions of the peptides determined by amino acid analysis (using norleucine as internal standard) and their radioactivities determined by scintillation counting; second, by direct determination of the radioactivity of the dimethyllysine peak eluted from the amino acid analyzer. The results are generally in good agreement.

Note that, in principle, the α-amino group might also be radiolabeled in the "cold–hot–cold" modification procedure. This would be detected as a new peak containing radioactivity by amino acid analysis, and the α-N,N-dimethylated peptide itself would not stain with fluorescamine (or react with Edman reagents). However, it is worth noting that at neutral pH α-amino groups are more readily reductively methylated than ϵ-amino groups[9] and, therefore, are likely to be modified in the native state (unless buried). Of the histones, H1, H2A, and H4 from chicken erythrocytes, and many other species, are α-N-acetylated. In H3 the amino-terminal tryptic peptide does not contain lysine, and the amino-termini of H2B and H5 react readily with "cold" reagent in the native structure and, thus, are not radiolabeled in the second modification step.

An *absolute* value for the degree of protection of a particular lysine in chromatin under the particular conditions of the "cold–hot–cold" modification may be obtained by comparison of its specific activity with that of radiolabeled ϵ-N,N-dimethyllysine derived from either α-N-acetyllysine or any readily available lysine-containing peptide whose amino terminus is blocked, or from which the α-N,N-dimethylated amino-terminal portion

[26] J. H. Seely and N. L. Benoiton, *Can. J. Biochem.* **48**, 1122 (1970).

may be easily cleaved. Using such a peptide, we found the specific activity of dimethyllysine to be 123,960 dpm/nmol.[4] Thus, a radiolabeled peptide resulting from the "cold–hot–cold" modification procedure and having, for example, one lysine and 12,396 dpm/nmol would be deemed 10% protected in the native structure. [The effective specific radioactivity of the [^{3}H]H·CHO determined in this way was 28 Ci/mol; the stock from the supplier (nominally at 70–80 Ci/mol) was therefore 56 Ci/mol reactive formaldehyde.]

To determine the specific radioactivity of individual lysines in a multilysine peptide, sequential (Edman) degradation of the peptide is essential. In the manual method,[27] the 4-N,N-dimethylaminoazobenzene 4'-thiohydantoin (DABTH) derivative of ε-N,N-[^{3}H]dimethyllysine, analyzed on polyamide-coated thin-layer sheets, may be detected by fluorography as described above for peptides on silica-coated sheets. A very rough estimate of the relative radioactivities of different lysines may thus be obtained. With automated sequencing, more accurate estimation is possible. For example, in a gas-phase sequencer, only a portion of the phenylthiohydantoin is diverted from the reaction vessel to the HPLC detection system; the rest (whose concentration will then be known) may be retrieved, and its specific radioactivity determined.

Representative results for histones H4 and H5 are shown in Tables I and II. For illustration, the specific radioactivities are determined in different ways in the two cases. The "average value" for the degree of protection of lysine, which assumes that radioactivity in a peptide is uniformly distributed between all the lysines, is convenient for an initial comparison of peptides. However, it could also be misleading because some lysines might be strongly radiolabeled and others not at all. Sequence analysis (e.g., Table II, final column) shows that although radioactivity is indeed distributed between various lysines of several multilysine radiolabeled peptides, in many cases approximately equally, one lysine in peptide 1 in Table II, for example, is much more strongly labeled than the other. This lysine, Lys-85 in H5, is the lysine of highest specific activity after the "cold–hot–cold" modification procedure and is assumed to be present at the interface between the globular domain of H5 and the nucleosome.[5]

Acknowledgment

Work in this laboratory was supported by the Science and Engineering Council of the U.K.

[27] J.-Y. Chang, D. Brauer, and B. Wittmann-Liebold, *FEBS Lett.* **93**, 205 (1978).

[20] Mapping DNA–Protein Interactions by Cross-Linking

By A. D. Mirzabekov, S. G. Bavykin, A. V. Belyavsky,
V. L. Karpov, O. V. Preobrazhenskaya, V. V. Shick,
and K. K. Ebralidse

Introduction

DNA–protein interactions attract much attention for they are involved in the structural organization of genomic DNA as well as in directing and regulating DNA function. This interest has stimulated the development of a number of new approaches; one of them is DNA–protein cross-linking (for review, see Ref. 1). A general strategy for using DNA–protein cross-linking to study DNA–protein complexes which has been devised in our laboratory is described in this chapter.

The cross-linking techniques presented here enable one to answer two major questions on different structural levels. First, what are the proteins associated with specific DNA regions and sequences? Second, what is the fine structure of the complexes, that is, which regions or monomer residues in the DNA and proteins are near one another in the complex and interact with each other? Different but compatible experimental approaches are applied to answer either of the above questions.

In order to characterize components of the cross-linked complex, hybridization with cloned DNA probes should be used for recognizing DNA sequences, whereas gel electrophoresis and/or specific antibodies are useful for identification of proteins. Gel electrophoresis is the method of choice for proteins that are present in the complex in a sufficient amount, for example, for histones. Thus, two-dimensional gel electrophoresis and hybridization with various probes, as described below, was used to demonstrate the reversible displacement of histones from transcribed gene regions[2] as well as an essential similarity in the arrangement of histones on DNA in core nucleosomes within transcribed and silent chromatin regions.[3] In the case of proteins that are much less abundant than histones, DNA cross-linked to these proteins can be isolated by the use of specific antibodies. In this way, the arrangement of HMG 14 and 17 proteins on

[1] J. Welsh and C. R. Cantor, *Trends Biochem. Sci.* **9**, 505 (1984).

[2] V. Karpov, O. Preobrazhenskaya, and A. Mirzabekov, *Cell* **36**, 423 (1984).

[3] V. Studitsky, A. Belyavsky, A. Melnikova, and A. Mirzabekov, *Mol. Biol. (Moscow)* **22**, 706 (1988).

English translation copyright © 1989
by Academic Press, Inc.

nucleosomal DNA[4,5] and the location of HMG 14/17, HMG 1/2, and histones H1 and H5 on transcribed and repressed DNA sequences[6] were determined.

The same approach has been also developed by Gilmour and Lis,[7] and Blanco et al.[8] to study the distribution of RNA polymerase, topoisomerase I, and HMG-T molecules within genomic DNA.

Two cross-linking methods have shown themselves as reliable and informative in our hands: (1) cross-linking proteins to DNA, partly depurinated under mild conditions,[9] and (2) cross-linking by ultraviolet (UV) radiation.[6] While UV irradiation allows cross-linking of proteins to DNA within whole cells, cross-linking to depurinated DNA works well on isolated nuclei, nucleosomes, and reconstituted complexes.

Cross-linking to depurinated DNA provides significant advantages over other methods for studying the fine structure of complexes. This method produces a single-stranded nick in DNA at the site of cross-linking and attaches the protein molecule only to the 5′-terminal fragment thus produced. By measuring the length of the fragment cross-linked to the protein, one can precisely locate the protein molecule on DNA of a certain size, as a distance of a given cross-linking site from the DNA 5′-terminus.[10,11] The basic principle of this method as well as the exploited chemical reactions are very much the same as in the Maxam–Gilbert procedure for sequencing guanine and adenine bases in DNA.[12,13] In this way, the arrangement of histones and HMG 14/17 proteins on DNA in different nucleosomes[11,14] as well as the binding sites of RNA polymerase subunits on promoter DNA[15] have been determined. The localization of cross-linked regions or amino acid residues within proteins can be found by labeling the protein with [32]P

[4] V. Shick, A. Belyavsky, A. Mirzabekov, V. Ermekova, and I. Belezky, *Proc. Acad. Sci. USSR* **274**, 1254 (1984).

[5] V. Shick, A. Belyavsky, and A. Mirzabekov, *J. Mol. Biol.* **185**, 329 (1985).

[6] A. Belyavsky et al., *Proc. Acad. Sci. USSR* (submitted).

[7] D. S. Gilmour and J. T. Lis, *Mol. Cell. Biol.* **5**, 2009 (1985).

[8] J. Blanco, J. C. States, and G. M. Dixon, *Biochemistry* **24**, 8021 (1985).

[9] E. Levina, S. Bavykin, V. Shick, and A. Mirzabekov, *Anal. Biochem.* **110**, 93 (1981).

[10] A. Mirzabekov, V. Shick, A. Belyavsky, and S. Bavykin, *Proc. Natl. Acad. Sci. U.S.A.* **75**, 4184 (1978).

[11] V. Shick, A. Belyavsky, and A. Mirzabekov, *J. Mol. Biol.* **139**, 491 (1980).

[12] W. Gilbert, A. Maxam, and A. Mirzabekov, *in* "Control of Ribosome Synthesis," Alfred Benzon Symposium IX (N. O. Kjeldgaard and O. Maaloe, eds.), p. 139. Munksgaard, Copenhagen, 1976.

[13] A. M. Maxam and W. Gilbert, *Proc. Natl. Acad. Sci. U.S.A.* **74**, 560 (1977).

[14] S. Bavykin, S. Usachenko, A. Lishanskaya, V. Shick, A. Belyavsky, I. Undritsov, A. Strokov, I. Zalenskaya, and A. Mirzabekov, *Nucleic Acids Res.* **13**, 3439 (1985).

[15] A. Chenchick, R. Beabealashvili, and A. Mirzabekov, *FEBS Lett.* **128**, 46 (1981).

at the cross-linked oligonucleotide[16] and applying standard methods for peptide analysis.[17] This approach was used to characterize and localize the cross-linking sites on segments of nucleosomal DNA and peptide regions of histone H4[17,18] or H5.[19]

We describe here an experimental procedure for induction and analysis of DNA–protein cross-links. We present the protocols for cross-linking proteins and DNA by dimethyl sulfate (DMS) methylation–depurination–reduction and by UV light. Then, methods for recovery of the cross-linked complexes by depleting them of the uncross-linked DNA and proteins are explained. Finally, methods for analysis of cross-links by gel electrophoresis diagonal techniques, DNA hybridization, immunological reactions, and peptide chemistry are described.

DNA–Protein Cross-Linking to Partly Depurinated DNA

The chief device for this "zero-length" DNA–protein cross-linking is the unmasking of the highly reactive aldehyde group of deoxyribose brought about by hydrolysis of the N-glycosidic bond.[9] Then any protein functional group, capable of reacting with aldehydes and localized closely to them in the DNA–protein complex, can be covalently bound to the DNA. The aldehydes are known to react with amino-terminal α-amino groups, ϵ-amino groups of lysines, and, recently, imidazole rings of histidines were shown to be reactive.[17]

The splitting of the glycosidic bond is facilitated by purine methylation with dimethyl sulfate at the N-7 position of guanine (facing the major groove of the DNA double helix) and, to a lesser extent, at the N-3 position of adenine (facing the minor groove).[20] Introduction of the methyl group into the DNA is not expected to perturb noticeably the nucleoprotein structure, at least not in the case of RNA polymerase–promoter DNA complexes[15] or nucleosomes[14] where much more bulky substituents can be tolerated.[21] Nor is the DNA methylation hindered by histones, as shown by kinetics measurements of methylation of DNA in chromatin[20,22] and of depurination of methylated bases.[20] However, in considering the application of the cross-linking technique to sequence-specific DNA-binding pro-

[16] E. Ebralidse, V. Tuneev, A. Melnikova, E. Goguadze, and A. Mirzabekov, Proc. Acad. Sci. USSR 265, 1272 (1982).

[17] K. Ebralidse, S. Grachev, and A. Mirzabekov, Nature (London) 331, 365 (1988).

[18] K. Ebralidse and A. Mirzabekov, FEBS Lett. 194, 69 (1986).

[19] D. Pruss, K. Ebralidse, and A. D. Mirzabekov, Mol. Biol. (Moscow) 22, 1108 (1988).

[20] A. Mirzabekov, D. San'ko, A. Kolchinsky, and A. Melnikova, Eur. J. Biochem. 75, 379 (1977).

[21] J. D. McGhee and G. Felsenfeld, Annu. Rev. Biochem. 49, 1115 (1980).

[22] J. D. McGhee and G. Felsenfeld, Proc. Natl. Acad. Sci. U.S.A. 76, 2133 (1979).

teins, one must pedantically account for possible protection/enhancement/interference artifacts.

Although all histones can be cross-linked to DNA at a proper level of methylation and depurination, such a complex is difficult to dissolve, and it is more convenient to use a lower level of cross-linking, produced by introduction of roughly 1 methyl group per 100–200 base pairs (bp). Depurination of methylated purines (which can be effected by simply elevating the temperature, usually to 37° or 45°) does not proceed to completion under the temporal conditions specified below; instead, some 20% of methylated bases are removed. Since the rate of removal of methylated guanines is some 10 times less than that of methylated adenines, the differences in methylation efficiencies are thus compensated for. The motive for choosing small extents of methylation–depurination is to minimize the influence of local structural perturbations on the structure as a whole.

The formation of the cross-link is accompanied by a DNA single-strand scission 3′ to the depurinated cross-linked site. It proceeds by 3′-phosphate elimination intramolecularly catalyzed by the $>C=N-$ function in the β position. This side reaction proves in fact to be advantageous in that it allows the alignment of the cross-linked proteins along the DNA.[10]

The cross-linking (but not the strand scission) can be reversed[11] at acidic pH, albeit incompletely. In practice, it is convenient for most applications to stabilize the cross-links by $>C=N-$ bond reduction with either sodium borohydride or sodium cyanoborohydride. The latter mild cross-linking agent is preferable in that it is selective toward $>C=N-$ (over $>C=O$) and can therefore be present during the aldehyde liberation–cross-linking step. Then the newly formed cross-links are immediately rendered irreversible by reduction, including cross-links with side chains which, as aldimines, might otherwise be too short-lived to survive prolonged incubation.[17]

When working with systems that can be reconstituted *in vitro,* one can methylate and sometimes depurinate the DNA prior to readdition of protein(s).[15] Here, the advantage is not only reduced reaction time (brief methylation and depurination steps at higher DMS concentrations and higher temperatures), but also less, if any, damage to proteins: note, for example, that DMS reacts readily with SH groups.

DNA – Protein Cross-Linking on Nucleosomes and Soluble Chromatin

1. Chill the nucleosomes or soluble chromatin prepared by an appropriate method (Kornberg *et al.,* this volume [1]) from 5 to 20 A_{260} units/ ml, in 25 mM cacodylate – NaOH, pH 7.2 (or in another thiol-free buffer of pH range 6.0–8.0 and having enough capacity to neutralize the products of DMS hydrolysis), 1 mM EDTA, 0.1 mM phenylmethylsulfonyl fluoride

(PMSF) in an ice-water bath. Add 1/10 volume of 50 mM DMS in 200 mM cacodylate–NaOH, pH 7.2, prepared prior to use by dissolving DMS (Merck, Darmstadt, FRG) (4.2 μl/ml) in ice-cold buffer with thorough vortexing for 30 sec. Incubate at 4° for 8 hr. Incubation can be also done at 20° for about 3 hr or at 37° for 1 hr.

2. Dialyze the methylated sample against 100 volumes of 20 mM HEPES–NaOH, pH 7.4, 1 mM EDTA, 1 mM PMSF in the cold overnight. Alternatively, exchange the medium by gel filtration.

3. Incubate the methylated sample at 37° for 24 hr or at 45° for 8 hr to perform depurination of methylated bases and cross-linking.

4. Chill the cross-linked sample in an ice-water bath. Add 1/10 volume of 200 mM NaBH$_4$, dissolved in ice-cold 100 mM HEPES–NaOH, pH 7.4, 0.2 mM EDTA immediately before use. Add a drop of isopentanol or n-octanol to prevent foaming. Incubate in an ice-water bath for 30 min in the dark.

5. Dialyze the sample against 100 volumes of 10 mM Tris-Cl, pH 8.0, 0.1 mM EDTA, 0.1 mM PMSF.

Notes. (1) DMS should be stored in small aliquots in sealed ampules. Do not use the reagent if pink coloration develops. (2) This protocol is routinely used for nucleosome structural studies and provides cross-linking of approximately 1 protein molecule per 200 bp of DNA. For "protein image" hybridization, it is advisable to use a higher degree of cross-linking, and methylation may be performed in 10–15 mM DMS. Check to confirm that no change in pH has occurred.

DNA–Protein Cross-Linking in Nuclei

1. Isolate nuclei as appropriate for a particular tissue (for examples, see Refs. 14, 23, and 24), avoiding primary and secondary amines (such as spermine or spermidine) and thiols (2-mercaptoethanol, dithiothreitol) in the isolation media, at least during the final stages of isolation. Suspend the nuclei at 10–50 A_{260} units/ml in, for example, MB[24] [60 mM KCl, 15 mM NaCl, 2 mM MgCl$_2$, 15 mM HEPES–NaOH, pH 7.4, 0.5 mM diisopropyl fluorophosphate (DFP)], or any other medium suitable for maintaining the integrity of nuclei. Ensure efficient inhibition of protease activity.

2. Follow the cross-linking protocol for nucleosomes (see previous section) but instead of dialysis or gel filtration (steps 2, 3, and 5), pellet the nuclei, discard the supernatants, and wash the pellets twice, for example, with MB.

---

[23] D. Rhodes and R. A. Laskey, this volume [27].
[24] J. P. Langmore and J. R. Paulson, *J. Cell Biol.* **96**, 1120 (1983).

Cross-Linking by Addition of Proteins to Partly Depurinated DNA

As discussed above, this protocol is well suited for DNA–protein complexes that can be reconstituted *in vitro*. Here, the cloned or synthetic DNA fragment is methylated and depurinated in the absence of protein, so that the latter cannot be damaged by methylation. The method is rapid and simple. However, when interpreting the results of cross-linking, one should not overlook the possibility that the binding equilibrium may be disturbed by arresting some kinetic steps. The procedure can be exemplified by the cross-linking of *Escherichia coli* RNA polymerase holoenzyme to *lac* operator DNA, which allowed localization of polymerase subunits on *lac* operator DNA.[15]

Protocol

1. Methylate a DNA fragment containing the *lac* UV5 operator fragment (100 μg/ml in 50 mM cacodylate–NaOH, pH 7.0, 50 mM NaCl, 0.1 mM EDTA) by adding DMS to 50 mM. Incubate at 20° for 5 min, then terminate the reaction by adding 2-mercaptoethanol to 100 mM. Recover the DNA by ethanol precipitation.

2. Heat the DNA in 10 mM cacodylate–NaOH, pH 7.0, 10 mM NaCl, 0.1 mM EDTA at 65° for 1 hr.

3. Increase the buffer strength to 50 mM cacodylate–NaOH, pH 7.0, 50 mM NaCl, 5 mM MgCl$_2$, 1 μg/ml heparin, and add RNA polymerase to 100 μg/ml. Incubate at 37° for 2 hr.

4. Reduce the cross-links with NaBH$_4$, as described for nucleosomes.

Cross-Linking by UV Light

Cross-linking of nucleoprotein complexes by UV light has some significant advantages over chemical cross-linking methods. The procedure is simple and rapid and can be completed within 10–20 min. It can be performed on intact cells. Because of its rapidity, it is a suitable method when dealing with especially labile and easily redistributed chromosomal components. On the other hand, a number of protein–protein and interstrand DNA cross-links are produced after long irradiation exposures, which may interfere with analysis of DNA–protein complexes. The chemistry of UV cross-linking is rather complex, and the reader is referred to Ref. 1 for further explanation.

Procedure

1. Place the material to be irradiated (~2.5 A_{260} units/ml) into a petri dish to a height of 0.5 mm. Chill the dish in an ice-water bath. Stir the

suspension gently on a magnetic stirrer. Flush the container with argon and cover it with a quartz plate. Irradiate with a Chromato-Vue C-61 transilluminator (Ultraviolet Products, San Gabriel, CA) for 10–15 min, providing 20 mW/cm^2 light intensity at the surface of the sample.

2. Collect the cells by centrifugation. Lyse the cells in a buffer containing 0.5% Nonidet P-40 (NP-40) or Triton X-100, and collect the nuclei by centrifugation. Crude nuclei can be lysed in 2% sarkosyl, and after shearing they can be used for isolation of DNA–protein complexes by CsCl gradient centrifugation or by other methods (see below).

Note. Prolonging the irradiation exposure time beyond that specified above will result in higher yields of cross-linking, paralleled by greater damage to DNA by UV light. If the cross-linked DNA is to be detected by DNA hybridization (see below), a longer exposure time will therefore not improve the sensitivity of detection.

Recovery of Cross-Linked Complexes

Cross-linked material has a tendency to aggregate. The formation of aggregates is promoted by various contaminants in the sample. The aggregates are difficult to handle, even to solubilize. Therefore, purification of cross-linked material facilitates its utilization for analysis. Also, the depletion from the sample of uncross-linked, "free" DNA and proteins increases the sensitivity of analysis of their cross-linked counterparts.

The choice of purification procedure depends on the source and the state of the cross-linked sample. Cross-linked whole cells and nuclei, contaminated by cytoplasmic debris, should be subjected to CsCl gradient centrifugation to remove uncross-linked proteins, lipids, carbohydrates, etc. Less complex samples, such as nucleosomes, can be depleted of free proteins by Cetavlon precipitation, followed by the removal of uncross-linked DNA by either hydroxyapatite chromatography, phenol–chloroform extraction, or KCl–sodium dodecyl sulfate (SDS) precipitation.

When manipulating samples cross-linked by the methylation–depurination procedure, keep in mind that low levels of depurination call for special means of protecting proteins from artifactual, repeated cross-linking via aldehydes that continue to be generated at the methylated sites. To protect the proteins, ensure the presence of amines, thiols, and detergents. Bearing in mind a tendency of cross-linked complexes to aggregate irreversibly, seek to minimize the number of steps involving precipitation of the cross-linked sample and avoid storage of cross-linked sample in the form of precipitate.

Centrifugation in CsCl Gradients[7,8]

1. Suspend the cross-linked nuclei in a medium containing 150 mM NaCl, 10 mM Tris-Cl, ph 8.0, 1 mM EDTA, 0.1% NP-40, 0.5 mM DFP, and 2% sarkosyl. Sonicate 10 sec at 22 kHz. Remove insoluble material by centrifugation (10,000 rpm, 15 min).

2. Layer the supernatant onto a preformed CsCl step gradient, consisting of 18.5 ml of CsCl at 1.75 g/cm^3, 6.0 ml of CsCl at 1.5 g/cm^3, and 3.5 ml of CsCl at 1.35 g/cm^3. Each layer of CsCl contains 0.5% sarkosyl, 1 mM EDTA, and 1 mM PMSF. Centrifuge at 40,000 rpm in a Ti 55.2 or Ti 50.2 fixed-angle rotor (Beckman) for 40 hr at 20°. Collect DNA-containing fractions by monitoring UV absorbance.

3. Dialyze samples against 0.1% sarkosyl, 10 mM Tris-Cl, pH 8.0, 1 mM EDTA. Recover the cross-linked complexes by adding 2.5 volumes of ethanol containing 0.3 M sodium acetate and then incubate at −20° overnight. Collect the precipitated DNA–protein complex by centrifugation for 30 min at 15,000 rpm in an SW27 rotor (Beckman).

Depletion of Uncross-Linked Proteins by Cetavlon Precipitation

The method is based on the fact that the Cetavlon salt of DNA precipitates quantitatively below 0.4 M NaCl and solubilizes fully above 0.5 M NaCl.[25] Proteins are soluble in this strongly cationic detergent.

1a. Cross-linked nuclei: Redissolve nuclear pellets at 1 mg DNA/ml in 1% Cetavlon (cetyltrimethylammonium bromide) (Serva, Heidelberg, FRG), 1 M NaCl, 5 M urea, 20 mM Tris-Cl, pH 8.0, 10 mM dithiothreitol (DTT), 2 mM EDTA, 0.1 mM PMSF. Sonicate very briefly, if solubilization is incomplete. Centrifuge to remove insolubilities. Precipitate the Cetavlon salt of DNA by diluting the supernatant with 4 volumes of water. Allow the precipitate to form at room temperature for 15 min. Centrifuge at 15,000 g for 30 sec or at 4000 g for 10 min. Discard supernatants.

1b. Cross-linked nucleosomes: To the solution of nucleosomes at 5–20 A_{260} units/ml in 20 mM Tris-Cl, pH 8.0, 10 mM DTT, 1 mM EDTA, 0.1 mM PMSF, add NaCl to 0.15 M and Cetavlon to ensure a 4- to 5-fold excess by weight over that of DNA. Allow to stand at room temperature for 15 min. Centrifuge as above and discard supernatants.

2. Wash the Cetavlon pellets of cross-linked nuclei or nucleosomes with 0.2% Cetavlon. Redissolve pellets in 3 M ammonium acetate or 4 M guanidine-HCl in the case of poor solubility. Recover the cross-linked

[25] V. Naktinis, N. Maleeva, D. San'ko, and A. Mirzabekov, *Biokhimiya* **42**, 1783 (1977).

DNA by precipitation with 3 volumes of 96% ethanol. Wash the pellets twice with 75% ethanol.

Depletion of Uncross-Linked Denatured DNA by Hydroxyapatite Chromatography in the Presence of SDS

The basis of this method is the tight binding of protein–dodecyl sulfate complexes by hydroxyapatite;[26] the complexes are desorbed at sodium phosphate concentrations above 0.16 M, while single-stranded DNA elutes below 0.16 M. Hence, cross-linked denatured DNA–protein complexes are retained at around 0.16 M sodium phosphate via their protein moieties while free single-stranded DNA is removed; thereafter, cross-linked complexes are recovered by elution with 0.5 M sodium phosphate.

Procedure

1. Make the cross-linked sample (20–30 A_{260} units/ml) 1% in SDS, 10 mM in DTT, 5 mM in sodium phosphate, pH 6.4. Heat in a boiling water bath for 1 min to denature DNA. Dilute with 10 volumes of water, and spin to pellet insolubilities. Load the supernatant onto a jacketed hydroxyapatite column (BioGel HTP, Bio-Rad Laboratories, Richmond, CA) thermostatted at 30°, at 60–400 μg DNA/ml bed volume. Elute the column with 10 volumes of 0.1% SDS, 5 mM sodium phosphate, pH 6.4, then with 10 volumes of 0.1% SDS, 0.15 M sodium phosphate, pH 6.4, to remove uncross-linked DNA, then with 0.1% SDS, 0.5 M sodium phosphate, pH 6.4, to elute the cross-linked complexes as a sharp peak.

2. Dialyze the recovered sample against 2 changes of 100 volumes of 0.1% SDS, 0.2 mM EDTA, pH 7.0. Precipitate with ethanol.

Extraction of Cross-Linked Complexes with Phenol–Chloroform

The method is based on the fact that proteins and DNA–protein complexes are partitioned into the organic phase and interphase while free single-stranded DNA remains in the aqueous phase during extraction.

Procedure

1. Dissolve cross-linked material at approximately 1 mg DNA/ml in 100 mM Tris-Cl, pH 8.0, 0.1% SDS.

2. Heat in boiling water for 2 min, cool to room temperature, and add an equal volume of phenol–chloroform (4:1). Mix by vortexing, and centrifuge to separate phases at 15,000 g. Remove upper aqueous layer.

[26] B. Moss and E. N. Rosenblum, *J. Biol. Chem.* **247**, 5194 (1972).

3. Add to the organic layer and interphase an equal volume of 1 M Tris-Cl, pH 8.0, 1% SDS. Mix by vortexing, centrifuge, and remove the upper aqueous layer again. Repeat this step twice.

4. Add to the organic phase and interphase 20 μg tRNA as a carrier and recover the cross-linked complex by ethanol precipitation. Centrifuge 10 min at 15,000 g. Wash pellets twice with cold 70% ethanol.

Depletion of Uncross-Linked Single-Stranded DNA by KCl Precipitation of Dodecyl Sulfate – Protein Complexes

This procedure makes use of the fact that potassium salt of dodecyl sulfate and its protein complexes are insoluble in water.[27]

Procedure

1. Dissolve the cross-linked sample in 0.5% SDS, 10 mM Tris-Cl, pH 8.0, 10 mM cysteamine-HCl, 0.2 mM EDTA, at 2 – 5 A_{260} units/ml, heat in the boiling water bath for 60 sec, and chill immediately in an ice-water bath.

2. Add 1/10 volume of 3 M KCl. Incubate the sample in an ice-water bath for 10 min, spin at 15,000 g for 30 sec, and remove supernatant by aspiration. Wash the pellets with a small volume of 0.2 M KCl, 10 mM Tris-Cl, pH 8.0, 10 mM cysteamine-HCl, 0.2 mM EDTA. Discard supernatant and redissolve the pellets in a small volume of 1% SDS, 10 mM Tris-Cl, pH 8.0, 10 mM cysteamine-HCl, 0.2 mM EDTA. Reprecipitate by adding 1/10 volume of 10 M ammonium acetate and 3 volumes of 96% ethanol; spin 1 min at 15,000 g. Discard supernatant and wash the pellet with 96% ethanol.

Note. Of the three procedures for depletion of uncross-linked single-stranded DNA, hydroxyapatite chromatography works most neatly. However, when processing a large number of samples, phenol – chloroform extraction or KCl – SDS precipitation are the methods of choice, owing to their simplicity and rapidity.

Analysis of DNA – Protein Cross-Links

Several experimental procedures to identify cross-linked proteins and DNA and locate the site of cross-linking within these molecules are described below.

[27] D. Bray and S. M. Brownlee, *Anal. Biochem.* **55,** 213 (1973).

*Identification of Cross-Linked Proteins and DNA by Diagonal Gel
Electrophoresis and "Protein Image" Hybridization*[2]

DNA cross-linked to different proteins of various electrophoretic mobilities can be separated and thus identified by two-dimensional gel electrophoresis techniques. Cross-linked DNA–protein complexes move in the
first direction according to the size of both DNA and protein components.
The mobility of the DNA molecule is decreased by the cross-linked protein. The extent of retardation depends on the molecular weight of the
cross-linked polypeptide. The proteins are then digested directly in the gel
with Pronase,[28] and the released DNA is fractionated in the second direction according to its size. As a result, two or more diagonals can be
observed on two-dimensional gels that contain uncross-linked DNA and
DNA fragments that had been cross-linked to different proteins. The
position of diagonals of protein–cross-linked DNA relative to that of
uncross-linked DNA is determined by the retardation imposed by the
cross-linked protein on the DNA during electrophoresis in the first direction. The retardation, and thus the molecular weight of cross-linked proteins, can be calibrated with cross-linked proteins of known size (see
Fig. 1).

The smaller the size of DNA relative to the size of cross-linked proteins,
the better the separation of the respective diagonals. However, an effective
separation can be achieved with short single-stranded DNA fragments
produced either by digestion with DNase I or during cross-linking to partly
depurinated DNA (see below). Separation into individual diagonals of
denatured DNA fragments, cross-linked to four or to all five histone
fractions, can be achieved.

DNA in the gel can be visualized by staining with ethidium bromide, by
autoradiography of ^{32}P-labeled material, or by hybridization with various
probes. The latter approach, named "protein image" hybridization,[2] allows
assessment of the presence of proteins (histones) on different DNA sequences in the genome. The experimental procedure for protein image
hybridization is described below.

Procedure

1. Adjust the cross-linked sample, for example, cross-linked nuclei,
depleted of uncross-linked single-stranded DNA and proteins, to 1% SDS,
7 M urea. Add 0.5–3 volumes of sample buffer containing 10 mM DTT,
15% Ficoll, 0.025% bromphenol blue (BPB) 62.5 mM Tris-Cl, pH 6.8, 1%

---

[28] D. W. Cleveland, S. G. Fisher, M. W. Kirschner, and U. K. Laemmli, *J. Biol. Chem.* **232**,
1102 (1977).

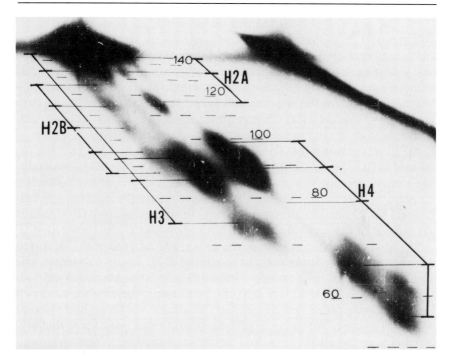

FIG. 1. Two-dimensional sequencing gel electrophoresis of single-stranded, ^{32}P-labeled DNA cross-linked to histones in sea urchin sperm chromatin core particles. Electrophoresis in the first dimension was carried out in 7 M urea under DNA denaturing conditions in a 17% polyacrylamide slab gel (165 × 365 × 0.6 mm). After digestion of histones with Pronase directly in the gel, electrophoresis was continued in the second dimension in a 15% polyacrylamide slab gel (300 × 400 × 1 mm) containing 7 M urea. The dashed lines show the position in the gel, and the figures give the length (in bases) of ethidium bromide-stained DNA fragments from DNase I digests of rat liver nuclei. The positions of ^{32}P-labeled DNA fragments cross-linked to different histones and arranged on separate diagonals were revealed by autoradiography and are indicated by solid lines. The extreme right diagonal contains un-cross-linked DNA fragments.

SDS, and 7 M urea, and heat the mixture in a boiling water bath for 1 min to denature DNA.

2. Load the sample onto a slab gel (200 × 200 × 0.6 mm), formulated as with the following components: stacking gel, 6% acrylamide, 0.1% $N,N,N'N'$-methylenebisacrylamide (MBA), 125 mM Tris-Cl, pH 6.8, 0.1% SDS, 7 M urea; resolving gel, 15% acrylamide, 0.1% MBA, 375 mM Tris-Cl, pH 8.8, 0.1% SDS, 7 M urea; reservoir buffer, 192 mM glycine, 25 mM Tris, pH 8.3, 0.1% SDS. Run at 10–15 V/cm until the dye reaches the bottom of the resolving gel.

3. Cut out a strip from the one-dimensional gel. Soak it twice in 30 volumes of 0.1% SDS, 62.5 mM Tris-Cl, pH 6.8, to remove urea. Prepour the two-dimensional gel (200 × 200 × 1.0 mm) constructed of the following: stacking gel, 6% acrylamide in the above-described stacking buffer, but without urea; resolving gel, 15–20% acrylamide as above, with urea, leaving room on top for the one-dimensional gel strip. Polymerize the gel strip to the top of this gel using 6% acrylamide in stacking buffer without urea. Leave room at the top to load the Pronase-containing buffer.

4. Load the two-dimensional gel with 1.5 ml of sample buffer (1% SDS, 10 mM DTT, 15% Ficoll, 0.025% BPB, 62.5 mM Tris-Cl, pH 6.8) containing 1 mg of Pronase (Calbiochem). Run in the above reservoir buffer at a constant voltage of 100 V until the BPB dye enters the stacking gel. Disconnect for 30 min, then resume the run. Terminate when the dye reaches the bottom of the resolving gel.

5. After electrophoresis, soak the gel in 200 ml H_2O with 2 g mixed ion-exchange resin. Shake for 1 hr to remove SDS. Remove the resin, and equilibrate the gel with transfer buffer (10 mM Tris–acetate, 5 mM sodium acetate, 0.5 mM EDTA, pH 7.8) for 15 min. Assemble the sandwich in the following order: Scotchbrite pad, Whatman 3 MM, gel and Hybond nylon membrane, Whatman 3 MM, Scotchbrite pad. Make sure that Hybond-N is on the anode (+) side. Transfer for 4 hr at 10 V/cm.

6. Remove Hybond-N, wash in water, and air-dry. Using a standard UV transilluminator (256 nm), irradiate the membrane as described in Ref. 29.

7. Hybridize with ^{32}P-labeled DNA probe, as described in Ref. 29.

Immunoaffinity Analysis of Cross-Linked Complexes with Antibodies to Specific Proteins[4-6]

Monospecific antibodies to a protein of interest, if available, make it possible to find out which genomic sequences interact with that particular protein and to detect its exact location on these sequences. The cross-linked complex(es) is purified on the basis of affinity to a specific antibody. The cross-linked DNA can be identified by hybridization with specific probes. If the cross-links were produced by the methylation–depurination methods, the binding sites for the protein along the DNA can be mapped by determining the length(s) of cross-linked denatured DNA, as described in the next section.

The task of mapping the exact DNA binding sites of a protein within an affinity-purified cross-linked complex along a DNA fragment requires that

[29] G. M. Church and W. Gilbert, *Proc. Natl. Acad. Sci. U.S.A.* **81,** 1991 (1984).

the boundaries of the cross-linked DNA fragments be well defined. In the case of HMG 14 and 17, these boundaries are the ends of the nucleosomal core particle generated by micrococcal nuclease digestion.[4,5] In other cases, the boundaries could be produced by restriction endonucleases either before or after immunoaffinity isolation.

Procedure

1. Having removed uncross-linked proteins by CsCl centrifugation (see above), solubilize the cross-linked sample in 0.2% SDS at a concentration of about 100 A_{260} units/ml. Add the following components: NP-40 (Fluka, Buchs, Switzerland) to 1%, sodium deoxycholate (Merck) to 1%, NaCl to 2.15 M, EDTA to 5 mM, Tris-Cl to 50 mM (pH 7.4), bovine serum albumin (BSA, Calbiochem, San Diego, CA) to 1 mg/ml (RIPA-M buffer).

2. Add 50 μg of affinity-purified rabbit antibody for every 100 A_{260} units of sample DNA. Incubate for 10 hr at 4°. Add 30 μl (bed volume) of a suspension of staphylococcal protein A–Sepharose CL-4B, prewashed in water and equilibrated with the same buffer as formulated above. Incubate 8–16 hr at 4° with gentle shaking. Pellet the Sepharose beads by centrifugation and transfer the beads to a smaller tube. Wash 10 times by shaking with 20–50 volumes of the same buffer for several minutes, then centrifuge. Wash 2 times with TEN buffer (50 mM Tris-Cl, 5 mM EDTA, 0.15 M NaCl, pH 7.4) for 3 min. Remove the liquid carefully.

3. To the slurry of immunosorbent, add 90 μl of elution buffer (2% SDS, 5 mM EDTA, 100 mM DTT, pH 8.0). Heat the sample on a boiling water bath for 3 min. If DNA cross-linked to antigen is needed in the double-stranded form (e.g., for subsequent digestion with restriction endonucleases), boiling should be replaced by heating at 65° for 20–30 min with shaking. Centrifuge at 10,000 g for 1 min. Collect the eluate. Repeat the elution once again and combine both eluates. As a rule, immunoaffinity purification must be repeated at least once.

4. To verify the purity and monospecificity of immunofractionation, take a small aliquot of the CsCl centrifugation-purified cross-linked sample, iodinate with Na[125]I by the ICl method (described below), perform the immunofractionation as described above, hydrolyse the DNA (see below), and analyze the liberated protein by conventional gel electrophoresis in the presence of SDS[30] with subsequent autoradiography.

5. Since protein location on specific sequences in the genome is determined by hybridization with cloned probes (below), ascertain that the immunoaffinity isolate contains only cross-linked DNA. Take a small aliquot (5–10 μl bed volume) of the protein A–Sepharose CL-4B slurry to

[30] U. K. Laemmli, *Nature (London)* **227,** 680 (1970).

which (antigen-DNA)–antibody complex has been bound. Wash the immunosorbent by centrifugation in 200 μl 100 mM bicine–NaOH, pH 9.5, 10 mM MgCl$_2$, 0.1 mM ZnCl$_2$ several times for 5 min each time. Add 20 μl of the same buffer, about 0.05 units of calf intestinal phosphatase (Sigma, St. Louis, MO), and incubate at 37° for 1 hr. Centrifuge for 1 min and wash the beads 3–5 times in RIPA-M as described above. Wash 2–3 times in kinase buffer (100 mM bicine–NaOH, pH 9.0, 10 mM MgCl$_2$, 10 mM DTT). Add to the beads 10–50 μCi of [γ-^{32}P]ATP in kinase buffer in a volume of 20 μl and several units of T4 polynucleotide kinase. Wash several times with RIPA-M and 2 times with TEN buffer. Elute the sample from the beads with 20–50 μl of elution buffer by heating the beads on a boiling water bath for 2–3 min. Centrifuge. Repeat the elution step and combine both eluates. Run two-dimensional DNA electrophoresis (as described below) to evaluate the content of free DNA.

6. To identify the genomic DNA that has been cross-linked to affinity-purified protein, treat the eluate collected from immunosorbent with proteinase K to digest cross-linked proteins, extract with phenol, treat with RNase, and extract and recover DNA by ethanol precipitation as described.[31] Dissolve the DNA in 30 μl of 40 mM Tris-Cl, 20 mM sodium acetate, 1 mM EDTA, pH 7.4. Denature and dot DNA on a wet nylon membrane (zeta probe, Bio-Rad), preequilibrated with the same buffer. Blot the membrane between two sheets of Whatman 3 MM paper, dry under vacuum, irradiate with UV light, and hybridize to ^{32}P-labeled probe as described.[29]

7. To map the sites of cross-linking of affinity-purified protein along the DNA, introduce the ^{32}P label into the 5' ends of the DNA fragment (described below), and follow the protocol for two-dimensional DNA gel electrophoresis.

Notes. (1) All antibodies or antisera used should be carefully checked for the absence of any cross-reaction with other proteins. This can be done by ELISA, Western blotting, or protein A–Sepharose-mediated immunoprecipitation.[32] In general, the antibody used must recognize denatured proteins, and cross-linking should not interfere with antigen binding. (2) For antibodies of very high affinity, elution of the cross-linked complex from immunosorbent (step 3 above) should be performed with 8 M guanidine-HCl instead of 2% SDS. (3) If the cross-linked DNA has a high molecular weight, employ formalinized *Staphylococcus aureus* cells as an immunosorbent[33] instead of protein A–Sepharose CL-4B.

[31] T. Maniatis, E. F. Fritsch, and J. Sambrook, "Molecular Cloning." Cold Spring Harbor Lab., Cold Spring Harbor, New York, 1982.

[32] This series, Vol. 92.

[33] R. G. D. McKay, *J. Mol. Biol.* **145,** 471 (1981).

Mapping Protein Cross-Linking Sites on a DNA Fragment[10,11,14]

The use of the depurinated DNA cross-linking method provides an opportunity to map the locations of cross-linked proteins on DNA fragments. This task is facilitated by splitting DNA at the site of cross-linking and the attachment of the protein molecule only to the 5'-terminal fragments thus produced. Measurements of the length of cross-linked DNA fragments and mapping of protein binding sites on DNA can be done by gel electrophoresis in two different ways. Gel electrophoresis in the first dimension (identical in both systems) separates the cross-linked protein – single-stranded DNA complexes depleted of uncross-linked DNA strands and proteins. The cross-linked complexes are made up of a particular protein plus single-stranded DNA fragment extending from the defined 5' end of the DNA fragment up to the site of cross-linking.

The design of the "DNA version" of two-dimensional electrophoresis requires the removal of the protein component of the complex, which can be effected by Pronase as described for the diagonal technique (see above). The run in the second dimension results in partitioning of single-stranded DNA fragments into several diagonals, each containing DNA fragments initially attached to a particular protein. The positions of the spots within each diagonal, when calibrated with DNA fragments of known size, enable one to assign each protein to certain sites along the DNA fragment (Fig. 1).

In the "protein version" of two-dimensional gel electrophoresis, the DNA component of the cross-linked complex is chemically degraded after the first dimension run, and the released ^{125}I-labeled proteins are separated from each other in the second dimension. In a two-dimensional gel, proteins are arranged as spots on different horizontal lines. The spots of cross-linked proteins are shifted from those of uncross-linked ones by a distance proportional to the length of the cross-linked DNA fragment (Fig. 2).

The yield of cross-linking by this method as described is low, approximately 1 – 5% nuclear or nucleosomal protein. By raising the extent of methylation, the yield can be increased. This is permissible, for example, when evaluating the histone content in a particular genomic region by the diagonal technique (see above). However, when aligning proteins along the DNA strands, it is important to set the yield reasonably low to avoid "shielding" of the cross-links distant to the reference site by cross-linking intervening proteins to the same DNA molecule. Keep in mind that low yields of cross-linking necessitate exhaustive depletion from cross-linked samples of uncross-linked proteins and DNA strands.

Two-Dimensional "DNA" Gel Electrophoresis

1. Label the cross-linked complexes with ^{32}P at the 5' ends of the component DNA by incubation with 50 – 200 μCi of [γ-^{32}P]ATP and 2

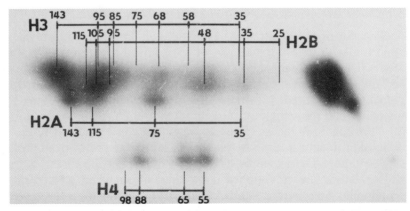

FIG. 2. Two-dimensional gel electrophoresis of ^{125}I-labeled histones cross-linked to DNA in chicken erythrocyte chromatin core particles. The cross-linked complex was electrophoresed in the first dimension in a 15% polyacrylamide slab gel (200 × 200 × 0.6 mm) containing 7 M urea. After hydrolysis of DNA in the gel, electrophoresis in the second dimension was performed in a 15% polyacrylamide slab gel (200 × 400 × 1 mm). The spots of uncrosslinked histones are seen at the extreme right of the gel. The number at each spot indicates the size of DNA fragments cross-linked to histones as described in Fig. 1.

units polynucleotide kinase in 50 mM Tris-Cl, pH 8.0, 10 mM MgCl$_2$, 10 mM DTT at 37° for 30 min. Stop by chilling and adding EDTA to 10 mM.

2. Adjust the cross-linked sample to 1% SDS, 7 M urea and add 0.5–3 volumes of sample buffer containing 10 mM DTT, 15% Ficoll, 0.025% BPB, 62.5 mM Tris-Cl, pH 6.8, 1% SDS, and 7 M urea. Heat the mixture in a boiling water bath for 1 min to denature DNA. Run two-dimensional gel electrophoresis as described for two-dimensional diagonal gel electrophoresis above, but supplement the Pronase mix loaded onto the two-dimensional gel with 100–500 μg of nonradioactive single-stranded DNA fragments of known lengths (e.g., denatured, deproteinized products of DNase I digestion of nuclei[34]). Terminate the run in the second direction when the dye has migrated out of the gel.

3. Extract SDS and urea from the gel with 70% alcohol. Wash the gel in water. Stain DNA with ethidium bromide (0.001 mg/ml) and mark the positions of standard DNA fragments under UV light. Wash the gel with 5% glycerol for 5 min, dry under vacuum, and autoradiograph.

[34] A. Prunell, R. D. Kornberg, L. Lutter, A. Klug, M. Levitt, and F. H. C. Crick, *Science* **204**, 855 (1979).

Two-Dimensional "Protein" Electrophoresis

1. Label the proteins within the cross-linked complexes by the iodine monochloride method.[35] Mix the cross-linked material (~ 1 A_{260} unit), in 10 μl of 1% SDS, 0.1 M sodium borate, pH 8.5, with 0.1 – 1 mCi of $Na^{125}I$ (10 – 20 Ci/mg), add 7 μl of 0.5 mM ICl, and vortex immediately [the 0.5 mM solution of ICl is prepared by mixing 0.05 ml of stock solution (13.4 mM KI, 6.6 mM KIO_3, 2.5 M NaCl, 1.0 M HCl) with 1.95 ml of 0.1 M NaCl].

2. Adjust the cross-linked sample to 1% SDS, 7 M urea, add 0.5 – 3 volumes of sample buffer containing 10 mM DTT, 15% Ficoll, 0.025% BPB, 62.5 mM Tris-Cl, pH 6.8, 1% SDS, 7 M urea, and heat the mixture in a boiling water bath for 1 min to denature DNA.

3. Load onto the first-dimensional gel, composed of the following: stacking gel, 6% acrylamide, 0.1% MBA, 125 mM Tris-Cl, pH 6.8, 0.1% SDS, 7 M urea; resolving gel, 15% acrylamide, 0.2% MBA, 375 mM Tris-Cl, pH 8.8, 0.1% SDS, 7 M urea; reservoir buffer, 192 mM glycine, 25 mM Tris, pH 8.3, 0.1% SDS. Run until the dye reaches the bottom of the resolving gel.

4. Cut out a strip from a first-dimensional gel, soak, using a magnetic stirrer, successively in 30 volumes of 66% formic acid, then in two changes of 66% formic acid, 2% (w/v) diphenylamine. Incubate the strip in this solution at 70° for 20 min. Wash the strip in 3 changes of 66% formic acid, 5 mM cysteamine-HCl, 5 changes of 5 mM cysteamine-HCl, until the pH is almost neutral, and finally in 0.1% SDS, 62.5 mM Tris-Cl, pH 6.8. Heat the strip in this buffer in a boiling water bath for 1 min. Place the gel strip horizontally over a preformed 6% acrylamide stacking gel and 15% acrylamide resolving gel formulated as stated above with or without urea. Seal the strip using 6% acrylamide, 0.2% MBA in 0.1% SDS, 125 mM Tris-Cl, pH 6.8, with TEMED and ammonium persulfate. After the run, soak the gel successively in 70% alcohol, water, and for 5 min in 5% glycerol. Dry in vacuum and autoradiograph.

Mapping and Identification of Polypeptide Domains Cross-Linked to DNA[16,17]

The strategy is to remove the DNA moiety of the cross-linked complex so as to leave a defined nucleotide tag covalently linked to protein. This is achieved by acidic hydrolysis of DNA in the presence of diphenylamine which proceeds via complete depurination/5',3'-elimination.[36] The DNA

---

[35] M. A. Contreras, W. F. Bale, and I. L. Spar, this series, Vol. 92, p. 277.
[36] B. K. Burton and G. B. Peterson, *Biochem. J.* **75**, 17 (1960).

remnant on the protein is then a modified deoxyribose with either a 5'-phosphate (if a purine residue happens to occur 5' to the site of cross-linking) or mono- or oligopyrimidine nucleotides.[16] These mono- and oligopyrimidine residues, if treated with phosphatase and trimmed by micrococcal nuclease to mononucleotides, can incorporate radioactive phosphate in a reaction catalyzed by polynucleotide kinase. Individual labeled proteins can be fractionated by polyacrylamide gel electrophoresis or by reversed-phase (RP) liquid chromatography, then fragmented chemically or enzymatically with specific proteases, and the tagged products patterned by gel electrophoresis or RP HPLC. Individual cleavage products, nucleotide peptides, can be purified by RP HPLC and subjected to amino acid analysis and mass spectrometry for identification of the cross-linked peptide(s).

The above phosphatase–nuclease–kinase–protease gel electrophoresis procedure could be applied to cross-linked histones eluted from each of the spots of the second-dimensional gel of protein gel electrophoresis (see above) thus allowing patterning of histone molecule domains involved in interactions with nucleosomal DNA at each turn of the double helix[18] (Fig. 3).

Patterning of DNA-Binding Domains in Cross-Linked Proteins
Hydrolysis of DNA in the cross-linked complex

1. Pellet the cross-linked complex, freed of uncross-linked proteins and DNA. Redissolve the pellet in freshly prepared 2% diphenylamine in 70% formic acid at a concentration less than 10 A_{260} units/ml, incubate at 70° for 20 min, dilute with an equal volume of 100 mM cysteamine-HCl, extract 3 times with water-saturated ethyl ether, vortex 30 sec, spin 2 sec, and discard upper phase (spare interphase, if present). Dilute with 5 vol-

---

FIG. 3. "Three-dimensional" gel electrophoresis with patterning of ^{32}P-labeled nucleotide peptides representing histone H4 contact sites in the nucleosomal core particle. First and second dimension gels were run as described in the legend to Fig. 2. The H4 spots corresponding to histone H4 cross-linking to either strand of core particle DNA at a distance of 55 (lanes 5, 6, and 7), 65 (lanes 3 and 4), and 88 (lane 1) nucleotides from the 5' ends were excised, freed of ^{125}I label,[18] trimmed with phosphatase and micrococcal nuclease, labeled with ^{32}P, digested with trypsin, and run in a glycine–acetic acid gel electrophoretic system as described in the text. The identical nucleotide peptide pattern at these three sites of cross-linking reflects the interaction of a single DNA-binding domain of H4 with these DNA segments juxtaposed at nucleosomal sites ±1.5 nucleotides. The major doublet band represents His-18–Arg-19, as determined by amino acid analysis and mass spectrometry.[17] The doublet is produced by the difference in ionization state of cytidine and thymidine, the former bearing an extra +1 charge at acidic pH.[16]

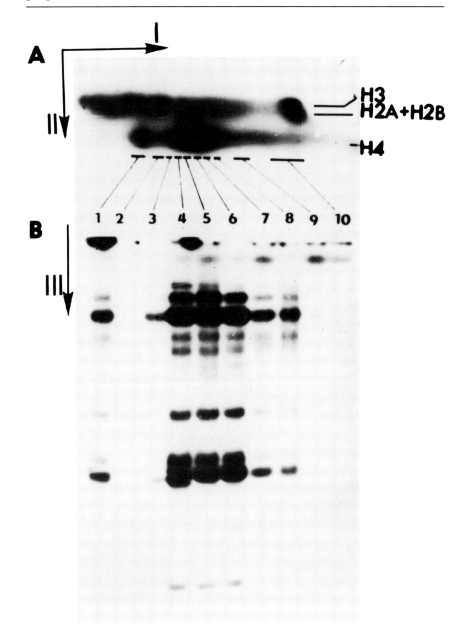

umes of water and freeze-dry. Wash the dried-down residues of acid hydrolysis with cold 15% (w/v) trichloroacetic acid (TCA), then with 0.01 N HCl in 95% acetone; dry under vacuum.

2. Incubate with 0.001 unit bacterial alkaline phosphatase (BAP, Pharmacia, Piscataway, NJ) in 20 μl 10 mM Tris-Cl, pH 8.0, 10 mM cysteamine-HCl, at 60° for 20 min. Precipitate the cross-linked protein by adding TCA to 20% (w/v) (ice-water bath, 3 min) and spin 30 sec at 10,000 g. Wash the pellets with ice-cold 0.01 N HCl in 95% acetone and dry under vacuum.

3. Incubate with micrococcal nuclease (0.1 mg/ml) (Sigma, grade VI) in 20 μl 25 mM bicine–NaOH, pH 9.0, 2 mM CaCl$_2$, 10 mM cysteamine-HCl, 0.1 mM DFP at 37° for 45 min. Terminate by chilling on ice and adding EDTA to 3 mM. Recover the protein by TCA precipitation and HCl–acetone washes.

Labeling of the Site of Cross-Linking with [^{32}P]ATP and Purification of Labeled Proteins

1. Incubate the cross-linked protein (0.1–20 μg) in 10 μl 50 mM Tris-Cl, pH 8.0, 10 mM MgCl$_2$, 10 mM DTT, 0.1 mM DFP, containing 20–50 μCi [γ-^{32}P]ATP and 2 units of T4-induced polynucleotide kinase, at 37° for 30 min. Terminate the reaction by adding EDTA to 10 mM and chilling on ice.

2. Add 5–7 μl 2% SDS, 20% Ficoll-400, 0.01% BPB. Incubate at 60° for 5 min and load onto a Laemmli gel,[30] supplementing the upper chamber buffer with 1 mM thioglycolic acid (Sigma). After the run, autoradiograph the gel for 5–30 min, locate the radioactive band(s) on the autoradiogram, and excise the respective gel slices. (Be careful not to macerate the gel.)

3. Incubate the slice(s) in 4–5 volumes of 0.1% Triton X-100, 10 mM Tris–thioglycolic acid, pH 8.0, 0.1 mM EDTA, 0.1 mM DFP at 37° for 12 hr, or at 65° for 2 hr. Remove the slice. Freeze-dry the eluates. Wash the dry pellets with 2% LiClO$_4$ in acetone twice, then with 0.01 N HCl in 95% acetone.

Protease Digestion of Protein Labeled with ^{32}P at the Site of Cross-Linking. Digest the eluted labeled protein with TPCK-treated trypsin (Serva) at 0.05 mg/ml in 0.2 M ammonium bicarbonate, 5 mM DTT, 0.1 mM EDTA, 0.5% Triton X-100 at 37° for 4 hr. Freeze-dry. For other enzymatical or chemical cleavages, see Ref. 37.

Polyacrylamide Gel Electrophoresis of Nucleotide Peptides in Tris–Glycine–EDTA Buffer (pH 8.6). This gel electrophoretic system is de-

---

[37] This series, Vol. 47.

signed for separation of usually small (2–35 amino acid residues) nucleotide peptides from tryptic digests of cross-linked proteins. For larger
fragments of, for example, CNBr-induced cleavage, conventional Laemmli
or acetic acid–urea gel systems are suitable.

> Mold: 200 × 200 mm glass plates, 0.4 mm spacers
> Gel: 20% acrylamide, 0.67% MBA, 12.5 mM Tris, 75 mM glycine,
>   0.1 mM EDTA
> Lower chamber buffer: 12.5 mM Tris, 75 mM glycine, 0.1 mM
>   EDTA
> Upper chamber buffer: 12.5 mM Tris, 75 mM glycine, 0.1 mM
>   EDTA, 0.3 mM thioglycolic acid
> Electrodes: "−" at the top, "+" at the bottom

Prerun either at a constant current of 10 mA/gel (chill the gel by a cold
air stream), discontinuing 30 min after the voltage reaches 1,000 V, or at a
constant voltage of 60 V overnight. Load samples of cross-linked protein
fragments (in 8% Ficoll/15% glycerol/dyes) in the smallest possible volume
(<5 μl). Run at 1,000 V (chill the gel by a cold air stream), until the BPB
has migrated 8 cm (~90 min). Dry the gel immediately under vacuum for
30 min with heating (60°, 15 min). Expose to X-ray film without intensifying screen.

*Polyacrylamide Gel Electrophoresis of Nucleotide Peptides in Glycine–
Acetate Buffer (pH 3.0)*

> Mold: as above
> Gel: 20% acrylamide, 0.67% MBA, 20 mM glycine, 0.2 M acetic acid
> Electrodes: "+" at the top, "−" at the bottom
> Chamber buffer (upper and lower) for preelectrophoresis: 20 mM
>   glycine, 0.2 M acetic acid

Prerun in the cold, either at a constant current of 20 mA, stopping 1 hr
after 1,000 V has been reached, or at a constant voltage of 60 V overnight.
Refill the electrode chambers with fresh glycine–acetate buffer, but supplement the upper chamber buffer with 1 mM cysteamine-HCl. Prerun for
further 30 min. Fill the outermost wells with 1 M sucrose, 0.05% methylene green. Load the samples in Ficoll/glycerol/dyes (see above). Run at
1,000 V in the cold until the methylene green migrates 12 cm (~90 min).
Dry and expose as above.

Identification of Nucleotide Peptides. For a protein of known amino
acid sequence, the determination of amino acid content of associated
nucleotide peptides would normally suffice for identification of the peptide
region. We employ conventional amino acid analysis with o-phthalalde-

hyde detection of the peptide moiety of purified individual nucleotide peptides after cross-linking on a preparative scale.

Isolation of individual nucleotide peptides is achieved by reversed-phase HPLC on silica gel C_{18} with a quasi-linear acetonitrile gradient.[17] Usually, several rechromatographies, done at different pH values, are required to obtain sufficient purity of nucleotide peptide. The correspondence between the isolated nucleotide peptide and the band position in the gel electrophoretic map is established by repeat electrophoresis of chromatographically purified nucleotide peptide, labeled with [32]P.

An independent means of checking the assignment is to determine the exact molecular mass. This is performed by mass spectrometry of the isolated nucleotide peptides, using the method of solvation of extracted ions at atmospheric pressure as described.[17]

Acknowledgments

The authors are greatly indebted to Dr. M. L. Pifer for critical comments on the manuscript and to Miss E. I. Novikova for assistance.

[21] Analysis of Nucleosome Positioning by Indirect End-Labeling and Molecular Cloning

By S. A. NEDOSPASOV, A. N. SHAKHOV, and G. P. GEORGIEV

Introduction

There is a growing body of evidence for the specificity of nucleosome positioning both *in vivo* and *in vitro* (reviewed in Refs. 1 and 2; see also Refs. 3 and 4). Differential flexibility of different DNA sequences has been suggested as a major determinant of this phenomenon (reviewed in Ref. 5) In this chapter, we present two approaches to the analysis of nucleosome positioning. The first is rather qualitative and gives an idea of nucleosome alignment (i.e., random versus specific) along gene-sized DNA segments in

[1] T. Igo-Kemenes, W. Hörz, and H. G. Zachau, *Annu. Rev. Biochem.* **51**, 89 (1982).

[2] J. C. Eissenberg, I. L. Cartwright, G. H. Thomas, and S. C. R. Elgin, *Annu. Rev. Genet.* **19**, 485 (1985).

[3] B. Neubauer, W. Linxweiler, and W. Hörz, *J. Mol. Biol.* **190**, 639 (1986).

[4] F. Thoma and R. T. Simpson, *Nature (London)* **315**, 250 (1985).

[5] A. A. Travers and A. Klug, *Philos. Trans. R. Soc. London, Ser. B* **317**, 537 (1987).

English translation copyright © 1989
by Academic Press, Inc.

experiments with total chromatin or chromatin reconstitution. The second is precise and determines unique or alternative nucleosome positions with sequencing gel resolution, which is important for critical evaluation of different theoretical models (reviewed in Ref. 5). Both methods are applicable with certain modifications for analysis of the organization of known genes in eukaryotic chromatin. Knowledge of the restriction map and availability of subcloned fragments are prerequisites for these studies.

Most of the experiments described in this chapter rely on the simian virus 40 (SV40) minichromosome as a convenient model system. Methods for SV40 chromatin preparation are given elsewhere in this volume.[6] For both of the methods described below, it is essential that minichromosomes be purified in soluble form, which makes sequential nuclease digestions feasible. In addition, the presence of several unique restriction nuclease sites on SV40 DNA and the availability of the complete nucleotide sequence are helpful for these experiments.

Nucleosome Positioning Derived from Chromatin Digestions and Indirect End-Labeling Analysis

Indirect end-labeling has been introduced independently to detect nuclease hypersensitive sites in chromatin[7] and to analyze nucleosome positioning.[8] The basis for this approach is related to the chemical sequencing method[9] and a procedure for mapping of restriction nuclease sites.[10] The method has also been used for genomic sequencing.[11]

Rationale

It is generally assumed that the sites of nuclease cleavage reflect the organization of DNA in chromatin. In particular, mapping of these sites gives information about nucleosome alignment. To accomplish such mapping, one needs a reference point. However, specific cuts (usually sites of restriction nuclease cleavage) cannot be easily and efficiently introduced into DNA in chromatin or in isolated nuclei. Thus, direct labeling, this is, creation of specific labeled sites in chromatin DNA that might serve as reference points for subsequent nonspecific nuclease cleavage, is not feasible.

[6] P. Oudet, E. Weiss, and E. Regnier, this volume [2].
[7] C. Wu, *Nature (London)* **286,** 854 (1980).
[8] S. A. Nedospasov and G. P. Georgiev, *Biochem. Biophys. Res. Commun.* **92,** 532 (1980).
[9] A. Maxam and W. Gibert, this series, Vol. 65, p. 499.
[10] H. O. Smith and M. L. Birnstiel, *Nucleic Acids Res.* **3,** 2387 (1976).
[11] G. Church and W. Gilbert, *Proc. Natl. Acad. Sci. U.S.A.* **81,** 1991 (1984).

Indirect end-labeling takes an opposite approach: nonspecific cleavages that reflect chromatin structure are introduced at the first step (Fig. 1A), while specific reference points (sites of appropriate restriction nuclease digestion) are created at the next step after isolation of pure DNA. Southern hybridization with appropriate DNA probes performed at the third step enables alignment of nuclease cleavages with respect to the restriction nuclease site. The sensitivity of this technique is high enough to analyze the chromatin organization of a single-copy gene of higher eukaryotes. In addition, this technique does not depend on any biochemical reactions (such as kinasing or filling in with DNA polymerase), and it works well for any method of DNA cleavage.

Cleavage of DNA in Chromatin

Micrococcal nuclease has been widely used as a probe for chromatin structure. It preferentially cleaves the linker between nucleosomes, generat-

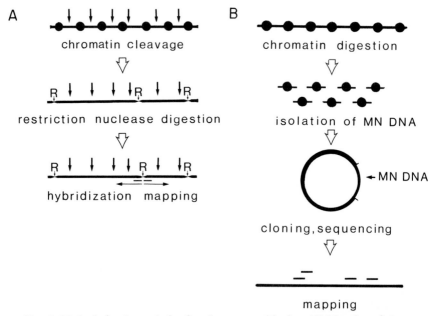

Fig. 1. Methods for the analysis of nucleosome positioning. (A) Mapping of cleavages introduced into chromatin DNA by nucleases or chemical agents using the indirect end-labeling technique. Two hybridization probes are shown below as solid bars. R corresponds to restriction endonuclease sites. Mapping is effective between neighboring restriction sites. (B) Cloning and sequencing of mononucleosomal DNA as an approach to map precise nucleosome positions. MN, Mononucleosomes (core particles).

ing the well-known DNA ladder at early stages of hydrolysis and the nucleosome core DNA later in digestion. In some cases, however, the sequence specificity of micrococcal nuclease[12,13] makes it difficult to identify those cleavage sites that are determined solely by chromatin organization. Although indirect end-labeling analysis, performed in parallel on chromatin and naked DNA, can give qualitative information on nucleosome arrangement,[14,15] a better biochemical probe for DNA in chromatin is nevertheless required.

Prunell and Kornberg[16] introduced an important modification for probing chromatin structure with micrococcal nuclease. In subsequent experiments,[17,18] exonuclease III (ExoIII) from *Escherichia coli* has been used to "trim" one strand of the external linker DNA. The DNA is then made blunt-ended by digestion with S_1 nuclease from *Aspergillus oryzae*. Another important development is the introduction of effective chemical reagents [such as methidiumpropyl-EDTA–Fe(II)] capable of cleaving in internucleosomal DNA in chromatin.[19,20] Recently, this technique, in combination with indirect end-labeling analysis, has given an indication of precise nucleosome positioning on an inactive single-copy gene.[21] Procedures for nonenzymatic cleavage of chromatin are described elsewhere in this volume.[20] A protocol based on micrococcal nuclease–ExoIII/S_1 digestions is given below.

Procedure

1. SV40 minichromosomes are extracted in isotonic buffer from the nuclei of infected cells (10^8 cells in a typical experiment) by procedures described elsewhere[6,22] and purified by sucrose gradient centrifugation. To follow yields at all stages and to be able to analyze agarose gels by fluoro-

[12] C. Dingwall, G. P. Lomonosoff, and R. A. Laskey, *Nucleic Acids Res.* **9,** 2659 (1981).
[13] W. Hörz and W. Altenburger, *Nucleic Acids Res.* **9,** 2643 (1981).
[14] A. Worcel, G. Gargiulo, B. Jessee, A. Udvardy, C. Louis, and P. Schedl, *Nucleic Acids Res.* **11,** 421 (1983).
[15] S. Cereghini, P. Herbomel, J. Jouanneau, S. Saragosti, M. Katinka, B. Bourachot, B. de Crombrugghe, and M. Yaniv, *Cold Spring Harbor Symp. Quant. Biol.* **47,** 935 (1983).
[16] A. Prunell and R. G. Kornberg, *Cold Spring Harbor Symp. Quant. Biol.* **42,** 103 (1978).
[17] T. Igo-Kemenes, A. Omori, and H. G. Zachau, *Nucleic Acids Res.* **8,** 5377 (1980).
[18] W. Linxweiler and W. Hörz, *Nucleic Acids Res.* **12,** 9395 (1984).
[19] I. L. Cartwright, R. P. Hertzberg, P. B. Dervan, and S. C. R. Elgin, *Proc. Natl. Acad. Sci. U.S.A.* **80,** 3213 (1983).
[20] I. L. Cartwright and S. C. R. Elgin, this volume [18].
[21] R. Benezra, C. R. Cantor, and R. Axel, *Cell* **44,** 697 (1986).
[22] A. N. Shakhov, S. A. Nedospasov, and G. P. Georgiev, *Nucleic Acids Res.* **10,** 3951 (1982).

graphy,[23] DNA is labeled with [^{3}H]thymidine starting from 20–30 hr post-infection. Specific activity is usually $10–50 \times 10^3$ cpm/μg of DNA.

2. Minichromosome fractions with DNA concentrations of 10–20 μg/ml in 5 mM triethanolamine (TEA)-HCl, pH 7.6, 0.1 mM EDTA, 15% sucrose are digested by micrococcal nuclease [EC 3.1.31.1; Worthington, Freehold, NJ, 15 units (U)/μg] in the presence of 0.5 mM CaCl$_2$ at various nuclease concentrations (1–10 U/ml) at 37° for 10–30 min. Reactions are terminated by adding EGTA to 1 mM and incubating on ice for 10–30 min. Aliquots are withdrawn at every stage to monitor DNA digestion by agarose gel electrophoresis followed by staining or fluorography.

3. For the ExoIII reaction, MgCl$_2$ is added to 3 mM Tris-HCl, pH 8.0, to 66 mM, dithiothreitol (DTT) to 1 mM. Digestion with ExoIII (New England Biolabs, Beverly, MA, 70 U/μl) at a concentration of 10–20 U/ml is carried out at 37° for 15 min. The reaction is terminated by adding EDTA to 10 mM, and sodium dodecyl sulfate (SDS) to 0.1%. DNA from the samples is purified by digestion with proteinase K (Merck, Darmstadt, FRG), phenol–chloroform extraction, and ethanol precipitation.[24]

4. For S$_1$ digestion, DNA samples are dissolved in 1 mM Tris-HCl, pH 7.5, 0.1 mM EDTA. Fivefold concentrated S$_1$ reaction buffer is then added to give final concentrations of 300 mM NaCl, 30 mM sodium acetate, pH 4.5, 0.1 mM ZnCl$_2$. Digestions with S$_1$ nuclease (P-L Biochemicals, Milwaukee, WI; 500,000 U/ml) at enzyme concentrations 50 U/ml are carried out at 20° for 30 min. Reactions are terminated by shifting the pH to 7.5 with 2 M Tris base and adding EDTA to 10 mM. DNA is purified by phenol–chloroform extraction and ethanol precipitation.

5. DNA samples are dissolved in 10 mM Tris-HCl, pH 7.5, 0.1 mM EDTA. Aliquots are withdrawn, and then DNA samples taken before and after ExoIII/S$_1$ digestions are analyzed by 2% agarose gel electrophoresis in TBE buffer.[24] After ethidium bromide staining or fluorography, samples corresponding to an optimal degree of micrococcal nuclease/ExoIII digestions (in the case of SV40 DNA, a few cleavages per viral genome) are chosen for further analysis. At this stage, oligonucleosomal DNA bands are shifted without being sharpened, while the mononucleosomal DNA band is sharpened and shifts to the core DNA position.

6. DNA from selected samples is digested with an appropriate restriction enzyme (for SV40 DNA, usually *Eco*RI, *Bam*HI, or *Bgl*I), and the

[23] R. A. Laskey, this series, Vol. 65, p. 363.
[24] T. Maniatis, E. F. Fritsch, and J. Sambrook, "Molecular Cloning." Cold Spring Harbor Lab., Cold Spring Harbor, New York, 1982.

fragments are separated on 1.6%, 30-cm-long, 1.5-mm-thick vertical gels.[25] With somewhat lower resolution, conventional thick horizontal agarose gels[26] can also be used.

7. Southern transfer to Gene Screen (New England Nuclear, Boston, MA) or Hybond-N (Amersham International plc, Amersham, UK) membranes is performed according to Maniatis et al.[24] and manufacturer's instructions.

8. Appropriate restriction nuclease fragments or subclones are labeled by nick-translation.[24] Alternatively, single-stranded DNA or RNA hybridization probes can be made after subcloning of these fragments into appropriate vectors.[27,28]

9. Hybridization is carried out as described.[24] The same filter is then probed with another labeled DNA fragment for mapping nuclease cleavages in the opposite direction.

Notes. (1) Chelation and readdition of divalent cations (steps 2–3) is a crucial point. We recommend checking the inhibition of micrococcal nuclease by EGTA and feasibility of sequential digestion with ExoIII/S_1 by making a pilot experiment with an excess of plasmid DNA added to chromatin preparation. (2) Single-stranded cleavages introduced into chromatin DNA by nucleases and chemicals can also be analyzed by the indirect end-labeling technique. In this case, alkaline agarose gel electrophoresis[24,26] is used.

Critical Comments

First, this protocol cannot be used directly for digestions of chromatin in the nuclei. In this case, nuclei are lysed after micrococcal nuclease digestion with 0.5 mM EDTA, then extracted and pelleted. Exonuclease III digestion can be then carried out with soluble oligonucleosomal chromatin from the supernatant. Second, at present, a good alternative method for total chromatin is a nonenzymatic cleavage.[20] To our knowledge, the best illustration of the indirect end-labeling approach so far is a demonstration of nucleosome positioning on single-copy genes.[21] This work used cleavage with methidiumpropyl-EDTA – Fe(II) and fragments with lengths of 0.4 – 1.0 kilobases (kb) as hybridization probes for genomic blots.

[25] S. A. Nedospasov, A. N. Shakhov, and G. P. Georgiev, *Mol. Biol. (Engl. Transl.)* **18**, 896 (1984).

[26] J. Favaloro, R. Treisman, and R. Kamen, this series, Vol. 65, p. 718.

[27] J. Messing, this series, Vol. 100, p. 20.

[28] D. A. Melton, P. A. Krieg, M. R. Rebagliati, T. Maniatis, K. Zinn, and M. R. Green, *Nucleic Acids Res.* **12**, 7035 (1984).

Molecular Cloning and Sequencing of Mononucleosomal DNA as a
Method to Determine Precise Nucleosome Positions

Rationale

To critically evaluate predictions from theoretical considerations concerning nucleosome positioning[5] and to reveal intrinsic features of the
nucleosomal DNA, mapping of nucleosomes with agarose gel resolution is
not sufficient. For precise mapping, especially in reconstruction experiments, the distance between the DNA ends of core particles and the
appropriate restriction nuclease site can be accurately measured by conventional techniques (e.g., see Ref. 3). Another approach, described below,
relies on random cloning and sequencing of mononucleosomal DNA
(Fig. 1B).

Initially, we applied this technique to map nucleosomes along SV40
DNA using micrococcal nuclease alone as a probe of chromatin structure.[29] Independently, Igo-Kemenes and co-workers[30] used a micrococcal
nuclease – ExoIII/S_1 procedure to map nucleosome positions on rat satellite
chromatin. After ExoIII/S_1 treatment, mononucleosomal DNA is more
homogeneous in size, thus allowing more accurate mapping. In addition,
earlier stages of micrococcal nuclease digestion can be used for mononucleosomal DNA preparation, which probably excludes artifacts similar to
that described by McGhee and Felsenfeld[31] (when bias in the nucleosomal
DNA content has been introduced owing to depletion of sequences preferentially cut by micrococcal nuclease).

What follows is a further modification of the procedure of Böck *et al.*[30]
Sequential digestions with micrococcal nuclease and ExoIII are performed
on soluble, purified SV40 minichromosomes or on reconstituted chromatin. Mononucleosomal DNA is cloned into M13-based vectors allowing
rapid chain termination sequencing.[32] Plaque hybridization[24] can be included as a convenient step before sequence analysis of the clones when
positioning on a particular DNA fragment is under analysis (e.g., in reconstitution experiments with the DNA insert cloned into a bacterial plasmid
vector).

Procedure

1. SV40 minichromosome preparation, digestion with micrococcal
nuclease, ExoIII, DNA purification, and digestion with S_1 nuclease are

[29] S. A. Chuvpilo, S. A. Nedospasov, A. N. Shakhov, and G. P. Georgiev, *Dokl. Akad. Nauk SSSR* **267**, 1268 (1982).
[30] H. Böck, S. Abler, X. Y. Zhang, H. Fritton, and T. Igo-Kemenes, *J. Mol. Biol.* **176**, 131 (1984).

carried out essentially as described above (steps 1–5 above). Alternatively, chromatin can be reconstituted *in vitro* from the DNA of interest and core histones and digested with micrococcal nuclease, ExoIII and then with S_1, as described above.

2. On analysis of RNase A-treated DNA on agarose gels, samples with optimal yields of mononucleosomes are chosen. In some cases, DNA samples corresponding to different stages of micrococcal nuclease digestion are combined (excluding very late stages when oligonucleosomal DNA is not visible and subnucleosomal DNA appears).

3. DNA is purified by preparative electrophoresis in 1.8–2% low melting temperature agarose (type VII, Sigma, St. Louis, MO) in Tris–acetate buffer.[24] The mononucleosomal DNA band is excised, and DNA is purified from agarose according to McMaster *et al.*[33]

4. Blunt-ended DNA (1 μg in a typical experiment) is ligated to an excess of appropriate phosphorylated synthetic linkers labeled with [γ-^{32}P]ATP. Kinasing, purification from the unbound linkers, redigestion with restriction enzyme, and final native acrylamide gel purification are carried out essentially according to Maniatis *et al.*[24] On autoradiography, mononucleosomal DNA with the attached linkers (octamers) gives a rather sharp band with a mobility corresponding to DNA of 160 base pairs (bp) (^{32}P-labeled *Msp*I digest of pBR322 with doublet bands of 147 and 160 bp is an excellent marker for this kind of experiment). The distinct mononucleosomal band is excised, and DNA is then eluted as described.[9]

5. A nucleosomal library is constructed in M13-based vectors (such as mp8 or mp10[27] digested with appropriate restriction enzyme and dephosphorylated as described[24]). Tranfection is made into JM103 cells, which are plated in the presence of isopropylthiogalactoside (IPTG) and 5-bromo-4-chloro-3-indolyl-β-D-galactoside (X-Gal) to distinguish recombinant clones. Where necessary, libraries are screened by plaque hybridization[24] with appropriate labeled DNA probes.

6. Minipreparations of single-stranded DNA are made from individual plaques in 1 ml cultures according to the described protocol.[34]

7. Identification and precise mapping of the inserts is based on conventional chain termination sequencing.[32] For analysis of mononucleosomal inserts, 6% acrylamide sequencing gels[9] are used (see Fig. 2A for illustration).

[31] J. D. McGhee and G. Felsenfeld, *Cell* **32,** 1205 (1983).
[32] F. Sanger, S. Nicklen, and A. R. Coulson, *Proc. Natl. Acad. Sci. U.S.A.* **74,** 5463 (1977).
[33] G. K. McMaster, P. Beard, H. D. Engers, and B. Hirt, *J. Virol.* **38,** 317 (1981).
[34] P. H. Schreirer and R. Cortese, *J. Mol. Biol.* **129,** 169 (1979).

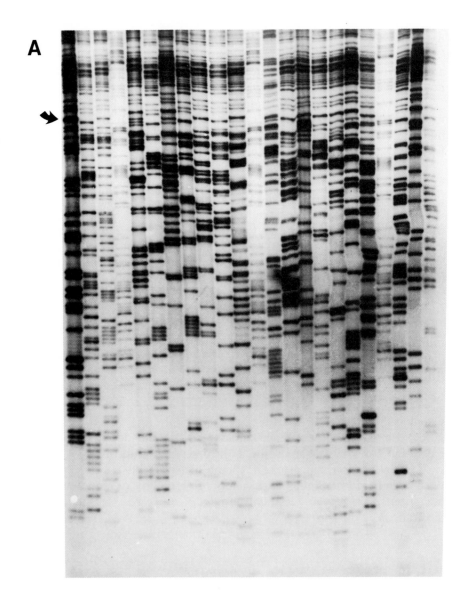

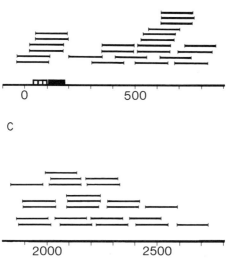

FIG. 2. Mapping of nucleosomes on SV40 DNA by partial sequencing of mononucleosomal clones. (A, facing page) Example of a 6% sequencing gel showing T-tracks of 20 individual mononucleosomal clones readily identifiable by the coding–decoding procedure.[35] Tracks 4, 12, and 20 (from left to right) contain marker DNA. The arrow indicates the distal end of the insertion in track 1 (T-doublet in the *Eco*RI linker). The upper parts of all tracks contain recognizable patterns of vector DNA above the ends of the insertions. (B) Mapping nucleosomes, prepared from the isolated minichromosomes, on the promoter–enhancer region of SV40 DNA. Only mononucleosomal DNA fragments of length 145 ± 3 bp are shown. Numbers correspond to nucleotides in the SV40 sequence. Note that the SV40 strain used contains a single copy of the 72-bp repeat (solid bar).[43,44] Open bars correspond to 21-bp repeats. (C) The same as in B, but for another part of the viral genome. The *Bam*HI site is nucleotide 2451 according to this numbering.

8. When the DNA sequence is known (as in the case of SV40 minichromosomes or in reconstitution experiments with known DNA) and analysis of a large number of individual clones is anticipated, the following approach is recommended.[35] Only T-tracks are made for each clone, thus allowing analysis of more clones in a single experiment. After autoradiography, the T-track patterns are converted into digital codes according to certain rules.[35] The whole DNA sequence is also converted by computer program into overlapping coded segments. All these compiled codes are sorted and printed out in the form of a simple catalog. The insert is then

[35] E. I. Golovanov and S. A. Nedospasov, *Comp. Appl. Biosci.* 2, 297 (1986).

identified according to the code of its T-track in this catalog with assignment of nucleotide number in the sequence. The problem of precise mapping of the borders of the insert by one sequencing track with nucleotide accuracy can be solved using appropriate sequencing markers (Fig. 2A).

Notes. (1) Mononucleosomal libraries can also be constructed in plasmid vectors, such as pUC13[36] or pUR222.[37] If inserts are cloned into an internal site of the polylinker, there is no need for the isolation of labeled fragments prior to sequencing.[37] Identification of inserts can be done by Maxam–Gilbert sequencing[9] of minipreparations of cloned DNA. For DNA of known sequence, identification of nucleosome positions only by G-tracks followed by coding–decoding procedure[35] is recommended. Alternatively, the inserts can be analyzed by chain termination sequencing[32] directly on plasmic minipreparations using standard synthetic oligonucleotides as sequencing primers. (2) Dinucleosomal DNA can be analyzed by a similar approach. However, dinucleosomal DNA bands remain heterogeneous after ExoIII/S_1 treatment. Therefore, during two electrophoretic purifications (steps 3–4 of the procedure), size fractionation might occur. For analysis of dinucleosomal inserts, gradient 4.5% sequencing gels on thermostatted plates (Macrophor, LKB, Bromma, Sweden), made according to manufacturer's instructions, are recommended. (3) The same protocol can be applied for analysis of nucleosome positioning in reconstitution experiments with gene-sized DNA or with shorter fragments cloned into a plasmic vector. In the latter case, plaque or colony hybridization (at step 5) is performed. (4) Analysis by molecular cloning of nucleosome positioning on a single-copy gene in total chromatin of higher eukaryotes demands screening of a very large mononucleosomal library and is not feasible.

Critical Comments

With regard to this method, the most crucial question concerns possible artifacts of molecular cloning. There are at least two problems. First, certain DNA fragments corresponding to nucleosome positions can be lost during cloning, thus imposing a bias on the observed nucleosome distribution. Second, the relative frequencies of different nucleosome positions can be changed in the course of the cloning procedure, which can distort the statistical distribution (for discussion, see Ref. 38). Thus, wherever possible, alternative approaches should not be ignored. As for the first problem,

---

[36] J. Vieira and J. Messing, *Gene* **19**, 259 (1982).
[37] U. Rüther, M. Koenen, K. Otto, and B. Müller-Hill, *Nucleic Acids Res.* **9**, 4087 (1981).
[38] S. C. Satchwell, H. R. Drew, and A. A. Travers, *J. Mol. Biol.* **191**, 659 (1986).

in several cases alternative methods for the analysis of nucleosome positioning give results consistent with those derived from molecular cloning. Böck et al.[30] performed a number of control experiments based on restriction nuclease digestion of satellite chromatin and arrived at consistent results. A mononucleosomal library, constructed by the method described above with nuclei of SV40 infected cells, has been screened by hybridization for α-satellite sequences. A number of analyzed clones correspond to several of the alternative nucleosome positions previously determined in independent studies[39] and may serve as an internal control of the procedure.[40] On the other hand, distortion of the statistical weights of certain nucleosome positions during cloning certainly takes place. Still, one part of the problem, namely, distinguishing between a frequent position and "amplification" of a certain position during cloning, can be solved because of the microheterogeneity of the ends of individual clones corresponding to the same nucleosome position. In addition, as in the case of satellite chromatin,[30] microheterogeneity of the DNA sequence might be helpful.

At least in three studies significant numbers of nucleosomal DNA fragments were sequenced and determinants of rotational and translational nucleosome positioning were analyzed.[30,38,40] Analysis of 177 random mononucleosomal clones from chicken erythrocyte H1/H5 depleted chromatin[38] (mean length 145 ± 1.5 bp) obtained by the procedure similar to that described above revealed sequence-dependent preferences that correlated with rotational orientation of the DNA toward the histone octamer. No preferences in dinucleotide distribution along the DNA were reported. Analysis of 200 mononucleosomal clones derived from the isolated H1-containing SV40 minichromosomes[40] showed a complicated nucleosome distribution (see Fig. 2B and 2C for illustration; note that the SV40 genome sequence differs from the conventional one[43,44]). Careful analysis of a subset of 125 clones (mean length 145 ± 2 bp)[41] revealed weak periodical variations of dinucleotide frequencies resulting in a matrix for rotational preferences similar to that found in the previous study.[38] Taken together, the findings based upon cloning and sequencing of nucleosomal DNA directly contradict the predictions of the simple models for DNA bending in the nucleosome (for example, see Ref. 42).

In nucleosomal DNA derived from the SV40 minichromosomes, significant preferences to certain nucleotides and dinucleotides were found at

[39] X. Y. Zhang, F. Fittler, and W. Hörz, Nucleic Acids Res. 11, 4287 (1983).
[40] S. A. Nedospasov, A. N. Shakhov, and V. V. Kushnirov, unpublished data, 1984.
[41] R. A. Abagyan, A. V. Ulyanov, and S. A. Nedospasov, unpublished results, 1988.
[42] Y. B. Zhurkin, J. Biomol. Struct. Dyn. 2, 785 (1985).
[43] A. R. Buchman, L. Burnett, and P. Berg, in "DNA Tumor Viruses" (J. Tooze, ed.), 2nd Rev. Ed., p. 799. Cold Spring Harbor Lab., Cold Spring Harbor, New York, 1982.
[44] S. A. Nedospasov, A. N. Shakhov, and R. Sahli, Mol. Biol. (Engl. Transl.) 18, 889 (1984).

several characteristic distances from the dyad.[41] The data were used to develop an algorithm for predicting the fit of histone octamer to a DNA of an arbitrary sequence.[41] Böck *et al.*[30] have analyzed 49 mononucleosomal clones from rat satellite chromatin. The average insert length in this case is 142.5 ± 2 bp and might result from slight overdigestion with ExoIII and/or S_1. There are good statistics for a few alternative nucleosome positions on these simple sequences that have been confirmed by independent data.

Finally, although dinucleosomal DNA after ExoIII/S_1 treatment is not as well defined as is core particle DNA, cloning and analysis of dinucleosomal DNA might be helpful in revealing nucleosomal "frames" among overlapping mononucleosomal clones. This has been verified for some regions of SV40 minichromosomes.[40]

Acknowledgments

We are grateful to T. Igo-Kemenes for making his detailed protocol available to us. We thank Eu. I. Golovanov and V. V. Kushnirov for collaboration on the analysis of mononucleosomal clones and to R. A. Abagyan for helpful discussion. We are indebted to M. L. Pifer for critical reading of the manuscript. Experiments on cloning and sequencing in M13 have been started at Swiss Institute for Experimental Cancer Research. S.A.N. thanks B. Hirt and R. Sahli for support and helpful discussions.

[22] Crystallization of Nucleosome Core Particles

By D. Rhodes, R. S. Brown, and A. Klug

Introduction

After the discovery that the DNA in eukaryotic chromosomes is packaged by histones into a continuous chain of repeating structural units, or nucleosomes (reviewed in Ref. 1), it became evident that, in order to obtain definitive information, it was necessary to determine the structure of the nucleosome, or parts thereof, by X-ray diffraction of single crystals. Such a structure promises to reveal the fine details of how the DNA helix wraps around the histone octamer and how individual histones interact with the DNA and with each other.

In attempting to grow crystals, it is essential to have access to homogeneous and pure preparations of the molecule, or particle, of interest in tens of milligram quantities. The initial hint that it might be possible to prepare

[1] R. D. Kornberg, *Annu. Rev. Biochem.* **46,** 931 (1977).

Copyright © 1989 by Academic Press, Inc.
All rights of reproduction in any form reserved.

a defined complex from chromatin came from the observation by Noll and Kornberg in 1977[2] that, on further digestion of chromatin with micrococcal nuclease following the initial cleavage of the linker DNA between nucleosomes, the DNA is cut away until an obstacle to further digestion is reached. In this process, histone H1 is lost to leave a substructure of the nucleosome known as the nucleosome core particle. Lutter[3,4] built on this original observation by refining the conditions of micrococcal nuclease digestion (involving the prior stripping of H1) and introducing a simple column fractionation method. The purification protocol he developed reproducibly gives large quantities (10–100 mg) of pure nucleosome core particles that have a well-defined structure [146 ± 2 base pairs (bp) of DNA wrapped around a protein core consisting of two copies each of histones H2A, H2B, H3, and H4 (the histone octamer)] (see Lutter, this volume [13]). This development enabled us to investigate a large number of crystallization conditions, which in turn led to an understanding over the years of how to grow crystals of reproducible size and quality suitable for study by X-ray diffraction.

Testifying to the universality of the nucleosome core structure, the same X-ray diffraction patterns have been obtained from crystals grown from nucleosome cores prepared from a variety of species of chromatin: rat liver, chicken erythrocytes, beef kidney, calf thymus, sea urchin sperm, scallop sperm, and mouse myeloma cells. Most of our recent crystals have been grown from nucleosome core particles purified from chicken erythrocyte chromatin, because chicken blood is easily available and does not contain many proteases.

Crystallization Method

Nucleosome core crystals are grown using the very simple "direct addition method" in small, stoppered glass tubes. In this method, changes in the crystallization conditions are achieved by changing the salt mixture directly in the tube. In principle, as a consequence of the nature of the nucleosome core structure, in which most of the surface presented to the solvent is DNA, crystallization is achieved by the addition of millimolar amounts of metal ions such as Mn^{2+}, Mg^{2+}, or Co^{2+},[3,5] rather than the

[2] M. Noll and R. D. Kornberg, *J. Mol. Biol.* **109,** 393 (1977).

[3] J. T. Finch, L. C. Lutter, D. Rhodes, R. S. Brown, B. Rushton, M. Levitt, and A. Klug, *Nature (London)* **269,** 29 (1977).

[4] L. C. Lutter, *J. Mol. Biol.* **124,** 391 (1978).

[5] J. T. Finch, R. S. Brown, D. Rhodes, T. Richmond, B. Rushton, L. C. Lutter, and A. Klug, *J. Mol. Biol.* **145,** 757 (1981).

often used and harsher precipitating agents such as organic solvents (methylpentanediol, polyethylene glycol) or high salt (ammonium sulfate).[6] The divalent metal ions neutralize the negatively charged DNA surface and, thus, presumably promote DNA–DNA interactions and assembly of nucleosome core particles into a crystal lattice. Nucleosome core crystals have also been obtained using spermine as the precipitating agent.[7] As Mn^{2+} gives the better ordered crystals, here we describe only the crystallization protocol and results using $MnCl_2$ as the precipitant.

Crystallization Vials. Crystals are grown in small, flat-bottomed glass tubes with an internal diameter of 3–4 mm and a height of about 25 mm. The glass tubes are boiled in 1% (w/v) sodium dodecylsulfate (SDS), 2 mM EDTA, rinsed several times in distilled water, and dried thoroughly. After this, they are siliconized by immersion in dimethyldichlorosilane solution (BDH, Poole, UK), shaken to remove excess siliconizing solution, washed directly in several changes of water, and finally rinsed in ethanol and again dried thoroughly. The tubes are stoppered with size 3 or 4 mm rubber or silicone stoppers (Gallenkamp, London) that have been cleaned by boiling in water and then dried.

Stock Solutions

> Nucleosome cores (from chicken blood) at 7–8 mg/ml stored in 20 mM potassium cacodylate (pH 6.0), 50 mM KCl, and 1 mM EDTA (1 mg nucleosome cores = 10 A_{260} units read in 0.1 M NaOH).
> 0.5 M KCl
> 0.2 M MnCl$_2$
> 0.1 M potassium cacodylate (pH 6.0)

All reagents are BDH analytical grade. All solutions including nucleosome core preparations, should be passed through 45-μm cut-off disposable filters (Millex-HA, USA) and stored 4°.

Setting Up a Crystallization Experiment

A crystallization experiment involves setting up many tubes containing systematically increasing amounts of $MnCl_2$ and KCl to produce a phase diagram that is described in detail in the next section. Crystallization trials can be carried out in volumes of 25–100 μl and at a final concentration of nucleosome core particles ranging from 2 to 4 mg/ml. The largest crystals

[6] A. McPherson, "The Preparation and Analysis of Protein Crystals." Wiley, New York, 1982.

[7] E. C. Uberbacher and G. J. Bunick, *J. Biomol. Struct. Dyn.* **2,** 1033 (1985).

grow at a final concentration of 4 mg/ml and in 100-μl volumes. About 20 mg of a nucleosome core preparation is required to set up one coarse and one fine phase diagram with tubes containing 100 μl crystallization mix and a nucleosome core concentration of 4 mg/ml. If material is abundant, it is advisable to duplicate each crystallization condition.

An example of a 100 μl crystallization mix is given below.

Final concentrations	Mix
40 mM MnCl$_2$	20 μl 0.2 M stock stolution
60 mM KCl	7 μl 0.5 M stock solution
	(50 mM present in core solution)
10 mM potassium cacodylate	(20 mM present in core solution)
(pH 6.0)	
4 mg/ml nucleosome cores	50 μl at 8 mg/ml

$$\underline{23 \ \mu l \ H_2O}$$
Total volume $\overline{100 \ \mu l}$

All procedures are carried out at room temperature. The amounts of stock solutions required and water are dispensed directly into a siliconized glass tube. For accuracy, volumes are measured and dispensed with a Hamilton glass syringe of a suitable size (10, 25, or 50 μl). The nucleosome core is added last, and the crystallization solution mixed thoroughly by pipetting several times with a 50- or 100-μl Hamilton syringe. Often, when nucleosome cores mix with the crystallization buffer, a white precipitate forms immediately. This precipitate will, however, depending on the concentration of MnCl$_2$ and KCl present, redissolve or clear prior to crystal growth. The tubes are stored standing upright in racks and placed in an incubator at 27°–28°. The racks are made of polystyrene foam to insulate the tubes from rapid changes in temperature.

Crystallization Strategy: The Phase Diagram

The nucleosome core is one of the largest structures to have been determined by X-ray diffraction of single crystals[3,8] (with the exception of viruses, which are highly symmetrical). It has a molecular weight of 206,000, the mass being distributed equally between protein and DNA. For any given resolution the average diffracted intensity is weakened by a factor roughly proportional to the volume or size of the structure. Therefore, when working with large molecules it is essential to grow large crys-

[8] T. J. Richmond, J. T. Finch, B. Rushton, D. Rhodes, and A. Klug, *Nature (London)* **311**, 532 (1984).

tals, so that the crystal size does not limit the resolution of the data to be collected. Large crystals of nucleosome cores ($0.3 \times 0.3 \times 1.5$ mm) can be grown by careful manipulation of salt conditions in the phase diagram.

As increasing amounts of $MnCl_2$ are added to a solution containing nucleosome cores, the precipitation point will be reached, and small crystals can be grown around these "salting-out" conditions. However, if more $MnCl_2$ is added the precipitation boundary is passed, and the nucleosome core particle redissolves. It is these latter conditions around the "salting-in" edge that we have exploited for growing crystals. The exact concentration of $MnCl_2$ at the "salting-in" point varies with the concentration of KCl present, and the crystal shape and size vary with the proportions of the two salts. Therefore, we construct a phase diagram to study the effects of systematic variations in the concentrations of both divalent Mn^{2+} and monovalent K^+ on the habit of crystal growth.

The strategy for growing large crystals with any one nucleosome core preparation is to first do a trial, or coarse phase diagram (Fig. 1), to find approximate crystallization conditions (crystallization conditions may vary for different preparations of nucleosome cores, see below). Then a second, finer phase diagram is designed around the range of conditions that appear most promising. If we look at Fig. 1, in which each square represents a crystallization tube, we can see that as the KCl concentration is increased from 40 to 60 mM, the molarity of $MnCl_2$ that produces crystals

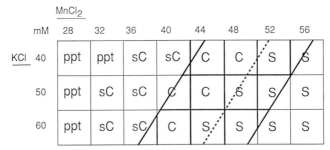

FIG. 1. Example of a coarse phase diagram using $MnCl_2$ as the precipitating agent. ppt, Precipitate; sC, small crystals; C, crystals; S, soluble. It can be clearly seen that the crystallization of nucleosome core particles is determined by the precise relationship between divalent and monovalent salt. The area at the center of the diagram marked by heavy lines indicates the range of conditions that should be repeated in the fine phase diagram, but using smaller increments in the molarity of the two salts. The solubility boundary is shown as a dashed line. The example of crystallization conditions given are for nucleosome cores used at a final concentration of 4 mg/ml. Note that the range of $MnCl_2$ that produces crystals will vary and is dependent on the average length of the DNA in a nucleosome core preparation. Other divalent metal ions (Mg^{2+}, Co^{2+}) can be used for growing crystals, but the molarity required for crystallization varies with the metal. KCl can be replaced by NaCl.

decreases from 48 to 40 mM, creating a boundary that runs diagonally across the phase diagram. Generally, the largest crystals are found along this diagonal and at salt concentrations just below solubility at the "salting-in" edge. To the left of the solubility boundary there is a tendency to go from tubes containing a few large crystals, to many smaller ones, and then to precipitates as the MnCl$_2$ concentration is reduced. Indicated in Fig. 1 is the range of conditions from the coarse phase diagram that should be covered when setting up the second crystallization experiment. In the fine phase diagram, however, the molarity of the two salts is increased in small steps, 1 or 2 mM for MnCl$_2$, 5 mM for KCl, and, consequently, the number of tubes containing large crystals will increase. Figure 2 shows an example of such a tube containing a single nucleosome core crystal.

The coarse phase diagram is examined after about 1 week. This is an adequate time for crystals to grow and, consequently, to give a preliminary guide of which conditions to use for the fine phase diagram. Depending on the salt conditions, crystals may appear in a few hours or after several days. The largest crystals grow over several weeks in tubes with only a few

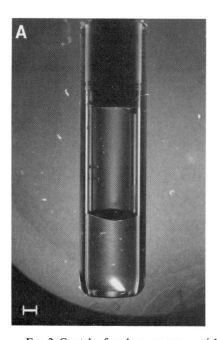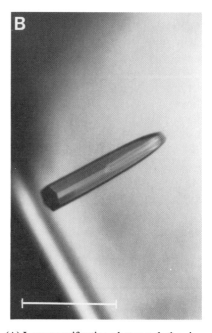

FIG. 2. Crystals of nucleosome core particles. (A) Low-magnification photograph showing a single crystal growing from the bottom of the crystallization vial. (B) High-magnification photograph taken at an angle to the long crystal axis to show the hexagonal cross section. Scale bar, 1 mm.

nucleation points. It is, therefore, logical that after precise crystallization conditions are set up, the tubes should not be disturbed for 2–3 weeks. The crystals have a prismatic shape that is hexagonal in cross section for the large crystals (Fig. 2B) and rhombic for the smaller ones, but both have the same crystal packing.

Factors That Affect Crystallization Conditions

Cleanliness. The rule of successful crystallization is cleanliness. The use of absolutely clean glassware and filtered solutions prevents bacterial growth and reduces the number of nucleation points to promote the growth of large, single crystals rather than many small ones.

Variability in Crystal Growth. It is also important to note that although crystallization is reproducible, that is, certain conditions will tend to produce larger crystals than others, there is also an inherent variability. In general, it is unlikely that two tubes containing the same nucleosome core preparation in identical crystallization conditions will produce the same number or size crystals. Therefore, it is inadvisable to restrict the proposed crystallization conditions to a range so narrow it does not allow for a certain amount of variability in habit of crystal growth.

Variability in DNA Length. The mean length of the DNA of different nucleosome core preparations varies as a consequence of the way in which nucleosome cores are produced. The extent of micrococcal nuclease digestion cannot be controlled precisely, and this results in nucleosome cores containing a mean DNA length that varies between different preparations (between 146 ± 2 and 148 ± 2 bp). The mean length of DNA affects both the amount of $MnCl_2$ required for crystallization and, more seriously, the unit cell dimensions (see below). The position of the solubility boundary (Fig. 1) at, for instance, 50 mM KCl varies for different preparations between 40 and 54 mM $MnCl_2$. Most typically, the DNA is around 146 ± 2 bp in length, but if it is 147 bp, less $MnCl_2$ is required to obtain crystals. If the DNA is 148 bp or longer, crystals grow poorly or not at all.

Temperature. As indicated earlier, the largest crystals are grown in an incubator at 27–28°. It is also possible to grow crystals in the cold. At 4° the solubility boundary for a nucleosome core preparation moves about 4 mM $MnCl_2$ to the left in the phase diagram, and crystals take longer to grow.

Rate of Crystal Growth. It is generally true that crystals which grow slowly also grow bigger and have a more perfect hexagonal rod shape. However, many of the crystals grow partly hollow, and we have not been able to overcome this problem routinely. The problem is more pronounced with some nucleosome core preparations than others and seems to occur when the "salting-in" border has been estimated at too low a concentrate of $MnCl_2$, which results in rapid crystal growth.

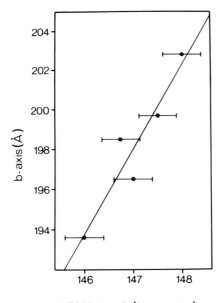

DNA length (base pairs)

FIG. 3. Variations in the dimensions of the crystallographic *b* axis with the mean length of the DNA on the nucleosome core particle. (Reprinted, with permission, from *Cold Spring Harbor Symp. Quant. Biol.* **47**, 498. Copyright © 1983 Cold Spring Harbor Laboratory.)

Unit Cell Dimensions and Crystal Packing

In early crystallization experiments several divalent metals, particularly Mg^{2+} and Co^{2+}, were used as precipitating agents. These metals produce crystals containing different, but related, packing interactions between nucleosome core particles. Most contacts are DNA–DNA, between core particles in a hexagonal arrangement. $MgCl_2$ produces crystals that diffract out to spacings of 20 Å with unit cell dimensions of $a = 110$ Å, $b = 192$ Å, and $c = 340$ Å^3; $CoCl_2$ gives an even larger unit cell. $MnCl_2$ produces the better crystals that diffract to about 5 Å and have the smallest unit cell with sides $a = 111$ Å, b varying between 193 and 203 Å, and $c = 110$ Å.[5,8,9]

This variability in *b* axis dimensions is related to the variability in the mean length of the DNA in different preparations of nucleosome cores from one chromatin source. (The origin of this variability in DNA length is discussed above.) Figure 3 shows how the dimension of the *b* axis varies monotonically with the average length of the DNA present in different crystals.[9] One base pair difference corresponds to about 3.5 Å in the length

[9] T. J. Richmond, J. T. Finch, and A. Klug, *Cold Spring Harbor Symp. Quant. Biol.* **47**, 493 (1982).

of the *b* axis. Clearly then, the ends of the DNA in the particles are involved in packing interactions and cause the *b* axis to expand or contract depending on whether the DNA is long or short. (For details on packing interactions, see Ref. 5.) Because many crystals are needed for the collection of each data set, this variability in unit cell dimensions presented a serious problem. The problem was partly overcome by making use of an organic solvent, hexane-1,6 diol to shrink the unit cell dimensions to a constant size.[8,9]

Conclusions

The structure of the nucleosome core particle containing mixed sequence DNA was solved to 20 Å resolution in 1977[3] and to 7 Å resolution in 1984.[8] It has the shape of a disk 57 Å thick and 110 Å in diameter. The 146 bp of DNA are bent nonuniformly into $1\frac{3}{4}$ turns of a left-handed superhelix around the histone octamer. The central turn of the superhelix is held by the H3–H4 tetramer, and most of the histone–DNA interactions occur on the inside of the superhelix in the minor groove of the DNA double helix.

The crystallization conditions described in this chapter have also been used to obtain crystals of reconstituted nucleosome cores[10] that, by virtue of containing a specific-sequence DNA fragment of precise length, are better ordered and diffract to spacings of 3 Å.[11] This promises to reveal a higher resolution structure.

Acknowledgments

The crystallization procedures described evolved over the years with the help of our colleagues, in particular, B. Rushton, J. Finch, and T. Richmond.

[10] D. Rhodes, *Nucleic Acids Res.* **6**, 1806 (1979).
[11] T. J. Richmond, M. A. Searles, and R. T. Simpson, *J. Mol. Biol.* **199**, 161 (1988).

Section III

Histones: Preparation and Analysis

[23] Isolation and Characterization of Histones

By C. VON HOLT, W. F. BRANDT, H. J. GREYLING, G. G. LINDSEY,
J. D. RETIEF, J. DE A. RODRIGUES, S. SCHWAGER, and B. T. SEWELL

Introduction

In chromatin, the histone polypeptide chains H2A, H2B, H3, and H4 are associated to form an octameric protein that in turn interacts with DNA to form the nucleosomal core. The association of the core particle with the fifth histone type, the histone H1, results in the nucleosome,[1] the fundamental repeat unit of the elementary chromatin fiber.

On the association of the histone octamer with the DNA, ionic bonds are established between the basic amino acid side chains, exposed on the surface of the octamer, and the adjacent phosphate groups of the DNA. Two types of ionic bonds can be realized in this interaction, namely, between phosphate and the ϵ-amino group of lysine or the guanidinium residue of arginine. The more resonant guanidinium–phosphate bond forms a stronger bond between DNA and histone than the lysyl–phosphate pair. The isolation of the component histone–histone complexes and subunits is determined by the relative contribution of the lysyl and arginyl residues to protein–DNA interaction. This is why the arginine-rich polypeptide chains of H3 and H4 are more difficult to dissociate from the DNA than H2A, H2B, and H1. In the latter three, the lysyl residues are envisaged to be predominantly responsible for the stabilization of the protein–DNA interaction. The ease of dissociation of any particular isohistone (histone variant) of the H1 or H2A and H2B classes from the DNA is governed by its amino acid sequence. Thus, the higher positive charge density in sea urchin sperm H1[2,3] and H2B[4-6] causes them to dissociate from the nucleosome at sodium chloride concentrations greater

[1] R. D. Kornberg, *Annu. Rev. Biochem.* **46**, 931 (1977)

[2] W. N. Strickland, M. Strickland, P. C. de Groot, C. von Holt, and B. Wittmann-Liebold, *Eur. J. Biochem.* **104**, 559 (1980).

[3] W. N. Strickland, M. Strickland, W. F. Brandt, C. von Holt, A. Lehmann, and B. Wittmann-Liebold, *Eur. J. Biochem.* **104**, 567 (1980).

[4] M. Strickland, W. N. Strickland, W. F. Brandt, and C. von Holt, *Eur. J. Biochem.* **77**, 263 (1977).

[5] W. N. Strickland, M. Strickland, W. F. Brandt, and C. von Holt, *Eur. J. Biochem.* **77**, 277 (1977).

[6] M. Strickland, W. N. Strickland, W. F. Brandt, C. von Holt, B. Wittmann-Liebold, and A. Lehmann, *Eur. J. Biochem.* **89**, 443 (1978).

Copyright © 1989 by Academic Press, Inc.
All rights of reproduction in any form reserved.

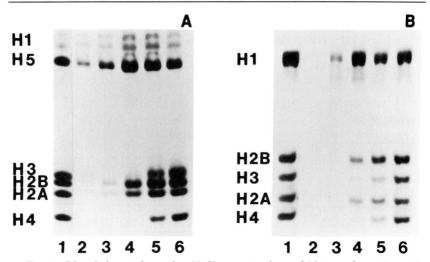

FIG. 1. Dissociation at increasing NaCl concentrations of histones from chromatin. Chicken erythrocyte and sea urchin sperm chromatin recovered after digestion with micrococcal nuclease was dialysed against various NaCl concentrations in 10 mM Tris-HCl, pH 7.5, 2 mM EDTA, 1 mM PMSF, 0.02% sodium bisulfite for 16 hours, the chromatin was separated by centrifugation, the supernatant dialyzed, freeze-dried and the histones subjected to polyacrylamide gel electrophoresis in the presence of sodium dodecyl sulfate (SDS–PAGE). (A) Chicken erythrocyte histones; (B) sea urchin sperm histones, lanes 1–6, total acid histones extracted at 0.6, 0.8, 1.0, 1.5, and 2 M NaCl, respectively.

than those applicable to the homologous polypeptides in avian erythrocyte chromatin (Fig. 1).

Within the core particle, the four types of polypeptide chains, H2A, H2B, H3, and H4, form an octamer.[7] Controlled variation of ionic conditions allows dissociation of this octamer from the DNA, either intact (see Isolation of Natural Histone Complexes) or in successive stages as a number of defined histone–histone complexes. Under more drastic conditions, they are released as the randomly structured component polypeptide chains (see Purification of Histone Subunits).

The intact octamer in solution and in the absence of DNA can also be cleaved under nondenaturing conditions into a tetramer of the composition (H3–H4)$_2$ and two dimers of the composition (H2A–H2B). The tetramer and the dimer can be further dissociated under denaturing conditions to result in randomly coiled component polypeptide chains. Alternatively, all the histones can be dissociated in a single-step procedure under denaturing conditions and recovered as individual denatured polypeptide chains.

[7] J. O. Thomas and R. D. Kornberg, *Proc. Natl. Acad. Sci. USA* **72**, 2626 (1975).

The full reversal of these pathways of dissociation is possible. Conditions can be defined leading to the *in vitro* reconstitution of a histone octamer (see Assembly of Core Particles from Natural and Reconstituted Octamers) in which both the natural composition and conformation are fully restored, as shown by the identical crystallization properties of the natural and reassembled octamer (see Reconstitution of Natural and Hybrid Octamers, Confirmation of Octameric Structure). This technology also allows the construction of hybrid octamers. These octamers may contain either specific, naturally occurring isohistones or histone analogs and derivatives, prepared through semisynthetic amino acid residue substitution (see Preparation of Histone Derivatives) or site-directed mutagenesis and/or recombinant DNA procedures. Assembly of such octamers into core particles on genomic and unique DNA as well as on synthetic, sequence-defined DNA provides the means to investigate histone–DNA interactions at single base pair resolution (see Assembly of Core Particles from Natural and Reconstituted Octamers).

The testing of the competence for octamer and core particle formation and the identification of any modulation of core particle properties resulting from a hybrid octamer is the only critical test to establish the functional competence of a histone or isohistone subunit. In view of the absence of any other known function of the histones, beside their structural role, this methodology thus provides a very powerful tool to investigate structure–function relation of any individual histone subunit or its derivative.

Isolation of Nuclei and Nucleoprotein

It may be desirable to isolate the histones in their natural conformation to investigate histone–histone or histone–DNA interaction. For such purposes, it is essential to isolate initially nuclei, as source for the octamer, in which the natural structure of chromatin is maintained as faithfully as possible. The isolation of individual histone subunits may be required solely for structural characterization or for inquiries into their biosynthesis, postsynthetic modification, and involvement in aspects of transcription. In the latter category such investigation may have been undertaken on intact cells or isolated nuclei.

In order to isolate individual histone subunits, ideally the purification of nuclei should always precede the isolation of the histones. This, however, often proves difficult because suitable protocols are not available for the particular cell type under investigation. In such cases nucleoproteins must be prepared rapidly under conditions which remove interfering cytoplasmic and nuclear protein and which simultaneously inhibit proteolysis. No specific provisions have to be made in these protocols to maintain the natural conformation of histone–DNA or histone–histone complexes because the subunits are ultimately isolated from the nucleoprotein under denaturing conditions. The preparation of nuclei from chicken erythrocytes, sea urchin embryos, and sea urchin sperm cells, described here, yields either chromatin suitable for the isolation of natural chromatin complexes or histone subunits labeled in metabolic experiments.

Isolation of Erythrocyte Nuclei from Chicken Blood

The method described has also been used successfully to isolate erythrocyte nuclei form the blood of shark, crocodile, and frog.

Solutions

Anticoagulant: 16 mM citric acid, 89 mM trisodium citrate, 16 mM sodium dihydrogen phosphate, 0.13 M glucose

Saline–citrate: 0.14 M NaCl, 10 mM trisodium citrate

Lysis buffer: 15 mM Tris, 65 mM KCl, 65 mM NaCl, 5 mM 2-mercaptoethanol, 0.15 mM spermine, 0.5 mM spermidine, 0.2 mM ethylenediaminetetraacetic acid (EDTA), 0.5 mM ethylene glycol bis(β-aminocthyl ether)-N,N'-tetraacetic acid (EGTA), 0.1 mM phenylmethylsulfonyl fluoride (PMSF) (from a stock solution), 1% (w/v) Nonidet LE, pH 7.5

PMSF: 100 mM in dimethyl sulfoxide (DMSO)

Washing buffer: as lysis buffer but without Nonidet LE

Glycerol: analytical grade

Procedure. Blood from mature hens is collected from the abattoir in 140 ml anticoagulant per liter blood. The suspension is cooled on ice and filtered through four layers of cheesecloth. Erythrocytes are sedimented (5000 g, 10 min, 4°). All subsequent operations are performed at the same temperature. The supernatant is carefully removed together with the white layer of leukocytes on top of the pellet. The erythrocytes are suspended in 5 volumes of NaCl–citrate and centrifuged as before. The washing procedure is repeated 3 times.

Washed cells are suspended in lysis buffer equivalent in volume to the original blood volume and quickly processed to the next step. The precise time for lysis varies, depending on the variety of chicken and probably nutritional status, and must be established in a pilot experiment. Overexposure to the lysis buffer leads to the bursting of the nuclear membrane and manifests itself in clumping of the nuclei in subsequent operations.

Nuclei are washed by repeated centrifugation and resuspension in 5 volumes of washing buffer until the supernatant is free of hemoglobin. It is important that, for all centrifugation steps following lysis, the rotor is quickly brought up to 5000 g and then immediately decelerated. This avoids too dense packing of the very fragile nuclei. Nuclei can be stored for several months at $-20°$ as a suspension in 50% (v/v) glycerol in washing buffer. Such stored nuclei are processed further after sedimenting and repeated washing with the buffer required for the particular subsequent methodology. Such nuclei are particularly suitable for the preparation of the histone octamer.

Isolation of Functionally Intact Nuclei from Sea Urchin Embryos

Sea urchin embryo cultures can easily be synchronized and thus lend themselves to investigation into functional changes of chromatin structure in the course of genome activity. *In vitro* investigations into the involvement of histones in functional aspects of chromatin, performed in nuclei, require functionally intact nuclei. Nuclei isolated according to the given protocol exhibit full transcriptional activity.[8]

Solutions

Buffer A (10× concentrated stock solution): 650 mM KCl, 150 mM NaCl, 1.5 mM spermine, 5 mM spermidine, 150 mM Tris-HCl, pH 8.0

KCl solution: 0.55 mM KCl

Sucrose solution 1: 0.25 M sucrose, 10 mM Tris-HCl, pH 8.0, 0.1 mM EDTA

Sucrose solution 2: 0.15 mM spermine, 0.5 mM spermidine, in sucrose solution 1

Sucrose solution 3: 2.0 M sucrose in buffer A, 10 mM 2-mercaptoethanol, 0.1 mM PMSF (from PMSF stock solution)

Storage buffer: 25% (v/v) Glycerol in 1× buffer A, 10 mM 2-mercaptoethanol, 0.1 mM PMSF, 0.2 mM EDTA, 0.2 mM EGTA

PMSF stock solution: 100 mM in DMSO

SDS: 10% (w/v) sodium dodecyl sulfate

Isolation of Nuclei from Embryos That Have Passed through the Hatching Blastula Stage up to Late Gastrula. In the investigated species, *Parechinus angulosus,* the hatching blastula up to gastrula stages of development refer to embryos between 9 and 20 hr after fertilization. The embryo culture is immobilized by the addition of 0.3 ml SDS/liter seawater, and the embryos are washed 3 times with KCl solution. After each wash they are sedimented (6000 g, 5 sec). The final sediment is suspended in 2–3 volumes of sucrose solution 1 and spun (6000 g, 2 min). For embryos harvested from 1 liter of a 4% culture, 80 ml is required.

[8] G. F. Morris and W. F. Marzluff, *Biochemistry* 22, 645 (1983)

The sediment is resuspended in the same volume of sucrose solution 2 and homogenized with 10 strokes of the tightly fitting Dounce homogenizer. Microscopic control with phase-contrast optics should show almost all nuclei liberated. To this homogenate, sufficient sucrose solution 3 is added to adjust the sucrose concentration to 1.25 M, that is, 1.33 volumes sucrose solution 3 per volumes of homogenate. The solution is spun (50,000 g, 50 min).

The pellet is suspended in storage buffer and spun (6000 g, 2 min). This pellet is finally resuspended in a small volume of storage buffer and transferred to an Eppendorf vial; it can be stored for up to 1 month in liquid nitrogen without losing its transcriptional activity.

Isolation of Nuclei from Prehatching Stages. From *P. angulosus,* intact nuclei can be isolated prior to hatching from embryos grown without fertilization membrane. The following solutions are utilized for removal of the fertilization membrane:[9]

3-Amino-1,2,4-triazole (ATA): 1 M (stock solution)
Pronase: 100 mg/ml H_2O
Antibiotics: 100 mg penicillin, 50 mg streptomycin/ml H_2O

To a 4% (v/v) suspension of sea urchin eggs in seawater, ATA is added to a final concentration of 1 mM. The eggs are fertilized, the formation of the fertilization membrane is confirmed by light microscopy, and then the fertilized eggs are incubated for 15 min at 20° on a rotary shaker. Eggs are then allowed to settle. To speed up the latter step, larger cultures should be divided into batches of approximately 250 ml. The sediment is taken up in the same volume of filtered seawater, the embryos allowed to settle, and the procedure repeated.

The sediments of washed eggs are then taken up in filtered seawater, combined, and made up to one-half the original volume, that is, an 8% culture. To this is added Pronase to a final concentration of 50 μg/ml. After about 20 min of incubation on the rotary shaker at 20°, the fertilization membrane has disappeared. The precise time may vary from experiment to experiment and must be determined through light microscopic observation. The cell suspension is then allowed to settle, as before, and washed twice with filtered seawater to remove Pronase. The culture is adjusted to the original volume (i.e., 4%), antibiotics are added to a dilution of 1 : 1000, and the culture is grown to the desired stage. Under these conditions, the synchrony of the development, as judged by light microscopic criteria, is maintained.

The harvested embryos are then processed for nuclei essentially as described for embryos past the hatching stage. However, possibly as a result

---

[9] A. G. Carroll and H. Ozaki, *Exp. Cell Res.* **119,** 307 (1979).

of the shorter cell cycle in the earlier stages of development, the nuclei are much more fragile. This requires the following modifications: for 6-hr embryos, only 5 strokes of the tight-fitting pestle of the Dounce homogenizer should be used; for 4 hr embryos, this is reduced to 3 strokes.

Rapid Isolation of Sea Urchin Embryo Nuclei with Detergent

(a) For the isolation of histones following *in vivo* experiments on intact cells or embryos, for example, postsynthetic modifications or isohistone synthesis program, the speed of isolation of the nucleosomal proteins is of higher priority than functional integrity of nuclei. The following fast procedure[10] yields nuclei with partially disintegrated nuclear envelopes and variable functional properties, but with a minimum of proteolytic degradation of nucleosomal proteins.

(b) For the isolation of core particles, mono, and oligonucleosomes, the Hewish and Burgoyne buffer A[11] containing spermine and spermidine is used to stabilize the nuclei during lysis with detergent rather than $MgCl_2$. This prevents the activation of Mg^{2+}-dependent endonucleases. Nonidet LE is preferentially used in this method as it does not influence the determination of DNA concentration at 260 nm as does Triton X-100.

Solutions

Ca^{2+},Mg^{2+}-free seawater: 48 mM KCl, 11 mM NaCl, 2.5 mM NaHCO$_2$, 0.5 mM EDTA

SDS: 10% (w/v)

Sucrose: 0.5 M sucrose, 10 mM Tris-HCl, pH 7.5

Buffer A (10× concentrated stock solution): 150 mM NaCl, 650 mM KCl, 1.5 mM spermine, 5 mM spermidine, 150 mM Tris-HCl, pH 7.4

Lysis buffer 1: 1% (w/v) Triton X-100, 20 mM NaCl, 10 mM MgCl$_2$, 20 mM Tris-HCl, pH 7.5, 0.2 mM PMSF

Lysis buffer 2: 2× buffer A, 20 mM EDTA, 2.5 mM EGTA, 0.5% (w/v) Nonidet LE, 10 mM 2-mercaptoethanol, 0.2 mM PMSF

Washing buffer: 0.25 M sucrose, 1× buffer A, 0.2 mM EDTA, 0.2 mM EGTA, 5 mM 2-mercaptoethanol, 0.2 mM PMSF

Storage buffer: 25% glycerol, 1× buffer A, 0.2 mM EDTA, 0.2 mM EGTA, 5 mM 2-mercaptoethanol, 0.1 mM PMSF

Procedure. The embryos, at the desired stage of development, are immobilized through the addition of SDS (0.3 ml/liter seawater) and allowed to settle. This step may be omitted for young and small cultures, which can

[10] L. D. Keichline and P. M. Wassarman, *Biochemistry* **18**, 214 (1979).
[11] D. R. Hewish and L. A. Burgoyne, *Biochem. Biophys. Res. Commun.* **52**, 504 (1973).

be centrifuged directly at 500 g for 1 min. The sediment is washed 5 times with Ca^{2+},Mg^{2+}-free seawater. The embryos are suspended in sucrose solution, at a volume ratio of 1 : 20 (embryos : solution), and homogenized with 10 strokes of the loosely fitting pestle of the Dounce homogenizer. This leads to complete disintegration of the embryos into a cell suspension.

For the (a) procedure, an equal volume of lysis buffer 1 is added to lyse the cells. Lysis is monitored by light microscopy to avoid overexposure to the detergent which leads to disintegration of the nuclei. The nuclei are collected via centrifugation (2000 g, 10 min). Such nuclei can be stored in liquid nitrogen. For the (b) procedure, an equal volume of lysis buffer 2 is added, and again lysis is monitored microscopically. Resulting nuclei are then washed twice in washing buffer (2000 g, 10 min). Nuclei are taken up in a small volume of storage buffer and stored in liquid nitrogen.

Sea Urchin Sperm Nuclei

The procedure is based on methods described by Zentgraf and Franke[12] and Spadafora et al.[13]

Solutions

Buffer A ($5\times$ concentrated stock solution): 0.5 M NaCl, 1.25 M sucrose, 50 mM Tris-HCl, pH 7.4
MgCl$_2$ stock solution: 100 mM MgCl$_2$
Washing buffer: $1\times$ buffer A, 0.5 mM MgCl$_2$
Lysis buffer: $1\times$ buffer A, 1.5% (w/v) Nonidet LE, 0.5 mM MgCl$_2$, 0.2 mM PMSF
KCl: 0.5 M

Procedure. Sperm is collected from male sea urchins after intracoelomic injection of 0.5 M KCl. The individuals are inverted, and the sperm is collected in plastic trays. The sperm is taken up in filtered seawater, filtered through plastic mesh (pore size 250 μm), and centrifuged (2000 g, 5 min). The sediment is washed 3 times in filtered sea water and resedimented after each wash (2000 g, for 5 min). The sediment is then taken up in washing buffer and homogenized (Dounce homogenizer, loosely fitting pestle). The homogenate is spun (2000 g, 5 min). This process is repeated 3 times.

The pellet is suspended in lysis buffer and homogenized as before. The suspension is shaken 5 min and centrifuged (800 g, 10 min). The pellet is washed twice in buffer A to remove the Nonidet and detached sperm tails. This step is controlled microscopically. It may be necessary to repeat the

[12] H. W. Zentgraf and W. Franke, *J. Cell Biol.* **99**, 272 (1984).
[13] C. Spadafora, M. Bellard, L. Compton, and P. Chambon, *FEBS Lett.* **69**, 281 (1976).

lysis step, but at a concentration of 0.75% (w/v) Nonidet in lysis buffer until all tails have been removed. For acid extraction of histones, these nuclei may be stored at $-70°$ in 50% (v/v) glycerol in buffer A. For investigations involving core particles or DNA characterization, the nuclei used should be fresh.

Isolation of Deoxyribonucleoprotein

The most common procedures to isolate and purify the protein component of nucleoprotein have been described elsewhere (this series, Vol. XII, Part B, [96] and [97]). Some methods for preparation of nucleoprotein from less conventional cells are described here.

Yeast Nucleoprotein. The following solutions are required:

Solution A: 50 mM Na_3PO_4, 1 mM $MgCl_2$, 40 mM $NaHSO_3$, 1 mM PMSF (from stock solution), pH 6, adjusted with 1 M HCl
Solution B: 50 mM Tris-HCl, pH 8.0, 1 mM $MgCl_2$, 40 mM $NaHSO_3$, 1 mM PMSF
Sucrose 1: 1.8 M sucrose in solution B
Sucrose 2: 1 M sucrose in solution B
Saline–citrate: 10 mM trisodium citrate, 140 mM NaCl, 3 mM EDTA
Triton solution: 0.2% (w/v) Triton X-100 in solution B

It is very difficult to isolate reproducibly, in good yields, undamaged nuclei from pressed bakers' yeast. The procedure described by Tonino and Rozijn,[14] however, with some modifications, yields a nucleoprotein fraction suitable as starting material for the isolation of histones under denaturing conditions.[15] To begin, 100 g of pressed bakers' yeast is suspended in 200 ml of solution A. The cells are pelleted and the procedure repeated. The washed cells are then resuspended in 200 ml of solution A and disintegrated in a ball mill (Braun, Melsungen, FRG) for 3 min with 0.5-mm-diameter glass beads under cooling with liquid carbon dioxide.

The broken cells are centrifuged (2000 g, 40 min). Several layers of insoluble material collect at the bottom of the centrifuge tube. The top layer, which contains nucleoprotein, is removed and suspended in a small volume of solution B. This suspension is homogenized with a Dounce homogenizer and subsequently spun (2000 g, 30 min). The nucleoprotein pellet is taken up in solution B containing 0.25 M sucrose, homogenized (Potter type instrument), and the suspension layered over a discontinuous

[14] G. J. M. Tonino and T. H. Rozijn, *Biochim. Biophys. Acta* **124**, 427 (1966).
[15] W. F. Brandt, K. Patterson, and C. von Holt, *Eur. J. Biochem.* **110**, 67 (1980).

gradient made from sucrose 1 and sucrose 2 solutions. The nucleoprotein collects at the interface on centrifugation (48,000 g, 80 min). The nucleoprotein is then homogenized in solution B (from this stage onward, PMSF may be omitted from solution B) and centrifuged (20,000 g, 10 min). The pellet is homogenized and centrifuged sequentially (20,000 g, 10 min) in the following solutions: Triton solution followed by saline–citrate and then finally in absolute ethanol. The precipitate can be stored at −20°.

Wheat Germ Deoxyribonucleoprotein.[16] Prepare the following solutions:

> Saline–citrate: 140 mM NaCl, 10 mM trisodium citrate, 10 mM EDTA, 10 mM NaHSO$_3$, 0.5 mM PMSF (from stock solution), 0.5% (v/v) thiodiglycol (TDG), 0.25% (w/v) Triton X-100, 0.1% (v/v) 1-octanol
> Solubilizing solution: 2 M NaCl, 0.1% (v/v) TDG, 1 mM PMSF (from stock solution)
> PMSF stock solution: 100 mM PMSF in DMSO

Freshly prepared germ from wheat *(Triticum aestivum)* can be stored at 4° in a closed container containing silica gel until further processing. All operations are performed at 4°. Seventy-five grams of wheat germ is homogenized in 250 ml saline–citrate in a blender at full speed for 1 min. The homogenate is centrifuged (20,000 g, 20 min). The sticky brown top layer (the crude chromatin) is scraped off from the underlying starch and cell debris and placed in a beaker on ice. The starch–debris layer is rehomogenized in saline–citrate in the blender and centrifuged. The chromatin layer is again scraped off and pooled with that from the previous run. Then 250 ml of saline–citrate is added, and the crude chromatin is rehomogenized for 30 sec. This is then centrifuged (27,000 g, 10 min). This washing step is repeated 4 times or until the supernatant is clear. The combined washed pellets contain crude nucleoprotein.

One hundred twenty-five milliliters of solubilizing solution is added to the washed, impure nucleoprotein and homogenized 30 sec in the blender. The blender is rinsed with 50 ml of solubilizing solution, and the combined homogenate stirred for 15 hr at 4°. The viscous homogenate is centrifuged (20,000 g, 40 min), and the slightly turbid supernatant is slowly poured into 2 liters of cold distilled water. Randomly associated deoxyribonucleoprotein (DNP) is rendered insoluble. This DNP is wound onto a glass rod, placed into centrifuge tubes and spun (27,000 g, 10 min). It can be stored at −20° for several months.

[16] J. de A. Rodrigues, W. F. Brandt, and C. von Holt, *Biochim. Biophys. Acta* **578**, 196 (1979).

Wheat Leaf Mesophyll Cell Deoxyribonucleoprotein. Prepare the following solutions:

Grinding solution: 50 mM sucrose, 50 mM Tris-HCl, pH 6.3, 3 mM CaCl$_2$, 10 mM NaHSO$_3$, 0.2 mM PMSF (from stock solution), 0.5% (v/v) thiodiglycol (TDG), 0.15% (w/v) Triton X-100, 0.1% (v/v) 1-octanol

Saline–citrate: as for wheat germ nucleoprotein isolation (above)

Washing solution: 50 mM Tris-HCl, pH 6.3, 0.1% (v/v) TDG, 10 mM NaHSO$_3$, 0.2 mM PMSF

Extraction solution: 0.5 mM EDTA adjusted to pH 7.5 with Tris, 0.1% (v/v) TDG

PMSF stock solution: 100 mM in DMSO

When the leaves of young wheat plants are about 10 cm long, the tips are harvested, leaving most of the stems behind. Wheat leaves can be stored in a polyethylene bag at −20° until required. Fifty grams of wheat leaf tissue is cut into lengths of approximately 5 cm, placed in a blender, and homogenized at full speed in 250 ml of grinding solution for 25 sec. Over 80% of the mesophyll cells, but few epidermal cells, are destroyed during this treatment, as judged by light microscopy.

The suspension is squeezed through four layers of cheesecloth. The debris on the cheesecloth is reblended in approximately 100 ml grinding solution and filtered as before. The combined filtrates are centrifuged (4000 g, 20 min), and the supernatant discarded. The combined pellets are washed twice in 100 ml grinding medium and centrifuged (4000 g, 10 min). The pellet is resuspended in 100 ml saline–citrate buffer and stirred for 30 min to lyse the nuclei. The extract is centrifuged (2000 g, 10 min), and the pellet is resuspended in 100 ml washing medium and centrifuged (2000 g, 10 min). Extraction solution is added to the crude chromatin pellet (5 ml per 50 g starting material). The nucleoprotein is homogenized with 10–15 strokes of a tight-fitting pestle in a Dounce homogenizer and then left 15 min for the nucleoprotein to dissolve. This solution is centrifuged (2000 g, 10 min), and the pellets are reextracted in the same way. The combined supernatants, which contain the solubilized deoxyribonucleoprotein, are immediately processed to extract histones.

Isolation of Natural Histone Complexes

Isolation of natural histone complexes is described for chicken histones, as less experience is available on isolation of natural complexes from other organisms and alternative standard protocols have not yet been fully developed. Histones are separated from the DNA in three ways: (1) via ultracentrifugation,[17] (2) through elution from hydroxyapatite which retains the DNA,[18] and (3) by removal of the DNA as DNA–protamine complexes.[19]

Isolation of Natural Octamers via Ultracentrifugation

Solutions

Extraction buffer: 750 mM NaCl, 10 mM Tris-HCl, pH 7.4, 0.2 mM PMSF (from stock solution)

Dissociation buffer: 2 M NaCl, 10 mM Tris-HCl, pH 7.4, 0.2 mM PMSF

Sodium chloride: 4 M NaCl

PMSF stock solution: 100 mM in DMSO

Materials. Ultrafiltration membrane (Amicon PM10 or equivalent)

Procedure. Estimates of DNA content in nuclei, as well as in insoluble and soluble chromatin fractions, are made spectrophotometrically at 260 nm after adding sufficient 4 M NaCl to adjust the sample aliquot to 2 M NaCl and pelleting the insoluble material (OD$_{260\,nm}$, 20 = 1 mg DNA/ml).

The natural histone octamer complex is isolated essentially as described by Ruiz-Carrillo and Jorcano.[17] All operations are carried out at 4°. Washed nuclei are extracted (4 mg DNA/ml) with extraction buffer for 1 hr to remove H1 and H5. The residual nucleohistone sediments on centrifugation at 143,000 g in a 60 Ti rotor for 16 hr. The resulting pellet is homogenized (Ultra-Turrax) in dissociation buffer and, after adjusting the DNA concentration to 4 mg/ml with the same buffer, is left for 1 hr on ice. The DNA is pelleted by centrifugation as before. The supernatant, containing the octamer, is concentrated to approximately 12 mg/ml via ultrafiltration. The protein content in the supernatant is estimated on the basis of the DNA concentration in the initial suspension of nuclei, under the assumption of a histone:DNA ratio of 1:1. The octamer is purified from this solution via exclusion chromatography on a Sephadex G-100 column, equilibrated and eluted with dissociation buffer (see Fig. 2).

Histone Complex Isolation via Hydroxyapatite Adsorption Chromatography

The exposure of a partial digest of chromatin to increasing concentrations of NaCl at neutral pH results in sequential dissociaton, first, of the

[17] A. Ruiz-Carrillo and J. L. Jorcano, *Biochemistry* **18**, 760 (1979).

[18] R. H. Simon and G. Felsenfeld, *Nucleic Acids Res.* **6**, 689 (1979).

[19] D. R. van der Westhuyzen and C. von Holt, *FEBS Lett.* **14**, 333 (1971).

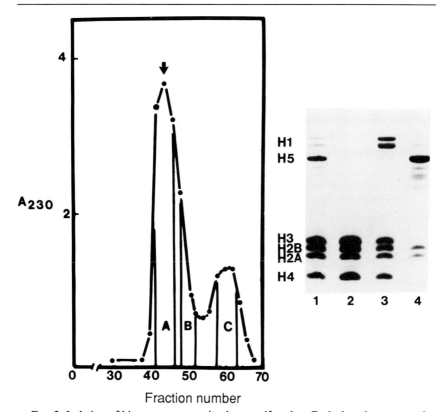

FIG. 2. Isolation of histone octamers via ultracentrifugation. Exclusion chromatography on Sephadex G-100 of the histone complexes dissociated from DNA (143,000 g supernatant) utilized the following: column, 2.5×95 cm; eluant, dissociation buffer; fraction volume, 3 ml; pressure head, 60 cm H_2O. Lanes 1–4 in the SDS–PAGE slab correspond to histone standard and fractions A–C in that order. [From H. J. Greyling, S. Schwager, B. T. Sewell, and C. von Holt, *Eur. J. Biochem.* **137**, 221–226 (1983). With permission of the Federation of European Biochemical Societies.]

histones H1 and H5 followed by the H2A–H2B dimer and, finally, the H3–H4 tetramer.[20] If, during this process, the DNA is immobilized through adsorption onto hydroxyapatite,[18] the core histones are eluted from the matrix as a mixture of the above complexes well separated from the histones H1 and H5,[20] provided the load for a 4.5×10 cm hydroxyapatite column does not exceed 100 mg DNA. With increasing DNA load, separation of the core histones from H1 and H5 deteriorates and further

[20] H. J. Greyling, S. Schwager, B. T. Sewell, and C. von Holt, *Eur. J. Biochem.* **137**, 221 (1983).

purification of the core histones is required. Dialysis of the eluted histone complexes against 2 M NaCl at pH 7.4 yields the reassociated octamer as the final product.[20]

Alternatively, dialysis of the complexes isolated by hydroxyapatite chromatography against 2 M NaCl, pH 5.0, yields the H3–H4 tetramer and the H2A–H2B dimer.[21] Subsequent exclusion chromatography allows purification of the complexes free of DNA and residual H1 and H5, which may not have been completely separated in the hydroxyapatite step if the hydroxyapatite column had been overloaded.

Solutions

Digestion buffer: 50 mM Tris-HCl, pH 7.4, 25 mM KCl, 4 mM MgCl$_2$, 1 mM CaCl$_2$, 0.2 mM PMSF (from stock solution)

EDTA stock solution: 250 mM

Extraction buffer: 10 mM Tris-HCl, pH 7.4, 0.25 mM EDTA, 0.2 mM PMSF

Micrococcal nuclease: stock solution of, e.g., 20 Worthington units/μl

Equilibration buffer: 10 mM sodium phosphate, pH 7.4 (or 10 mM Tris-HCl, pH 7.4), 0.1 mM PMSF

Elution buffer: 3 M NaCl, 10 mM sodium phosphate, pH 7.4, 0.2 mM PMSF

Octamer buffer: 2 M NaCl, 10 mM sodium phosphate, pH 7.4, 0.2 mM PMSF

Di–tetramer buffer: 2 M NaCl, 10 mM sodium phosphate, pH 5.0, 0.2 mM PMSF

PMSF stock solution: 100 mM in DMSO

Glycerol: analytical grade

Materials

Hydroxyapatite (see Ref. 22)

Sephadex G-100

Ultrafiltration membrane (Amicon P10 or equivalent)

Procedure. For the partial digestion of chromatin, nuclei are washed repeatedly in digestion buffer and suspended (5 mg DNA/ml) in digestion buffer. Micrococcal nuclease is added (40 units/mg DNA), thoroughly mixed, and digestion is then allowed to proceed at 37° for 30 min. The reaction is terminated by the addition of EDTA to 5 mM. The digest is

---

[21] H. J. Greyling, J. P. Hapgood, B. T. Sewell, and C. von Holt, *Eur. J. Biochem.* **161,** 133 (1986).

[22] C. Bernardi, this series, Vol. 21, p. 95.

centrifuged (6000 g, 5 min), and soluble chromatin is extracted from the pellet by homogenization in extraction buffer with a loose pestle in a Dounce homogenizer and dialysis against the same buffer for 16 hr at 4°. Soluble chromatin is recovered in the supernatant after centrifugation (10,000 g, 20 min).

For hydroxyapatite chromatography, soluble chromatin (~ 100 mg DNA in 20 ml of extraction buffer) is loaded on a hydroxyapatite column (4.5 × 10 cm), preequilibrated with equilibration buffer. Unbound material is eluted with 1 column volume of the same buffer. The column is then eluted with elution buffer (Fig. 3). From this eluate, either the reconstituted octamer or a mixture of dimers and tetramers can be recovered as follows.

Octamer isolation. The core histone mixture is concentrated by ultrafiltration to a concentration of approximately 1–2 mg/ml and then dialyzed against 20 volumes octamer buffer at 4° for 12 hr. Octamers are then separated from nonreconstituted material by exclusion chromatography on Sephadex G-100 equilibrated and eluted with octamer buffer (Fig. 4). Fractions are pooled and concentrated via ultrafiltration. Octamers in 2 *M* NaCl can be diluted with an equal volume of glycerol and stored at −20° for several months. To use such stored octamers in other experiments, they

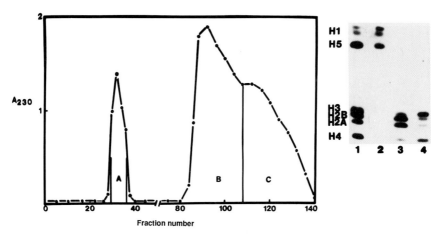

FIG. 3. Elution of histones from hydroxyapatite. The micrococcal nuclease digest of chromatin (100 mg DNA) was loaded onto the column and the column washed with 1 volume 10 m*M* sodium phosphate, 0.2 m*M* PMSF, pH 7.4. Histones were eluted with 3 *M* NaCl in the same buffer. Fractions of 2.5 ml were analyzed by SDS–PAGE. Lane 1, histone standards; lanes 2–4 correspond to fractions A–C. [From H. J. Greyling, S. Schwager, B. T. Sewell, and C. von Holt, *Eur. J. Biochem.* **137**, 221–226 (1983). With permission of the Federation of European Biochemical Societies.]

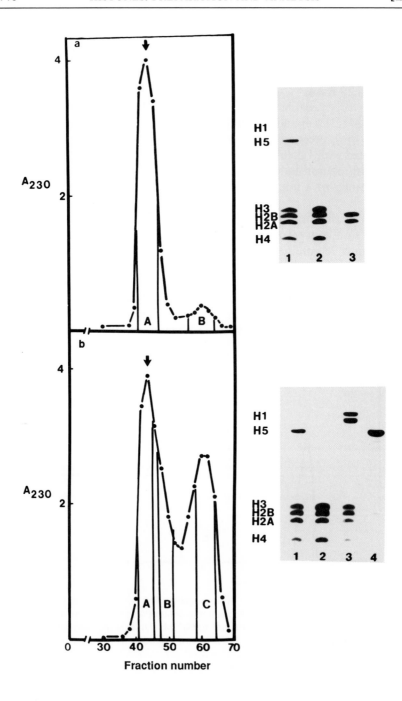

are dialyzed against the electrolyte solution dictated by the particular experiment and then further processed.

Dimer and tetramer isolation. The hydroxyapatite eluate containing core histones is concentrated as above and then dialyzed against 20 volumes of di–tetramer buffer at 4° for 12 hr. Dimers and tetramers are separated from each other by exclusion chromatography on a Sephadex G-100 column, equilibrated and eluted with di–tetramer buffer (Fig. 5). Fractions are pooled and concentrated via ultrafiltration.

Isolation of Dimers and Tetramers by the Protamine Displacement Method

Protamine successfully competes *in vitro* for DNA at sodium chloride concentrations at which histones remain fully dissociated from the DNA.[19] This allows the complete removal of DNA from a solution containing natural histone complexes and DNA. However, the sodium chloride concentration required to form an insoluble DNA–protamine complex is such that the histone octamer is dissociated[20] into tetramers and dimers. Initially, the displacement of histones by protamine had been executed at pH 5.4.[19] In chicken erythrocyte chromatin, however, a very active protease, probably of lysosomal origin, operates at this pH.[23] It is therefore preferable to perform the dissociation of dimers and tetramers at pH 7.0.[20]

Solutions

Dissociation buffer: 2 M NaCl, 10 mM Tris-HCl, pH 7.4, 0.2 mM PMSF (from stock solution)

Dialysis buffer: 150 mM NaCl, 10 mM Tris-HCl, pH 7.4, 0.2 mM PMSF

Protamine solution: 20 mg protamine/ml in dissociation buffer (contaminants are removed from crude protamine sulfate by precipitating the reagent at 4° overnight from a room temperature, saturated, unbuffered aqueous solution)

Elution Buffer: 50 mM sodium acetate, pH 5.0, 5 mM sodium bisulfite

[23] G. G. Lindsey, P. Thompson, and C. von Holt, *FEBS Lett.* **135**, 81 (1981).

FIG. 4. Isolation of histone octamers. (a) Reconstituted histone octamers from histones eluted from hydroxyapatite. (b) Reconstituted histone octamers from protamine-displaced histones. Conditions of chromatography were as in Fig. 2. Lanes 1–4 in the SDS–PAGE slab correspond to standard histones and fractions A–C, respectively. [From H. J. Greyling, S. Schwager, B. T. Sewell, and C. von Holt, *Eur. J. Biochem.* **137**, 221–226 (1983). With permission of the Federation of European Biochemical Societies.]

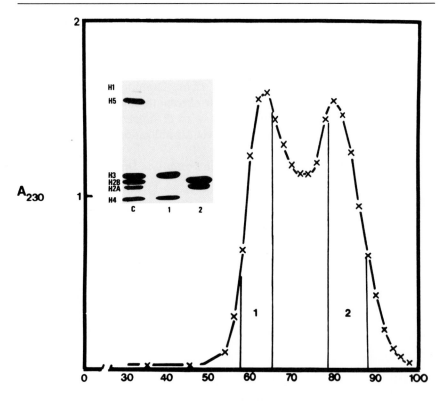

Fraction number

FIG. 5. Isolation of histone dimers and tetramers. Exclusion chromatography was performed on the hydroxyapatite eluate (see Fig. 3) concentrated and dialyzed against di–tetramer buffer, using the following: column, 2.5 × 90 cm; eluant, di–tetramer buffer; fraction volume, 3 ml; pressure head, 60 cm. On the SDS–PAGE slab, lanes C, 1, and 2 represent histone standards and chromatographic fractions 1 and 2, respectively. [From H. J. Greyling, J. P. Hapgood, B. T. Sewell, and C. von Holt, *Eur. J. Biochem.* **161**, 131–138 (1986). With permission of the Federation of European Biochemical Societies.]

Materials

Sephadex G-50
Sephadex G-100

Procedure. The histone pairs H2A – H2B and H3 – H4 are isolated by a modification of the method of van der Westhuyzen and von Holt.[19] All operations are executed at 4°. Nuclei are dissolved (4 mg DNA/ml) in dissociation buffer and allowed to equilibrate for 1 hr. Equal volumes of the viscous solution of lysed nuclei and protamine solution are mixed together and dialyzed against 15 volumes of dialysis buffer until all the DNA is precipitated as a deoxyribonucleic acid–protamine complex. The suspension is centrifuged (10,000 g, 20 min), and the supernatant, containing histone and excess protamine, is concentrated by ultrafiltration. The concentrate (10 mg histone/ml) is then fractionated into histone and protamine by gel filtration on a column of Sephadex G-50, equilibrated and eluted with dialysis buffer. The histone complex-containing fractions are again concentrated to 10–12 mg/ml by ultrafiltration, and dimers and tetramers may be separated from each other by exclusion chromatography on Sephadex G-100 (Fig. 5) at pH 7.4.

Alternatively, the dimers and tetramers may be separated on Sephadex G-100 with elution buffer at pH 5.0 to yield fractions A and B.[19] The H3 and H4 tetramer can be precipitated from fraction A at 0.7 saturation with ammonium sulfate, leaving H1 in the supernatant. The dimer in fraction B is recovered by ultrafiltration.

Purification of Histone Subunits

For many, though not all, investigated species, two successive gel-exclusion chromatographic procedures resolve the individual four core histones and the histone H1, extracted under denaturing conditions. Nevertheless, this fractionation procedure separates the histones, while producing good yields, into subfractions that can serve as suitable starting material for further fractionation.

Initial chromatography on BioGel P-60 in 20 mM HCl usually resolves H1–H5 from either the pair H3–H2A or H3–H2B and the remaining unpaired histone H4, as the case may be in histone mixtures from different organisms. The latter two pairs may then be partially or completely resolved by further exclusion chromatography under different conditions. Conservative cutting of fractions and recycling, using either chromatographic procedure, can produce almost pure fractions that still, however, contain the various posttranscriptionally modified derivatives of each of the respective histones. In addition, depending on the source of the histones, the fractions thus produced may contain various isohistones. Further subfractionation can be achieved through ion-exchange chromatography, reversed-phase HPLC, or preparative gel electrophoresis.

The presence of contaminating cytoplasmic or nuclear acidic proteins, particularly in histones isolated from rapidly dividing or transcriptionally very active cells (e.g., liver, yeast, sea urchin embryo), or the presence of acidic polysaccharides (myxomycetozoa) causes very poor separation of the histones because of the formation of ill-defined complexes with those contaminants. Elution order on exclusion chromatography, extractability in Johns procedure,[24] or mobility on SDS – PAGE or acid – urea – PAGE, in themselves, do not offer unequivocal proof of the identity of a particular histone. For example, sea urchin sperm H2B is extracted in Johns procedure, together with H3, and comigrates in urea gels with H3. A similar situation pertains to yeast histones[25] and cycad pollen[26] where H2A and H2B elute with H3 in the presence of 50 mM NaCl, whereas in acid only, H2A and H2B elute prior to H3.[15]

Extraction of Histone Subunits from Nucleoprotein

  Solutions

  Hydrochloric acid: 0.25 M HCl
  Sulfuric acid: 0.25 N H$_2$SO$_4$
  Perchloric acid: 10% (w/v) HClO$_4$
  Reinecke salt solution: 1.6 g NH$_4$[Cr(NH$_3$)$_2$(SCN)$_4$]·H$_2$O/100 ml
    H$_2$O, (saturated at 4°)
  Tris buffer: 50 mM Tris-HCl, pH 8.5
  Acetone: analytical grade
  Acidic acetone: acetone: 1 M HCl (98:2, v/v)

Histone subunits are extracted from nucleoproteins under denaturing conditions with dilute HCl, H$_2$SO$_4$, or HClO$_4$. The choice of acid depends on the nature of the steps planned for the subsequent purification. A convenient extraction volume is 1 ml for every 3 mg of DNA, although histone may be recovered from considerably more diluted extracts (0.1 mg or less/ml) either after freeze-drying or as the Reineckates.

Histone Perchlorates. Histone perchlorates exhibit characteristic solubilities that form the basis of fractionation procedures.[24] However, the histone perchlorates do not dissociate easily once redissolved, and partial dissociation causes nonhomogeneous and uncharacteristic charge densities on the surface of the histones. This in turn may lead to blurred electrophoretograms, often with very poor separation of the subunits. Similarly, ion-exchange chromatography of perchlorates may not result in the correct separation. It is therefore essential to convert the perchlorates to the histone chlorides via precipitation with acetone at −20° from a solution of 0.25 M HCl or via extensive dialysis of the histone perchlorates, dissolved in 0.25 M HCl, against distilled water at 4°.

Histone Chlorides. Histone chlorides dissociate more easily, and excess acid present after incomplete dialysis easily evaporates on freeze-drying. In

[24] E. W. Johns, *Biochem. J.* **104,** 78 (1967).
[25] W. F. Brandt and C. von Holt, *FEBS Lett.* **65,** 386 (1976).
[26] W. F. Brandt and C. von Holt, *FEBS Lett.* **51,** 84 (1975).

our hands extraction with hydrochloric acid, followed either by dialysis against water and freeze-drying or precipitation with acetone, has become the method of choice.

Histone Sulfates. If the histones are extracted with sulfuric acid, the acetone precipitation of the histone sulfates at $-20°$ is more successful because of their lower solubility relative to the chlorides. However, it is important to remove even traces of sulfuric acid from the precipitate because these will char the proteins during freeze-drying and on storage. Histone sulfates are best freeze-dried from aqueous solutions only after very extensive dialysis against distilled water or after reprecipitation with acetone from a concentrated solution in 0.25 M HCl and precipitation with acidic acetone. Histone sulfates often give blurred bands on electrophoresis in acetic acid–urea gels, probably because of incomplete dissociation.

Histone Reineckates. If acetone fails to precipitate histones from very dilute solution, they can still be successfully isolated as Reineckates[27] as follows. An equal volume of saturated Reinecke salt solution is added to the histone solution (pH 1–2) at 4°. The precipitate, formed after 30 min, is collected via centrifugation, and the supernatant is discarded. The histone Reineckate precipitate is washed twice with 1.0 ml Tris buffer to remove contaminating proteins and then suspended in 1.0 ml 0.25 M HCl. An equal volume of saturated Reinecke salt is added to reprecipitate some dissociated histone Reineckate at 4°. The precipitate is collected after 5 min and washed once with ice-cold water (further washing with water results in histone losses) and then extracted twice with acidic acetone. This dissociates the histone Reineckate, extracts the acetone-soluble Reinecke acid, and leaves free histone hydrochloride as a white insoluble precipitate. The precipitate is washed with acidic acetone, dried under vacuum, and stored at $-20°$ until further processing.

Tissue proteases are often coextracted with the histones. Proteolytic degradation in subsequent purification steps (exclusion and ion-exchange chromatography) can in many cases be inhibited by the addition of a mixture of 10 mM sodium bisulfite, 1 mM PMSF, and 1 mM tosyl-lysine chloromethylketone (TLCK) to all eluants.

Exclusion Chromatography

Separation of the individual histone subunits H2A, H2B, H3, H4, and the histone H1 as denatured, randomly coiled polypeptides is achieved via exclusion chromatography (BioGel P-60 and Sephadex G-100) on columns of 80–100 cm length with diameters of 1.5–14 cm. The sample load

[27] N. O. Lindh and B. L. Brantmark, *Methods Biochem. Anal.* **14**, 79 (1976).

may be 15–20 mg/cm² of column cross-section area.[28] Owing to the absence of tryptophan and the low content of tyrosyl and phenylalanyl residues, the histone concentrations in the eluate are monitored between 206 and 230 nm.

The best separation is achieved at temperatures between 20° to 25°. Despite the random conformation of the polypeptide chains under the chromatographic conditions, variation of the sodium chloride concentration in the HCl eluant leads to specific and reproducible aggregations for histones from a given source. This may be exploited for the isolation of individual subunits (Fig. 7). Given below is the two-step procedure for the isolation of histone subunits.[28-30]

Solutions

Eluant 1: 20 mM HCl, 50 mM NaCl (or as chosen from Fig. 7)
Eluant 2: 20 mM HCl
Eluant 3: 50 mM sodium acetate, 5 mM sodium bisulfite, pH 5.1
Application solvent: the chosen eluant, 8 M urea, 1% (v/v) 2-mercaptoethanol
Urea: 8 M
Tris: 50 mM Tris-HCl, pH 7.0
2-Mercaptoethanol: analytical grade

Materials

BioGel P-60, 100–200 mesh
Sephadex G-100, 100–200 mesh
o-Iodosobenzoate, analytical grade
Ultrafiltration membrane (Amicon P10 or equivalent)

[28] C. von Holt and W. F. Brandt, *Methods Cell Biol.* **16,** 205 (1977).
[29] E. L. Böhm, W. N. Strickland, M. Strickland, B. H. Thwaits, D. R. van der Westhuyzen, and C. von Holt, *FEBS Lett.* **34,** 217 (1973).
[30] D. R. van der Westhuyzen, E. L. Böhm, and C. von Holt, *Biochim. Biophys. Acta* **359,** 341 (1974).

FIG. 6. Exclusion chromatography of histones on BioGel P-60. The eluant was 20 mM HCl, 50 mM NaCl, except for yeast and wheat histones which were eluted with 20 mM HCl. The lane numbers in the electrophoretograms correspond to the chromatogram fraction numbers. A and B refer to unfractionated histones from calf and the respective organism. PAGE was performed in acid–urea gels in the absence of Triton X-100, except for the histones from wheat which were run in the presence of Triton at 7.6 M urea. The proteins move cathodically to the bottom of the lane. [From C. von Holt and W. F. Brandt, *Methods Cell Biol.* **16,** 205–225 (1977). With permission of Academic Press.]

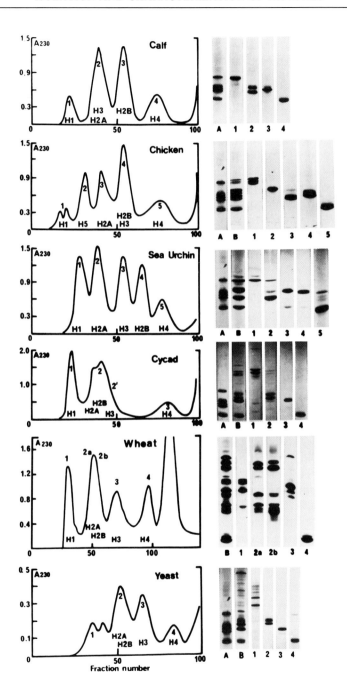

BioGel Chromatography. The sample in 8 *M* urea is adjusted with a minimum of Tris to pH 7.0, made 1% (v/v) with respect to mercaptoethanol, and left for 1 hr at room temperature. After acidification to pH 2, the appropriate sample application solvent is added to yield a sample volume compatible with the column dimensions. The sample is applied to the column and eluted with eluant 1 or 2.

The BioGel run results in different elution patterns depending on histone source (Fig. 6) and sodium chloride concentration (Fig. 7). Chromatographically identifiable fractions are pooled, dialyzed against distilled water at 4°, and freeze-dried. Histones H3 and H4, intended for octamer reconstruction, are not freeze-dried but concentrated via ultrafiltration. After addition of an equal volume of glycerol, they can be stored at −20°.

Sephadex Chromatography. Fractions recovered after BioGel chromatography, which still contain mixtures of histones, in particular the H2B–H3 and H2A–H3 pair, are redissolved in urea and reduced as before, if PAGE reveals the presence of H3 or H4 dimers depending on the source. They are then transferred to application solvent made up from eluant 3 and applied to a Sephadex column. The column is developed with eluant 3 to resolve the H3–H2A pair (calf thymus histones) or the H3–H2B pair (chicken erythrocytes) or other combinations as they arise from the BioGel runs of histone mixtures from other sources. Fractions are pooled according to the elution profile, dialyzed, and freeze-dried as before (Fig. 8). PAGE analysis of all the recovered fractions will aid in the decision as to which samples have to be recycled through the appropriate procedure.

Purification of H3 and H4 through Dimerization.[28,31] The histone H3-containing fractions (for histones H3 with a single cysteine residue) from the BioGel P-60 column are pooled and concentrated by ultrafiltration to 20 mg/ml. Solid recrystallized urea is added to 4 *M* and the pH adjusted to pH 7 with Tris. *o*-Iodosobenzoate in 4 *M* urea, pH 7, is added, 25 μg/mg protein, and the reaction is allowed to proceed for 1 hr at room temperature in the dark. The entire reaction mixture is then loaded onto a Sephadex G-100 column equilibrated with eluant 3 (without bisulfite). The fractions containing dimerized histone H3 are pooled (Fig. 9); H3 monomer is obtained from the dimer by the addition of excess 2-mercaptoethanol in Tris buffer and subsequent dialysis. The sample is then concentrated by ultrafiltration.

Sea urchin sperm histone H4 may elute from a BioGel P-60 column with a small H2B contamination. The sea urchin sperm histone H4 fractions contain two isohistones,[32] one of which is characterized by the re-

[31] W. F. Brandt and C. von Holt, *FEBS Lett.* **14**, 338 (1971).
[32] G. G. Lindsey and C. von Holt, submitted.

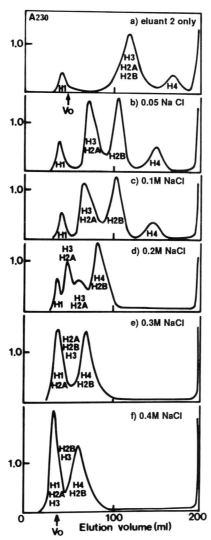

FIG. 7. Gel exclusion chromatography of histones on BioGel P-60. Calf thymus histones were developed with eluant 2 containing various sodium chloride concentrations. [From C. von Holt and W. F. Brandt, *Methods Cell Biol.* **16,** 205–225 (1977) and E. L. Böhm, W. N. Strickland, M. Strickland, B. H. Thwaits, D. R. van der Westhuyzen, and C. von Holt, *FEBS Lett.* **34,** 217–221 (1973). With permission of the Federation of European Biochemical Societies.]

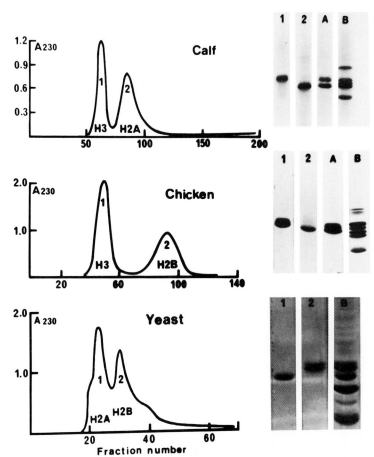

FIG. 8. Gel exclusion chromatography of histone pairs. Histone pairs not separated under conditions in Fig. 6 were rechromatographed (calf and chicken on Sephadex G-100 with eluant 3 and yeast on BioGel P-60 with eluant 1). PAGE was performed in acid–urea gels. The proteins move cathodically to the bottom of the lane. (From Refs. 15 and 28–30.)

placement of the threonine residue at position 73 (calf) by cysteine.[33] Dimerization of the histone H4-containing fraction from the BioGel P-60 column with o-iodosobenzoate in a procedure identical to that for H3 allows the separation of the cysteine-containing variant from the variant not containing cysteine and from histone H2B contamination if present.

[33] M. Strickland, W. N. Strickland, W. F. Brandt, and C. von Holt, *FEBS Lett.* **40,** 346 (1974).

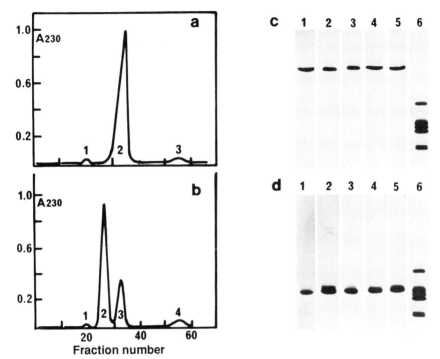

FIG. 9. Purification of H3 from various organisms via dimerization. Histone H3 mono-mer fraction 2 eluted from Sephadex G-100 with eluant 2 (a) is rechromatographed after dimerization (b) to remove contaminating histone H2A. (c) PAGE was performed in acid–urea gels at 2.5 M urea. Lanes 1–5, dimers of histone H3 from cycad pollen *(Encephalartos caffer)*, mollusk testis cells *(Patella granatina)*, sea urchin sperm *(Parechinus angulosus)*, shark erythrocytes *(Poroderma africanus)*, and chicken erythrocytes *(Gallus domesticus)* in that order. Lane 6 displays total calf thymus histone. (d) Monomers generated from the dimer: the remaining heterogeneity is due to fractional acetylation. Proteins move cathodic-ally to the bottom of the gels. [From C. von Holt and W. F. Brandt, *Methods Cell Biol.* **16**, 205–225 (1977) and W. F. Brandt and C. von Holt, *FEBS Lett.* **14**, 338–342 (1971). With permission of the Federation of European Biochemical Societies and Academic Press.]

Ion-Exchange Chromatography of Histone Subunits

Two anion-exchange matrices are in use, CM-cellulose and CM-Trisa-cryl M. The reproducibility of chromatograms on a given CM-cellulose matrix is inferior to that achieved with Trisacryl. In addition, CM-cellulose from different suppliers differs considerably. The general conditions de-scribed for the ion-exchange chromatography apply to both types of ma-trices.

The histones are developed from the ion-exchanger with a sodium chloride gradient either in the absence or in the presence of urea. However, the increasing sodium chloride concentration in the eluant leads to an increase in the expression of the hydrophobic properties of the histones, which in certain of the isohistone mixtures leads to aggregation and deterioration of separation. In such cases, the gradient must be run in the presence of urea to counteract the aggregation. Generally, the isohistones of the H1 type are better separated in the absence of urea, whereas in the case of most H2B and H2A isotypes, the presence of urea often improves the separation. As the histone sample is always small in relation to the vast excess of urea, even minor cyanate contamination in urea can lead to random blockage of amino groups in the histones. If this is to be avoided, the sodium acetate buffer system should be replaced by a 50 mM acetic acid–Tris buffer[28,34] to mop up cyanate contamination in the urea.

Solutions

EDTA (sodium salt): 10 mM in water
Acid: 0.5 M HCl
Alkali: 0.5 M NaOH
Eluant 1A: 50 mM sodium acetate, pH 4.4–5.5 (adjusted with HCl) or 50 mM acetic acid–Tris, pH 4.5–5.5
Eluant 2A: same as eluant 1A with 6–8 M urea (pH adjusted with Tris–acetate)
Eluant 1B: 50 mM acetate, (as eluant 1A), 1 M NaCl
Eluant 2B: 50 mM acetate, (as eluant 2A), 1 M NaCl
Eluant 4: 0.2 M HCl
Desalting solvent: 10 mM HCl

Materials

CM-cellulose (Whatman)
CM-Trisacryl M (LKB; Bromma, Sweden)
Sephadex G-25

Procedure. The CM matrix is stirred for 1 hr in EDTA, fines are removed through filtration on a Büchner funnel with a suitable nylon mesh, and washed successively with acid, water (until neutral), alkali, water. This is followed by equilibration with excess eluant 1A or 2A of the chosen pH. The matrix is then suspended in the same eluant, degassed under vacuum, and the column (1.5 × 10 cm) packed. Eluant 1A or 2A is then pumped through the column until inflow and eluate have identical pH. The protein sample (25 mg) is applied in all cases in eluant 2A containing 1% (v/v) mercaptoethanol. Two column volumes of either eluant 1A or 2A are pumped through the column before commencement

---

[34] A. Henschen and P. Edman, *Biochim. Biophys. Acta* **263**, 351 (1972).

of gradient elution. The gradient of increasing NaCl concentration is produced in a gradient former containing in the two 150-ml vessels either the eluant pair 1A and 1B or 2A and 2B. At the end of the gradient, eluant 4 is pumped through the column to elute the more basic H3, if present. The protein concentration in the eluate is monitored at 230 nm. Selected fractions are pooled and freeze-dried. These fractions are then desalted on a 2.5 × 30 cm Sephadex G-25 column with desalting solvent. Protein-containing fractions are freeze-dried again for further processing.

Histone Subunits from Various Organisms

Yeast Histones. Histones are extracted from yeast nucleoprotein with 0.25 M HCl. The extract is made 80% (v/v) with respect to acetone and precipitated overnight at −20°. Alternatively, the extract is dialyzed against distilled water at 4° and subsequently freeze-dried. Fractionation of total acid-extracted yeast histones on BioGel P-60 yields the H2A–H2B group and the histones H3 and H4 (Fig. 6). The latter is purified on Sephadex G-75 (Fig. 10). The H2A–H2B group is again subjected to BioGel P-60 exclusion chromatography, but now using eluant 1 (Fig. 8), yielding electrophoretically homogeneous H2A. The H2B fraction, however, proves to be heterogeneous. It is therefore more economical to subject the H2A–H2B pair, as eluted in the first BioGel P-60 run, to CM-cellulose ion-exchange chromatography. This yields two H2A and two H2B isohistones[15] (Fig. 11).

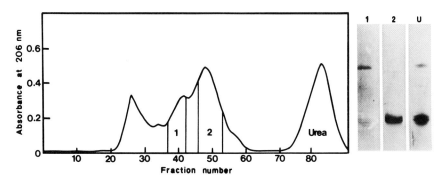

Fig. 10. Purification of yeast histone H4. Crude H4 eluted from BioGel P-60 with eluant 2 (Fig. 6) is recycled on Sephadex G-75 with the same eluant to result in the removal of an electrophoretically slow moving contaminant. Numbering of lanes corresponds to chromatographic fractions; U, crude fraction. [From W. F. Brandt, K. Patterson, and C. von Holt, *Eur. J. Biochem.* **110**, 67–76 (1980). With permission of the Federation of Biochemical Societies.]

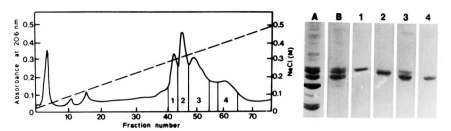

Fig. 11. Ion-exchange chromatography of the yeast H2A–H2B pair, eluted from BioGel P-60 (Fig. 6) on CM-cellulose. PAGE lanes 1–4 correspond to the respective chromatographic fractions; A, total yeast histones; B, H2A–H2B pair eluted from BioGel P-60 (Fig. 6). [From W. F. Brandt, K. Patterson, and C. von Holt, *Eur. J. Biochem.* **110,** 67–76 (1980). With permission of the Federation of European Biochemical Societies.]

Sea Urchin Sperm Histones. The procedure described has been used for the isolation of histones from *Parechinus angulosus.* Only limited experience is available for the histones from other echinoderms. Sperm nuclei are suspended for 1 hr at 4° in distilled water, 1 : 5 (v/v), and then acidified to 0.25 *M* HCl to extract the histones for 1 hr at 4°. After centrifugation, these are precipitated from the supernatant in 80% acetone at −20° or freeze-dried after dialysis against distilled water. The histones are dissolved in application solvent, applied to a BioGel P-60 column, and developed with eluant 1 (Fig. 6). The chromatographic fractions are combined, dialyzed, and freeze-dried.

The histones H2B can also be isolated by a modified Johns extraction procedure, the details of which have been described elsewhere.[4] This H2B-enriched fraction is further purified by exclusion chromatography on Bio-Gel P-60 (Fig. 6).

The crude histone H2B fraction isolated by either of these procedures is heterogeneous, containing several isohistones H2B. The nature and composition of this H2B subset differ in various investigated species.[35,36] The sperm chromatin from a single animal contains several isohistones H2B.[35] The histones H2B are further fractionated via ion-exchange chromatography (Fig. 12).

Histones H2A and H3 are isolated via BioGel P-60 exclusion chromatography[4,28] as homogeneous fractions. The histone H4 contains two isohistones exhibiting either cysteine or threonine at position 73.[33] The separation of these two isohistones via exclusion chromatography after dimerization is described above.

[35] C. von Holt, W. N. Strickland, W. F. Brandt, and M. Strickland, *FEBS Lett.* **100,** 201 (1979).
[36] M. Strickland, W. N. Strickland, and C. von Holt, *FEBS Lett.* **135,** 86 (1981).

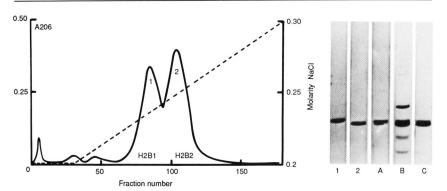

FIG. 12. Ion-exchange chromatography of sea urchin sperm H2B fraction eluted from BioGel P-60 (Fig. 6) on CM-cellulose (NaCl gradient in eluants 1A and 1B in 6 M urea). PAGE lanes 1 and 2 correspond to chromatographic fractions; lane A, artificial mixture of fraction 1 and 2 in the ratio 1:2; lane B, crude H2B; lane C, H2B eluted from BioGel P-60 (Fig. 6). [From M. Strickland, W. N. Strickland, W. F. Brandt, and C. von Holt, *Eur. J. Biochem.* **77**, 263–275 (1977). With permission of the Federation of European Biochemical Societies.]

The histone H1 is extracted with perchloric acid.[37] To this end, nucleo-protein is suspended in distilled water (1:5, w/v), allowed to stand for 1 hr at 4°, and adjusted under stirring to 10% (w/v) perchloric acid. The precipitate is pelleted. The H1 in the supernatant is converted, if required, to the chloride (see above), dialyzed, and freeze-dried. The H1 thus isolated from *P. angulosus* is electrophoretically homogeneous. Alternatively, the H1 can be isolated via BioGel P-60 exclusion chromatography (Fig. 6) from the total acid extract.

Sea Urchin Embryo Histones. Histones are extracted either as total histones from nuclei or nucleoprotein with 0.25 M HCl (see above) or as subfractions using Johns procedure.[38] Embryo chromatin contains a series of closely related isohistones of the H1, H2A, and H2B classes.[39,40] Therefore, the dimerization of H3 prior to exclusion chromatography simplifies the elution pattern from BioGel P-60. (For the dimerization of H3 see above.) The dialyzed and freeze-dried histone mixture still containing the H3 dimer is applied to BioGel P-60 without 2-mercaptoethanol in the application solvent and developed with eluant 1.

The H3 dimer separates partially from the H2A–H2B pair (Fig. 13).

[37] W. N. Strickland, H. Schaller, M. Strickland, and C. von Holt, *FEBS Lett.* **66**, 322 (1976).
[38] E. W. Johns, *Methods Cell Biol.* **16**, 183 (1977).
[39] W. F. Brandt, W. N. Strickland, M. Strickland, L. Carlisle, D. Woods, and C. von Holt, *Eur. J. Biochem.* **94**, 1 (1979).
[40] W. F. Brandt and C. von Holt, *Biochim. Biophys. Acta* **537**, 177 (1978).

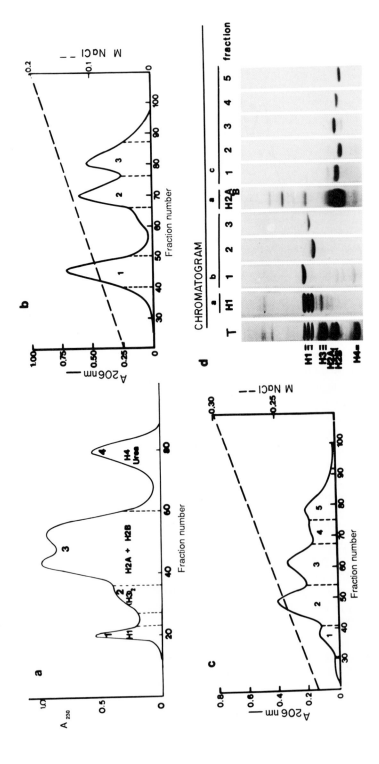

This H2A–H2B fraction is isolated, dialyzed, freeze-dried, and then subjected to CM-cellulose ion-exchange chromatography (Fig. 13). A number of H2A and H2B isohistones are separated from each other, which on sequencing, however, still reveal heterogeneity in some positions. The histones H1 recovered from the BioGel-P60 exclusion chromatogram (Fig. 13) are further fractionated via CM-cellulose ion-exchange chromatography to result in a number of isohistones (Fig. 13), some again still heterogeneous as revealed by sequencing.[41]

Wheat Germ Histones. Wheat germ histones constitute a very complex mixture of isohistones of the H1, H2A, and H2B classes in addition to the H3 and H4. Within each group, variations in both size and charge mean that comparatively elaborate separation procedures are required. A combination of exclusion chromatography, ion-exchange separation, and HPLC allows the isolation of some pure isohistones and, in addition, the isolation of mixtures of very closely related polypeptides.

Histones H3 and H4. A group separation is achieved on BioGel P-60 (Fig. 6). The H3 and H4 fractions are recycled through the same procedure and yield electrophoretically pure fractions; the former proved to be homogeneous on sequencing.[42]

Histones H1. The H1 fraction recovered from BioGel exclusion chromatography is subjected to CM-Trisacryl ion-exchange chromatography. Only poor separation into a number of isohistones is achieved[43] (Fig. 14).

The large histone variants $H2B_{(1)}$ and $H2B_{(2)}$. The large H2B variants recovered with fraction 2a after BioGel P-60 chromatography[44] (Fig. 6) are subjected to reversed-phase HPLC on Ultrapore RPSC C_3 (Beckman) employing an acetonitrile gradient containing 0.1% (v/v) heptafluorobu-

[41] P. de Groot, W. N. Strickland, W. F. Brandt, and C. von Holt, *Biochim. Biophys. Acta* **747**, 276 (1983).
[42] A. M. Modro, J. de A. Rodrigues, W. F. Brandt, and C. von Holt, *Biol. Chem. Hoppe Seyler* **369**, 193–197 (1988).
[43] W. F. Brandt and C. von Holt, *FEBS Lett.* **194**, 282 (1986).
[44] W. F. Brandt, J. de A. Rodrigues, and C. von Holt, *Eur. J. Biochem.* **173**, 547–554 (1988).

FIG. 13. Purification of sea urchin late gastrula histones. (a) Gel exclusion chromatography (BioGel-P60) of acid-extracted total histones after dimerization of histone H3. (b) Ion-exchange chromatography of the histone H1 fraction from a on CM-cellulose (NaCl gradient in eluants 1A and 1B, pH 5.5). (c) Ion-exchange chromatography of the H2A–H2B fraction from a on CM-cellulose (NaCl gradient in eluants 2A and 2B, pH 4.5). (d) PAGE was performed in acetic acid–urea gels (2.5 *M* urea). Lanes correspond to chromatographic fractions of the respective chromatograms; lane T, total gastrula histones. [From W. F. Brandt, W. N. Strickland, M. Strickland, L. Carlisle, D. Woods, and C. von Holt, *Eur. J. Biochem.* **94**, 1–10 (1979). With permission of the Federation of European Biochemical Societies.]

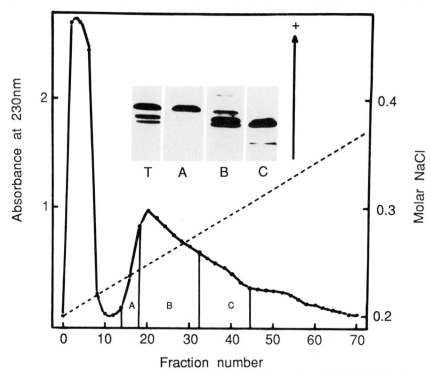

Fig. 14. Ion-exchange chromatography of wheat histone H1. The H1 fraction from exclusion chromatography (Fig. 6) was rechromatographed on CM-Trisacryl (NaCl gradient in eluants 2A and 2B, pH 5.3). PAGE was performed in SDS gels, with fractions moving anodically to the top of the slab. [From W. F. Brandt and C. von Holt, *FEBS Lett.* **194**, 282–286 (1986). With permission of the Federation of European Biochemical Societies.]

tyric acid (Fig. 15). The first run separates the large H2B isohistones from the corresponding H2A isohistones. The H2B fraction is recycled through the same procedure to resolve H2B$_{(1)}$ and H2B$_{(2)}$. The large H2A isohistones require further purification via CM-cellulose ion-exchange chromatography.

The large histone variants H2A$_{(2)}$ and H2A$_{(3)}$. The large histone H2A variants are separated only incompletely by HPLC. Ion-exchange chromatography of the gel exclusion chromatography fraction 2a (Fig. 6) resolves both H2A$_{(2)}$ and H2A$_{(3)}$[45] (Fig. 16). The final purification is achieved via rechromatography of the H2A$_{(2)}$ and H2A$_{(3)}$ fractions on BioGel P-60 with eluant 1.[45]

[45] J. de A. Rodrigues, W. F. Brandt, and C. von Holt, *Eur. J. Biochem.* **173**, 555–560 (1988).

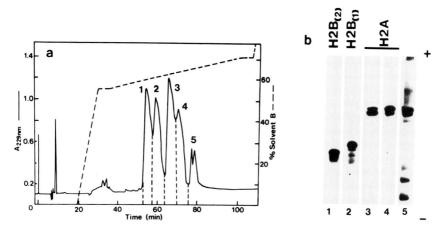

FIG. 15. Reversed-phase HPLC of wheat histones H2A and H2B. (a) Gel exclusion chromatography fraction 2a (Fig. 6) dissolved in eluant A containing 1% (v/v) heptafluorobutyric acid was fractionated on a Beckman Ultrapore RPSC 5-μm column. Eluant A, 0.1% (v/v) heptafluorobutyric acid in water; eluant B, 0.1% (v/v) heptafluorobutyric acid in 70% (v/v) acetonitrile; flow rate, 2 ml/min; pressure, 960 psi; temperature, ambient. (b) PAGE, acetic acid, 2.5 M urea–Triton X-100.

The small H2A and H2B variants. The fraction 2b from the BioGel P-60 run contains the small H2A and H2B histones. These are separated from each other on CM-cellulose ion-exchange chromatography[46] (Fig. 17). However, in themselves they are still heterogeneous. Partial separation of H2A$_{(1)}$ is achieved on subsequent gel exclusion chromatography on BioGel P-60 with eluant 1. The small H2B variants are recovered only as a mixture.

Preparative Polyacrylamide Gel Electrophoresis

Purification of single subunits can be achieved via gel electrophoresis when only small amounts of complex isohistone mixtures are available. The histone subunit complement of sea urchin gastrulas constitutes a complex mixture of basic proteins, which are only incompletely separated by elaborate and time-consuming column chromatographic techniques with comparatively large losses of material. The cause for such loss is often the need for repeated recycling and a succession of chromatographic operations in order to produce reasonably pure fractions. This may make it virtually impossible to recover enough subunit material for further investigations. Preparative gel electrophoresis of the Triton X-100 complexes of

[46] J. de A. Rodrigues, W. F. Brandt, and C. von Holt, *Eur. J. Biochem.* **150,** 499 (1985).

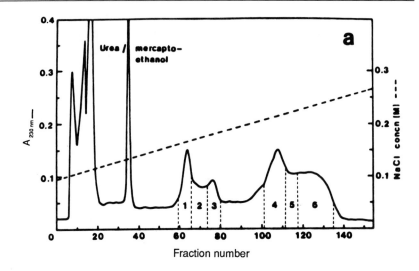

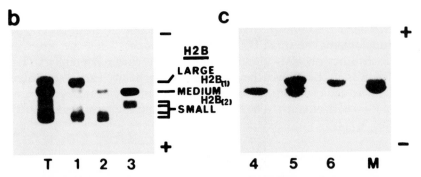

FIG. 16. Ion-exchange chromatography of wheat histone H2A and H2B variants. (a) Gel exclusion chromatography fraction 2a (Fig. 6) in sample application buffer was applied to a CM-cellulose column and developed with eluants 2A and 2B (pH 5.3, 6 M urea). (b) SDS-PAGE lanes 1–3 correspond to the chromatographic fractions and T to the starting material, fraction 2a (Fig. 6). (c) Acetic acid–urea PAGE with Triton X-100 at 7.6 M urea: lanes 4–6 correspond to the chromatographic fractions 4–6 and M to the mixture of fractions 4 and 6.

histones constitutes a simple and convenient method to isolate the less common isohistone subunits from a complex mixture.[47]

Solutions

Solutions for PAGE: see below (SDS–PAGE)
Formic acid 1: 70% (v/v) formic acid

[47] S. Schwager, W. F. Brandt, and C. von Holt, *Biochim. Biophys. Acta* **741,** 315 (1983).

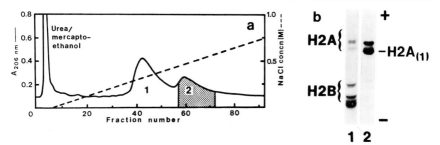

FIG. 17. Ion-exchange chromatography of wheat histone H2A and H2B variants. (a) Gel exclusion chromatography fraction 2b (Fig. 6) in sample application buffer was applied to a CM-cellulose column and developed with eluants 1A and 1B (pH 5.0, 6 M urea). (b) Acetic acid–urea PAGE with Triton X-100 at 7.6 M urea. [From J. de A. Rodrigues, W. F. Brandt, and C. von Holt, *Eur. J. Biochem.* **150**, 499–506 (1985). With permission of the Federation of European Biochemical Societies.]

Formic acid 2: 40% (v/v) formic acid
H_2SO_4: 0.4 N
Amido Black: 0.1% (w/v), in 7% (v/v) acetic acid
Destainer: 7% (v/v) acetic acid

Procedure. Gel casting is done as described below. Preparative gel slabs of $1.7 \times 120 \times 200$ mm are polymerized in front of a fluorescent light screen. With the given riboflavin concentration in the gels and the intensity of the light screen (four 40-W tubes) the polymerization is usually complete within 30 min.

From an analytical transverse urea gradient (0–9 M) electrophoresis, the optimal urea concentration is selected that produces sufficiently large differences in the mobility of histone subunit–Triton complexes to make electrophoretic separation possible (Fig. 18). Once this urea concentration has been established, preparative slab gels with that concentration are run either for the unfractionated histone subunits or mixtures enriched in particular subunits. The histone sample (10–15 mg) is dissolved in application buffer and incubated for 30 min at 30° to reduce disulfide bonds, and after adjusting the pH to 2 it is applied to the slot extending 100 mm across the slab. Electrophoresis is for 16–17 hr at 10 mA constant current. The gels are stained with Amido Black. The position of the desired subunits on the gel is identified through staining.

After destaining, fractions from four slabs are cut out from the slabs on an illumination box and homogenized with an Ultra Turrax in formic acid 1 and extracted by shaking at 5° overnight. The extract is separated from the gel particles via filtration (Whatman 1 filter paper), and the solution is passed through a Dowex 1-X8 (50–100 mesh) column (1 × 20 cm) equili-

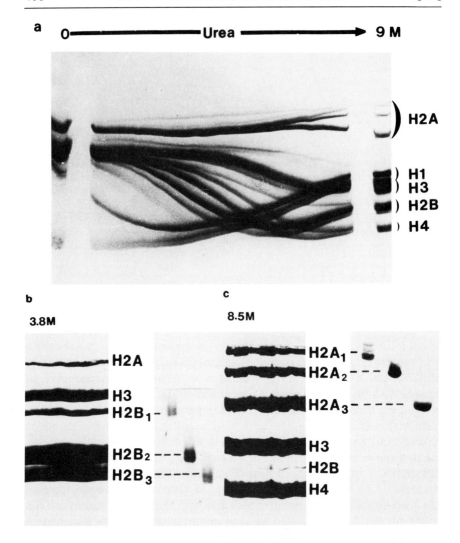

FIG. 18. Polyacrylamide gel electrophoresis of sea urchin embryo histones. (a) Total sea urchin embryo histones (*P. angulosus,* late gastrula) in a transverse urea gradient in a 6 m*M* Triton X-100 gel. Optimal separation for H2A and H2B isohistones is achieved at 8.5 and 3.8 *M*, respectively. (b) Preparative gel slab electrophoresis of an H2B-enriched fraction eluted from BioGel P-60. Histone, 1 mg/cm slot, was applied to a 3.8 *M* urea–Triton gel. Purified isohistones $H2B_1$, $H2B_2$, and $H2B_3$ are indicated next to the preparative slab. (c) Preparative slab gel electrophoresis of an H2A, H3, and H4 enriched fraction. Histone, 1 mg/cm slot was applied to a 8.5 *M* urea–Triton gel. Purified isohistones $H2A_1$, $H2A_2$, and $H2A_3$ are indicated next to the preparative slab. [From S. Schwager, W. F. Brandt, and C. von Holt, *Biochim. Biophys. Acta* **741**, 315–321 (1983).]

brated with formic acid 2.[48] The Amido Black is retained by the column, and the protein is eluted with formic acid 2. The eluate and the washings are combined and dialyzed against deionized water and freeze-dried.

Alternatively, bands containing the desired fractions can be electroeluted with high yields using the Biotrap device (Schleicher and Schüll, Dassel, FRG). The freeze-dried samples are taken up in 3 ml 0.02 N H_2SO_4 and precipitated with 6 volumes of acetone. The precipitate is subsequently washed 2–3 times with acetone, redissolved in water, and freeze-dried to result in a white powder. Approximately 50 mg of histones is processed on four slabs with an average yield of 35–40%.

Chemical Characterization of Histone Subunits

Determination of the Relative Molecular Weight (M_r) of Histone Subunits by Polyacrylamide Gel Electrophoresis

A first approximation of the M_r value is often done by SDS–PAGE using molecular weight standards falling in the range of histone subunits, in the same way that M_r is determined for many globular proteins. However, basic proteins bind SDS in an anomalous fashion[49] with the result that the thus determined M_r values are misleading. Even the use of histone subunits with known molecular weights as standards for the determination of the M_r value of an unknown isohistone subunit often does not resolve this dilemma (Fig. 19).

The relative molecular weight of subunits can, nevertheless, be determined with a fair degree of reliability in the absence of detergent provided that, under the electrophoretic conditions used, all intrinsic anionic charges of the polypeptide are completely eliminated and the subunit is fully unfolded.[50] Under such conditions, the mobility should be determined only by the size of the polypeptide chain, its positive charge, and the particular, although ill-defined, conditions in the polyacrylamide gel. If the mobility of the subunit H4 with known size and charge is used as a reference to account for the gel conditions, it can be shown that after correction of the observed mobility of any histone, in terms of the standard H4, the relationship between charge, size, and mobility holds (Table I [Ref. 51 cited therein] and Fig. 19). Thus the M_r value of a histone subunit with

[48] H. Wada and E. F. Snell, *Anal. Biochem.* **46**, 548 (1972).

[49] K. Weber and M. Osborn, *in* "The Proteins" H. Neurath and R. L. Hill, eds., pp. 179–223. Academic Press, New York, 1975.

[50] W. F. Brandt, B. T. Sewell, and C. von Holt, *FEBS Lett.* **194**, 273 (1986).

[51] M. O. Dayhoff, "Atlas of Protein Sequence and Structure," Vol. 5, Suppl. 3 (National Biomedical Research Foundation, Washington, D.C., 1978).

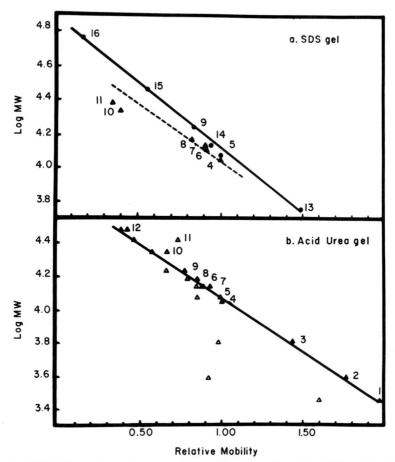

Fig. 19. Relation between electrophoretic mobility of proteins and peptides and their M_r values. (a) Mobility of histones (▲) and globular proteins (●) in the presence of SDS. The numbers correspond to proteins listed in Table I; 13, 14, 15, and 16 represent insulin, ribonuclease, carbonate dehydratase, and bovine serum albumin, respectively. (b) Mobility of proteins and peptides in 8 M urea–0.9 M acetic acid gels (15% acrylamide) [S. Panyim and R. Chalkley, *Arch. Biochem. Biophys.* **130**, 337–346 (1969)]. (△) Relative mobility (histone H4 = 1), (▲) relative mobility divided by the relative cationic charge using histone H4 as reference (see Table I). The proteins and peptides correspond to those given in Table I. [From W. F. Brandt, B. T. Sewell, and C. von Holt, *FEBS Lett.* **194**, 273–277 (1986). With permission of the Federation of European Biochemical Societies.]

unknown sequence can be determined with reasonable accuracy provided its amino acid composition has been established:

$$Mp \text{ (corrected)} = \frac{Mp \text{ (observed)}}{M(\text{H4})} \times \frac{B(\text{H4})}{Bp}$$

TABLE I
ELECTROPHORETIC MOBILITIES OF VARIOUS PROTEINS AND PEPTIDES IN
ACIDIC 8 M UREA GELS, THEIR M_r VALUES, AND POSITIVE CHARGE
DENSITIES

Protein	$M_r{}^a$	Cationic charge density[b]	Relative electrophoretic mobility[c]	
			Observed	Corrected[d]
1. Melittin	2,850	21.1	1.75	1.95
2. H2A peptide	3,900	12.7	0.94	1.76
3. H2B CN-2	6,840	17.5	1.07	1.41
4. H4	11,280	23.9	1.00	1.00
5. Cytochrome c	11,000	20.5	0.85	0.99
6. H2B	13,770	33.2	0.85	0.89
7. H2A	14,000	21.4	0.83	0.93
8. H3	15,340	22.2	0.79	0.85
9. Myoglobin	17,000	20.9	0.67	0.77
10. H1 (sea urchin sperm)	21,850	28.1	0.67	0.57
11. H1	26,500	37.7	0.71	0.45
12. (H3)$_2$	30,680	22.2	0.39	0.42

[a] M_r values have been taken from Dayhoff[51] and that for H2A peptide (tryptic core of wheat histone H2A)[16,46], H2B CN-2 (the carboxy-terminal CNBr fragment of wheat H2B$_{(1)}$),[35] and H1 and its CNBr peptides calculated from the sequence.[3] Histones used were isolated from calf thymus except where otherwise indicated.

[b] Cationic charge density was calculated from the amino acid composition, that is, the sum of the mol% Lys, His, Arg, and the N terminus divided by the sum of M_r values of the constituent amino acids (i.e., charge per 10,000 Da). This figure is in most proteins close to the mol% basic amino acids present.

[c] Electrophoretic mobility was measured relative to histone H4 in 15% acrylamide–8 M urea gels, pH 2.3.

[d] Corrected mobility was calculated by dividing the relative mobility by the cationic charge using the charge of histone H4 as a reference.

where Mp is the observed mobility of the protein, M(H4) the mobility of H4, B(H4) the mol% basic amino acids per 10,000 Da in H4, and Bp the mol% basic amino acids per 10,000 Da in protein X. From the thus corrected mobility, the M_r value is determined by Mp (corrected) $= A + B \log M_r X$ (see Table II).

The microblotting techniques described in below allow the determination of the amino acid composition of a gel electrophoretic fraction directly

TABLE II

ESTIMATES OF M_r FROM CORRECTED ELECTROPHORETIC MOBILITIES[a]

Protein	M_r	Cationic charge density	Relative electrophoretic mobility		M_r^b estimated	Error[c] (%)
			Observed	Corrected		
Protamine	4,300	48.0	2.70	1.35	6,900	+61.0
H1 CN-1 (sperm)	21,200	38.7	1.05	0.65	20,000	−5.6
H1 CN-2 (sperm)	5,300	34.0	1.92	1.34	7,000	+30.1
H5 (chicken)	21,600	34.4	0.75	0.52	24,300	+12.5
H2B$_3$ (sperm)	16,500	26.0	0.81	0.76	16,900	+2.4
H2B$_2$ (sperm)	16,000	25.6	0.83	0.78	16,400	+2.5
H3 CN-1 (chicken)	10,300	25.3	1.15	1.09	10,300	0.0
H2B$_1$ (sperm)	16,100	24.8	0.81	0.78	16,400	+1.9
H4 (chicken)	11,300	23.9	1.00	1.00	11,700	+3.5
H4 (acetyl)$_1$	11,400	23.1	0.97	1.00	11,700	+2.6
H4 (acetyl)$_2$[d]	11,400	22.2	0.95	1.02	11,400	+0.0
H2A$_1$ (wheat)	15,500	22.6	0.79	0.83	15,200	−1.9
H3 (chicken)	15,300	22.2	0.79	0.85	14,700	−3.9
Ribonuclease	13,700	16.7	0.70	1.00	11,700	−14.6
Carbonate dehydratase	29,200	12.3	0.36	0.65	20,000	−31.5
H3 (CN-2)$_2$ (chicken)	6,600	12.1	0.86	1.69	4,100	−37.8
Trypsinogen	24,000	8.8	0.35	0.95	12,700	−47.0

[a] Sperm histones have been isolated from the sea urchin *Parechinus angulosus*. CN, cyanogen bromide cleavage fragment. For details regarding columns 2–5, see footnotes *a–d* of Table I.

[b] The data in Table I fit the equation M_r (corrected) $= A + B \log M_r X$ with $A = 4.729$, $B = 0.66$ ($P^2 = 0.997$, maximum absolute residual $= 0.031$ by regression analysis). This equation was used to estimate the M_r values of various proteins.

[c] Error in the M_r determination is the difference between the estimated and the actual molecular weight, expressed as a percentage.

[d] Histone H4 is heterogeneous with respect to the number of ϵ-N-acetylated lysine residues present.

from the gel. For proteins in the size range of histone subunits, $1-2\ \mu g$ suffices to give a reasonably accurate approximation of the amino acid composition.

When the charge density of standards deviates considerably from that of the subunits under investigation, a significant error in the M_r determination occurs (Table II). In such cases, either more suitable standards have to be chosen or a useful correction factor can be derived from Fig. 20. PAGE for M_r determinations of histone subunits is performed in acetic acid–urea gels.

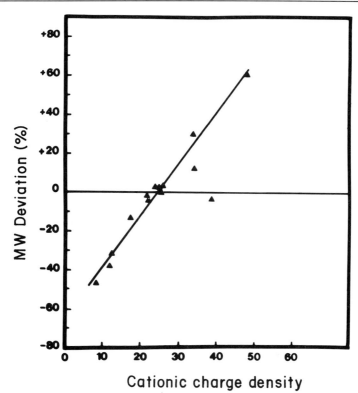

Cationic charge density

FIG. 20. Relation between the error in the estimated M_r and cationic charge density (see footnotes in Table I). The data were taken from Table II. The error in the M_r was calculated using the formula $[M_r \text{ (estimated)} - M_r]/M_r \times 100$. [From W. F. Brandt, B. T. Sewell, and C. von Holt, *FEBS Lett.* **194**, 273–277 (1986). With permission of the Federation of European Biochemical Societies.]

Determination of Electrophoretic Characteristics

SDS–PAGE.[52] The following solutions are prepared:

Acrylamide stock solution: 30% (w/v) acrylamide in H_2O, 0.8% (w/v) N,N′-methylenebisacrylamide in H_2O

Running gel buffer: 1.125 *M* Tris-HCl, pH 8.8, 0.3% (w/v) SDS

Stacking gel buffer: 0.375 *M* Tris-HCl, pH 6.8, 0.3% (w/v) SDS

Sample buffer: 62.5 m*M* Tris-HCl, pH 6.8, 2% (w/v) SDS, 10% (v/v) glycerol, 5% (v/v) 2-mercaptoethanol, 0.001% (w/v) bromphenol blue

Electrode buffer (10×): 0.25 *M* Tris-HCl, pH 8.3, 1.92 *M* glycine, 1% (w/v) SDS

[52] U. K. Laemmli, *Nature (London)* **227**, 680 (1970).

Persulfate: 10% (w/v) ammonium persulfate
TEMED
Staining solution: 0.25% Coomassie Brilliant Blue, 50% (v/v) methanol, 10% (v/v) acetic acid in water
Destaining solution: 25% (v/v) ethanol, 10% (v/v) acetic acid in water

The 15% separating gel ($120 \times 200 \times 1.5$ mm) is prepared by mixing the following solutions: 15 ml acrylamide stock solution, 10 ml running gel buffer, 20 μl TEMED, 4.68 ml water. Polymerization is initiated by adding 0.3 ml persulfate. The gel monomer solution is poured up to a level that leaves enough space for a 1.5-cm stacking gel. The gel is overlayered with water to ensure a straight interface. The solution is allowed to polymerize; setting time is about 15 min. The water is removed.

The 3% acrylamide stacking gel is made up from the following solutions: 1 ml acrylamide stock solution, 3.33 ml stacking gel buffer, 5.35 ml H_2O, and 20 μl TEMED. Polymerization is initiated by addition of 0.3 ml persulfate. The stacking gel is cast, a comb resulting in 0.5-cm wells is inserted, and the gel is allowed to polymerize.

The sample is dissolved at 2 mg/ml in the application buffer, incubated for 2 min in boiling water, and $10-20$ μg applied to the gel. Electrophoresis is carried out at 90 V constant voltage overnight, until the bromphenol blue has run off. The gels are stained in Coomassie Brilliant Blue and destained.

Acetic Acid–Urea Gels with[53]/without[54] Triton X-100, Transverse 0–9 M Urea Gradient. The following solutions are employed:

Sample buffer: 50 mM Tris-HCl, pH 8.5, 1% (v/v) 2-mercaptoethanol, 8 M urea, 0.001% (w/v) Pyronin Y (tracking dye)
Acrylamide: 15% (w/v) acrylamide, 0.1% (w/v) N,N-methylenebisacrylamide, 5.4% (v/v) acetic acid, 6 mM Triton X-100 [omitted for M_r determination (see above); for protein electroblotting Nonidet LE is substituted for Triton X-100], 0.5% (v/v) TEMED, 0.1% (v/v) thiodiglycol; urea is added to give appropriate molarity.
Riboflavin: 1 mg/ml
Electrolyte: 0.9 M acetic acid

For M_r determination, the Triton X-100 is omitted from the above mixture and urea included to give a molarity of 8 M.

For the casting of transverse urea gradients, the following procedure applies. A linear $0-9$ M urea slab gel is cast with the aid of a gradient former containing 8 ml each of the acrylamide and the acrylamide plus 9 M urea solutions. Riboflavin (50 μl) is added to each solution and mixed

[53] A. Zweidler, *Methods Cell Biol.* **17,** 223 (1978).
[54] S. Panyim and R. Chalkley, *Arch. Biochem. Biophys.* **130,** 337 (1969).

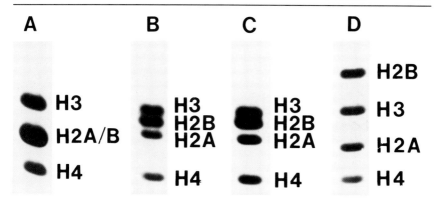

FIG. 21. SDS–PAGE of acid-extracted core histones from (A) sea urchin embryo, (B) calf thymus, (C) chicken erythrocytes, (D) sea urchin sperm.

well. The stirrer is activated, and simultaneously the stopcock is opened and the pump switched on. The polymerization mixture is pumped into the slab gel shell from the bottom starting with the 9 M urea acrylamide solution. Illumination is kept subdued to avoid polymerization while pouring. The pump delivery speed should not be higher than 1 ml/min to avoid turbulence with mixing of layers. After casting, the inlet tube is clamped off, and the gel placed in front of an illumination box containing four 40-W fluorescent tubes.

Polymerization takes about 15 min.

After setting, the gel is turned 90°, the spacer removed, and a wide sample well is poured across the gel using the acrylamide solution. The sample is dissolved in buffer and kept at room temperature for 30 min. Just before application, the pH is lowered to 2 with dilute HCl. One hundred microliters of sample solution (2 mg/ml) is applied to the gel. Acetic acid at 0.9 M is used as electrolyte. Electrophoresis is carried out until the Pyronin Y has run off completely (about 6–8 hr at 10 mA). The slabs are stained for a minimum of 3 hr in Coomassie Brilliant Blue and destained until the background is clear. Typical SDS (Fig. 21) and acetic acid–urea/Triton PAGE (Fig. 22) results, displaying the varying complexities arising from the presence of either postsynthetic modification or isohistones, are given below.

Blotting of Histones onto Glass Fiber Membranes

Histone subunits, separated on analytical acid–urea or acid–urea/ Triton PAGE gels, can be further characterized by determination of their amino acid composition.[55] In addition, if the terminal amino group is not

---

[55] W. F. Brandt and C. von Holt, in "Advanced Methods in Protein Microsequence Analysis" (B. Wittmann-Liebold et al., eds.), pp. 161–178. Springer-Verlag, Berlin, 1986.

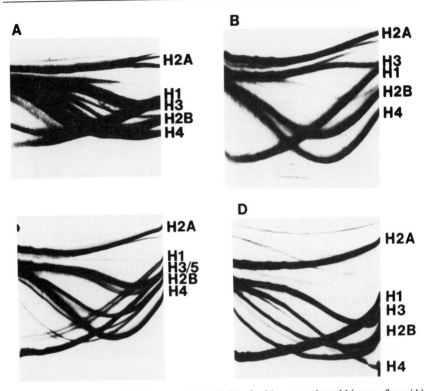

FIG. 22. Transverse 1–8 M urea gradient PAGE of acid-extracted total histones from (A) sea urchin embryo, (B) calf thymus, (C) chicken erythrocytes, (D) sea urchin sperm.

blocked, a number of sequential amino-terminal amino acid residue determinations can be accomplished to identify the amino-terminal sequence, or the fingerprinting pattern resulting from enzymatic or chemical fragmentation can be prepared.[55] To this end the PAGE gels are first blotted onto a glass fiber membrane for further processing, either for amino acid analysis or amino-terminal automated Edman degradation, by typical methods described elsewhere.[56] For sequencing purposes, Triton X-100 is replaced by Nonidet LE because the former contains amino-reactive contaminants blocking the amino termini of proteins.

Composition and sequence data are derived from 0.2–4 μg histones, that is, less than 1 nmol of protein. However, a single human fingerprint on a glass fiber membrane may contain nanomoles of amino acids/cm².[57]

[56] This series, Vol. 91, Part I, Section VI.
[57] P. B. Hamilton, *Nature (London)* **205**, 284 (1965).

If amino acids are present on the membranes, it is necessary to clean them, after which they should be handled only with gloves and tweezers.

Solutions

Membrane scrubbing solutions:
Trifluoroacetic acid: analytical grade
Hydrochloric acid: 10 M HCl
Protein stain (Coomassie): 0.2% (w/v) Coomassie Brilliant Blue, 25% (v/v) methanol, 9% (v/v) acetic acid (1–5 min staining)
Destainer: 25% (v/v) methanol
Electrolyte: 0.7% (v/v) acetic acid
Triton: 0.5% (w/v) Triton X-100 in electrolyte [for amino-terminal group determination and sequencing, Triton is replaced by 0.5% (w/v) Nonidet LE]

All aqueous solutions are made in double-distilled water.

Materials

Glass fiber membranes: Whatman GF/C or GF/F

Procedure. One of the following three methods will rid glass fiber membranes of contaminants: (1) pyrolysis at 500° for 5 hr; (2) Immersion in 50–100% trifluoroacetic acid for 1–5 hr, 22°; (3) Immersion in 10 M HCl, 1–5 hr, 50–60°. Following the wet treatment, the membranes are washed in double-distilled water until neutral and then dried at 100°.

The electroblotting equipment is essentially that described by Bittner *et al.*[58] Special attention has to be paid to the symmetry of the electrodes to achieve even transfer. The faithfulness of the transfer is equally improved by the careful sandwiching of the PAGE gel and the glass fiber membrane, avoiding the trapping of air bubbles. Sandwiching is best done by submerging the PAGE gel, membrane, and mounting gear in the transfer buffer. Electrotransfer with the given electrolyte is usually achieved at 4–7 V/cm within 1–4 hr.

Alternatively, histones can be electroblotted from gels onto glass fiber membranes sandwiched between graphite electrode plates at 5 V for 2 hr.[59] The layers of the sandwich are assembled in the following order: graphite anode, filter paper soaked in electrolyte, polyacrylamide gel, glass fiber membrane 1 and 2, filter paper soaked in electrolyte, graphite cathode. This sandwich is placed on crushed ice with the cathode facing the ice.

For underivatized membranes, the binding capacity is of the order of 20 μg protein/cm^2 or approximately 2 μg per band of 0.2 × 0.5 cm. If the

[58] M. Bittner, P. Kupferer, and C. F. Morris, *Anal. Biochem.* **102**, 459 (1980).
[59] J. Kyhse-Andersen, *J. Biochem. Biophys. Methods* **10**, 203 (1984).

primary PAGE gel contains more protein per band, two successively sand-wiched membranes will trap excess histones. Acid urea gels with up to 9 M urea with or without Triton X-100 or Nonidet LE in the gel are first conditioned in electrolyte for 1 hr and then electroblotted toward the cathode with the same solution within.

Staining of membranes is done with Coomassie Brilliant Blue (Fig. 23). The histone blots can be used in a variety of protein chemical procedures, such as determination of the amino acid composition, chemical or enzy-matic fragmentation to establish peptide maps, and, ultimately, amino acid sequence determination of peptides thus produced or of the unfrag-mented histone or isohistone. Detailed description of these procedures is beyond the scope of this chapter (see Ref. 56).

For amino acid analysis, the stained bands are excised from the fiber membranes and hydrolyzed directly, or the stained protein is eluted with

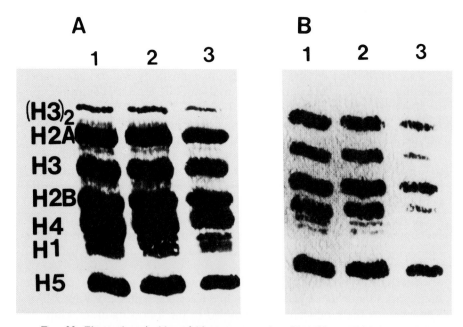

FIG. 23. Electrophoretic blot of histones onto glass fiber filters. Chicken erythrocyte histones were separated on polyacrylamide slab gels (see text) under the following conditions: urea, 3.8 M; 0.5% w/v Nonidet LE to replace Triton X-100 in the acrylamide gel. The protein bands were electroblotted to the glass fiber filters (GF/C) using graphite electrodes. The protein was stained with Coomassie Brilliant Blue. The amount of histone applied to the gel was 60 (lane 1), 40 (2), and 30 μg (3), respectively. (A and B) First and second glass filter layer, respectively. The second layer becomes necessary to trap histones escaping through the first membrane.

TABLE III
AMINO COMPOSITION OF HISTONES RECOVERED BY ELECTROBLOTTING FROM 3.8 M UREA–NONIDET LE POLYACRYLAMIDE GELS

	\multicolumn{12}{c}{Amount (mol%)}
Band:	\multicolumn{2}{c}{1}	\multicolumn{2}{c}{2}	\multicolumn{2}{c}{3}	\multicolumn{2}{c}{4}	\multicolumn{2}{c}{5}	\multicolumn{2}{c}{6}
Histone:	\multicolumn{2}{c}{H2A}	\multicolumn{2}{c}{H3}	\multicolumn{2}{c}{H2B}	\multicolumn{2}{c}{H4}	\multicolumn{2}{c}{H1}	\multicolumn{2}{c}{H5}
Asp	7.1	(6.9)[c]	4.5	(4.1)	5.2	(4.8)	5.4	(4.9)	2.1	(2.5)	2.4	(1.7)
Thr	2.7	(3.8)	6.4	(6.8)	5.9	(6.4)	5.9	(6.9)	3.5	(5.6)	2.1	(3.2)
Ser	2.7	(3.0)	3.2	(3.8)	6.4	(11.2)	1.6	(2.0)	5.0	(5.6)	7.9	(11.9)
Glu	9.3	(9.2)	13.1	(11.3)	8.9	(8.0)	7.3	(5.9)	5.0	(3.7)	5.5	(4.3)
Pro[a]	0.0	(3.8)	0.0	(4.5)	0.0	(4.8)	0.0	(1.0)	0.0	(9.2)	ND[d]	(4.7)
Gly	11.2	(10.7)	6.4	(5.3)	7.6	(5.6)	17.5	(16.7)	7.1	(7.2)	4.7	(5.3)
Ala	14.4	(13.0)	14.7	(13.5)	10.1	(10.4)	8.2	(6.9)	25.5	(24.3)	15.7	(16.3)
Cys[b]	0.0		0.0	(1.5)	0.0		0.0		0.0		0.0	
Val	5.5	(6.1)	4.8	(4.5)	6.9	(7.2)	9.1	(8.8)	4.3	(5.4)	3.9	(4.2)
Met	0.0		0.6	(1.5)	0.5	(1.5)	0.5	(1.0)	0.0		0.0	(0.4)
Ile	4.1	(4.6)	5.1	(5.3)	4.9	(4.8)	5.2	(5.9)	0.7	(1.5)	2.4	(3.2)
Leu	11.2	(12.2)	8.3	(9.0)	4.7	(4.4)	7.3	(7.8)	3.5	(4.5)	3.9	(4.7)
Tyr[b]	0.0	(2.3)	0.0	(2.3)	0.0	(4.0)	0.0	(3.9)	0.0	(0.9)	0.0	(1.2)
Phe	0.8	(0.8)	1.9	(3.0)	1.0	(1.6)	1.4	(2.0)	0.7	(0.9)	0.2	(0.6)
His[b]	0.0	(3.0)	0.0	(1.5)	0.0	(2.4)	0.0	(2.0)	0.0		0.0	(1.9)
Lys	9.8	(11.4)	7.3	(9.7)	14.0	(16.0)	8.8	(10.8)	19.0	(26.5)	18.1	(23.6)
Arg	8.0	(9.2)	10.6	(13.5)	5.5	(6.4)	12.0	(13.7)	1.0	(1.8)	7.9	(12.4)

[a] No hypochlorite was added.

[b] Probably destroyed by oxidation during hydrolysis. Coomassie stained bands (chicken erythrocyte histones) containing about 2 μg protein were eluted with trifluoroacetic acid into hydrolysis tubes and hydrolyzed in 6 M HCl–1% phenol at 150° for 90 min. No correction for destruction or incomplete hydrolysis was made.

[c] Values in parentheses have been calculated from the sequence or are from Ref. 38.

[d] Not determined.

trifluoroacetic acid into the hydrolysis tube. The eluate is dried in a stream of nitrogen. Hydrolysis is best done by the gas-phase procedure[60] and amino acid analysis by an OPA-postcolumn procedure, described elsewhere.[61] Typical amino acid composition data can be derived from as little as 2 μg histone (Table III). Identification of a histone by its amino-terminal sequence is possible with 2 μg histone, an amount which is carried per band in only a slightly overloaded gel (Fig. 24).

[60] B. A. Bidlingmeyer, S. A. Cohen, and T. L. Tarwin, *J. Chromatogr.* **336,** 93 (1984).
[61] D. G. Klapper, *in* "Methods in Sequence Analysis" (M. Elzinga, ed.), pp. 509–515. Humana Press, Clifton, New Jersey, 1982.

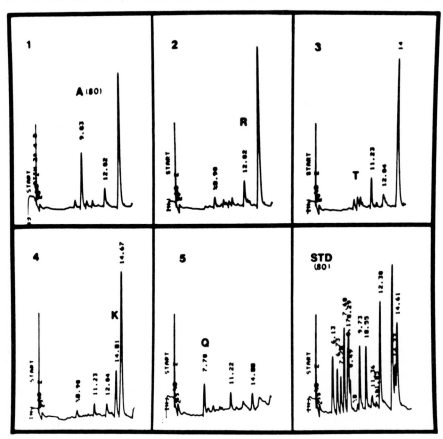

Fig. 24. Amino-terminal sequence analysis of histone H3 from a glass filter blot. A glass fiber disk suitably sized for the gas-phase sequencer cartridge was cut from a filter blot (Fig. 23). The disk contained 4 μg protein. About 50% of the phenylthiohydantoin (PTH) amino acid of each degradation cycle was used for HPLC analysis of the PTH derivatives. The yield of alanine in step 1 was 60%; internal standard, 500 pmol norleucine. The HPLC chromatograms[55] of the first 5 cycles (1–5) correspond to the sequence Ala-Arg-Thr-Lys-Gln. No polybrene was added to the disk. (With permission of Springer Verlag.)

Preparation of Histone Derivatives

The investigation of subunit interaction within the octamer or of the octamer with its DNA, as well as crystallographic explorations of the octamer or the core particle at the X-ray crystallographic and electron microscopic levels, should be greatly aided by the availability of well-defined octamer or core derivatives. Such derivatives have been prepared by derivatizing easily accessible amino acid side chains, on the intact or temporarily denatured core particle, with side group-specific reagents.[62,63] In this section, an alternative route to the production of octamer derivatives is described. The following protocols detail the methodology to prepare well-defined derivatives of octamer subunits from which, together with unlabeled complementary subunits and by using the appropriate methodology, one can reconstruct unequivocally characterized derivatives of the octamer. Such octamers can subsequently be assembled on random or unique DNA to core particles (see below).

Fluorescent Labeling of Chicken Erythrocyte Histone H3 and Sea Urchin Sperm Histone H4

Solutions

Labeling buffer: 0.1 M Tris-HCl, pH 8.3, 4 M urea
Dialysis solution: 0.25 M HCl
2-Mercaptoethanol: analytical grade

Materials

Iodoacetylfluorescein (IAF)
N-Iodoacetyl-N-(5-sulfo-1-naphthyl)ethylenediamine (IAEDANS)
Ultrafiltration membrane (Amicon PM10 or equivalent)

Procedure. Chicken erythrocyte histone H3 eluted from a BioGel P-60 column, together with H2B and some H2A, requires further purification prior to fluorescent labeling. This purification involving dimerization of the cysteinyl residues in H3 is described above. Fluorescent labeling of chicken erythrocyte histone H3[64] by carboxymethylation with N-iodoacetyl-N-(5-sulfo-1-naphthyl)ethylenediamine (IAEDANS) is carried out in labeling buffer. One mole of label in the same buffer is added per mole cysteine, and the reaction is allowed to proceed at room temperature for 20 min in the dark. The reaction is quenched by the addition of excess (100 mol/mol) mercaptoethanol. The labeled protein is dialyzed against 0.25 M HCl followed by dialysis against water, both at 4°. Finally, the protein is recovered by ultrafiltration (Amicon PM10 membrane). The labeling of chicken erythrocyte histone H3 with IAEDANS occurs under these conditions stoichiometrically and specifically at the cysteine residue. This is

[62] A. E. Dieterich, R. Axel, and C. R. Cantor, *Cold Spring Harbor Symp. Quant. Biol.* **42,** 199 (1977).
[63] A. E. Dieterich, R. Axel, and C. R. Cantor, *J. Mol. Biol.* **129,** 587 (1979).
[64] G. G. Lindsey, P. Thompson, L. Pretorius, and C. von Holt, *FEBS Lett.* **192,** 230 (1985).

shown by the inability of the labeled protein to dimerize after o-iodosoben-zoate treatment.[64]

Since labeling the cysteine residues of other histones, for example, H4 with different probes (see below), may be neither stoichiometric nor specific, labeling must be followed by dimerization with o-iodosobenzoate and Sephadex G-100 chromatography (eluant 2, see above) to remove any unlabeled or nonspecifically labeled histone. The latter two will dimerize and elute earlier than the desired derivative.

Fluorescent labeling of the cysteine-containing isohistone H4 from sea urchin sperm (see above) by carboxymethylation with 5-iodoacetylfluorescein (IAF) is carried out as described for IAEDANS labeling of chicken erythrocyte histone H3. Dimerization and Sephadex G-100 chromatography (eluant 2, see above) are required after labeling to separate the cysteine-labeled histone H4 from unlabeled and nonspecifically labeled histone H4. The reason for the increased nonspecific labeling as compared to IAEDANS is probably due to the high hydrophobicity of the fluorescein probe leading to multiple nonspecific interactions between probe and protein.

Gold Labeling of Histone H2A–H2B Dimers

Those amino groups that are easily accessible in the octamer can be labeled with a carboxyl group carrying probe after carbodiimide activation. From such labeled octamers are isolated labeled dimers. The latter, together with a complementary unlabeled tetramer, are then reconstructed to an octamer labeled in its dimer component.

Gold-labeled dimer can be inserted into a reconstituted octamer to achieve heavy metal labeling for crystallographic and electron microscopic purposes.[21] The procedure described yields dimers with approximately three aurothiomalate molecules per dimer.

Solutions

Aurothiomalate activation buffer: 20 mM sodium phosphate, pH 5.0
Histone octamer coupling buffer: 2 M NaCl, 0.2 M sodium phosphate, pH 7.4, 0.2 mM PMSF (from stock solution, see below)
Eluant: 2 M NaCl, 10 mM sodium phosphate, pH 5.5, 0.1 mM PMSF

Materials

Sephadex G-100
Aurothiomalate
1-Ethyl-3-(3-dimethylamino)propylcarbodiimide hydrochloride

Procedure. Activation is carried out by the method of Davis and Preston.[65] Equimolar (1,2-dicarboxyethylthio)gold, disodium salt (aurothiomalate), and 1-ethyl-3-(3-dimethylaminopropyl)carbodiimide hydrochloride (EDC) in aurothiomalate activation buffer are mixed with reactants at a final concentration of 50 mM and activation is allowed to proceed for 90 sec. An appropriate volume of the activation reaction mixture is added to the octamer solution (1 mg/ml in coupling buffer) to yield a final concentration of 6 mM for aurothiomalate. The reaction is allowed to proceed for 16 hr with constant stirring at room temperature after which the unreacted aurothiomalate is removed by exclusion chromatography on Sephadex G-50. The pH of the eluant causes the dissociation of octamers to H2A–H2B dimers and H3–H4 tetramers. The histone fraction is pooled, concentrated by ultrafiltration (Amicon PM10 membrane) to 14 mg/ml, and loaded onto a Sephadex G-100 column and eluted with eluant. The dimer fraction is pooled and used for reconstitution of labeled octamer.[21]

Preparation of Cys[110]-Dethio-Histone H3 (Ala[110]-Histone H3)[66]

Cys[110]-dethio-histone H3 can be inserted into a reconstituted octamer hybrid prepared from chicken erythrocyte H2A,H2B and the cysteine-containing isohistone H4 from sea urchin sperm[32] in order to prepare an octamer with thiol groups in H4 only.[66]

Solutions

Nickel acetate: 1% (w/v) in water
Histone H3: 7 mg/ml in 6 M guanidinium chloride, pH 7.0, adjusted with 0.2 M NaOH

Procedure. One gram of sodium borohydride is added to 600 ml nickel acetate solution at room temperature. The black amorphous precipitate of Raney nickel is collected by filtration on Whatman No. 1 paper and washed with 4 liters distilled water. Care must be taken not to allow the precipitate to become dry. The entire washed precipitate is then added to 4 ml of the histone H3 solution. The reaction proceeds for 72 hr at room temperature under H$_2$ using a pH-Stat (0.2 M HCl in the syringe) to maintain pH 7.0. Fresh Raney nickel is added every 12–16 hr. The reaction is terminated by centrifugation (7000 g, 30 min). The dethio-histone H3 in the supernatant is dialyzed, first against 0.25 M HCl and then against water. This solution is subjected to the dimerization procedure

[65] M. T. B. Davis and J. F. Preston, *Anal. Biochem.* **116,** 402 (1981).
[66] H. J. Greyling, B. T. Sewell, and C. von Holt, *Eur. J. Biochem.* **171,** 721–726 (1987).

described above to remove unconverted H3 via exclusion chromatography. The purified Cys[110]-dethio-histone H3 is used for reconstitution, yielding crystalline tubes, by the methodology described in the next section.

Reconstitution of Natural and Hybrid Octamers

Reconstitution of the histone octamer may be attempted by two general methodological approaches.[20] First, the ability of the octamer to reassociate from its natural products of dissociation, namely, the H2A–H2B dimer and H3–H4 tetramer, under suitable conditions provides a general strategy for the reconstitution of octamers from salt-extracted, undenatured histone complexes. However, when it is desirable to reconstitute from individually purified histones, an alternative strategy is used. Procedures employed to purify individual histones require denaturing conditions such as high urea concentrations and dilute hydrochloric acid (see Purification of Histone Subunits). Therefore, experimental procedures for the reconstitution of octamers from histones denatured during isolation constitutes the second approach.

Octamer Reconstitution from Salt-Extracted Complexes and the Modified H2A–H2B Dimer

Octamer reconstitution from undenatured products of octamer dissociation requires an increase in ionic strength and/or an increase in pH to yield the solution conditions under which octamer formation is favored. Reconstituted octamers are then purified by Sephadex G-100 chromatography to yield the octamer free of contaminating H1, H5, and excess H2A–H2B dimers.[20,21]

Solutions

Dialysis buffer: 2 M NaCl, 10 mM Tris-HCl, pH 7.4, 0.2 mM PMSF (from stock solution)
Sephadex G-100 buffer: 2 M NaCl, 10 mM Tris-HCl, pH 7.4, 0.1 mM PMSF
PMSF stock solution: 100 mM PMSF in DMSO

Materials

Ultrafiltration membrane (Amicon PM10 or equivalent)

Procedure. For octamer reconstitution from histone complexes isolated by hydroxyapatite chromatography (see Isolation of Natural Histone Complexes), the core histone fractions are pooled and concentrated by ultrafiltration to 10–12 mg/ml and dialyzed for 24 hr against dialysis buffer before application to the Sephadex G-100 column, equilibrated and eluted with the Sephadex G-100 buffer. For octamer reconstitution from tetramers and dimers, produced by the protamine displacement procedure (see above), the fractions containing the two histone complexes are combined and dialyzed against dialysis buffer for 24 hr at 4°. The dialyzed sample is concentrated by ultrafiltration, and Sephadex G-100 chromatography is carried out, as above. Reconstitution from modified H2A–H2B dimers is carried out by subjecting stoichiometric amounts of modified dimer, prepared as described in Preparation of Histone Derivatives, together with natural tetramer to the same reconstitution procedure.

Octamer Reconstitution from Acid-Extracted Individual Subunits

The reconstitution of octamers from individually purified chicken erythrocyte histones requires the initial disaggregation of the individual polypeptides before reconstitution. The formation of nonspecific aggregates during reconstitution is minimized if the histones are dissolved first in high concentrations of urea at protein concentrations not exceeding 0.5 mg/ml. This procedure presumably results in the total unfolding and disaggregation of the polypeptide chains before subsequent renaturation and complex formation during dialysis against 2 M NaCl.[20,64]

Octamer reconstitution from individually purified histones, with the exception of H3, can be performed with protein samples stored at −20° as freeze-dried fractions after isolation.[20] Freeze-drying of H3 causes its structure to become permanently altered,[20,67] and this polypeptide must therefore be used either immediately after purification or storage in the presence of 50% (v/v) glycerol at −20° [final concentrations of the storage solution for H3 : 5 mM HCl, 1% (v/v) 2-mercaptoethanol, 50% (v/v) glycerol].

Solutions

Histone solvent: 8 M urea, 0.25 M HCl, 1% (v/v) 2-mercaptoethanol
Dialysis buffer: 2 M NaCl, 10 mM Tris-HCl, pH 7.4, 0.2 mM PMSF (from stock solution)
Elution buffer: 2 M NaCl, 10 mM Tris-HCl, pH 7.4, 0.1 mM PMSF
PMSF stock solution: 100 mM PMSF in DMSO

Materials

Sepharose 6B
Ultrafiltration membrane (Amicon PM10 or equivalent)

[67] G. G. Lindsey, P. Thompson, L. Pretorius, L. R. Purves, and C. von Holt, *FEBS. Lett.* **155**, 301 (1983).

Procedure. For octamer reconstitution, freeze-dried histones H2A, H2B, and H4 are dissolved separately (0.5 mg/ml) in freshly prepared histone solvent. Histone H3 is brought to a concentration of 0.5 mg/ml in the same solution by ultrafiltration. The solutions are allowed to stand for 16–24 hr at 4°, and then equal amounts are mixed before dialysis against dialysis buffer for 24 hr at 4°. The sample is concentrated to 10–12 mg/ml by ultrafiltration as before and applied to a Sepharose 6B column equilibrated and eluted with elution buffer. From this eluate, the octamer is recovered (Fig. 25).

The reconstitution of hybrid histone octamers, that is, octamers in which a structural variant has replaced one erythrocyte histone, is executed by the same methodology. The reconstitution of hybrid octamers from stoichiometric amounts of (1) sea urchin sperm $H2B_1$ and erythrocyte H3, H2A, and H4 and (2), sea urchin sperm H4 and Cys^{110}-dethio-histone H3 and erythrocyte H2A and H2B has been performed by this methodology.[66] Sperm H4, erythrocyte H3, and Cys^{110}-dethio-histone H3 (see above) are not to be freeze-dried during isolation and are made 0.5 mg/ml in histone solvent by ultrafiltration prior to reconstitution.

Confirmation of Octameric Structure

The octameric structure of the reconstitutes is confirmed by cross-linking with dimethyl suberimidate and by the crystallisation of the putative octamer to the helical tubes described by Klug *et al.*[68]

Cross-Linking with Dimethyl Suberimidate. Dimethyl suberimidate cross-links proteins via the ε-amino groups of the lysyl residues. The cross-linked products are analyzed by SDS–PAGE. If an octamer is present in solution, the cross-linking proceeds rapidly through the di-, tetra-, and hexamer to the octamer, with a distinct pause at this level (Fig. 26). Further cross-linking then proceeds to the 16-mer followed by higher multiples of the octamer, which, however, no longer enter the polyacrylamide gel. Since the rate of the cross-linking reaction depends on the local concentration of the nucleophilic amino group on the protein surface, subtle differences in the nature of the octamer may be observed by comparison of the cross-linking kinetics at different pH values (Fig. 27).

The following solutions are required:

Dialysis buffer A: 2 *M* NaCl, 50 m*M* sodium borate, pH 9.0, 8.5, or 8.0

---

[68] A. Klug, D. Rhodes, J. Smith, J. T. Finch, and J. O. Thomas, *Nature (London)* **287,** 509 (1980).

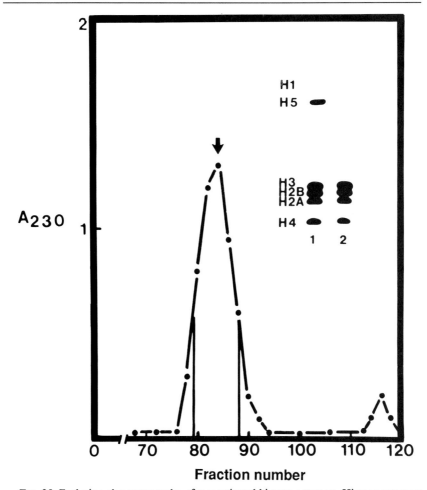

FIG. 25. Exclusion chromatography of reconstituted histone octamers. Histone octamers were reconstituted from individually purified histones H2A, H2B, H3, and H4. Column, Sepharose 6B, 2.5 × 90 cm; eluant, elution buffer; fraction volume, 3 ml; pressure head, 60 cm. SDS–PAGE slab: lane 1, histone standard; 2, octamer fraction. Arrow denotes elution volume of natural octamer. [From H. J. Greyling, S. Schwager, B. T. Sewell, and C. von Holt, *Eur. J. Biochem.* **137**, 221–226 (1983). With permission of the Federation of European Biochemical Societies.]

Cross-linking buffer: 2 *M* NaCl, 250 m*M* sodium borate, pH 9.8, 9.25, or 8.8

Reagent A: 10 mg dimethyl suberimidate-HCl/ml of cross-linking buffer yielding a final pH of 9.0, 8.5, and 8.0 (prepared immediately prior to use)

Glycine: 250 m*M*

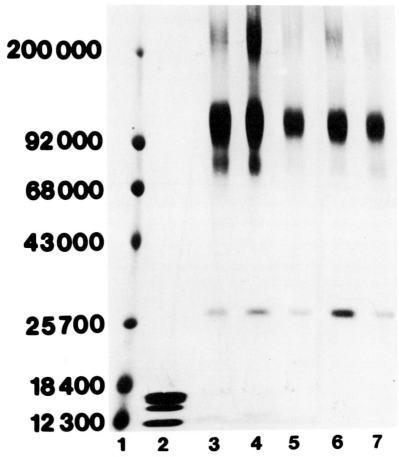

FIG. 26. SDS–PAGE of natural and reconstituted histone octamers after cross-linking with dimethyl suberimidate. Molecular weight standards (lane 1) in order of decreasing electrophoretic mobility are as follows: cytochrome c, β-lactoglobulin, α-chymotrypsinogen, ovalbumin, bovine serum albumin, phosphorylase b, and heavy chain of myosin. Lane 2, core histones; lane 3, hydroxyapatite-eluted octamer; lane 4, natural octamer (ultracentrifuged); lane 5, octamer reconstituted from acid-extracted histones; lane 6, octamer reconstituted from individually purified core histones; lane 7, octamer reconstituted from protamine-displaced histones. [From H. J. Greyling, S. Schwager, B. T. Sewell, and C. von Holt, *Eur. J. Biochem.* **137**, 221–226 (1983). With permission of the Federation of European Biochemical Societies.]

Histone octamer, which may or may not have been stored in 50% (v/v) glycerol, is dialyzed against 100 volumes of dialysis buffer of the desired pH. After dialysis the protein concentration is adjusted to 2 mg/ml. Equal volumes of the protein solution and the reagent are mixed at room temperature to give an octamer to reagent A ratio of 1 : 5 (w/w), and the reaction is

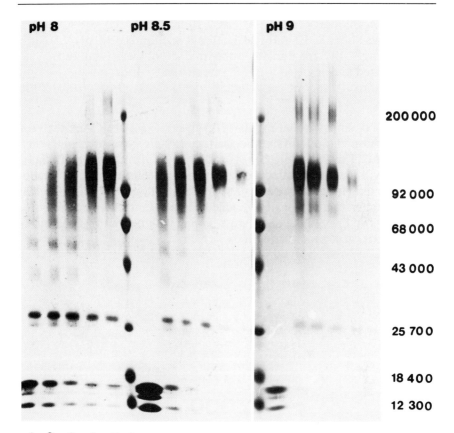

FIG. 27. SDS–PAGE of natural octamers after cross-linking with dimethyl suberimidate at different pH. Cross-linking time: 1 min (lanes 1, 8, 15); 3 min (lanes 2, 9, 16); 5 min (lanes 3, 10, 17); 10 min (lanes 4, 11, 18); and 15 min (lanes 5, 12, 19). Lanes 6 and 13 contain molecular weight standards in order of decreasing electrophoretic mobility: cytochrome c, β-lactoglobulin, α-chymotrypsinogen, ovalbumin, bovine serum albumin, phosphorylase b, and heavy chain of myosin. Lanes 7 and 14 contain standard core histones.

allowed to proceed for 1, 3, 5, 10, and 15 min. On mixing of the reagent with the cross-linking buffer, the pH changes to 9.0, 8.5, and 8.0, respectively. The reaction is stopped at the appropriate time by the addition of glycine to a final concentration of 25 mM to quench excess reagent. The reaction mixture is chilled on ice and dialyzed at 4° overnight against distilled water.

To cross-link poly(glutamic acid)-stabilized octamers, the protocol is different. Prepare the following solutions:

Sodium chloride–Tris: 0.2 M NaCl, 10 mM Tris-HCl, pH 8.0
Dialysis buffer B: 10 mM NaCl, 50 mM sodium borate, pH 9.5
Reagent B: 12.5 mg dimethyl suberimidate/ml dialysis buffer B
  (prepared and adjusted to pH 9.5 immediately prior to use)
Poly(Glu): 10 mg/ml poly(glutamic acid) (MW 50,000–100,000) in
  50 mM Tris-HCl, pH 8.0

Histone octamers stored in 50% (v/v) glycerol–1 M NaCl are dialyzed against sodium chloride–Tris, and the final protein concentration is then adjusted to 2.5 mg/ml. Poly(Glu) is added in 20-μl increments. Initially, the solution turns turbid after each addition but clears on mixing. Gradually the increments are increased until a volume of poly(Glu) equal to that of the octamer solution has been added. This solution is then ready for dialysis against dialysis buffer for cross-linking with reagent B. The ratio of octamer to reagent should be 1 : 10 (w/w) (Fig. 28).

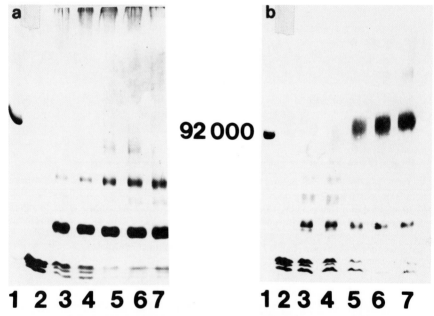

FIG. 28. SDS–PAGE of poly(glutamic acid)-stabilized histone octamers. SDS–PAGE was performed on natural octamers stabilized with poly(glutamic acid) at 10 mM NaCl, pH 9.5, and cross-linked with dimethyl suberimidate. (a) Controls without polyglutamic acid. (b) Octamer stabilized with poly(glutamic acid). Cross-linking times: 1 min (lane 3); 5 min (lane 4); 10 min (lane 5); 30 min (lane 6); and 60 min (lane 7). Lane 1, molecular weight standard phosphorylase b (MW 92,000); lane 2, core histones. Note that the linear gradient does not normally resolve the core histones.

For analysis by SDS–PAGE, equal volumes of the dialyzed reaction mixture and twice-concentrated SDS application buffer are mixed and applied to polyacrylamide gradient slabs (250 × 125 × 1.5 mm) containing a 5–20% linear vertical gradient of polyacrylamide running gel with an acrylamide:bisacrylamide ratio of 30:0.8. The stacking gel is 3%. Gels are run vertically at constant voltage of 100 V for 16 hr and stained with Coomassie Brilliant Blue. For other electrophoresis conditions, see above.

Crystallization of the Octamer. In assessing whether a faithful reconstitution has occurred, the severest test is that the crystal form of the reconstituted octamers should be the same as that of native octamers, particularly in the case of a multisubunit aggregate where the effect of a small perturbation of the structure may be expected to be magnified in the crystal. A crystallization procedure is described which leads to the formation of microscopic crystalline tubes first described by Klug et al.[68] Prepare the following solutions:

Solution 1: 2 *M* NaCl, 100 m*M* Tris-HCl, pH 7.7, 20% saturated ammonium sulfate at room temperature
Solution 2: same as solution 1 but 45% saturated ammonium sulfate at 4°

For crystallization, pure natural or reconstituted octamers at a concentration of 2–3 mg/ml are dialyzed for 16 hr at room temperature (22°) against solution 1. The dialyzed solution is clarified by centrifugation (2000 *g*, 2 min). The clarified sample is then dialyzed against solution 2 at 4°. Within 4 days, a shiny, birefringent precipitate forms, which contains the crystallization tubes.

For electron microscopy, prepare the specimens by diluting droplets of the tube-containing precipitate with an equal volume of 0.4% (w/v) freshly prepared uranyl acetate solution and place on a sheet of Parafilm. A carbon-coated electron microscope grid is suspended, carbon film side down, on the droplet. These steps are performed as quickly as possible. After 1 min, the grid is blotted with Whatman 542 filter paper and then resuspended, film side down, on a droplet of freshly prepared but now 0.2% (w/v) uranyl acetate solution. After 20 sec the grid is blotted dry as above and examined with the electron microscope.

With pure octamer at low magnification, typically, large numbers of crystalline tubes are seen (Fig. 29). At higher magnification the typical tube structure becomes evident (Fig. 29). Images of the tubes have optical diffraction patterns similar to those shown in Fig. 29 which can be indexed according to Klug et al.[68] Layer lines containing zero-order Bessel functions can usually be seen with a spacing of 1/65Å, and layer lines containing tenth and twentieth order Bessel functions can be seen in favorable

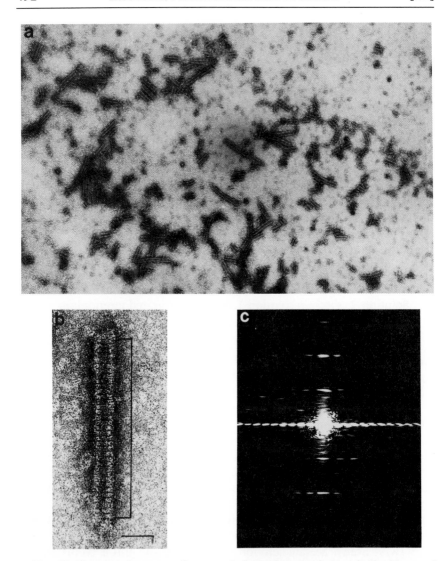

FIG. 29. Electron microscopy of octamer helical tubes negatively stained with uranyl acetate. Electron micrographs were recorded at the following magnifications. (a) ×7,500 and (b) ×37,500. The bar in b represents 50 nm. The optical transform (c) was recorded of a section (2 unit cells) of the helical tube as shown in b.

cases (i.e., those cases in which minimum deformation of the tubes has occurred).

Crystalline tubes have been produced from a number of octamers reconstituted from individually purified histones[20] and from hybrid octamers in which the SH-containing sea urchin H4 had replaced the vertebrate type H4[69] or in which, in addition to that replacement, Cys^{110}-dethio-histone H3 had been introduced.[66] H2A–H2B dimers, labeled covalently with aurothiomalate and reconstituted to octamers, have yielded crystalline tubes[21] as have octamers reconstituted from natural dimers and tetramers.[20]

Assembly of Core Particles from Natural and Reconstituted Octamers

Naturally, histone octamers are associated with DNA to form nucleosome core particles. Therefore, reassociation of reconstituted octamers with DNA constitutes the only functional test for the integrity of the octamer and provides, in addition, a powerful tool to study DNA–histone interactions.

In vitro assembly systems using high ionic strength buffers are widely used. The assembly of cores in the presence of poly(glutamic acid)[70] may, however, have a number of advantages. Under the assembly conditions, the histone octamer remains stabilized by the poly(glutamic acid) at low ionic strength in the absence of DNA.[71] This eliminates any rearrangements of the octamer and also the possibility that tetramers, dimers, or individual subunits may interact with the DNA to form spurious complexes. The use of rigorously purified, intact octamers under nondissociating conditions results in almost quantitative assembly of cores within a short period.[71] The poly(glutamic acid) can be removed from the products of assembly by gradient centrifugation.[71]

Assembly on Random DNA

Histone octamers are stable in either 2 M NaCl or in their storage solution of 1 M NaCl, 50% glycerol, pH 7.0. On initiation of assembly, the poly(glutamic acid) is introduced into this high ionic solution before progressive dilution to a moderate ionic strength of approximately 200 mM NaCl. The poly(glutamic acid) stabilizes the octamer at this ionic strength (Fig. 30). If the solution during the dilution becomes cloudy, this indicates the presence of free or degraded histones or other contaminants in the octamer preparation. Poly(glutamic acid) of 50,000–100,000 Da appears to promote assembly reactions much more effectively than molecules below 20,000 Da. The lowest poly(glutamic acid) to histone ratio that

[69] G. G. Lindsey and C. von Holt, unpublished observations, 1987.
[70] A. Stein, J. P. Whitlock, and M. Bina, *Proc. Natl. Acad. Sci. U.S.A.* **76**, 5000 (1979).
[71] J. D. Retief, B. T. Sewell, H. J. Greyling, S. Schwager, and C. von Holt, *FEBS Lett.* **167**, 170 (1984).

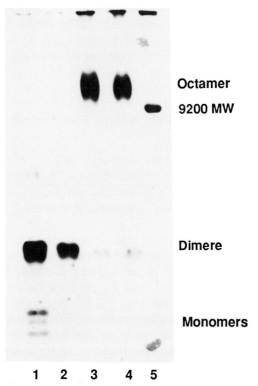

FIG. 30. SDS–PAGE of cross-linked histones. The polyacrylamide gradient was 5–20% from top to bottom. Core histones were cross-linked in the following solutions: lane 1, 200 mM NaCl; 2, 500 mM NaCl; 3, 2 M NaCl; 4, 200 mM NaCl plus poly(glutamic acid) [poly(glutamic acid): histone ratio 2:1 (w/w)]. Lane 5 contains phosphorylase b (M_r 92,000). [From J. D. Retief, B. T. Sewell, H. J. Greyling, S. Schwager, and C. von Holt, *FEBS Lett.* **167**, 170–175 (1984). With permission of the Federation of European Biochemical Societies.]

promotes assembly is 1:1, while 2:1 is optimal. Further increasing the ratio to 8:1 does not significantly improve the yield of cores.

The octamers introduced into the assembly reaction are almost quantitatively assembled on the DNA. At all investigated ratios "tight" dimers are formed.[71] At histone to DNA ratios higher than 1, extensive tracts of tight polycores are produced. At such high ratios excess octamers can also bind to the already assembled cores.[72] After addition of the DNA to the histone–poly(glutamic acid) mixture, the solution should remain clear for at least 1 hr. If it turns cloudy within a few minutes after the addition of the

[72] G. Voordouw and H. Eisenberg, *Nature (London)* **273**, 446 (1978).

DNA with the formation of a precipitate, the histone to DNA ratio is too high. Assembly reactions at histone to DNA ratios between 0.5 and 0.8 proceed smoothly. Cores prepared from octamers, isolated by various methods and also reassembled from their individually purified components or dimer and tetramer subunits, yield DNase I digestion nuclease cutting patterns that are indistinguishable from those produced from natural cores.[71]

Solutions

Poly(glutamic acid): 10 mg/ml poly(glutamic acid) (MW 50,000–100,000), 2 mM EDTA, 50 mM Tris-HCl, pH 7.6

Histone octamer: 10 mg/ml in 50% (w/v) glycerol, 1 M NaCl, 1 mM 2-mercaptoethanol, 10 mM phosphate buffer, pH 7.0

350 mM NaCl: 350 mM NaCl, 1 mM EDTA, 10 mM Tris-HCl, pH 7.6

Assembly buffer: 70 mM Tris-HCl, pH 7.6, 1 mM EDTA

DNA stock: 3.5 mg DNA/ml, prepared from a nuclease digest of chromatin of approximately greater than 1000 base pairs, 10 mM Tris-HCl, pH 7.6, 1 mM EDTA

Dialysis buffer: 20 mM ammonium acetate, 0.2 mM EDTA, 2 mM 2-mercaptoethanol, 5 mM Tris-HCl, pH 7.6

20 mM NaCl: 20 mM NaCl, 2 mM EDTA, 20 mM Tris-HCl, pH 7.6

CaCl$_2$: 100 mM CaCl$_2$, 10 mM Tris-HCl, pH 7.6

Micrococcal nuclease: 1 mM CaCl$_2$, 10 mM Tris-HCl, pH 8.0, 20 Worthington units nuclease/μl

EDTA: 500 mM, pH 8.0

Procedure. Forty microliters of histone octamer is added to 80 μl poly(glutamic acid) in a glass centrifuge tube. Then 430 μl of 350 mM NaCl solution is added followed by 1.7 ml assembly buffer, after which 103 μl DNA solution is added. The mixture is incubated at 37° for 3 hr with gentle shaking. Assembly is essentially complete within 1 hr (Fig. 31).

To assess the characteristics of the assembled polycores, these may be digested to cores for further analysis. Polycores are dialyzed against dialysis buffer at 4° for 2 hr and adjusted to 1.4 mM Ca^{2+}. Micrococcal nuclease stock solution is added at 1.5 μl/350 μg DNA in the assembly mixture. The digestion is stopped at the appropriate times by excess EDTA (10 μl EDTA/330 μl digest). The redigested cores are applied to 5–20% (w/v) sucrose gradients, in 20 mM NaCl solution, and centrifugd for 16 hr in a Beckman SW40 Ti rotor at 182,000 g. Such cores on DNase I digestion (see below) give typical 10-bp periodicity protection patterns (Fig. 32).

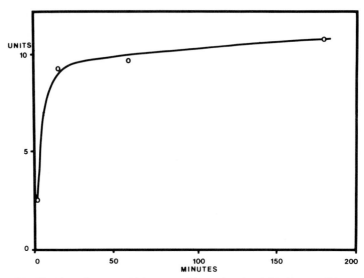

FIG. 31. Kinetics of core particles assembly under the following conditions: 800 µg poly(glutamic acid), 400 µg histone octamers, 400 µg DNA in 10 mM Tris-HCl (pH 8.0), 0.1 mM EDTA, 0.1 mM PMSF, 200 mM NaCl in a volume of 2 ml incubated at 37°. The reaction was stopped at the times indicated by the addition of CaCl$_2$ from a stock solution (to give an excess of 1 mM free calcium) and micrococcal nuclease, 1000 units/DNA. The nuclease reaction was terminated with excess EDTA after 1 min at 37°. The monomeric core particles produced were analyzed on a sucrose gradient, and the area under the core fraction was determined in arbitrary units. [From J. D. Retief, B. T. Sewell, H. J. Greyling, S. Schwager, and C. von Holt, *FEBS Lett.* **167,** 170–175 (1984). With permission of the Federation of European Biochemical Societies.]

Assembly on Unique DNA

The poly(glutamic acid) assembly system can easily be scaled down to accommodate small quantities of DNA. Using the plasmid pGV403, into which a unique piece of DNA to be investigated has been inserted, a simple and rapid method to map cores at single base pair resolution becomes available.[73]

The pGV403 plasmid[74] possesses a *Sma*I site for blunt-ended insertion. This site is flanked by two different *Tth*III 1 sites (Fig. 33). The *Tth*III 1 site has the general sequence

$$G—A—C—N\|N—N—G—T—C$$
$$C—T—G—N—N\|N—C—A—G$$

[73] J. D. Retief, B. T. Sewell, and C. von Holt, *Biochemistry* **26,** 4449 (1987).
[74] G. Volckaert, E. de Vleeschouwer, H. Blöcker, and R. Frank, *Gene Anal. Tech.* **1,** 52 (1984).

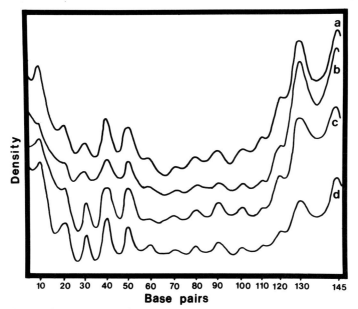

FIG. 32. Densitometer traces of DNase I digestions of 5' end-labeled cores. Nucleosome cores were assembled on partially digested chicken erythrocyte DNA (20 units micrococcal nuclease/mg DNA for 4 min at 37°). After assembly the cores were further trimmed to 145 bp by micrococcal nuclease digestion. The resultant cores were isolated on a 5–20% sucrose gradient, 5' end-labeled, and digested with DNase I. The DNA fragments were separated by denaturing DNA gel electrophoresis, and the gels were dried and autoradiographed. Densitometer traces were recorded after 15 sec of DNase I digestion (0.05 unit DNase I/μg DNA at 37°). The following octamers were used for the assembly reactions:[71] (a) natural chicken erythrocyte cores; (b) octamers extracted according to Ruiz-Carillo and Jorcano[17] (see text); (c) octamers extracted from cores immobilized on hydroxyapatite (see text); (d) octamers assembled from acid-extracted, individually isolated histones (see text).

whereas the *Tth*III 1 site upstream of *Sma*I is

$$G—A—C—T\|A—A—G—T—C$$
$$C—T—G—A—T\|T—C—A—G$$

and the *Tth*III 1 site downstream of *Sma*I is

$$G—A—C—T\|C—A—G—T—C$$
$$C—T—G—A—G\|T—C—A—G$$

Restricting these *Tth*III 1 sites yields a single-base 5' overhang at the four ends of the two fragments produced. Each 5' overhang contains one unique base, A, T, C, and G, respectively. When the corresponding recessed ends are filled in with a single radioactively labeled nucleotide, only one of the four generated ends becomes labeled. The fragment containing

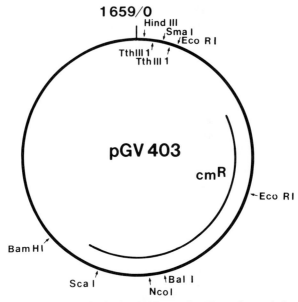

FIG. 33. pGV403. The plasmid size is 1659 base pairs. The main restriction endonuclease cutting sites are indicated. cm^R, Chloramphenicol-resistance gene.

the insert is flanked by G and A 5′ overhangs. This allows the inserted fragment to be labeled uniquely at either end by using [α-³²P]dCTP or [α-³²P]dTTP for the fill-in reaction. Since the single-base 5′ overhangs produced by the vector will not be labeled, the vector DNA does not have to be removed and can act as carrier DNA in the system.[73]

The use of the pGV403 plasmids removes the methodological difficulties imposed by other methods of preparing end-labeled DNA for purposes of core assembly. The plasmids, with any desired insert, can be produced and purified in milligram quantities on cesium chloride gradients. The labeling protocols have been tuned to produce sufficient amounts of uniquely end-labeled DNA rapidly. Since only a single restriction enzyme digest at 65° and a fill-in reaction using the Klenow fragment are needed, DNA degradation and exposure to radioactivity during preparation are reduced to a minimum.

At low histone to DNA ratios the assembly of a small number of cores that will not be sterically hindered from occupying preferred positions on the DNA can be expected. Incubating the assembled products at 37° increases the mobility of the cores, which should allow them to find areas of preferred binding.

Preparation of Plasmids. pGV403 plasmids (Amersham) are supplied pretreated with *Sma*I and dephosphorylated. Ligation of the insert and transformation into HB101 host cells with subsequent propagation are carried out according to Maniatis *et al.*[75] The presence of inserts in the pGV403 plasmids is verified by colony hybridizations on Hybond N (Amersham) using a suitable nick-labeled probe. The positive colonies are picked, and the plasmids are amplified and subsequently purified on cesium chloride gradients, according to Ref. 75.

End-Labeling of DNA Standards. A suitable size standard for the region 38–624 bp is a *Hpa*II digest of pBR322. Digestion and labeling via a Klenow fill-in reaction can be done in a virtually single-step operation, thus avoiding the more elaborate procedure of 5' end-labeling with a kinase reaction. The following solutions are required:

TE buffer:[75] 1 mM EDTA, 10 mM Tris-HCl, pH 8.0
Reaction buffer: 20 mM MgCl$_2$, 20 mM 2-mercaptoethanol, 40 mM
    Tris-HCl, pH 7.6
pBR322 stock: 0.6 mg/ml solution of pBR322 plasmid in TE buffer
dCTP: 2.5 mM in water
dGTP: [α-^{32}P]dGTP (5 μCi/μl, 3000 Ci/mmol)

To end-label DNA standards, the 3' fill-in reaction of protruding 5' ends with the Klenow fragment of DNA polymerase is used. The nucleotide chosen for the labeling reaction should be the last one to be filled in to prevent the formation of labeled fragments smaller than the nominal fragment length. Six-tenths microgram of *Hpa*II-digested pBR322 is dissolved in a mixture of 1 μl TE buffer, 1 μl 2.5 mM dCTP, 1 μl [α-^{32}P]dGTP, 7 μl reaction buffer, and 1 unit Klenow fragment of DNA polymerase. The reaction is allowed to proceed at 16° until incorporation of the nucleotides ceases <10 min), as monitored by trichloroacetic acid (TCA) precipitation on glass fiber filters. Desalting is carried out on spun columns.[75]

Digestion and Labeling of pGV403 Plasmids. The following solutions are required:

TE buffer: 1 mM EDTA, 10 mM Tris-HCl, pH 8.0[75]
Plasmid stock solution: 6 mg/ml pGV403 DNA in TE buffer
Digestion buffer: 70 mM MgCl$_2$, 600 mM NaCl, 70 mM 2-mercaptoethanol, 100 mM Tris-HCl, pH 8.0
Klenow buffer: 20 mM MgCl$_2$, 20 mM 2-mercaptoethanol, 40 mM
    Tris-HCl, pH 7.6

[75] T. Maniatis, E. F. Fritsch, and J. Sambrook, "Molecular Cloning." Cold Spring Harbor Lab., Cold Spring Harbor, New York, 1982.

dCTP: $[\alpha^{32}P]$dCTP (5 μCi/μl, 3000 Ci/mmol)
Klenow: Klenow fragment of DNA polymerase
EDTA stock solution: 250 mM EDTA, pH 8.0

The reaction mixture contains 18 μl plasmid stock solution, 2 μl digestion buffer, and 21 units TthIII 1. The digestion proceeds overnight at 65°. Then 10 μl of the digest is added to 10 μl Klenow buffer, 2 μl $[\alpha\text{-}^{32}P]$dCTP, and 1 unit Klenow fragment. The reaction proceeds at 16° (normally 10 min) monitored by TCA precipitation on glass fiber filters. The reaction is terminated by the addition of 3 μl EDTA stock solution (same as stopped labeling mixture, below).

Assembly on Specific DNA Sequences. Prepare the following solutions:

Poly(glutamic acid) stock solution: 10 mg poly(glutamic acid)/ml 10 mM phosphate buffer, pH 6.8
Octamer stock solution: 6.2 mg chicken erythrocyte octamers/ml, 1 M NaCl, 50% (v/v) glycerol, 0.1 mM PMSF, 10 mM Tris-HCl, pH 7.6
NaCl solution: 1 M NaCl
TE buffer: 10 mM Tris-HCl, pH 8.0, 1 mM EDTA

One microliter of octamer stock solution and 1 μl 1 M NaCl are added to 2 μl poly(glutamic acid) solution in an Eppendorf tube. This droplet is mixed well by repeated pipetting. Of this mixture, 2 μl is transferred to the bottom of an Eppendorf tube containing 5 μl of TthIII 1-digested pGV403 derivative in stopped labeling mixture (see above). The volume of this assembly mixture is made up to 10 μl with TE buffer, and the mixture is incubated for 3 hr at 37°. This is followed by an equilibration period of 60 min at 20°, followed by 30 min at 16°.

DNase I Probing of Cores Assembled on Unique DNA. Successful assembly in a specific position must reveal, on DNase I probing, a protected area of the size expected from the core-associated DNA exhibiting an approximately 10-bp periodicity, typical for the DNase I digestion of one side-protected DNA[76] (Figs. 34 and 35). The following solutions are required:

MgCl$_2$ stock solution: 100 mM MgCl$_2$, 10 mM Tris-HCl, pH 7.6
DNase I stock solution: 100 units DNase I/μl, 10 mM CaCl$_2$, 10 mM Tris-HCl, pH 6.0 (to be diluted to 1 unit/μl immediately prior to use)
Extraction buffer: 1% (v/v) SDS, 300 mM sodium acetate, 1 mg/ml proteinase K, 250 mM EDTA

---

[76] L. C. Lutter, *J. Mol. Biol.* **124,** 391 (1978).

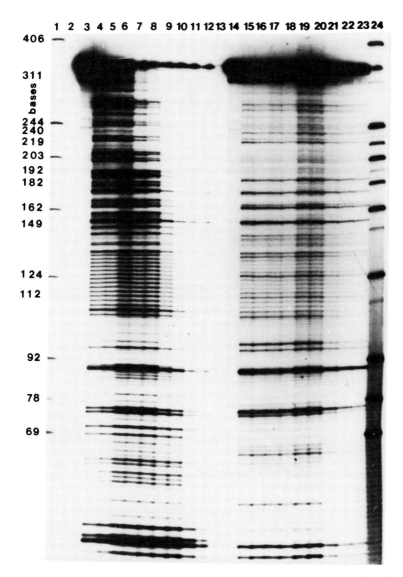

FIG. 34. Autoradiogram of a DNase I-digested fragment in pGV403. The plasmid was constructed by digesting the h22 histone gene quintet with *Hae*III and isolating the fragment containing the $(G-A)_{16}$ sequence. This fragment was digested with *Alu*I, and the $(G-A)_{16}$-containing sequence was isolated, treated with phosphatase, and inserted into the *Sma*I site of pGV403. The fragment was labeled with [α-^{32}P]dCTP and digested with DNase I. Lanes 3–12, free DNA digested with 1.25 units of DNase I/μg DNA; lanes 14–23, DNA assembled at 0.6:1 histone to DNA ratio digested at 12.5 units of DNase I/μg DNA. The digestions were carried out at 16° for 0, 15, 30, 60, 90, 120, 240, 480, 720, and 1200 sec. Lanes 1, 2, and 24 contain pBR322 digested with *Hpa*II and labeled with [α-^{32}P]dGTP by a fill-in reaction. The DNA was analyzed on a 6% polyacrylamide denaturing DNA gel. [From J. D. Retief, B. T. Sewell, and C. von Holt, *Biochemistry* **26**, 4449–4453 (1987). With permission of the American Chemical Society.]

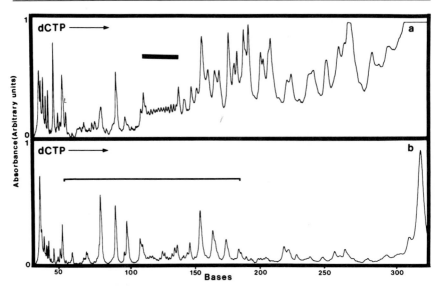

Fig. 35. Densitometer scan and autoradiogram of a DNase I-digested fragment in pGV403 (Fig. 34). The assembly frame is denoted with a bracket and the $(G-A)_{16}$ sequence by a solid bar. (a) DNase I digestion of free DNA; (b) DNase I digestion of core particle associated DNA. [From J. D. Retief, B. T. Sewell, and C. von Holt, *Biochemistry* **26**, 4449–4453 (1987). With permission of the American Chemical Society.]

Phenol: equilibrated against 100 mM Tris-HCl, pH 8.0, 0.2% (v/v)
   2-mercaptoethanol
Chloroform–isoamyl alcohol: 24:1 (v/v)
Ethanol: 80% (v/v)

Three microliters of $MgCl_2$ stock solution and 10 units DNase I are added to the assembly products. After appropriate time intervals at 16°, aliquots are removed and placed in 20 μl extraction buffer, followed by incubation at 20° for 20 min. DNA is then isolated by two phenol extractions with 20 μl phenol each and three chloroform extractions with 20 μl chloroform–isoamyl alcohol (24:1, v/v), followed by ethanol precipitation and washing with 80% (v/v) ethanol. The dried pellets are suspended in sequencing gel sample buffer and run on horizontally poured denaturing DNA gels according to the method of Maxam and Gilbert.[77]

[77] A. M. Maxam and W. Gilbert, this series, Vol. 65, pp. 499.

Acknowledgments

The work referred to from the authors' laboratory was supported by grants from the Foundation of Research Development of the Council for Scientific and Industrial Development (Republic of South Africa) to C.v.H.

Appendix: Histone Sequences

Amino acid sequences of histones are represented by the single-letter notation. They have been visually aligned to demonstrate the most obvious sequence similarities. Of the protein sequences, only those that have been unambiguously determined by chemical and enzymatic means are shown. Thus, sequences derived by positioning of residues based on composition of peptides and homology consideration have been omitted. An asterisk indicates that the protein sequence was deduced from the DNA sequence. All sequences are compared to the bovine histones, and differences are underlined with the exception of the H1 histones. Heterogeneity is identified by two single-letter notations placed in one sequence position. Modified amino acids that have been determined in the original reference are shown. Modified sites are indicated by placing symbols next to the amino acid residues as follows: amino-terminal acetylation sites ($<$); ϵ-N-acetyl lysine residues, complete or partial ($-$); mono, di, and tri ϵ-N-lysine methylation, partial or complete ($+$); and phosphorylation ($\circ$).

Within one histone group individual sequences are identified by arabic numerals to allow identification over more than one printed page and in the bibliographic references. The arabic numerals in parentheses at the carboxy-terminal end equals the total number of residues in the protein. Partial sequences have been incorporated at the end of the list. If amino- or carboxy-terminal amino acids have not been positioned this is indicated by (. . .); unsequenced positions are indicated likewise.

HISTONES H1 AND H5

```
                              10         20         30         40         50         60         70         80         90        100        110        120
 1  Rabbit          H1.3  <SE-APAETAA PAPAEKSPAK KKKAAKKPGA GAAKRKAAGP PVSELITKAV AASKERNGLS LAALKKALAA GGYDVEKNNS R-IKLGLKSL VSKGTLVETK GTGASGSFKL
 2  Chicken         *H1    SETAPVAAPA VSAPGAKAAA -KKP-KKAAG GAKPRKPAGP SVTELITKAV SASKERKGLS LAALKKALAA GGYDVEKNNS R-IKLGLKSL VSKGTLVQTK GTGASGSFKL
 3                  *H1    SETAPAAAPD APAPGAKAAA -KKP-KKAAG GAKARKPAGP SVTELITKAV SASKERKGLS LAALKKALAA GGYDVEKNNS R-IKLGLKSL VSKGTLVQTK GTGASGSFRL
 4  Duck            *H1    SETAPVAAPA VSAPGAKAAG       GSKARKPAGP SVTELITKAV SASKERKGLS LAALKKALAA GGYDVEKNNS R-IKLGLKSL VGKGTLVQTK GTGASGSFKL
 5  Frog            *H1   AETASTETTPA APPAEPKQKK KKQQPKKAAG GAKAKKPSGP SASELIVKSV SASKERGGVS LAALKKALAA GGYDVERNNS R-LKLALKAL VTKGTLTQVK GSGASGSFKL
 6                  *H1   TETAATETTPA APPAEPKQKK K-QQPKKAAG GAKAKKPSGP SASELIVKSV SASKERGGVS LAALKKALAA GGYNVERNNS R-LKLALKAL VTKGTLTQVK GSGASGSFKL
 7  Trout           H1          <AEVAPAPAA AAPAKPKKK AAAKPKKSGP AVGELAGKAV AASKERSGVS LAALKKSLAA GGYDVEKNNS R-VKIAVKSL VIKGTLVETK GTGASGSFKL
 8                  *H1          AEVAPAPAA AAPAKAPKKK AAAKPKKAGP SVGELIVKAV SASKERSGVS LAALKKSLAA GGYDVEKNNS R-VKIAVKSL VIKGTLVQTK GTGASGSFKL
 9  Fruit fly       *H1   SDSA--VATS ASPVAAPPA- ------TVEK KVVQKKASGS AGTKAKKASA TPSH.RGGSS LLAIKKYITA TYKCDAQKLA PFIKKYLKSA VVNGKLIQTK GKGASGSFKL
10  Polychaete annelid H1a                   AR RRKTAAAAHP PVATMVVTAI LGLKDRKGSS MVAIKKYIAA N-YRVDVARL APFIRKFIRK AV-----KQT KGTGASGSFR
11                  H1b                       AR RKR--AATHP PVATAVVAAI LGLKDRKGSS MVAIKKYLAA N-YRVDVARL GPFIRRFVRK AVAEGMLVQN K-----GSFR
12  Sea urchin      H1    PGS PQKRAASPRK SPRKSPKKSP RKASASPRRK AKRARASTHP PVLEMVQQAAI TAMKERKGSS AAKIKSYMAA N-YRVDMNVL APHVRRALRN GVASGALKQV TGTGASGRFR
13                  *H1            AEKNSSKK VTTKKPAAHP RVSKRPASHP PYSDMIAAA. TELKDRNGSS LQAIKKY1AT N-FDVQMDRQ LLFIKRALKS GVEKGKLVQT KGKGASGSFK
14                  *H1           SAAKPKTA KKARAAPAHP KASKKSTDHP KYSDMIVAAI PTSQMVVAAI TALKERGGSS NQAIKKY1AA N-YKVDINKQ ATFIKRALKA GVANGTLVQV KGKGASGSFK
15  Human           *H1°            TENSTS--- -APAAKPKRA KASKKSTDHP KYSDMIVAAI QAEKNRAGSS RQSIQKYIKS HYKVGENADS QIKLSIKRLV TT-GVLKQTK GVGASGSFRL
16  Duck            *H5            TDSPIPA-- PAPAAKPKRA RAEKSRPASHP TYSEMIVAAI RAEKSRGGSS RQSIQKYVKS HYKVGQHADL QIKLAIRRLL TT-GVLKQTK GVGASGSYRL
17  Goose           H5             TDSPIPA-- PAPAAKPKRA RAPRKPASHP TYSEMIAAAI TYSEMIAAAI RADKSRGGSS RQSIQKYVKS HYKVGQHADL QIKLAIRRLL AA-GVLKQTK GVGASGSFRL
18  Chicken         H5             TESLVLS-- PAPAKPKRVK ASRRS-ASHP TYSEMIAAAI RAEKSRGGSS RQSIQKYIKS HYKVGHNADL QIKLSIRRLL AA-GVLKQTK GVGASGSFRL
19  Ciliate protozoan *H1  APRSSTSKSA TREKKDHKKA PIKKAIAKKD TKPTPTKGKA ASASTTPVKK DVTPVKADTK KKIHKTKTMK ETVSDAKKTV HAAAGDKKLS KKRPAKEAAK KAINPGKKAA

PARTIAL SEQUENCES

20  Pigeon          H5     TESPIPVPA PAPAAKPKPK RVSKRPASHP PYSDMIAAA. ...........
21  Sea urchin      H1    AAS PQKRAASPRK SPKKSPRKSP KKK--SPRKR KARSAA--HP PVIDMITAAI AAQKERRGSS VAKIQSYIAA KYRCDIN-AL NPHIRRALKN QVKSGALKQV SGVGATGRFR
22  Bovine          H1.1  <SETAPAAPAA APPAEK--TP VKKKAAKKPA GA-RRKASGP PVSELITKAV AASKERSGVS LAALKKALAA AGYDVEKNNS R-IKLGLKSL VSKGTLVQTK GTGASGSF..
23  Rabbit          H1.4  <SE-APAETAA PAPAKSPAKT PVKARKKKSA GAAKRKASGP PVSELITKAV AASKERSGVS LAALKKALAA AGY....
24  Sea urchin      H1.1              ....KK VTTKKPAAHP P...MVTTAI TELK[F/E]RNGSS N-FDVEMDRQ LIKIKKALKS GVEKGKLVQT KGTGASGSFK
25                  H1.2                     ....MVVAAI TALKERGGSS ......MKKQ SVFIKKALKS GVEKGTLVQV KGKGASGSFK
26                  H1.3                     ....MVVAAI TALKERGGSS NQALKKYKAA N.....
27                  H1        PGS PQKRAASPRK SPRKGSPKKS P........ ....MIRAAI TAMKERGGSS VAKIKSYIAA N-YRVNMTNL QPHIRRALRS GVASGALKQV TGTGATGRFR
28                  *H1             TDTAKK VTQKKPAAHP PAAEMVTTAI KELKERKGSS RQAIANYIKA H-FDVEIDQQ LVFIKKALRS GVAKGTLVQT KGTGASGSFK
29  Wheat           H1.1              ....MVSEAI AALKEREGSS EFAIGKKKE. ...
30                  H1.3              ....MVSEAI TALKERTGS. .........MLT QIKKLVAAGK LTK....
31  Maize           H1      ....YAEMISEAI TSLKERTGSS ZYAIAKFVED KHKQKLPXDF ...........YKL
32  Frog            *H1    $TDV$ETP $PL.......... .......... ....NS R-LKLALKAL VTKETLLQVK GSGASGSFKL
```

```
          130        140        150        160        170        180        190        200        210        220        230        240        250        260

 1  DKKAASGEAK PKPKKAGAAK PKKPAGATPK KP-KKAAGAK KAV--KKTPK KAPKPKAAAK PKVAKPKSPA KVAKSPKKAK AVKPKAAKPK APKPKAAKAK KTAAKKKK                                              (213)
 2  NKKPGETKEK ATKKKPAAKP KKPAAKKPAA AA-KKPKKAA AVK--KSPKK AKKPAAAATK KAAKSPKKAT KAGRPKKTAK SPAKAKAVKP KAAKSKAAKP KAAKAKKAAT KKK                                          (217)
 3  NKKPGEVKEK APRKRATAAK PKKPAAKKPA AAAKKPKKAA AVK--KSPKK AKKPAAAATK KAAKSPKKAA KAGRPKKAAK SPAKAKAVKP KAAKPKATKP KAAKAKKTAA KKK                                          (218)
 4  NKKPGETKEK ATKKKPAAKP KKPAAKKPAS AA-KKPKKAA AVK--KSPKK AKKPAAAATK KAAKSPKKAA KAGRPKKAAK SPAKAKAVKP KAAKPKAAKP KAAKAKKAAP KKK                                          (217)
 5  NKKQLETKVK AVAKKKLVAP KAAKPVTAKK KP-KSPKKPK KVSAAAAKSP KKAKKPVKAA KSPKKPKAVK SKKVTKSPAK KATKPKAAKA KIAKPKIAKA KAAKGKKAAA KK                                           (221)
 6  NKKQLETKVK AVAKKKLVAP KAAKPVAAKK KP-KSPKKPK KVSAAAAKSP KKAKKPVKAA KSPKKPKAVK PKKVTKSPAK KATKPKAAKA KIAK-----A KAAKGKKAAA KK                                           (215)

 7  NKKAVEAKKP AKKAAAPKAK KVAAKKPAAA KKPKKVAAKK AVAAKKSPKK AKKPATPKKA AKSPKKATKA AKPKAAKPKK AAKSPKKVKK PAAAKK                                                             (194)
 8  NKKAVEAKKP AKKAAAPKAK KVAAKKPAAA KKPKKVAAKK AVAAKKSPKK AKKPATPKKA AKSPKKVKKP AAAAKKAAKS PKKATKAAKP KAAKPKAAKA KKAAPKKK                                               (206)
 9  SASAKKEKDP KAKSKVLSAE KKVQSKKVAS KKIGVSSKKT AVGAADKKPK AKKAVATKKT AENKKTEKAK AKDAKKTGII KSKPAATKAK VTAAKPKAVV AKASKAKPAV SAKPKKTVKK ASVSATAKKP KAKTTAAKK           (255)

10  VNKTAVPKKK KAAKKPKAKK VKKPKSAAKK KTNRARKPKT -KKNRN                                                                                                                  (121)
11  VNKTALPKKK KAAKKPKAKK VKKPKSAAKK KTNRARAPKT -KKNRN                                                                                                                  (119)

12  VGAVAKPKKA KKTSAAAKAK KAKAAAAKKA R$^R_K$AKAAAKRK AALAKKKAAA AKRKAAAKAK KAKKPKKKAA KKAKKPAKKS PKKKKKAKRS PKKAKKAAGK RKPAAKKARR SPRKAGKRRS PKKARK                   (248)

13  VNVQAAKAQA SEKAKKEKEK AKLLAQREKA KEKGCSEEGE TAEGSRPKKV KAAPKKAKKP VKKTTEKKEK KKTPKKAPKK PAAKSTPRK TPKKAAAKKP KTAKPKKPAX KKAAKSK                                       (204)
14  LGKVKAGKTE AQKARAAAKK AKLAAKKKEQ KEKKAAKTKA RKEKLAAKKA AKKAAKKVKK PAAKAKKPAK KAAKKPAAKK AAKKPAAKKP AKKAAKKPAA KKAAKPAKKA AKKPAKKAA KK                                (209)
15  AKSDEPKKSV A-FKKTKK-E IKKVATPKKA SKPKKAASKA PTKKPKATPV KKAKKLLAAT PKKAKK-PKT VK-AKPVKAS KPKKAKPVKP KAKSSAKRAG KKK                                                   (193)
16  AKGDKAKKSP A-GRKKKKKA ARRSTSPKKA ARPRKARSPA -KKPKAA--A RKARKKSRAS PKKAKK-PKT VK-AKSLKTS KVKKAKKSRP RAKSGARKSP KKK                                                   (193)
17  AKGDKAKRSP A-GRKKKKKA ARKSTSPKKA ARPRKARSPA -KKPKAA--A RKARKKSRAS PKKAKK-PKT VK-AKSLKTS KPKKARRSKP RAKSGARSKP KKK                                                   (193)
18  AKSDKAKRSP --G--KKKKA VRRSTSPKKA ARPRKARSPA -KKPKAT--A RKARKKSRAS PKKAKK-PKT VK-AKSRKAS KAKKVKRSKP RAKSGARKSP KKK                                                   (189)
19  AQPKSTKKEV KKDNKTAKKE TKKDHKPAKK EAKKETKPAK KDAKKSSKPA KKN                                                                                                         (163)

PARTIAL SEQUENCES

20  .....
21  VGAVKRSAAS ANKLKATREK ARARAKKA KAAARRKAAA -AKRKAAAAK RRAAKKARKA KAKP.....
22  .....
23

24  LNVQAAKAQA AEKAKKEKEK AKLQAQREKA KAKAAAAKKEK QQKATAAKKA KAAPKKKPAK KAVEKKEKKK TPKKATAATK KATPKTVT.. ..
25  LGKKPAAGRT DA.....

26  .....
27  VG....
28  L......... ......... ......... ......... ......... ......... .........SK PAAKKPAAKK PAAKVIAKSK
29  .........

30
31  PSXTX$$$K. ....
32  NKKQLQSKDK AAKKKAPLAA KTKKPAARAK KAPKSPKKPK KVSA-AAKSP KVVKKPAKAA LAAKSPKKNK AAKPKKATKS PAKKTAVKPK TAAAKSPAKA KVAKKKAAP KKK                                          (163)
```

Histones H1 and H5

Seq.	Organism	Source	Reference
1	ORYCTOLAGUS CUNICULUS	THYMUS	HSIANG M., LARGMAN C.R., COLE R.D.; UNPUBLISHED RESULTS, CITED BY: COLE R.D.; (IN) THE MOLECULAR BIOLOGY OF THE MAMMALIAN GENETIC APPARATUS, VOL.1, TS'O P.O.P., ED., PP.93-104, ELSEVIER/NORTH-HOLLAND, AMSTERDAM, NEW YORK, 1977.
2	GALLUS GALLUS	DNA	SUGARMAN B.J., DODGSON J.B., ENGEL J.D.; J. BIOL. CHEM. 258:9005-9016(1983).
3	GALLUS GALLUS	DNA	COLES L.S., WELLS J.R.E.; NUCL. ACIDS RES. 13:585-594(1985).
4	CAIRINA MOSCHATA	DNA	TONJES R., DOENECKE D.; J.MOL. EVOL. 25:361-370(1987).
5,6	XENOPUS LAEVIS	DNA	TURNER P.C., ALDRIDGE T.C., WOODLAND H.R., OLD R.W.; NUCL. ACIDS RES. 11:4093-4107(1983).
7	SALMO GAIRDNERII	TESTIS	MACLEOD A.R., WONG N.C.W., DIXON G.H.; EUR J. BIOCHEM. 78:281-291(1977).
8	SALMO GAIRDNERII	DNA	MEZQUITA J., CONNOR W., WINKFEIN R.J., DIXON G.H.; J. MOL. EVOL. 21:209-219(1985).
9	DROSOPHILA MELANOGASTER	DNA	MURPHY T.J., BLUMENFELD M.; NUCL. ACIDS RES. 14:5563-5563(1986).
		DNA	GOLDBERG M.L.; UNPUBLISHED (1979) THESIS, UNIVERSITY OF STANFORD, U.S.A.
10,11	PLATYNEREIS DUMERILII	SPERM	KMIECIK D., SELLOS D., BELAICHE D., SAUTIERE P.; EUR. J. BIOCHEM. 150:359-370(1985).
12	PARECHINUS ANGULOSUS	SPERM	STRICKLAND W.N., STRICKLAND M., BRANDT W.F., VON HOLT C., LEHMANN A., WITTMANN-LIEBOLD B.; EUR. J. BIOCHEM. 104:567-578(1980).
13	STRONGYLOCENTROTUS PURPURATUS	DNA	LEVY S., SURES I., KEDES L.; J. BIOL. CHEM. 257:9438-9443(1982).
14	LYTECHINUS PICTUS	DNA	KNOWLES J.A., CHILDS G.J.; NUCL.ACIDS RES. 14:8121-8133(1986).
15	HOMO SAPIENS	DNA	DOENECKE D., TONJES R.; J. MOL. BIOL. 187:461-464(1986).
16	CAIRINA MOSCHATA	DNA	DOENECKE D., TONJES R.; J. MOL. BIOL. 178:121-135(1984).
17	ANSER ANSER	ERYTHROCYTE	YAGUCHI M., ROY, SELIGY V.L.; BIOCHEM. BIOPHYS. RES. COMM. 90:1400-1406(1979).

18	GALLUS GALLUS	ERYTHROCYTE	BRIAND G., KMIECIK D., SAUTIERE P., WOUTERS D., BORIE-LOY O., BISERTE G., MAZEN A., CHAMPAGNE M.; FEBS LETT. 112:147-151(1980).
	GALLUS GALLUS	DNA	KRIEG P.A., ROBINS A.J., D'ANDREA R., WELLS J.R.E.; NUCL. ACIDS RES. 11:619-627(1983).
	GALLUS GALLUS	DNA	RUIZ-CARRILLO A., AFFOLTER M., RENAUD J.; J. MOL. BIOL. 170:843-859(1983).
19	TETRAHYMENA THERMOPHILA	DNA	WU M., ALLIS C.D., RICHMAN R., COOK R.G., GOROVSKY M.A.; PROC. NATL. ACAD. SCI. U.S.A. 83:8674-8678(1986). NO ALIGNMENT HAS BEEN ATTEMPTED.
20	COLUMBA LIVIA	ERYTHROCYTE	YAGUCHI M., ROY C., DOVE M., SELIGY V.; BIOCHEM. BIOPHYS. RES. COMM. 76:100-106(1977).
21	ECHINOLAMPAS CRASSA	SPERM	STRICKLAND W.N., STRICKLAND M., VON HOLT C.; BIOCHIM. BIOPHYS. ACTA 700:127-129(1982).
22	BOS TAURUS	THYMUS	LIAO L.W., COLE R.D.; J. BIOL. CHEM. 256:3024-3029(1981).
23	ORYCTOLAGUS CUNICULUS	THYMUS	RALL S.C., COLE R.D.; J. BIOL. CHEM. 246:7175-7190(1971).
24	PARECHINUS ANGULOSUS	EMBRYO	DE GROOT P., STRICKLAND W.N., BRANDT W.F., VON HOLT C.; BIOCHIM. BIOPHYS. ACTA 747:276-283(1983).
25	PARECHINUS ANGULOSUS	EMBRYO	DE GROOT P., STRICKLAND W.N., BRANDT W.F., VON HOLT C.; BIOCHIM. BIOPHYS. ACTA 747:276-283(1983).
26	PARECHINUS ANGULOSUS	EMBRYO	DE GROOT P., STRICKLAND W.N., BRANDT W.F., VON HOLT C.; BIOCHIM. BIOPHYS. ACTA 747:276-283(1983).
27	SPHAERECHINUS GRANULARIS	SPERM	STRICKLAND W.N., STRICKLAND M., VON HOLT C., GIANCOTTI V.; BIOCHIM. BIOPHYS. ACTA 703:95-100(1982).
28	PSAMMECHINUS MILIARIS	DNA(CLONE H22)	SCHAFFNER W., KUNZ G., DAETWYLER H., TELFORD J., SMITH H.O., BIRNSTIEL M.L.; CELL 14:655-671(1978).
29	TRITICUM AESTIVUM	EMBRYO	BRANDT W.F., VON HOLT C.; FEBS LETT. 194:282-286(1986).
30	TRITICUM AESTIVUM	EMBRYO	BRANDT W.F., VON HOLT C.; FEBS LETT. 194:282-286(1986).
31	ZEA MAYS	TASSELS	HURLEY C.K., STOUT J.T.; BIOCHEMISTRY 19:410-416(1980).
32	XENOPUS LAEVIS	DNA	TURNER P.C., ALDRIDGE T.C., WOODLAND H.R., OLD R.W.; NUCL. ACIDS RES. 11:4093-4107(1983).

Histones 2A

```
                            1      10      20      30      40      50      60      70

 1 Bovine        H2A.1        <SGRGKQGGKA RAK---A--K TRSSRAGLQF PVGRVHRLLR KGNYAERVGA GAPVYLAAVL
 2 Human         H2A.1        <S̊GRGKQGGKA RAK---A--K TRSSRAGLQF PVGRVHRLLR KGNYAERVGA GAPVYLAAVL
 3               H2A.2        <SGRGKQGGKA RAK---A--K TRSSRAGLQF PVGRVHRLLR KGNYAERVGA GAPVYLAAVL
 4               H2A.3        <SGRGKQGGKA RAK---A--K S̲RSSRAGLQF PVGRVHRLLR KGNYAERVGA GAPVYMAAVL
 5               H2A.4        <SGRGKQGGKA RAK---A--K S̲RSSRAGLQF PVGRVHRLLR KGNYAERVGA GAPVYMAAVL
 6 Rat           H2A.α        <SGRGKQGGKA RAK---A--K S̲RSSRAGLQF PVGRVHRLLR KGNYAERVGA GAPVYLAAVL
 7               H2A.B₁       <SGRGKQGGKA RAK---A--K TRSSRAGLQF PVGRVHRLLR KGNYAERVGA GAPVYLAAVL
 8               H2A.B₂       <SGRGKQGGKA RAK---A--K S̲RSSRAGLQF PVGRVHRLLR KGNYAERVGA GAPVYLAAVL
 9 Mouse         *H2A          SGRGKQGGKA RAK---A--K TRSSRAGLQF PVGRVHRLLR KGNYS̲ERVGA GAPVYLAAVL
10 Chicken       H2A.1        <SGRGKQGGKA RAK---A--K S̲RSSRAGLQF PVGRVHRLLR KGNYAERVGA GAPVYLAAVL
11 Frog          *H2A          SGRGKQGGKT̲ RAK---A--K TRSSRAGLQF PVGRVHRLLR KGNYAERVGA GAPVYLAAVL
12 Trout         *H2A          SGRGKT̲GGKA RAK---A--K TRSSRAGLQF PVGRVHRLLR KGNYAERVGA GAPVYLAAVL
13 Nematode      H2A          XSGRGK-GGKA KT̲-GG̲K̲A--K S̲RSSRAGLQF PVGRL̲HR̲ILR KGNYAQ̲RVGA GAPVYLAAVL
14 Cuttlefish    H2A          <SGRGK-GGK̲V KGK---S̲--K TRSSRAGLQF PVGRI̲HRLLR KGNYAQ̲RVGA GAPVYLAAVM̲
15 Starfish      H2A          <SGRGK-GGKA RAK---A--K SR̲S̲RAGLQF PVGRVHR̲FLR KGNYAE̲RVGA GAPVY̲L̲AAVM̲
16 Sea urchin    *H2A-1        SGRGK-GK̲AK̲ GT̲K---S̲--K TRSSRAGLQF PVGRVHR̲FL̲K KGNYGS̲RVGA GAPVYLAAVL
17               *H2A-2        SGRGK-GA̲KA KS̲K---A--K S̲RSSRAGLQF PVGRVHR̲FL̲K KGNYGN̲RVGA GAPVYLAAVL
18               *H2A-3        SGRGK-GA̲KA KG̲K---A--K S̲RSSRAGLQF PVGRVHR̲FLR KGNYAN̲RVGA GAPVYLAAVL
19               *H2A          SGRGK-S̲GKA RT̲K---A--K S̲RSSRAGLQF PVGRVHR̲FLR KGNYAK̲RVGG GAPVYM̲AAVL
20               H2A          <SGRGK-GA̲KG̲ K̲AK---A--K S̲RSSRAGLQF PVGRVHR̲FLR KGNYAN̲RVGA GAPVYLAAVL
21               H2A          <SGRGK-GA̲KA KG̲K---A--K S̲RSSRAGLQF PVGRVHR̲FLR KGNYAN̲RVGA GAPVYLAAVL
22               *H2A          SGRGK-S̲GKA RT̲K---A--K TRSSRAGLQF PVGRVHR̲FLR KGNYAK̲RVGG GAPVYM̲AAVL
23 Marine worm   H2A          <SGRGK-GGKA KG̲K---S̲--K S̲RSSRAGLQF PVGRI̲HRLLR KGNYAER̲I̲GA GAPVYLAAVM̲
24 Wheat         H2A₍₁₎        AGR-K-GG̲D̲- RKK---A--V̲ TRSV̲KAGLQF PVGRI̲GRYL̲K KGR̲YAQ̲RVGS̲ GAPVYLAAVL
25               H2A₍₂₎       <MD̲ᴳS̲KᴬKK VAᴬK̲KFGG-P̲ RKK---S̲--V̲ TᴷR̲S̲IKAGLQF PVGRI̲GRYL̲K KGR̲YAQ̲RVGS̲ GAPVYLAAVL
26 Ciliate protozoan H2A(1)   <ST̲T̲G̲K̲-GGKA KG̲K̲T̲ASSKQV̲ SR̲SARAGLQF PVGRI̲SRFL̲K HG̲R̲YS̲ER̲I̲G̲T̲ GAPVYLAAVL
27               H2A(2)       <ST̲T̲GK-GGKA KG̲K̲T̲ASSKQV̲ SR̲SARAGLQF PVGRI̲SRFL̲K NG̲R̲YS̲ER̲I̲G̲T̲ GAPVYLAAVL
28 Yeast         *H2A1/2       SG-GK-GGKA -G̲-S̲AAKASQ̲ SR̲SAKAGL̲TF PVGRVHRLLR R̲GNYAQ̲RI̲G̲S̲ GAPVYL̲T̲AVL
29               *H2A1/2       SG-GK̲S̲GGKA ---A̲V̲AKSAQ̲ SR̲SA̲ᵏ̲AGLAF PVGRVHRLLR KGNYAQ̲RVGA GAPVYLAAVL
30 Chicken       *H2AF         AG̲- GKA̲GKD̲S̲GKA K̲AK---A--V̲ SR̲SQ̲RAGLQF PVGRI̲HRHL̲K̲T̲RT̲T̲SHG̲RVGA I̲A̲AVYS̲AAI̲L̲
```

PARTIAL SEQUENCES

```
31 Human         *H2A          SGRGKQGGKA RAK---A--K TRSSRAGLQF PVGRVHRLLR K..YS̲ERVGA GAPVYLAAVL
32 Fruit fly     *H2A          SGRGK-GGK̲V̲ KG̲K---A--K S̲RSN̲RAGLQF PVGRI̲HRLLR KGNYAERVGA GAPVYLAAVM̲
33               *H2A          SGRGK-GGK̲V̲ KG̲K---A--K S̲RSN̲RAGLQF PVGRI̲HRHQ̲R̲ KGNYAERVGA GAPVYLAAVM̲
```

```
               80        90       100       110       120       130       140      150       160
1  EYLTAEILEL AGNAARDNKK TRIIPRHLQL AIRNDEELNK LLGKVTIAQG GVLPNIQAVL LP---KKTES HHKA-KGK                        (129)

2  EYLTAEILEL AGNAARDNKK TRIIPRHLQL AIRNDEELNK LLGKVTIAQG GVLPNIQAVL LP---KKTES HHKA-KGK                        (129)

3  EYLTAEILEL AGNAARDNKK TRIIPRHLQL AIRNDEELNK LLGRVTIAQG GVLPNIQAVL LP---KKTES HHKA-KGK                        (129)

4  EYLTAEILEL AGNAARDNKK TRIIPRHLQL AIRNDEELNK LLGKVTIAQG GVLPNIQAVL LP---KKTES HHKA-KGK                        (129)

5  EYLTAEILEL AGNAARDNKK TRIIPRHLQL AIRNDEELNK LLGKVTIAQG GVLPNIQAVL LP---KKTES H-KA-KGK                        (129)

6  EYLTAEILEL AGNAARDNKK TRIIPRHLQL AIRNDEELNK LLGKVTIAQG GVLPNIQAVL LP---KKTES HHKA-KGK                        (129)

7  EYLTAEILEL AGNAARDNKK TRIIPRHLQL AIRNDEELNK LLGRVTIAQG GVLPNIQAVL LP---KKTES HHKA-KGK                        (129)

8  EYLTAEILEL AGNAARDNKK TRIIPRHLQL AIRNDEELNK LLGRVTIAQG GVLPNIQAVL LP---KKTES HHKA-KGK                        (129)

9  EYLTAEILEL AGNAARDNKK TRIIPRHLQL AIRNDEELNK LLGRVTIAQG GVLPNIQAVL LP---KKTES HHKA-KGK                        (129)

10 EYLTAEILEL AGNAARDNKK TRIIPRHLQL AIRNDEELNK LLGKVTIAQG GVLPNIQAVL LP---KKTDS H-KA-KAK                        (128)

11 EYLTAEILEL AGNAARDNKK TRIIPRHLQL AVRNDEELNK LLGGVTIAQG GVLPNIQFVL LP---KKTES S-KSAKSK                        (129)

12 EYLTAEILEL AGNAARDNKK TRIIPRHLQL AVRNDEELNK LLGGVTIAQG GVLPNIQAVL LP---KKTE- --KAVKAK                        (127)

13 EYLAAEVLEL AGNAARDNKK TRIAPRHLQL AVRNDEELNK LLAGVTIAQG GVLPNIQAVL LP---KKTGG DKE                             (126)

14 EYLAAEVLEL AGNAARDNKK SRIIPRHLQL AIRNDEELNK LLSGVTIAQG GVLPNIQAVL LP---KKTQ- --KA--AK                        (124)

15 EYLAAEILEL AGNAARDNKK TRINPRHLQL AIRNDEELNK LLSGVTIAQG GVLPNIQAVL LP---KKTA- --KA--AK                        (124)

16 EYLAAEILEL AGNAARDNKK SRIIPRHLQL AVRNDEELNK LLGGVTIAQG GVLPNIQAVL LP---KKTA- ----KASK                        (124)

17 EYLAAEILEL AGNAARDNKK SRIIPRHLQL AVRNDEELNK LLGGVTIAQG GVLPNIQAVL LP---KKTG- ----KSA                         (123)

18 EYLAAEILEL AGNAARDNKK TRIIPRHLQL AIRNDEELNK LLGGVTIAQG GVLPNIQAVL LP---KKTGS --KS--SK                        (125)

19 EYLTAEILEL AGNAARDNKK SRIIPRHLQL AVRNDEELNK LLGGVTIAQG GVLPNIQAVL LP---KKTG- -----KSS                        (123)

20 EYLAAEILEL AGNAARDNKK TRIIPRHLQL AIRNDEELNK LLGGVTIAQG GVLPNIQAVL LP---KKTGS --KS--SK                        (125)

21 EYLAAEILEL AGNAARDNKK TRIIPRHLQL AIRNDEELNK LLGGVTIAQG GVLPNIQAVL LP---KKTGS --KS--SK                        (125)

22 EYLTAEILEL AGNAARDNKK SRIIPRHLQL AVRNDEELNK LLGGVTIAQG GVLPNIQAVL LP---KKTA- -----KSS                        (123)

23 EYLAAEVLEL AGNAARDNKK TRIIPRHLQL AIRNDEELNK LLSGVTIAQG GVLPNIQAVL LP---KKTQ- -----KSK                        (123)
```

```
24 EYLAAEVLEL AGNAAKDNKK TRIIVPRHLLL AVRNDQELGR LLAGVTIAHG GVIPNINSVL LP---KKSPA AAEKEA---K SPKKKTSTKS PKKKVAAKE   (145)

25 EYLAAEVLEL AGNAAKDNKK TRIIVPRHLLL AIRNDQELGR LLSGVTIAHG GVIPNINPVL LP---KKAAE KAEKAGAAPK SPKKIT--TKS PKKA        (151)

26 EYLAAEVLEL AGNAAKDNKK TRIVPRHILL AIRNDEELNK LMANTITADG GVLPNINPML LPSKSKKTES RGQASQDI                          (137)

27 EYLAAEVLEL AGNAAKDNKK TRIVPRHILL AIRNDEELNK LMANTITADG GVLPNINPML LPSKTKKTSE AEH                              (132)

28 EYLAAEILEL AGNAARDNKK TRIIPRHLQL AIRNDDELNK LLGNVTIAQG GVLPNIHQNL LP---KKSAK ATTAKASQEL                        (131)

29 EYLAAEILEL AGNAARDNKK TRIIPRHLQL AIRNDEELNK LLGHVTIAQG GVVPNINAHL LP---KTSGKQ-SQ TGKPSQEL                       (130)

30 EYLTAEVLEL AGNASKDLKV KRITPRHLQL AIRGDEELDS LI-KATIAGG GVIPHIHKSL IG---KKGQQ K-TA                             (127)
```

PARTIAL SEQUENCES

```
31 EYLTAEILEL AGNAARDNKK TRIIPRHLQL AIRNDEELNK LLGRVTIAQG GVLPNIQAVL LP---KKTES HHKA-KGK

32 EYLAAEVLEL AGNAARDNKK TRIIPRHLQL AIRNDEELNK LLSGVTIAQG GVLPNIQAVL LP---KKTE- --K....

33 EYLAAEVLEL AGNAARDNKK ....
```

Histones H2A

Seq.	Organism	Source	Reference
1	BOS TAURUS	THYMUS	YEOMAN L.C., OLSON M.O.J., SUGANO N., JORDAN J.J., TAYLOR C.W., STARBUCK W.C., BUSCH H.; J. BIOL. CHEM. 247:6018-6023(1972).
2-5	HOMO SAPIENS	SPLEEN	HAYASHI T., OHE Y., HAYASHI H., IWAI K.; J. BIOCHEM. 88:27-34(1980).
6-8	RATTUS NORVEGICUS	CHLOROLEUKEMIC TUMORS	LAINE B., SAUTIERE P., BISERTE G.; BIOCHEMISTRY 15:1640-1645(1976).
9	MUS MUSCULUS	DNA	LIU T-J., LIU L., MARZLUFF W.F.; NUCL. ACIDS RES. 15:3023-3039(1987).
10	GALLUS GALLUS	ERYTHROCYTE	LAINE B., KMIECIK D., SAUTIERE P., BISERTE G.; BIOCHIMIE 60:147-150(1978).
		DNA	D'ANDREA R.,HARVEY R., WELLS J.R.E.; NUCL.ACIDS RES. 9:3119-3128(1981).
		DNA	WANG S.W., ROBINS A.J., D'ANDREA R., WELLS J.R.E.; NUCL. ACIDS RES. 13:1369-1387(1985).
11	XENOPUS LAEVIS	DNA	MOORMAN A.F.M., DE BOER P.A.J., DE LAAF R.T.M., DESTREE O.H.J.; FEBS LETT. 144:235-241(1982).
12	SALMO GAIRDNERII	DNA	CONNOR W., STATES J.C., MEZQUITA J., DIXON G.H.; J. MOL. EVOL. 20:236-250(1984).
13	CAENORHABDITIS ELEGANS	NEMATODE	VANFLETEREN J.R., VAN BUN S.M., VAN BEEUMEN J.J; BIOCHEM. J. 243:297-300(1987).
14	SEPIA OFFICINALIS	TESTIS	WOUTERS-TYROU D., MARTIN-PONTHIEU A., BRIAND G., SAUTIERE P., BISERTE G.; EUR. J. BIOCHEM. 124:489-498(1982).
15	ASTERIAS RUBENS	GONADS	MARTINAGE A., BELAICHE D., DUPRESSOIR T., SAUTIERE P.; EUR. J. BIOCHEM. 130:465-472(1983).

16-18	PSAMMECHINUS MILIARIS	DNA	BUSSLINGER M., BARBERIS A.; PROC. NATL. ACAD. SCI. U.S.A. 82:5676-5680(1985).
19	PSAMMECHINUS MILIARIS	DNA	SCHAFFNER W., KUNZ G., DAETWYLER H.,TELFORD J., SMITH H.O., BIRNSTIEL M.L.; CELL 14:655-671(1978).
20	PSAMMECHINUS MILIARIS	GONADS	WOUTERS D., SAUTIERE P., BISERTE G.; EUR. J. BIOCHEM. 90:231-239(1978).
21	PARECHINUS ANGULOSUS	SPERM	STRICKLAND W.N., STRICKLAND M.S., DE GROOT P.C., VON HOLT C.; EUR. J. BIOCHEM. 109:151-158(1980).
22	STRONGYLOCENTROTUS PURPURATUS	DNA	SURES I., LOWRY J., KEDES L.H.; CELL 15:1033-1044(1978).
23	SIPUNCULUS NUDUS	ERYTHROCYTE	KMIECIK D., COUPPEZ M., BELAICHE D., SAUTIERE P.; EUR. J. BIOCHEM. 135:113-121(1983).
24	TRITICUM AESTIVUM	GERM	RODRIGUES J. DE A., BRANDT W.F., VON HOLT C.; EUR. J. BIOCHEM. 150:499-506(1985).
25	TRITICUM AESTIVUM	GERM	RODRIGUES J.DE A., BRANDT W.F., VON HOLT C.; EUR. J. BIOCHEM. 173:555-560(1987).
26,27	TETRAHYMENA PYRIFORMIS	MACRONUCLEUS	FUSAUCHI Y., IWAI K.; J. BIOCHEM. 93:1487-1497(1983).
28	SACCHAROMYCES CEREVISIAE	DNA	CHOE J., KOLODRUBETZ D., GRUNSTEIN M.; PROC. NATL. ACAD. SCI. U.S.A. 79:1484-1487(1982).
29	SCHIZOSACCHAROMYCES POMBE	DNA	CHOE J., SCHUSTER T., GRUNSTEIN M.; MOL. CELL. BIOL. 5:3261-3269(1985).
30	GALLUS GALLUS	DNA	HARVEY R.P., WHITING J.A., COLES L.S., KRIEG P.A., WELLS J.R.E.; PROC. NATL. ACAD. SCI. USA 80:2819-2823(1983).
31	HOMO SAPIENS	DNA	ZHONG R., ROEDER R.G., HEINTZ N.; NUCL. ACIDS RES. 11:7409-7425(1983).
32	DROSOPHILA MELANOGASTER	DNA	GOLDBERG M.L.; UNPUBLISHED (1979) THESIS, UNIVERSITY OF STANFORD, U.S.A.
33	DROSOPHILA VIRILIS	DNA	DOMIER L.L., RIVARD J.J., SABATINI L.M., BLUMENFELD M.; J. MOL. EVOL. 23:149-158(1986).

```
                                 1    10    20    30    40    50    60    70

 1  Human, bovine, rat  H2B.1                         PEPAKSAP-A PKK-GSKKA- --VTKAQKKD GKKRKRSRKE
 2  Human               *H2B                          PEPAKSAP-A PKK-GSKKA- --VTKAQKKD GKSAAH-RKE
 3  Mouse               *H2B                          PEPAKSAP-A PKK-GSKKA- --LTKAQKKD GKKRKRSRKE
 4                      *H2B(A)                        PEPAKSAP-A PKK-GSKKA- --VTKAQKKD GKKRKRSRKE
 5                      *H2B(B)                        PEPTKSAP-A PKK-GSKKA- --VTKAQKKD GKKRKRSRKE
 6  Chicken             H2B                            PEPAKSAP-A PKK-GSKKA- --VTKTQKKG DKKRKKSRKE
 7                      *H2B                           PEPAKSAP-AS PKK-GSKKA- --VTKTQKKG DKKRKKSRKE
 8                      *H2B                           PEPAKSAP-A PKK-GSKKA- --VTKTQKKG DKKRKRARKE
 9  Frog                *H2B                           PEPAKSAP-A AKK-GSKKA- --ATKTQKKD GKKRRKTRKE
10  Trout               *H2B                           PEPAKSAP-- -KK-GSKKA- --VTKTAGKG GKKRKSRKE
11                      H2B                            PEPAKSAP-- -KK-GSKKA- --VTKTAGKG GKKRKSRKE
12  Fruit fly           H2B                            PPKTSGKA AKKAG-K-AQ KNITKT---D -KKKRKRKE
13  Nematode            H2B                            APPKPSK-- -K-GAKKAK- --KTVTKPKD GKKRKKARKE
                                                            A      A
14  Limpet              H2B                            PPKVSSKG AKKAG-K-A- K-AARS-G-D -KKRKRRRKE
15  Sea urchin          *H2B                           AP-TA-QV AKK-GSKKAV K-GTKTAX-G GKKRNRKRKE
16                      *H2B(1)                        AP-TA-QV AKK-GSKKAV K-APRPS--G GKKRNRKRKE
17                      *H2B-1       PSQ RS-----PTK RSPTKRSPQK GA-------G KGGKGSKRGG K-ARRRGG-A AVRRRRRRE
18                      *H2B                           AP-TA-QV AKK-GSKKAV K-APRPS--G GKKRNRKRKE
19                      *H2B                           AP-TG-QV AKK-GSKKAV K-PPRAS--G GKKRHRKRKE
20                      *H2B-1                         PAKAQAA GKK-GSKKA- K-APRPS--G DKKRRRKRKE
21                      *H2B-1       PSQ KS-----PTK RSPTKRSPQK GG-------- KGAKRGGKA- G-KRRRG--V AVKRRRRRRE
22                      H2B(1)       PSQ KSPTKRSPTK RSPTKRSPQK GG-------- KGGKGAKRGG K-AGKRRRGV QVKRRRRRRE
23                      H2B(2)       P-- RSPAKTSPRK GSPRKGSP-- ---SRKASPK RGGKGAKRAG K-GGRRRR-V -VKRRRRRRE
24                      H2B(3)       P-- RSPAKTSPRK GSPRKGSPRK GSPSRKASPK RGGKGAKRAG K-GGRRRR-V -VKRRRRRRE
25                      H2B-2             PRSPSK SSPRKGSPRK GSPR-KGSPK RGGKGAKRAG K-GGRRRN-V -VKRRRRRRE
26  Starfish            H2B                    (Di-CH3)PPKPSGKG QKKAG-K-A- K-GAPR-T-D -KKRRKRKE
27                      H2B                    X-PPK-SGKG QKKAG-K-A- K-GAPR-S-D -KKRRRKRKE
28  Yeast               *H2B(1/2)              S-SAAK-AEKKPAS KAPAEKKPAA KKTST--S-T-D GKKRSK-A-RKE
                                                SA                                    V       V
29                      *H2B1                  SAAEKKPAS KAPAGK--AP RDTM---KSA DKKRGKNRKE
30  Wheat               H2B(2)       X-PNK KPAAENKVEK AAEKTPAGKK PKAEKRLP-A GKT-ASKEAG GEGKTRGRKK GSKAKKGV-E
31  Ciliate protozoan   H2B                 (Tri-CH3)APKKAP AAA--EKK-- --VKKAPTTE -KKNKKKRSE
32                      *H2B                           APKKAP AAAA-EKK-- --VKKAPTTE -KKNKKKRSE
33                      *H2B                           APKKAP AATT-EKK-- --VKKAPTTE -KKNKKKRSE

PARTIAL SEQUENCES

34  Sea urchin          *H2B-2                         PAK-QTS GKGAKKAGKA K--GRPAG-A SKTRRRKRKE
35                      *H2B-2                                              ....RN VVKRRRRRRE
36                      H2B-2        PKS-PSK SS-----PRK GSPRKGSPRK GSPK-RGGK- ----GA-KRA GKGG-RR... .
37                      H2B          P-- KSPSKGSPRK GSPRKGS--- --PTRRGA-- -GGKGAKRAG K-GGRRRT-X VAKRRXXRRE
38                      H2B               AP-----TGQ VAKKGS.... .......... .........K KAVKAPRPSG GKKRBXKRKE
39  Crocodile           H2B                            PEPAKSAP-A PKK-GSKKA- --VTKTQKKG DKKR......
```

512

```
                  80        90       100       110       120        130       140       150       160

1   ŠYSVYVYKVL KQVHPDTGIS SKAMGIMNSF VNDIFERIAG EASRLAHYNK RŠTITŠREIQ TAVRLLLPGE LAKHAVSEGT KAVTKYTSSK   (125)

2   SYSIYVYKVL KQVHPDTGIS SKAMGIMNSF VNDIFERIAG EASRLAHYNK RSTITSREIQ TAVRLLLPGE LAKHAVSEGT KAVTKYTSAK   (124)

3   SYSVYVYKVL KQVHPDTGIS SKAMGIMNSF VNDIFERIAG EASRLAHYNK RSTITSREIQ TAVRLLLPGE LAKHAVSEGT KAVTKYTSSK   (125)

4   SYSVYVYKVL KQVHPDTGIS SKAMGIMNSF VNDIFERIAS EASRLAHYNK RSTITSREIQ TAVRLLLPGE LAKHAVSEGT KAVTKYTSSK   (125)

5   SYSVYVYKVL KQVHPDTGIS SKAMGIMNSF VNDIFERIAG EASRLAHYNK RSTITSREIQ TAVRLLLPGE LAKHAVSEGT KAVTKYTSSK   (125)

6   SYSIYVYKVL KQVHPDTGIS SKAMGSMNSF VNDIFERIAG EASRLAHYNK RSTITSREIQ TAVRLLLPGE LAKHAVSEGT KAVTKYTSSK   (125)

7   SYSIYVYKVL KQVHPDTGIS SKAMGIMNSF VNDIFERIAG EASRLAHYNK RSTITSREIQ TAVRLLLPGE LAKHAVSEGT KAVTKYTSSK   (125)

8   SYSIYVYKVL KQVHPDTGIS SKAMSIMNSF VNDIFERIAG EASRLAHYNK RSTITSREIQ TAVRLLLPGE LAKHAVSEGT KAVTKYTSSK   (125)

9   SYAIYVYKVL KQVHPDTGIS SKAMSIMNSF VNDVFERIAG EASRLAHYNN RSTITSREIQ TAVRLLLPGE LAKHAVSEGT KAVTKYTSAK   (125)

10  SYAIYVYKVL KQVHPDTGIS SKAMGIMNSF VNDIFERIAG ESSRLAHYNK RSTITSREIQ TAVRLLLPGE LAKHAVSEGT KAVTKYTSSK   (123)

11  SYAIYVYKVL KQVHPDTGIS SKAMGIMNSF VNDIFERIAG ESSRLAHYNK RSTITSREIQ TAVRLLLPGE LAKHAVSEGT KAVTKYTSSK   (123)

12  SYAIYIYKVL KQVHPDTGIS SKAMSIMNSF VNDIFERIAA EASRLAHYNK RSTITSREIQ TAVRLLLPGE LAKHAVSEGT KAVTKYTSSK   (122)

13  SYSVYIYRVL KQVHPDTGVS SKAMSIMNSF VNDVFERIAŠ EASRLAHYNK RSTIŠSREIQ TAVRLILPGE LAKHAVSEGT KAVTKYTSSK   (122)
                                                       A

14  SYSIYIYKVL KQVHPDTGVS SKAMSIMNSF VNDIFERIAA EASRLAHYNK RSTITSREIQ TAVRLLLPGE LAKHAVSEGT KAVTKYTSSK   (121)

15  SYGIYIYKVL KQVHPDTGIS SRAMVIMNSF VNDIFERIAG ESSRLAQYNK KSTISSREIQ TAVRLILPGE LAKHAVSEGT KAVTKYTTSK   (123)

16  SYGIYIYKVL KQVHPDTGIS SRAMVIMNSF VNDIFERIAG ESSRLAQYNK KSTISSREIQ TAVRLILPGE LAKHAVSEGT KAVTKYTTSK   (122)

17  SYGIYIYKVL KQVHPDTGIS SRGMSIMNSF VNDVFERVAA EASRLIKYNR RSTVSSREIQ TAVRLLLPGE LAKHAVSEGT KAVTKYTTSR   (139)

18  SYGIYIYKVL KQVHPDTGIS SRAMIIMNSF VNDIFERIAG ESSRLAQYNK KSTISSREIQ TAVRLILPGE LAKHAVSEGT KAVTKYTTSK   (122)

19  SYGIYIYKVL KQVHPDTGVS SRAMIIMNSF VNDIFERIAG EASRLTQYNK KSTISSREIQ TAVRLLLPGE LAKHAVSEGT KAVTKYTTAK   (122)

20  SYGIYIYKVL KQVHPDTGIS SRAMSIMNSF VNDVFERIAA EASRLAHYNK KSTITSREVQ TAVRLLLPGE LAKHAVSEGT KAVTKYTTSK   (122)

21  SYGIYIYKVL KQVHPDTGIS SRAMSVMNSF VNDVFERIAS EAGRLTTYNR RNTVSSREVQ TAVRLLLPGE LAKHAVSEGT KAVTKYTTSR   (136)

22  SYGIYIYKVL KQVHPDTGIS SRAMSVMNSF VNDIFERIAA EAGRLTTYNR RSTVSSREVQ TAVRLLLPGE LAKHAVSEGT KAVTKYTTSR   (144)

23  SYGIYIYKVL KQVHPDTGIS SRAMSVMNSF VNDVFERIAG EASRLTSANR RSTVSSREIQ TAVRLLLPGE LAKHAVSEGT KAVTKYTTSR   (143)

24  SYGIYIYKVL KQVHPDTGIS SRAMSVMNSF VNDVFERIAS EASRLTSANR RSTVSSREIQ TAVRLLLPGE LAKHAVSEGT KAVTKYTTSR   (148)

25  SYGSYIYRVL KQVHPDTGIS SRGMSVMNSF VNDVFERIAG EASRLCQANR RRTISSREIQ TAVRLLLPGE LAKHAVSEGT KAVTKYTTSR   (142)

26  SYGIYIYKVM KQVHPDTGIS SRAMSIMNSF VNDIFERIAA EASRLAHYNK KSTITSREVQ TAVRLLLPGE LAKHAVSEGT KAVTKYTTSK   (121)

27  SYGIYIYKVM KQVHPDTGIS SRAMSIMNSF VNDIFERIAA EASRLAHYNK KSTITSREVQ TAVRLLLPGE LAKHAVSEGT KAVTKYTTSK   (120)

28  IYSSYIYKVL KQTHPDTGIS QKSMSILNSF VNDIFERIAT EASKLAAYNK KSTISAREIQ TAVRLILPGE LAKHAVSEGT RAVTKYSSSIQA (130)

29  IYSSYIYRVL KQVHPDTGIS NQAMPILNSF VNDIFERIAT EASKLAAYNK KSTISSREIQ TAVRLILPGE LAKHAVIEGT KSVTKYSSSAQ  (125)

30  IYKIYIFKVL KQVHPDIGIS SKAMSIMNSF INDIFEKLAG ESAKLARYNK KPTITSREIQ TSVRLVLPGE LAKHAVSEGT KAVTKFTSS-   (149)

31  IFAIYIFKVL KQVHPDVGIS KKAMNIMNSF INDSFERIAL ESSKLVRFNK RRTLSSREVQ TAVKLLLPGE LARHAISEGT KAVTKFSSSIN  (120)

32  IFAIYIFKVL KQVHPDVGIS KKAMNIMNSF INDSFERIAL ESSKLVRFNK RRTLSSREVQ TAVKLLLPGE LARHAISEGT KAVTKFSSSIN  (121)

33  IFAIYIFKVL KQVHPDVGIS KKAMNIMNSF INDSFERIAL ESSKLVRFNK RRTLSSREVQ TAVKLLLPGE LARHAISEGT KAVTKFSSSSN  (121)

PARTIAL SEQUENCES

34  SYGIYIYKVL KQVHPDTGIS SKAMSIMNSF VNDVFERIAG EASRLAHY.. ..

35  SYGIYIYKVL KQVHPDTGIS SRGMSVMNSF VNDVFERIAG EASRLTSANR RSTISSREIQ TAVRLLLPGE LAKHAVSEGT KAVTKYTTARR

36  ....

37  SYGIY..... .......... ...MSIMNSF VNDVFERIAA EASRLŜHĈNR RSTISSRXI. .....................AR
                                                          A  A
                                                          Y  Y

38  .......... .......... ...MSIMNSF VNDĪFERIAG EASRL...
                                        V

39  .......... .......... ......MNSF VNDIFERIAG EASRLAHYNK RSTITSR... .
```

Histones H2B

Seq.	Organism	Source	Reference
1	HOMO SAPIENS	SPLEEN	OHE Y., HAYASHI H., IWAI K.; J. BIOCHEM. 85:615-624(1979).
	BOS TAURUS	THYMUS	IWAI K., HAYASHI H., ISHIKAWA K.; J. BIOCHEM. 72:357-367(1972).
	RATTUS NORVEGICUS	THYMUS	MARTINAGE A., MANGEAT P., SAUTIERE P., MARCHIS-MOUREN G., BISERTE G.; BIOCHIMIE 61:61-69(1979).
2	HOMO SAPIENS	DNA	ZHONG R., ROEDER R.G., HEINTZ N.; NUCL. ACIDS RES. 11:7409-7425(1983).
3	MUS MUSCULUS	DNA	SITTMAN D.B., GRAVES R.A., MARZLUFF W.F.; NUCL. ACIDS RES. 11:6679-6697(1983).
4,5	MUS MUSCULUS	DNA	LIU T-J., LIU L., MARZLUFF W.F; NUCL. ACIDS RES. 15:3023-3039(1987).
6	GALLUS GALLUS	ERYTHROCYTE	VAN HELDEN P. D., STRICKLAND W.N., STRICKLAND M., VON HOLT C.; BIOCHIM. BIOPHYS. ACTA 703:17-20(1982).
7	GALLUS GALLUS	DNA	HARVEY R.P., ROBINS A.J., WELLS J.R.E.; NUCL. ACIDS RES. 10:7851-7863(1982).
		DNA	GRANDY D.K., ENGELS J.D., DODGSON J.B.; J. BIOL. CHEM. 257:8577-8580(1982).
8	GALLUS GALLUS	DNA	GRANDY D.K., DODGSON J.B.; NUCL. ACIDS RES. 15:1063-1080(1987).
9	XENOPUS LAEVIS	DNA	MOORMAN A.F.M., DE BOER P.A.J., DE LAAF R.T.M., DESTREE O.H.J.; FEBS LETT. 144:235-241(1982).
10	SALMO GAIRONERII	DNA	CANDIDO E.P.M., DIXON G.H.; PROC. NATL. ACAD. SCI. USA 69:2015-2019(1972).
			WINKFEIN R.J., WAYNE C., MEZQUITA J., DIXON G.H.; J. MOL. EVOL. 22:1-19(1985).
11	SALMO TRUTTA	TESTIS	KOOTSTRA A., BAILEY G.S.; BIOCHEMISTRY 17:2504-2510(1978).
12	DROSOPHILA MELANOGASTER	EMBRYO	ELGIN S.C.R., SCHILLING J., HOOD L.E.; BIOCHEMISTRY 18:5679-5685(1979).
			GOLDBERG M.L; UNPUBLISHED (1979) THESIS, UNIVERSITY OF STANFORD, U.S.A.
13	CAENORHABDITIS ELEGANS	NEMATODE	VANFLETEREN J.R., VAN BUN S.M., DELCAMBE L.L., VAN BEEUMEN J.J.; BIOCHEM. J. 235:769-773(1986).
14	PATELLA GRANATINA	GONAD	VAN HELDEN P.D., STRICKLAND W.N., BRANDT W.F., VON HOLT C.; EUR. J. BIOCHEM. 93:71-78(1979).
15	STRONGYLOCENTROTUS PURPURATUS	DNA	SURES I., LOWRY J., KEDES L.H.; CELL 15:1033-1044(1978).
16	STRONGYLOCENTROTUS PURPURATUS	DNA	LIEBERMANN D., HOFFMAN-LIEBERMANN B., WEINTHAL J., CHILDS G., MAXSON R., MAURON A., COHEN S.N., KEDES L.; NATURE 306:342-347(1983).
17	STRONGYLOCENTROTUS PURPURATUS	DNA	LAI Z-C., LIEBER T., CHILDS G.; NUCL. ACIDS RES. 14:9218(1986).
18	PSAMMECHINUS MILIARIS	CLONE H19	BUSSLINGER M., PORTMANN R., IRMINGER J.C., BIRNSTIEL M.L.; NUCL. ACIDS RES. 8:957-977(1980).
19	PSAMMECHINUS MILIARIS	CLONE H22	BUSSLINGER M., PORTMANN R., IRMINGER J.C., BIRNSTIEL M.L.; NUCL. ACIDS RES. 8:957-977(1980).
			SCHAFFNER W., KUNZ G., DAETWYLER H., TELFORD J., SMITH H.O., BIRNSTIEL M.L.; CELL 14:655-671(1978).
20,21	PSAMMECHINUS MILIARIS	DNA	BUSSLINGER M., BARBERIS A.; PROC. NATL. ACAD. SCI. U.S.A. 82:5676-5680(1985).

22	PARECHINUS ANGULOSUS	SPERM	STRICKLAND M., STRICKLAND W.N., BRANDT W.F., VON HOLT C.; EUR. J. BIOCHEM. 77:263-275(1977).
23	PARECHINUS ANGULOSUS	SPERM	STRICKLAND W.N., STRICKLAND M., BRANDT W.F., VON HOLT C.; EUR. J. BIOCHEM. 77:277-286(1977).
24	PARECHINUS ANGULOSUS	SPERM	STRICKLAND M., STRICKLAND W.N., BRANDT W.F., VON HOLT C., WITTMANN-LIEBOLD B., LEHMANN A.; EUR. J. BIOCHEM. 89:443-452(1978).
25	LYTECHINUS PICTUS	DNA	LAI., Z-C., CHILDS G.; NUCL. ACIDS RES. 17:6845-6856(1987).
26	ASTERIAS RUBENS	GONAD	MARTINAGE A., BRIAND G., VAN DORSSELAER A., TURNER C.H., SAUTIERE P.; EUR. J. BIOCHEM. 147:351-359(1985).
27	MARTHASTERIAS GLACIALIS	SPERM	STRICKLAND M.S., STRICKLAND W.N., VON HOLT C.; EUR. J. BIOCHEM. 106:541-548(1980).
28	SACCHAROMYCES CEREVISIAE	DNA	WALLIS J.W., HEREFORD L.,GRUNSTEIN M.; CELL 22:799-805(1980).
29	SCHIZOSACCHAROMYCES POMBE	DNA	CHOE J., SCHUSTER T., GRUNSTEIN M.; MOL. CELL. BIOL. 5:3261-3269(1985).
30	TRITICUM AESTIVUM	GERM	BRANDT W.F., RODRIGUES J. DE A., VON HOLT C.; EUR J. BIOCHEM. 173:547-554(1988).
31	TETRAHYMENA PYRIFORMIS	MACRONUCLEUS	HAYASHI H., NOMOTO M., IWAI K.; PROC.JAPAN ACAD. 56(SER B):579-584(1980).
32,33	TETRAHYMENA THERMOPHILA	DNA	NOMOTO M., IMAI N., SAIGA H., MATSUI T., MITA T.; NUCL. ACIDS RES. 15:5681-5697(1987).
34	PSAMMECHINUS MILIARIS	DNA	BUSSLINGER M., BARBERIS A.; PROC. NATL. ACAD. SCI. U.S.A. 82:5676-5680(1985).
35	PSAMMECHINUS MILIARIS	SPERM	BUSSLINGER M., BARBERIS A.; PROC. NATL. ACAD. SCI. U.S.A. 82:5676-5680(1985).
36	PSAMMECHINUS MILIARIS	SPERM	STRICKLAND M., STRICKLAND W.N., BRANDT W.F., VON HOLT C.; BIOCHIM. BIOPHYS. ACTA 536:289-297(1978).
37	ECHINOLAMPAS CRASSA	SPERM	STRICKLAND W.N., STRICKLAND M., VON HOLT C.; (UNPUBLISHED).
38	PARECHINUS ANGULOSUS	EMBRYO	BRANDT W.F., STRICKLAND W.N., STRICKLAND M., CARLISLE L., WOODS D., VON HOLT C.; EUR. J. BIOCHEM. 94:1-10(1979).
39	CROCODILUS NILOTICUS	ERYTHROCYTE	VAN HELDEN P., STRICKLAND W.N., BRANDT W.F., VON HOLT C.; BIOCHIM. BIOPHYS. ACTA 533:278-281(1978).

Histones H3

```
                      1    10        20        30        40        50        60        70
1   Bovine     H3.2   ARTKQTARKS TGGKAPRKQL ATKAARKSAP ATGGVKKPHR YRPGTVALRE IRRYQKSTEL LIRKLPFQRL
2              H3.1   ARTKQTARKS TGGKAPRKQL ATKAARKSAP ATGGVKKPHR YRPGTVALRE IRRYQKSTEL LIRKLPFQRL
3   Human      H3.3   ARTKQTARKS TGGKAPRKQL ATKAARKSAP STGGVKKPHR YRPGTVALRE IRRYQKSTEL LIRKLPFQRL
4             *H3.1   ARTKQTARKS TGGKAPRKQL ATKAARKSAP ATGGVKKPHR YRPGTVALRE IRRYQKSTEL LIRKLPFQRL
5   Mouse     *H3.1   ARTKQTARKS TGGKAPRKQL ATKAARKSAP ATGGVKKPHR YRPGTVALRE IRRYQKSTEL LIRKLPFQRL
6             *H3.3   ARTKQTARKS TGGKAPRKQL ATKAARKSAP ATGGVKKPHR YRPGTVALRE IRRYQKSTEL LIRKLPFQRL
7   Chicken    H3.2   ARTKQTARKS TGGKAPRKQL ATKAARKSAP ATGGVKKPHR YRPGTVALRE IRRYQKSTEL LIRKLPFQRL
8             *H3     ARTKQTARKS TGGKAPRKQL ATKAARKSAP STGGVKKPHR YRPGTVALRE IRRYQKSTEL LIRKLPFQRL
9   Duck      *H3     ARTKQTACKS TGRKAPRKQL ATKAAHKSAP AMGGVKKPHC YRPGTVALHE IHRYQKSTEL LICKLPFQRL
10  Rainbow trout *H3 ARTKQTARKS TGGKAPRKQL ATKAARKSAP ATGGVKKPHR YRPGTVALRE IRRYQKSTEL LIRKLPFQRL
11  Carp       H3     ARTKQTARKS TGGKAPRKQL ATKAARKSAP ATGGVKKPHR YRPGTVALRE IRRYQKSTEL LIRKLPFQRL
12  Shark      H3     ARTKQTARKS TGGKAPRKQL ATKAARKSAP ATGGVKKPHR YRPGTVALRE IRRYQKSTEL LIRKLPFQRL
13  Frog      *H3     ARTKQTARKS TGGKAPRKQL VTKAAKKCAP ATGGVKKPHR YRPGTVALRE IRRYQKSTEL LIRKLPFQRL
14  Sea urchin *H3    ARTKQTARKS TGGKAPRKQL ATKAARKSAP ATGGVKKPHR YRPGTVALRE IRRYQKSTEL LIRKLPFQRL
15  Maize     *H3     ARTKQTARKS TGGKAPRKQL ATKAARKSAP ATGGVKKPHR FRPGTVALRE IRKYQKSTEL LIRKLPFQRL
16  Wheat      H3     ARTKQTARKS TGGKAPRKQL ATKAARKSAP ATGGVKKPHR FRPGTVALRE IRKYQKSTEL LIRKLPFQRL
17  Pea        H3     ARTKQTARKS TGGKAPRKQL ATKAARKSAP ATGGVKKPHR FRPGTVALRE IRKYQKSTEL LIRKLPFQRL
18  Cycad      H3     ARTKQTARKS TGGKAPRKQL ATKAARKSAP ATGGVKKPHR FRPGTVALRE IRKYQKSTEL LIRKLPFQRL
19  Bread mold H3     ARTKQTARKS TGGKAPRKQL ASKAARKSAP STGGVKKPHR YKPGTVALRE IRRYQKSTEL LIRKLPFQRL
20  Yeast      H3     ARTKQTARKS TGGKAPRKQL ASKAARKSAP STGGVKKPHR YKPGTVALRE IRRFQKSTEL LIRKLPFQRL
21  Ciliate protozoan H3 ARTKQTARKS TGGKAPRKQL ATKAARKSAP ATGGVKKPHK FRPGTVALRE IRKYQKSTDL LIRKLPFQRL
22  Mouse     *H3.4   AHTKQTARKS TCGKAPRKQL ATKAAXKSAP STGGVKKSHR YRPDTVALLE IRRYXXSTEL LIHKLPFQRL
PARTIAL SEQUENCES
23  Fruit fly *H3     ARTKQTARKS TGGKAPRKQL ATKAARKSAP PTGGVKKPHR YRPGTVALRE IRRYQKSTEL LIRKLPFQRL
24  Sea urchin H3     ARTKQTARKS TGGKAPRKQL ATKAARKSAP ATGGVKKPHR YRPGTVAL...
25  Limpet     H3     ARTKQTARKS TGGKAPRKQL ATKAARKSAP ATGGVKKPHR YRPGTVAL...
26  Frog      *H3
27  Frog      *H3
```

```
              80        90       100       110       120       130

 1   VREIAQDFKT DLRFQSSAVM ALQEASEAYL VGLFEDTNLC AIHAKRVTIM PKDIQLARRI RGERA   (135)

 2   VREIAQDFKT DLRFQSSAVM ALQEACEAYL VGLFEDTNLC AIHAKRVTIM PKDIQLARRI RGERA   (135)

 3   VREIAQDFKT DLRFQSAAIG ALQEASEAYL VGLFEDTNLC AIHAKRVTIM PKDIQLARRI RGERA   (135)

 4   VREIAQDFKT DLRFQSSAVM ALQEACEAYL VGLFEDTNLC AIHAKRVTIM PKDIQLARRI RGERA   (135)

 5   VREIAQDFKT DLRFQSSAVM ALQEACEAYL VGLFEDTNLC AIHAKRVTIM PKDIQLARRI RGERA   (135)

 6   VREIAQDFKT DLRFQSSAVM ALQEASEAYL VGLFEDTNLC AIHAKRVTIM PKDIQLARRI RGERA   (135)

 7   VREIAQDFKT DLRFQSSAVM ALQEASEAYL VGLFEDTNLC AIHAKRVTIM PKDIQLARRI RGERA   (135)

 8   VREIAQDFKT DLRFQSAAIG ALQEASEAYL VGLFEDTNLC AIHAKRVTIM PKDIQLARRI RGERA   (135)

 9   VREIAQDFKT DLRFQSSAVM ALQEASEAYL VGLFEDTNLC AIHAKRVSIM PKDIQLTRRI RGERA   (135)

10   VREIAQDFKT DLRFQSSAVM ALQEASEAYL VGLFEDTNLC AIHAKRVTIM PKDIQLARRI RGERA   (135)

11   VREIAQDFKT DLRFQSSAVM ALQEASEAYL VGLFEDTNLC AIHAKRVTIM PKDIQLARRI RGERA   (135)

12   VREIAQDFKT DLRFQSSAVM ALQEASEAYL VGLFEDTNLC AIHAKRVTIM PKDIQLARRI RGERA   (135)

13   VREIAQDFKT DLRFQRSAVM ALQEASEAYL VALFEDTNLC AIHAKRVTIM PKDIQLARRI RGERA   (135)

14   VREIAQDFKT ELRFQSSAVM ALQEASEAYL VGLFEDTNLC AIHAKRVTIM PKDIQLARRI RGERA   (135)

15   VREIAQDFKT DLRFQSSAVA ALQEAAEAYL VGLFEDTNLC AIHAKRVTIM PKDIQLARRI RGERA   (135)

16   VREIAQDFKT DLRFQSSAVS ALQEAAEAYL VGLFEDTNLC AIHAKRVTIM PKDIQLARRI RGERA   (135)

17   VREIAQDFKT DLRFQSSAVS ALQEA SEAYL VGLFEDTNLC AIHAKRVTIM PKDIQLARRI RGERA   (135)
                                    A
18   VREIAQDFKT DLRFQSSAVS ALQEAAEAYL VGLFEDTNLC AIHAKRVTIM PKDIQLARRI RGERA   (135)

19   VREIAQDFKS DLRFQSSAIG LLQESVESYL VSLFEDTNLC AIHAKRVTIQ SKDIKLARRL RGERN   (135)

20   VREIAQDFKT DLRFQSSAIG ALQESVEAYL VSLFEDTNLA AIHAKRVTIQ KKDIKLARRL RGERS  (135)
                                   H-F-K-A  EL                     S V
21   VRDIA        DI RFQS   AYL ALQEAAEAYL VGLFEDTNLC AIHARRVTIM TKDMQLARRI RGERF   (135)
          M-M-K-N         Q-I-L
22   VQEIAQDFKT DLRFQSAAIG ALQEASEANR VGLFEDTNLC AIHAKCVTVI PKDIQLAHSI LGERA   (134)

PARTIAL SEQUENCES

23   VREIAQDFKT DLRFQSSAVM ALQEASEAYL VGLFEDTNLC AIHAK..... .........I RGERA

24

25

26   ..NFKT DLRFQSAAIG ALQEASEAYL VGLFEDTNLC AIHAKRVTIM PKDIQLAR.. .

27   ...FKT DLRFQSSAVM ALQEASEAYL VGLFEDTNLC AIHAKRVTIM PKDIQLARRI R...
```

HISTONES H3

Seq.	Organism	Source	Reference
1	BOS TAURUS	THYMUS	MARZLUFF W.F. JR., SANDERS L.A., MILLER D.M., MCCARTY K.S.; J. BIOL. CHEM. 247:2026-2033(1972).
	BOS TAURUS	THYMUS	PATHY L., SMITH E.L.; J. BIOL. CHEM. 250:1919-1920(1975).
2	BOS TAURUS	THYMUS	HOOPER J.A., SMITH E.L.; J. BIOL. CHEM. 248:3255-3260(1973).
3	HOMO SAPIENS	DNA	WELLS D., KEDES L.; PROC. NATL. ACAD. SCI. U.S.A. 82:2834-2838(1985).
		SPLEEN(PARTIAL SEQ.)	OHE Y., IWAI K.; J. BIOCHEM.(TOKYO) 90:1205-1211(1981).
4	HOMO SAPIENS	DNA	ZHONG,R.,ROEDER,G.R.,HEINTZ,N. NUCEIC ACID RES 11,7409-7425(1983)
		DNA(PARTIAL SEQ.)	CLARK S.J., KRIEG P.A., WELLS J.R.E.; NUCL. ACIDS RES. 9:1583-1590(1981).
5	MUS MUSCULUS	DNA	SITTMAN,D.B.,GRAVES,R.A.,MARZLUFF W.F.; NUCL. ACIDS RES. 11:6679-6697(1983).
6	MUS MUSCULUS	DNA	SITTMAN,D.B.,GRAVES,R.A.,MARZLUFF W.F.; NUCL. ACIDS RES. 11:6679-6697(1983).
7	GALLUS GALLUS	ERYTHROCYTES	BRANDT W.F., VON HOLT C.; EUR. J. BIOCHEM. 46:419-429(1974).
		DNA	WANG S-W., ROBINS A.J., D'ANDREA R., WELLS J.R.E.; NUCL. ACIDS RES. 13:1369-1387(1985).
8	GALLUS GALLUS	DNA	ENGEL J.D., SUGARMAN B.F., DODGSON J.B; NATURE 297:434-436(1982).
9	CAIRINA MOSCHATA	DNA	TONJES R., DOENECKE D.; GENE 39:275-279(1985).
10	SALMO GAIRDNERII	DNA	CONNOR W., STATES J.C., MEZQUITA J., DIXON G.H.; J. MOL. EVOL. 20:236-250(1984).
		TESTIS(PARTIAL SEQ.)	CANDIDO E.P.M., DIXON G.H.; PROC. NATL. ACAD. SCI. U.S.A. 69:2015-2019(1972).
11	LETIOBUS BUBALUS	GONADS	HOOPER J.A., SMITH E.L., SOMMER K.R., CHALKLEY R.; J. BIOL. CHEM. 248:3275-3279(1973).
12	PORODERMA AFRICANUS	ERYTHROCYTES	BRANDT W.F., STRICKLAND W.N., VON HOLT C.; FEBS LETT. 40:349-352(1974).
13	XENOPUS LAEVIS	DNA	MOORMAN A.F.M., DE BOER P.A.J., DE LAAF R.T.M., VAN DONGEN W.M.A.M., DESTREE O.H.J.; FEBS LETT.136:45-52(1981).
14	PSAMMECHINUS MILIARIS	DNA CLONE H19	BUSSLINGER M., PORTMANN R., IRMINGER J.C., BIRNSTIEL M.L.; NUCL. ACIDS RES. 8:957-977(1980).
		CLONE H22	SCHAFFNER W., KUNZ G., DAETWYLER H., TELFORD J., SMITH H.O., BIRNSTIEL M.L.; CELL 14:655-671(1978).
			BIRNSTIEL M., PORTMANN R., BUSSLINGER M., SCHAFFNER W., PROBST E., KRESSMANN A.; PROC. ALFRED BENZON SYMP. 13:117-132(1979).
	STRONGYLOCENTROTUS PURPURATUS	DNA	SURES I., LOWRY J., KEDES L.H.; CELL 15:1033-1044(1978).

518

#	Organism	Source	Reference
15	ZEA MAYS	DNA CLONE H3C2, H3C4	CHAUBET N., PHILIPPS G., CHABOUTE M.-E., EHLING M., GIGOT C.; PLANT MOL. BIOL. 6:253-263(1986).
16	TRITICUM AESTIVUM	DNA	TABATA T., FUKASAWA M., IMABUCHI M.; MOL. GEN. GENETICS 196:397-400(1984).
17	PISUM SATIVUM	EMBRYO	PATTHY L., SMITH E.L., JOHNSON J.; J. BIOL. CHEM. 248:6834-6840(1973).
18	ENCEPHALARTOS CAFFER	POLLEN	BRANDT W.F., VON HOLT C., FEBS LETT. 194:278-281(1986).
19	NEUROSPORA CRASSA		WOUDT L.P., PASTINK A., KEMPERS-VEENSTRA A.E., JANSEN A.E.M., MAGER W.H., PLANTA R.J. NUCL. ACIDS RES. 11:5347-5360(1983).
20	SACCHAROMYCES CEREVISIAE		BRANDT W.F., VON HOLT C.; EUR. J. BIOCHEM. 121:501-510(1982).
21	TETRAHYMENA PYRIFORMIS		HAYASHI T., HAYASHI H., FUSAUCHI Y. IWAI K.; J. BIOCHEM.(TOKYO) 95:1741-1749(1984).
22	MUS MUSCULUS	DNA CLONE MM531	SITTMAN D.B., CHIU I.-M., PAN C.-J., COHN R.H., KEDES L.H., MARZLUFF W.F.; PROC. NATL. ACAD. SCI. U.S.A. 78:4078-4082(1981).
23	DROSOPHILA MELANOGASTER	DNA	GOLDBERG M.L.; UNPUBLISHED (1979) THESIS, UNIVERSITY OF STANFORD, U.S.A.
24	PARECHINUS ANGULOSUS	SPERM	BRANDT W.F., STRICKLAND W.N., MORGAN M., VON HOLT C.; FEBS LETT. 40:167-172(1974).
25	PATELLA GRANATINA	SPERM	BRANDT W.F., STRICKLAND W.N., MORGAN M., VON HOLT C.; FEBS LETT. 40:167-172(1974).
26	XENOPUS LAEVIS	DNA	RUBERTI I., FRAGAPANE P., PIERANDREI-AMALDI P., BECCARI E., AMALDI F., BOZZONI I.; NUCL. ACIDS RES. 10:7543-7559(1982).
27	XENOPUS LAEVIS	DNA	W. BAINS PH.D THESIS 1982

HISTONES H4

			1	10	20	30	40	50

1 Bovine,pig,rat H4 <SGRGKGGKGL GKGGAKRH-R ǨVLRDNIQGI TKPAIRRLAR RGGVKRISGL

2 Human,mouse,chicken,frog *H4 SGRGKGGKGL GKGGAKRH-R KVLRDNIQGI TKPAIRRLAR RGGVKRISGL

3 Sea urchin *H4 SGRGKGGKGL GKGGAKRH-R KVLRDNIQGI TKPAIRRLAR RGGVKRISGL

4 Sea urchin H4 <SGRGKGGKGL GKGGAK̄RH-R ǨVLRDNIQGI TKPAIRRLAR RGGVKRISGL

5 Wheat *H4.1 SGRGKGGKGL GKGGAKRH-R KVLRDNIQGI TKPAIRRLAR RGGVKRISGL

6 Wheat *H4.2 SGRD̲KGGKGL GKGGAKRH-R KVLRDNIQGI TKPAIRRLAR RGGVKRISGL

7 Maize *H4 SGRGKGGKGL GKGGAKRH-R KVLRDNIQGI TKPAIRRLAR RGGVKRISGL

8 Pea H4 <SGRGKGGKGL GKGGAK̄RH-R KVLRDNIQGI TKPAIRRLAR RGGVKRISGL

9 Slime mold *H4 SGRGKGGKGL GKGGAKRH-R KVLRDNIQGI TKPAIRRLAR RGGVKRIS̲N̲T̲

10 Bread mold *H4 T̲GRGKGGKGL GKGGAKRH-R K̲I̲LRDNIQGI TKPAIRRLAR ·RGGVKRIS̲A̲M̲

11 Yeast *H4 SGRGKGGKGL GKGGAKRH-R K̲I̲LRDNIQGI TKPAIRRLAR RGGVKRISGL

12 Ciliate protozoan H4 AG-GK̄GGK̄GM GK̄VGAK̄RHS̲K̲R̲ R̲K̲S̲N̲K̲A̲S̲I̲E̲GI TKPAIRRLAR RGGVKRIS̲S̲F̲

13 Ciliate protozoan *H4 AG-GKGGKGM GKVGAKRHS̲R̲ K̲S̲N̲K̲A̲S̲I̲E̲GI TKPAIRRLAR RGGVKRIS̲S̲F̲

14 Ciliate protozoan *H4 AG-GKGGKG̲M̲ GKV̲GAKRHS̲R̲ K̲S̲N̲K̲A̲S̲I̲E̲GI TKPAIRRLAR RGGVKRIS̲S̲F̲

PARTIAL SEQUENCES

15 Ciliate protozoan H4 AG-GK̄GGK̄GM GK̄VGAK̄RHS̲R̲ K̲S̲N̲K̲A̲S̲I̲E̲GI TKPAIRRLAR RGGVKRIS̲S̲F̲

16 Rainbow trout H4 <SGRGK̄GGK̄GL GK̄GGAK̄RH-R ǨV...

17 Fruit fly *H4 TGRGKGGKGL GKGGAKRH-R KVLRDNIQGI TKPAIRRLAR RGGVKRISGL

18 Sea urchin H4   ..

```
                      60        70        80        90        100
    1     IYEETRGVLK VFLENVIRDA VTYTEHAKRK TVTAMDVVYA LKRQGRTLYG FGG    (102)

    2     IYEETRGVLK VFLENVIRDA VTYTEHAKRK TVTAMDVVYA LKRQGRTLYG FGG    (102)

    3     IYEETRGVLK VFLENVIRDA VTYCEHAKRK TVTAMDVVYA LKRQGRTLYG FGG    (102)

    4     IYEETRGVLK VFLENVIRDA VTYCEHAKRK TVTAMDVVYA LKRQGRTLYG FGG    (102)

    5     IYEETRGVLK IFLENVIRDA VTYTEHARRK TVTAMDVVYA LKRQGRTLYG FGG    (102)

    6     IYEETRGVLK IFLENVIRDA VTYTEHARRK TVTAMDVVYA LKRQGRTLYG FGG    (102)

    7     IYEETRGVLK IFLENVIRDA VTYTEHARRK TVTAMDVVYA LKRQGRTLYG FGG    (102)

    8     IYEETRGVLK IFLENVIRDA VTYTEHARRK TVTAMDVVYA LKRQGRTLYG FGG    (102)

    9     IYEETRGVLK IFLENVIRDA VTYTEHARRK TVTAMDVVYA LKRQGRTLYG FGG    (102)

   10     IYEETRGVLK TFLEGVIRDA VTYTEHAKRK TVTSLDVVYA LKRQGRTLYG FGG    (102)

   11     IYEEVRAVLK SFLESVIRDS VTYTEHAKRK TVTSLDVVYA LKRQGRTLYG FGG    (102)

   12     IYDDSRQVLK SFLENVVRDA VTYTEHARRK TVTAMDVVYA LKRQGRTLYG FGG    (102)

   13     IYDDSRQVLK SFLENVVRDA VTYTEHARRK TVTAMDVVYA LKRQGRTLYG FGG    (102)

   14     IYDDSRQVLK SFLENVVRDA VTYTEHARRK TVTAMDVVYA LKRQGRTLYG FGG    (102)

PARTIAL SEQUENCES

   15     IYDD...                    ...MDVVYA LKRQGRT...

   16

   17     IYEETRGVLK VFLENVIRDA VT...

   18     .YEETRGVLK VFLENVIRDA VTYCEHAKRK TVTAMDVVYA LKRQGRTLYG FGG
```

Histones H4

Seq.	Organism	Source	Reference
1.	BOS TAURUS (bovine)	THYMUS	OGAWA Y., QUAGLIAROTTI G., JORDAN J., TAYLOR C.W., STARBUCK W.C., BUSCH H.; J. BIOL. CHEM. 244:4387-4392(1969).
		NOVIKOFF HEPATOMA	QUAGLIAROTTI G., OGAWA Y., OGAWA Y., TAYLOR C.W., SAUTIERE P., JORDAN J., STARBUCK W.C., BUSCH H.; J. BIOL. CHEM. 244:1796-1802(1969). WILSON R.K., STARBUCK W.C., TAYLOR C.W., JORDAN J., BUSCH H.; CANCER RES. 30:2942-2951(1970).
	SUS SCROFA (pig)	THYMUS	SAUTIERE P., LAMBELIN-BREYNAERT M.-D., MOSCHETTO Y., BISERTE G.; BIOCHIMIE 53:711-715(1971).
	RATTUS NORVEGICUS (rat)	CHLOROLEUKEMIC TUMOR	SAUTIERE P., TYROU D., MOSCHETTO Y., BISERTE G.; BIOCHIMIE 53:479-483(1971).
2.	HOMO SAPIENS (human)	DNA	SIERRA F., STEIN G., STEIN J.; NUCL. ACIDS RES. 11:7069-7086(1983).
	MUS MUSCULUS (mouse)	DNA	SEILER-TUYNS A., BIRNSTIEL M.L.; J. MOL. BIOL. 151:607-625(1981).
	GALLUS GALLUS (chick)	DNA	WANG S.W., ROBINS A.J., D'ANDREA R., WELLS J.R.E.; NUCL. ACIDS RES. 13:1369-1387(1985).
	XENOPUS LAEVIS (frog)	DNA	MOORMAN A.F.M., DE BOER P.A.J., DE LAAF R.T.M., VAN DONGEN W.M.A.M., DESTREE O.H.J.; FEBS LETT. 136:45-52(1981). CLERC R.G., BUCHER P., STRUB K., BIRNSTIEL M.L.; NUCL. ACIDS RES. 11:8641-8657(1983).
3.	PSAMMECHINUS MILIARIS	DNA	BUSSLINGER M., PORTMANN R., IRMINGER J.C., BIRNSTIEL M.L.; NUCL. ACIDS RES. 8:957-977(1980).
	LYTECHINUS PICTUS	DNA	ROBERTS S.B., WEISSER K.E., CHILDS G.; J. MOL. BIOL. 174:647-662(1984). CHILDS G., NOCENTE-MCGRATH C., LIEBER T., HOLT C., KNOWLES J.A.; CELL 31:383-393(1982).
	STRONGYLOCENTROTUS PURPURATUS	DNA	KAUMEYER J.F., WEINBERG E.S.; NUCL. ACIDS RES. 14:4557-4576(1986).
4.	PSAMMECHINUS MILIARIS	GONADS	WOUTERS-TYROU D., SAUTIERE P., BISERTE G.; FEBS LETT. 65:225-228(1976).
		DNA	GRUNSTEIN M., DIAMOND K.E., KNOPPEL E., GRUNSTEIN J.E.; BIOCHEMISTRY 20:1216-1223(1981)

5.	TRITICUM AESTIVUM	DNA (TH011)	TABATA T., SASAKI K., IWABUCHI M.; NUCL. ACIDS RES. 11:5865-5875(1983).
6.	TRITICUM AESTIVUM	DNA (TH091)	TABATA T., IWABUCHI M.; GENE 31:285-289(1984).
7.	ZEA MAYS	DNA	PHILIPPS G., CHAUBET N., CHABOUTE M.-E., EHLING M., GIGOT C.; GENE 42:225-229(1986).
8.	PISUM SATIVUM	SEEDLING	DELANGE R.J., FAMBROUGH D.M., SMITH E.L., BONNER J.; J. BIOL. CHEM. 244:5669-5679(1969).
9.	PHYSARUM POLYCEPHALUM	DNA	WILHELM M.L., WILHELM F-X.; FEBS LETT. 168:249-254(1984).
10.	NEUROSPORA CRASSA	DNA	WOUDT L.P., PASTINK A., KEMPERS-VEENSTRA A., JANSEN A.E.M., MAGER W.H., PLANTA R.J.; NUCL. ACIDS RES.11:5437-5360(1983)
11.	SACCHAROMYCES CEREVISIAE	DNA	SMITH M.M., ANDRESSON O.S.; J. MOL. BIOL. 169:663-690(1983).
		PROTEIN	BRANDT W.F., PATTERSON K., VON HOLT C.; EUR. J. BIOCHEM. 110:67-76(1980).
12.	TETRAHYMENA PYRIFORMIS	H4.1,2	HAYASHI H., NOMOTO M., IWAI K.; J. BIOCHEM.(TOKYO) 96:1449-1456(1984).
13.	TETRAHYMENA THERMOPHILA	DNA	HOROWITZ S., BOWEN J.K., BANNON G.A., GOROVSKY M.A.; NUCL. ACIDS RES.15:141-160(1987).
14.	TETRAHYMENA THERMOPHILA	DNA	BANNON G.A., BOWEN J.K., YAO M-C., GOROVSKY M.A.; NUCL. ACIDS RES. 12:1961-1975(1984).
15.	TETRAHYMENA THERMOPHILA	PROTEIN	GLOVER C.V.C., GOROVSKY M.A.; PROC. NATL. ACAD. SCI. U.S.A. 76:585-589(1979).
16.	SALMO GAIRDNERII	DNA	CANDIDO E.P.M., DIXON G.H.; J. BIOL. CHEM. 246:3182-3188(1971).
17.	DROSOPHILA MELANOGASTER	DNA	GOLDBERG M.L.; UNPUBLISHED (1979) THESIS, UNIVERSITY OF STANFORD, U.S.A.
18.	PARECHINUS ANGULOSUS	SPERM	STRICKLAND M., STRICKLAND W.N., BRANDT W.F., VON HOLT C.; FEBS-LETT. 40:346-348(1974).

[24] Purification and Analysis of H1 Histones

By R. DAVID COLE

General Introduction

No current method gives complete resolution of all the primary structural variants of H1 histones. Therefore, the choice of an analytical procedure will depend on the long-range purpose of the experiment at hand. The path of wisdom is to choose the most convenient procedure that is adequate to attain the particular objective in mind. The choice will depend on whether one wishes to distinguish H1 as a class from other histone classes; to distinguish between the regular set of H1 variants, on the one hand, and special variants such as H1°, H1t, or H5, on the other; or to distinguish among the regular variants of H1. These three objectives run from the ultimate in simplicity and ease to an as yet unmet challenge. The preparation of H1 histones, in terms of the same distinctions, presents the same problems as the analysis. Therefore, this chapter, after setting out some considerations common to all the methods, will be divided into three parts, each addressed to one of these levels of resolution.

General Considerations

The preparation of H1 histone of 95% homogeneity, at the milligram level and higher, is quick, easy, and straightforward—a matter of selective extraction and precipitation. The main problem is the presence of proteases, which differ among cell types in the vigor of their activity and in their responsiveness to particular inhibitors. Usually these proteases can be inhibited by 50 mM NaHSO$_3$ (which also inhibits some phosphatases) or 250 μM phenylmethylsulfonyl fluoride (PMSF). In any case, speed in the early steps of isolation is helpful.

The preparation of H1 histone at the microgram level frequently gives low yields owing to adsorption on the surfaces of labware. This is probably the cause of many reported chromatographic losses. Usually, it is a matter of the highly cationic H1 adsorbing on the negative surface of glass or dialysis tubing, and so it can be minimized by silanization of glassware or by the use of plasticware. The deleterious effects of glass surfaces can be minimized by use of acidic media (pH ≤ 2) or moderate salt concentrations (~ 100 mM). The use of a companion protein, so effective in protecting other proteins, is of more limited value in the case of H1 because the H1 is such a strong competitor for negative surfaces. Although the adsorp-

METHODS IN ENZYMOLOGY, VOL. 170

Copyright © 1989 by Academic Press, Inc.
All rights of reproduction in any form reserved.

tion process is usually minimized by use of hydrophobic labware, occasionally H1 can be lost on hydrophobic surfaces if the H1 is neutralized by a polyanion such as DNA in the presence of salt (e.g., 100 mM), which strengthens hydrophobic interactions.

The accurate quantification of H1 concentrations in absolute terms is more difficult than may meet the eye. Fortunately, for most purposes, the relative concentrations of H1 variants or the ratio of H1 to other histones is all that is required. Where absolute concentrations are desired, however, it is necessary to resort to amino acid analysis following total acid hydrolysis of the protein preparation. Dry weights are suspect because H1, even more than most proteins, binds water and anions tenaciously, and even lyophilized powders may well be only half protein. Spectrophotometric and colorimetric methods are very sensitive to contaminants because the amino acid composition of H1 is so unusual. For example, since the $E_{280}^{1\%}$ of H1 is about one-tenth that of bovine serum albumin, use of $E_{280}^{1\%}$ for pure H1 when measuring the concentration of an H1 preparation contaminated by 10% serum albumin would produce an underestimation by a factor of 2. Analogous errors apply to colorimetric analyses such as the Lowry[1] and Bradford[2] methods. It should be clear, then, that knowledge of the extinction coefficient of pure H1 does not solve the problem unless the sample being assayed is of equal purity. Many preparations in the author's laboratory have been cross-checked by A_{230} and amino acid analysis with the result that H1 purified by ion exchange (as described following) consistently had an $E_{280}^{1\%}$ of 18.5, but material prepared by perchloric acid extraction and trichloroacetate precipitation (see below) fairly consistently had an $E_{280}^{1\%}$ value of 35, even though the two preparations looked the same in SDS gel electrophoresis.

The resolution of the variants within the regular set of H1 histones is a challenge because the variants are so nearly alike in composition at the same time that they are heavy-handed in their electrostatic interactions. It ought to be kept in mind, if the following protocols are to be modified, that although the two most important methods of resolving H1 variants, namely, electrophoresis and ion exchange, are thought of primarily as resolving on the basis of charge, the major factors in resolution of H1 variants are probably shape and size. For example, this may explain why Amberlite IRC-50 (BioRex 70), a carboxylic resin with a hydrophobic surface, gives much better resolution than does carboxylmethyl cellulose, a carboxylic resin with a more hydrophilic surface. In any case, the challenge

[1] O. H. Lowry, N. J. Rosebrough, A. L. Farr, and R. J. Randall, *J. Biol. Chem.* **193**, 265 (1951).
[2] M. M. Bradford, *Anal. Biochem.* **72**, 248 (1976).

of resolving all the H1 variants has not yet been fully addressed. The H1 histones of rabbit thymus have been separated into six fractions by chromatography or five by two-dimensional electrophoresis, but amino acid sequencing revealed a minimum of eight molecular species.[3] Similarly, five electrophoretically purifed H1 fractions from sea urchin were judged by amino acid sequencing to contain at least seven molecular species.[4] Apparently there are many more H1 genes in the chicken than the six fractions that have been resolved by ion exchange, although how many are expressed in adults is not clear.[5]

Separation of Histone from Core Histones and other Cellular Components

Introduction

H1 histone is more loosely bound to chromatin than are the other histones, and so it has been selectively extracted from lysed nuclei by acids, salts, protamines, tRNA, and ion-exchange resins. If one is interested in H1 without concern for other chromosomal components, there are advantages in direct, selective extraction of H1 from whole cells or tissues, not only in convenience but also in minimizing exposure of H1 to proteases. The protocol given below, and variations of it, are widely used by those whose interest is focused on H1, and, as mentioned in the notes, it can be applied to the preparation of H1 from mixtures containing the other histones. The protocol is generally used as a preparative procedure, although it could be used analytically. Better techniques for analytical purposes in distinguishing H1 from other histones are adaptations of the protocols in the two following major sections of this chapter, and are mentioned in the notes in those sections (see below).

Protocol [6]

1. Homogenize about 100 g tissue (or 100 ml packed cells) for 2 min in a blender at top speed with 500 ml 5% perchloric acid (0.74 N) at $0°-5°$ and filter through two layers of cheesecloth. Maintain a temperature of $0°-5°$ throughout the following steps.
2. Centrifuge at approximately 600 g for 30 min; save the supernatant and reextract the pellet by thorough resuspension in 250 ml 5%

[3] R. D. Cole, *Int. J. Peptide Protein Res.* **30**, 433 (1987).
[4] P. De Groot, W. Strickland, and C. von Holt, *Biochim. Biophys. Acta* **747**, 276 (1983).
[5] B. J. Sugarman, B. Dodgoon, and J. D. Engle, *J. Biol. Chem.* **258**, 9005 (1983).
[6] B. J. Smith, J. M. Walker, and E. W. Johns, *FEBS Lett.* **112**, 42 (1980).

perchloric acid. A Dounce homogenizer makes the resuspension easier.

3. Centrifuge at 600 g for 30 min; discard the pellet and combine the supernatant with the supernatant from step 2.

4. Adjust the pooled supernatant to about 0.1 N HCl by the addition of concentrated HCl; add 3 liters acetone; stir several hours (e.g., overnight) at 4° to precipitate H1. This precipitate is sometimes transparent and deposited on the walls of the container.

5. Centrifuge at 12,000 g for 60 min to collect a pellet of H1.

6. Dry the pellet by washing 2 times with acidified acetone (1 ml concentrated HCl per liter) and filtering; or dissolve the pellet in water, dialyze overnight, and store frozen or lyophilized.

Notes

(1) The yield from calf or steer thymus is about 400 mg, but thymus is unusual in having little cytoplasm. Typically, 100 g of other tissues would yield about 75 mg H1. Histone H5 is isolated with H1 when this procedure is used and yields about 60 mg H5 plus 15 mg H1 from 100 ml blood. When a pellet of about 5×10^5 packed HeLa cells is treated by this method, the yield approximates 50 mg H1. Whole histone can also be used as starting material in this procedure.[7]

(2) Trichloroacetic acid (5%) may be substituted[8] for perchloric acid. The former gives purer H1 through step 3, but at step 5 the two acids give the same result for the regular set of H1 variants. Trichloroacetic acid is fully effective for the extraction of H1° but does not extract H1t, the testes-specific variant.[9]

(3) When the method is scaled down, H1 precipitates sometimes form a transparent coat on the sides of the container, and so careful washing of the container is important if a pellet is not obvious.

(4) Lyophilized powders can be stored for years at room temperature without harm if they are kept dry. If samples are stored cold and opened when cold, moisture will condense in the protein powder, and the H1 will slowly deteriorate.

(5) This method does not require protease inhibitors if fresh tissues are homogenized directly in perchloric acid. The freezing of tissues releases proteases, and, as a safeguard, 50 mM NaHSO$_3$ or 250 μM PMSF should be incorporated in the first two steps. If isolated nuclei are used as starting

[7] R. H. Stellwagen and R. D. Cole, *J. Biol. Chem.* **243**, 4456 (1968).
[8] J. M. Kinkade, Jr., and R. D. Cole, *J. Biol. Chem.* **241**, 5790 (1966).
[9] S. M. Seyedin, R. D. Cole, and W. S. Kistler, *Exp. Cell Res.* **136**, 399 (1981).

material, their preparation ought to include the inhibitors. The relative and absolute effectiveness of different protease inhibitors varies between tissues.

Separation of Special Linker Histones from the Regular Set of H1 Variants

Introduction

Some linker histones, although considered H1 variants, either differ widely from the regular set of variants structurally (e.g., H5, H1°) or occur only in specific tissues (e.g., H5, H1t). Their uniqueness makes them attractive for attempts to identify functions of linker histones. Of course, when the functions of these special H1 variants are understood, the question may arise whether those functions are unique or are general among H1 histones. In any case, the unique character of these special H1 variants makes them easy to prepare and to distinguish analytically from the regular set of H1 variants. Below, a chromatographic protocol is given which is also useful preparatively.

Protocol

1. Gently suspend 3 g BioGel P-100 (Bio-Rad, Richmond, CA) in a large filter flask containing 1.5 liters 20 mM HCl, 50 mM NaCl, 0.02% (w/v) NaN$_3$ and allow the resin to swell for 3 days at 20°.
2. Degas by applying suction with gentle swirling.
3. Pack a column with the swollen gel to give a bed of 2 × 150 cm.
4. Wash the column with 1 liter 20 mM HCl, 50 mM NaCl, 0.02% NaN$_3$.
5. Dissolve the H1 sample in HCl/NaCl/NaN$_3$ solution, apply it to the column, and elute with the same HCl/NaCl/NaN$_3$ solution.
6. Maintain the flow rate at 6–10 ml/hr, and monitor the column by A_{230}. H1 will elute near 110 ml, H1° and H1t near 125 ml, and H5 near 132 ml.

Notes

(1) With low loads (~10 mg perchloric acid extract) applied in small volumes (e.g., 2 ml), peak widths will be 6 ± 1 ml. High loads result in overlapping peaks, but loads of 100–150 mg protein (perchloric acid extract) can be processed effectively by rechromatographing those fractions of the first run that are found to contain the special H1 variant. (2) Urea

(to 8 M) is useful in dissolving large amounts of protein into small volumes.

(3) The concentrations of HCl and NaCl can be varied widely, but the pH ought to be lower than 3 and the ionic strength ought to be at least 25 mM to suppress ion exchange by carboxyl groups in the gel. Other antibacterial agents may be substituted for NaN$_3$.

(4) This protocol may be adapted to the separation of H1 from the core histones by use of BioGel P-60[10] in place of BioGel P-100. Approximate elution positions will be as follows: H1 at 350 ml; H5 at 425 ml; H2A at 500 ml; H2B and H3 at 600 ml; H4 at 750 ml.

Resolution of the Regular Set of H1 Variants

Introduction

Two good methods are available for separating H1 variants from each other: two-dimensional gel electrophoresis and ion-exchange chromatography. Because the former is described elsewhere in this volume, only the latter is presented in this chapter. Although the chromatography gives marginally better resolution than the electrophoresis, neither resolves all the H1 variants.[3] The chromatography is better for preparative purposes and more easily quantified, but electrophoresis is much more convenient. Both methods are time-consuming.

The chromatographic system given here and minor variations of it, have been used in many laboratories with good results. With one exception, no other chromatographic system has given comparable resolution, probably because this system achieves a balance between ion-exchange and hydrophobic processes, as mentioned in the general introduction to this chapter. The one exception just noted is a high-pressure liquid chromatography[11] that uses a reversed-phase (and therefore hydrophobic) column. Although other HPLC chromatographies for histones failed to resolve H1 variants, this HPLC column gave resolution of H1 variants comparable to two-dimensional electrophoresis and the ion-exchange protocol presented below. It is not, however, described procedurally in this chapter because there has not yet been sufficient experience published to test its intricacies and reproducibility. Its further application and development must be watched because it might well become the method of choice because of its speed and convenience. It could be developed for both analytical and preparative purposes.

[10] D. R. van der Westhuyzen, C. L. Bohm, and C. von Holt, *Biochim. Biophys. Acta* **359**, 341 (1974).

[11] M. Kurokawa and M. C. MacLeod, *Anal. Biochem.* **144**, 47 (1985).

Protocol[12]

1. Prepare 200 ml 50% guanidinium chloride (GuCl) by dissolving 100 g GuCl in 100 ml 0.1 M phosphate buffer, pH 6.8. From this stock and more phosphate buffer, make 50 ml 40% GuCl, 100 ml 13% GuCl, and 500 ml 8% GuCl. For accurate work, measure the concentration of the GuCl by refractometer (for 40% GuCl, $n =$ 1.4014; for 12%, $n = 1.3549$; for 8%, $n = 1.3483$).

2. Suspend 25 g Amberlite IRC-50 (Mallinkrodt, Paris, KY) or BioRex 70 (Bio-Rad, Richmond, CA), 200–400 mesh, in 50 ml water in a 100-ml graduated cylinder, allow the resin to settle 10 min, and decant fines; resuspend the resin in 50 ml 2 M HCl for 15 min and then suction-filter the resin. Resuspend the resin in 50 ml H_2O, 15 min, and filter. Similarly, wash the resin successively with 50 ml 2 M NaOH, 50 ml H_2O, 50 ml 2 M HCl, 50 ml H_2O. Resuspend the resin in 50 ml 2 M NaCl, and titrate with 0.5 M NaOH until pH 7 ± 0.1 is maintained for 10 min. Filter off the NaCl solution and resuspend the resin in 20 ml 8% GuCl–phosphate buffer.

3. Pack the resin in a column to give a bed 1×15 cm, and wash the column with 100 ml of 8% GuCl–phosphate buffer.

4. Appy 2–5 mg H1 histone (e.g., perchloric acid extract) in as much as 3 ml 8% GuCl–phosphate buffer, and elute with a linear gradient from 8 to 12% GuCl using 50 ml each of the 8% GuCl–phosphate and the 12% GuCl–phosphate. Maintain the flow near 1 ml/hr. Regenerate the column by washing with 30 ml 40% GuCl–phosphate and reequilibrate it with 100 ml 8% GuCl–phosphate. The effluent is most conveniently monitored by A_{230}. The last H1 variant will usually be off the column by 11.5% GuCl.

Notes

(1) The protocol given is for a typical analytical scale. It can be made suitable for smaller amounts of histone by use of fluorescamine for monitoring the effluent or by radiolabeling the H1 *in vivo*[13] or *in vitro*.[14]

(2) Resolution deteriorates from peak broadening as substantial increases in loading or flow rate occur. Because this is gradient elution rather than isocratic, resolution is not very sensitive to column length, per se, and excellent resolution is achieved on 15-cm columns if the protein loading is kept low. (3) High loads are accommodated by increasing the length of the

[12] M. Bustin and R. D. Cole, *J. Biol. Chem.* **244**, 5286 (1969).
[13] F. Caron and J. O. Thomas, *J. Mol. Biol.* **146**, 513 (1981).
[14] D. L. Bates and J. O. Thomas, *Nucleic Acids Res.* **9**, 5883 (1981).

column or increasing the diameter; the flow rate can then be increased in proportion to the volume of the resin bed. Resolution suffers slightly as the diameter of the column is increased, but it is still good for columns of 2 × 60 cm to which 100–150 mg H1 may be applied and eluted at 10 ml/ hr with a 3-liter gradient.

(4) H1 phosphorylation can be analyzed with the protocol given if ^{32}P labeling is used. Usually, phosphorylation causes a minor decrease in the elution volume of an H1 histone, but because this chromatography depends profoundly on protein conformation, phosphorylation at some sites has a larger effect on elution volume than it does at other sites. An analogous situation exists in the effects of phosphorylation on the mobility of H1 in denaturing gel electrophoresis.[15] The difference in elution volume of any H1 variant and that of its phosphorylated form can be exaggerated by use of an isocratic forerun of 200 ml of 8% GuCl–phosphate before beginning the gradient elution.[16]

(5) The GuCl used must be highly purified to allow A_{230} monitoring. For modest sized columns, it is well worth buying the expensive, commercially available, ultrapure GuCl. Otherwise, the GuCl can be purified by a 6-day procedure, described in detail elsewhere.[17] Another option for large-scale columns is to monitor the effluent turbidimetrically[18] or by the Lowry method. If the Lowry method is used, aliquots, especially at the upper end of the elution gradient, must be no greater than 0.4 ml to avoid the precipitation of some components of the reagents.

(6) The GuCl–phosphate buffers are chemically stable, and large volumes may be prepared and stored at room temperature. If buffers are stored at 5° they must be warmed to room temperature to prevent the release of tiny bubbles within the resin bed, with a resultant increase in peak width. Because 12% GuCl is only 1 M, it is not an effective denaturant for most proteins, and therefore the stored buffers must be protected by an antibiological agent such as 0.02% NaN_3. Although resolution for gradient elution analysis can usually be increased by the use of more shallow gradients, a limit is reached eventually because peak broadening finally overwhelms the advantage of greater separation of the peak positions. When tested,[12] the gradient given in the protocol above was found to have reached that limit.

(7) The protocol above can be modified for the separation of H1 from the other histone classes simply by use of a steeper gradient.[18] Elute with a

[15] P. C. Billings, J. W. Orf, D. K. Palmer, D. A. Talmage, C. G. Pan, and M. Blumenfeld, *Nucleic Acids Res.* **6**, 2151 (1979).
[16] L. R. Gurley, R. A. Walters, and R. A. Tobey, *J. Biol. Chem.* **250**, 3936 (1975).
[17] T. G. Spring and R. D. Cole, *Methods Cell Biol.* **14**, 227 (1977).
[18] P. S. Rasmussen, K. Murray, and J. M. Luck, *Biochemistry* **1**, 79 (1962).

60 ml linear gradient of 8–14% GuCl–phosphate and then 20 ml 40% GuCl–phosphate. Elution positions will be approximately as follows: H1, 10 ml; H2A, 35 ml; H2B, 45 ml; H3 plus H4, 67 ml.

[25] Analysis of Histone Subtypes and Their Modified Forms by Polyacrylamide Gel Electrophoresis

By RONALD W. LENNOX and LEONARD H. COHEN

Introduction

Although only 5 classes of histones exist, namely, H4, H3, H2A, H2B, and H1 (which we consider here to include H5 and H1°), more than 10 times this number of forms of the histones can be separated by gel electrophoresis. Some of this multiplicity results from the existence of several subtypes or isoforms of H1, H2A, H2B, and H3. The rest arises from various posttranslational modifications which affect principally the charge, in the case of acetylation or phosphorylation, or the mass, in the case of ubiquitination, of the parent histone molecules. Histone subtypes and their modified forms have been the subject of recent reviews that consider their occurrence, biochemistry, and putative roles in chromatin structure.[1-5]

The electrophoretic techniques used in this laboratory to study histones are variations and combinations of three methods. The first is electrophoresis in acetic acid–urea gels, of the type first described by Panyim and Chalkley,[6] which separates molecules, to a large extent, on the basis of charge. This method allows the separation of modified and unmodified forms of the histones and also the partial separation of the various subtypes of H1 and their phosphorylated forms.[7] The second technique is electro-

[1] R. W. Lennox and L. H. Cohen, *in* "Histone Genes, Structure, Organization and Regulation" (G. S. Stein, J. L. Stein, and W. F. Marzluff, eds.), p. 373. Wiley, New York, 1984.

[2] A. Zweidler, *in* "Histone Genes, Structure, Organization and Regulation" (G. S. Stein, J. L. Stein, and W. F. Marzluff, eds.), p. 339. Wiley, New York, 1984.

[3] C. von Holt, P. De Groot, S. Schwager, and W. F. Brandt, *in* "Histone Genes, Structure, Organization and Regulation" (G. S. Stein, J. L. Stein, and W. F. Marzluff, eds.), p. 65. Wiley, New York, 1984.

[4] R. S. Wu, H. T. Panusz, C. L. Hatch, and W. M. Bonner, *CRC Crit. Rev. Biochem.* **20,** 201 (1986).

[5] R. W. Lennox and L. H. Cohen, *in* "Chromatin and Chromosome Structure" (K. W. Adolp, ed.), p. 33. CRC Press, Boca Raton, Florida, 1987.

[6] S. Panyim and R. Chalkley, *Arch. Biochem. Biophys.* **130,** 337 (1969).

[7] R. W. Lennox, R. G. Oshima, and L. H. Cohen, *J. Biol. Chem.* **257,** 5183 (1982).

Copyright © 1989 by Academic Press, Inc.
All rights of reproduction in any form reserved.

phoresis in Triton–acetic acid–urea (TAU) gels to separate the various subtypes of H2A, H2B, and H3 and their modified forms.[8-10] The third is electrophoresis in gels containing sodium dodecyl sulfate (SDS), which separates principally on the basis of mass[11] and, in one or two instances, also separates certain phosphorylated forms of H1 from other phosphorylated forms and from the parent molecule,[7,12] presumably because these phosphorylations affect the conformation of the protein. Although electrophoresis by any of these methods alone is often used for particular purposes, resolution of all the forms of histones found in cells or tissues requires electrophoresis in two dimensions, the first in acid–urea gels, with or without Triton X-100, and the second in SDS gels.

Stock Solutions

60% (w/v) acrylamide: 180 g acrylamide (Bio-Rad, Richmond, CA) is dissolved in 133.2 ml H_2O, filtered through glass-fiber filter paper (GF/C, Whatman, Clifton, NJ), and stored protected from light.

2.0% (w/v) methylenebisacrylamide (MBA): 5 g MBA (Bio-Rad) is dissolved in H_2O to a final volume of 250 ml and filtered and stored as for acrylamide.

5% (v/v) acetic acid: 1 liter glacial acetic acid is added to 19 liters H_2O.

50 mg/ml protamine sulfate (from herring, grade III, Sigma, St. Louis, MO) in H_2O: Store in 0.5-ml aliquots at $-70°$.

5 M cysteamine: 28.4 g cysteamine-HCl is dissolved in H_2O to a final volume of 50 ml and stored in 0.5-ml aliquots at $-70°$.

1 M Tris base: 121.1 g Tris base (Sigma) is dissolved in H_2O to a final volume of 1 liter.

0.5 M Tris-HCl, pH 6.8: 30.25 g Tris base is dissolved in H_2O and brought to pH 6.8 with HCl and to a final volume of 500 ml.

20% (w/v) SDS: 100 g SDS (Bio-Rad) is made up in a final volume of 500 ml.

4× SDS separating gel buffer,[11] 1.5 M Tris-HCl, pH 8.8, 0.4% (w/v) SDS: 181.7 g Tris base is brought to pH 8.8 with HCl in a volume of 980 ml, and 20 ml 20% (w/v) SDS is then added.

4× SDS stacking gel buffer,[11] 0.5 M Tris-HCl, pH 6.8, 0.4% (w/v) SDS: 30.3 g Tris base is brought to pH 6.8 with HCl in a volume of 490 ml, and 10 ml 20% (w/v) SDS is then added.

5× SDS electrode buffer,[11] 0.125 M Tris, 0.96 M glycine, pH 8.3,

[8] C. R. Alfageme, A. Zweidler, A. Mahowald, L. H. Cohen, *J. Biol. Chem.* **249**, 3729 (1974).
[9] L. H. Cohen, K. M. Newrock, and A. Zweidler, *Science* **190**, 994 (1975).
[10] A. Zweidler, *Methods Cell Biol.* **17**, 223 (1978).
[11] U. K. Laemmli, *Nature (London)* **227**, 680 (1970).
[12] M. Blumenfeld, *Biochem. Genet.* **17**, 163 (1979).

0.5% (w/v) SDS: 302.5 g Tris base and 1440 g glycine (Bio-Rad) are dissolved in 19.5 liters H_2O, and then 500 ml 20% (w/v) SDS is added.

$2\times$ SDS sample buffer, 125 mM Tris-HCl, pH 6.8, 2% (w/v) SDS, 10% (v/v) glycerol (Fisher, Fair Lawn, NJ), 5% (v/v) 2-mercaptoethanol (Sigma), 0.002% bromphenol blue (Bio-Rad): 0.76 g Tris base dissolved in H_2O is brought to pH 6.8 with HCl, and 5 ml glycerol, 5 ml 20% (w/v) SDS, and 2.5 ml 2-mercaptoethanol are added and the volume brought to 50 ml. Finally, 1 mg bromphenol blue (the amount that adheres to a pipette or spatula) is added as tracking dye.

0.1% (w/v) Coomassie blue staining solution: 3.3 g Coomassie blue R (Bio-Rad) is dissolved in 750 ml methanol–150 ml glacial acetic acid, filtered through a glass fiber filter, and added to 2.2 liters of H_2O.

Destaining solution (5 : 16 : 1): 5 liters methanol and 1 liter glacial acetic acid are added to 16 liters H_2O.

Butanol saturated with H_2O: butanol and H_2O are mixed, and each is replenished when necessary.

20% (w/v) ammonium persulfate: 1 g ammonium persulfate (Bio-Rad) is dissolved in 5 ml H_2O. This solution is made fresh at least once a week.

10% (w/v) Triton X-100: 10 g Triton X-100 (Research Plus, Bayonne, NJ) is dissolved in H_2O to a final volume of 100 ml.

25% (w/v) perchloric acid (PCA): to 70% (w/w) PCA add 1.8 volumes of H_2O.

5% (w/v) perchloric acid: to 70% (w/w) PCA add 13 volumes of H_2O.

0.4 M H_2SO_4: add 22.2 ml H_2SO_4 to H_2O and dilute to a final volume of 1 liter.

2.2 M H_2SO_4: add 122.2 ml H_2SO_4 to H_2O and dilute to a final volume of 1 liter.

120% (w/v) trichloroacetic acid (TCA): add 144 ml H_2O to 500 g TCA.

0.2% (w/v) Pyronin Y: dissolve 0.2 g Pyronin Y in 100 ml H_2O.

Histone Preparation from Isolated Nuclei

The methods for histone preparation from isolated nuclei are based on the work of Johns[13] and Chalkley.[14] To a suspension of nuclei (0.5–2 mg

[13] E. W. Jones, *Methods Cell Biol.* **16**, 183 (1977).

[14] D. Oliver, K. R. Sommer, S. Panyim, S. Spiker, and R. Chalkley, *Biochem. J.* **129**, 349 (1972).

DNA/ml), in nuclear isolation medium or 0.1 M NaCl, add an equal volume of cold 0.4 M H_2SO_4 or 0.1 volume of 2.2 M H_2SO_4. After at least 20 min in the cold, centrifuge for 5 min at 5000 g. Reextract the pellet with half the original volume of 0.2 M H_2SO_4. To the combined extracts, add 3 volumes of absolute ethanol, let stand covered for at least 20 min on ice, then centrifuge at 10,000 g for 10 min. Wash the pellets once with 75% ethanol and 3 times with absolute ethanol; dry with a stream of dry nitrogen or in a vacuum desiccator. Seal tightly and store dry. The histones can be dissolved in water and stored at $-70°$, but if the solution is to be stored for any substantial period, 1–5% (w/v) thiodiglycol should be included to prevent oxidation of methionyl residues. (This is not needed for H1 proteins of most mammals, which lack methionine.[9])

If desired, H1 and HMG proteins can be separated from the core histones by adding 0.25 volume of 25% PCA to the H_2SO_4 extract with stirring. Let stand 20 min in the cold and spin down the core histones. To the supernatant fraction, add 0.2 volume 120% TCA (w/v), keep cold for 20 min, and centrifuge down H1 and HMG proteins. Wash the core histone pellet and the H1 pellet as above. Where only H1 is of interest, it can be obtained directly from cells and some tissues by extraction with 5% PCA, followed by precipitation with 20% TCA. However, this procedure is not applicable to all tissues (e.g., sea urchin embryos, *Drosophila* embryos and adults[15]).

Homogenization of cells or tissues in 5% (w/v) PCA rapidly inactivates proteases or phosphatases. However, it is prudent to have inhibitors of various interfering enzymes present when preparing nuclei for histone isolation: for example, aprotinin (100 units/ml) and phenylmethylsulfonyl fluoride (PMSF) (0.1 mM) to inhibit proteases, 50 mM sodium bisulfite[16] or 1–5 mM *p*-chloromercuribenzoate (pCMB)[17] to inhibit phosphatases, and 10 mM sodium butyrate to inhibit deacetylases.[18]

Acid – Urea Gels for Fractionating H1

Method

Electrophoresis of H1 in acetic acid – urea gels separates phosphorylated forms according to the number of phosphate groups and also separates some H1 subtypes.

[15] L. H. Cohen, unpublished observations, 1975.
[16] L. R. Gurley, R. A. Walters, and R. A. Tobey, *J. Biol. Chem.* **250,** 3936 (1975).
[17] J. R. Paulson, *Eur. J. Biochem.* **111,** 189 (1980).
[18] P. Pantazis and W. M. Bonner, *J. Biol. Chem.* **256,** 4669 (1981).

1. Assemble two glass plates 50 cm long by 25 cm wide, one of which is notched, with Teflon spacers 0.5 mm thick along the sides and across the bottom and clamp together with binder clips. Seal the edges with molten 1% (w/v) agarose (Seakem ME, FMC Corp., Rockland, ME) in H_2O. One of the glass plates is 0.5 cm thick, the other 1 cm thick, or both may be 1 cm thick.

2. Prepare 60 ml of gel solution containing 15% acrylamide, 0.1% MBA, 2.5 M urea, and 5% (v/v) acetic acid by combining 9.21 g urea [Bethesda Research Laboratories (BRL), Gaithersburg, MD], 15 ml 60% (w/v) acrylamide, 3.1 ml 2% (w/v) MBA, 3 ml glacial acetic acid, and 0.3 ml TEMED (Bio-Rad) and bringing the volume to 59.4 ml with H_2O. Degas the solution under vacuum. Add 0.6 ml 20% (w/v) ammonium persulfate (APS) and mix well. Pour the gel solution between the plates, taking care to avoid the formation of bubbles, until the surface is 2.0–2.5 cm from the top of the notched plate. Pipette butanol saturated with H_2O onto the surface of the gel, and allow to polymerize for 60–90 min.

3. Pour off the butanol, removing the dregs from the gel surface with the edge of a paper towel.

4. Prepare an upper gel solution by combining 1.5 g urea, 1 ml 60% (w/v) acrylamide, 0.2 ml 2% (w/v) MBA, 0.5 ml glacial acetic acid, 0.05 ml TEMED, and H_2O to a final volume of 9.9 ml, and degas it under vacuum. To this solution [2.5 M urea, 6% (w/v) acrylamide, 0.04% (w/v) MBA, and 5% (v/v) acetic acid] add 0.1 ml 20% (w/v) APS, mix well, and apply with a pipette to the surface of the 15% gel to a height of around 1 cm. Overlay with butanol, and allow to polymerize for 60–90 min.

5. Remove the butanol, the clamps, and the bottom spacer and then clamp the gel to the back plate of the lower reservoir. Fasten the upper reservoir to the top of the plates with clamps.

6. Fill both reservoirs with 5% (v/v) acetic acid, taking care that no air bubbles are trapped at the bottom of the gel.

7. Carefully rinse the gel surface, and apply 1 ml of overlay solution containing 2.5 M urea (150 mg), 5% (v/v) acetic acid (50 μl), and 0.02% (w/v) Pyronin Y (100 μl of 0.2%).

8. Connect electrodes (cathode to bottom, anode to top) and apply 500 V overnight.

9. Disconnect the power supply, rinse the gel surface, and replace the electrode solutions with fresh 5% (v/v) acetic acid.

10. Apply 1 ml scavenge solution containing 0.5 ml 5 M cysteamine, 150 mg urea, 50 μl acetic acid, and 0.1 ml 0.2% (w/v) Pyronin Y to the gel surface, and electrophorese at 750 V for 90 min. Disconnect the power supply, rinse the gel surface, apply another 1 ml of scavenge solution, and electrophorese at 600 V for 90 min.

11. Disconnect the power supply, rinse the gel surface, and apply 1 ml of solution containing 0.5 ml 50 mg/ml protamine sulfate, 150 mg urea, 50 μl acetic acid, and 0.1 ml 0.2% (w/v) Pyronin Y. Electrophorese at 500 V for 30–60 min.

12. Disconnect the power supply and replace the electrode solutions with fresh 5% (v/v) acetic acid. Carefully insert the comb between the plates until the tips of the teeth extend 2–4 mm into the upper loading gel. Using a 5- or 10-μl syringe, apply the protein samples in 2.5 M urea, 5% (v/v) acetic acid, 25 mg/ml protamine sulfate, and 0.02% (w/v) Pyronin Y to the surface of the loading gel between the teeth of the comb.

13. Apply voltage of 500 V for 15–30 min to electrophorese the samples into the gel. Turn off the power and remove the comb.

14. Electrophorese at 600 V for 42 hr.

15. Switch off power, discard electrode solutions, and disconnect plates from reservoirs.

16. With spatulas or a long knife, separate the glass plates.

17. Cut off the top two-thirds of the gel and discard. Transfer the bottom one-third (after marking a corner for orientation) to staining solution.

18. Incubate in stain with rocking or shaking for 20–60 min then destain in 5 : 16 : 1 destaining solution.

19. After destaining, gels can be photographed, prepared for fluorography or autoradiography, stained again with silver, or prepared for electrophoresis in a second dimension.

Notes. Notes are numbered to correspond to the steps described above.

1. These gels are thinner than those generally used for protein electrophoresis, with several advantages: (a) less heat is generated, and the heat is better dissipated, permitting higher voltages and shorter runs; (b) bands and spots are smaller, and thus detection sensitivity greater; (c) staining, destaining, and drying times are all reduced; (d) gels with high polyacrylamide content are less likely to be damaged during drying; (e) thin gels are no more fragile, and in some ways easier to handle, than thicker gels; and cost is reduced. If thicker gels are necessary, for example, for preparative purposes, the voltage should be proportionally reduced. It seems to be necessary for at least one plate to be 1 cm thick. If both plates are 0.5 cm thick, it is usually difficult to load strips of the first gel onto the two-dimensional gels because they are too thick. It is thought that in the one-dimensional gel, the weight of the acrylamide solution causes the plates to bow slightly, but the problem has not been investigated. Plates are cleaned with FL-70 detergent (Fisher), rinsed in warm tap water, distilled H_2O, and 95% ethanol, and air-dried. Spacers and combs are cut from Teflon sheets,

which may be obtained from distributors of plastic products. Glass plates are obtained from local glass companies.

4. We do not cast wells in gels this thin (0.5 mm) because the partitions swell as urea leaches out of them during preelectrophoresis and obstruct the wells. Consequently, a low-percentage loading gel is used that is easily penetrated by the comb and allows rapid entry of the proteins.

5. In order to accommodate gels of various heights, we use separate top and bottom electrophoresis buffer reservoirs. The reservoirs are similar in design to those in the Studier electrophoresis chambers,[19] but wider and without the plexiglass connection between the two reservoirs. The upper reservoir is modified by milling a U-shaped groove (width 0.5 cm, depth 0.4 cm) on the face that contacts the gel plate. A piece of Tygon or silicon tubing (OD 0.5 cm, ID 0.1 cm) is placed in this groove as a gasket to prevent leakage from the reservoir. This plate is also wider than the reservoir itself, so that the gel plates can be clamped to it.

7–11. Preelectrophoresis serves to remove APS from the gel. The cysteamine scavenge serves to neutralize free radicals, including those on the gel matrix. The protamine sulfate binds to any free carboxyl groups in the gel matrix and diminishes binding (and smearing) of histones. Bands on gels treated with herring protamine (Sigma, type III) are sharper than those on gels treated with salmon protamine (Sigma, Type X).

12. Samples may be loaded in a volume of 0.1–8.0 μl with no effect on resolution. The amount of protein in each band can drastically alter resolution, there being little leeway in the amount that can be loaded. Depending on the number of components, between 0.2 and 1.5 μg may be loaded.

18, 19. Phosphorylated forms of H1 migrate more slowly than the parent unmodified forms. Where it is known that no modified forms are present, the first dimension may resolve all forms of H1. Generally, however, a second dimension is needed to resolve all the subtypes and their phosphorylated forms. Gels stained with Coomassie blue may be incubated in 50% (v/v) methanol and then stained with silver by the method of Boulikas et al.[20] The H1 histones generally stain dark brown; H1a of some species stains black.

Electrophoresis of H1 in SDS-Containing Gels

Although electrophoresis in SDS-containing gels generally separates proteins according to their molecular weight, histones, which make substantial contributions to the charge of their complexes with SDS, behave anomalously. Histone H1 with a true molecular weight in the range 20K–

[19] F. W. Studier, J. Mol. Biol. **79**, 237 (1973).
[20] W. Wray, T. Boulikas, V. P. Wray, and R. Hancock, Anal. Biochem. **118**, 197 (1981).

25K, for example, migrates as a group of bands of apparent molecular weight 30K–35K. As phosphorylations do not appreciably affect the molecular weight of H1, their effects on migration in gels of this type are usually indiscernible, but particular phosphorylations do affect the mobility of certain subtypes in gels containing SDS.[7,12] It is thought that these phosphorylations alter the conformation of H1 in a way that is not overcome by SDS.

Electrophoresis of H1 in gels containing SDS as the second dimension resolves all of the H1 subtypes detected thus far and most of their phosphorylated forms. One-dimensional electrophoresis in minigels containing SDS is also used for assaying the purity or integrity of H1, determining the uptake of radioactive precursors, or monitoring the elution of subtypes during chromatography.

Electrophoresis in gels containing 18% polyacrylamide provides good separation of core histones but poor resolution of H1 proteins. Twelve percent gels are better for separating mouse H1, and 15% gels are better for separating rat and human H1. Other concentrations of acrylamide may be more useful in separating H1 of other species. The recipes given in Table I are for preparing various volumes of 12, 15, or 18% gels. Approximately 60 ml of gel solution is sufficient for one second-dimensional gel, 40 ml for eight minigels.

Second-Dimensional Gels for H1

1. Assemble two plates, 35 cm long by 25 cm wide by 0.5 cm (one of them notched), with 0.75-mm-thick Teflon spacers along the sides and across the bottom. Clamp them together with binder clips and seal the edges with 1% (w/v) agarose.

2. Prepare 60 ml gel solution using recipes in Table I and pour it between the plates to approximately 2 cm from the top of the notched plate. Overlay with butanol and allow to polymerize for 1 hr.

3. Pour off the butanol and remove the dregs with a paper towel.

4. Prepare 10 ml stacking gel using the recipes in Table I (sufficient for two gels) and pour to a height of 1 cm. Overlay with butanol and allow to polymerize for 1 hr.

5. Immediately begin to prepare the one-dimensional gel samples for loading onto the second-dimensional gels. Place the gel on a clean glass plate and carefully cut strips of each lane containing the stained H1 bands, using a razor blade or sharp knife and a straight edge as guide. Incubate each strip in 50 ml 100 mM Tris base, 1% (v/v) 2 mercaptoethanol with gentle shaking in a fume hood for 30 min.

6. Discard this solution and replace it with 50 ml 50 mM Tris base, 1% (v/v) 2-mercaptoethanol. Incubate for 15 min.

TABLE I
PREPARATION OF SDS GELS

Final acrylamide concentration (% w/v):	Resolving gels						Stacking gel
	12		15		18		15
Volume (ml)	40	60	40	60	40	60	10
60% acrylamide (ml)	8	12	10	15	12	18	0.84
2.0% MBA (ml)	6.8	9.6	8	12	9.6	14.4	0.58
4× buffer (ml)	10	15	10	15	10	15	2.5
TEMED (μl)	25	50	25	50	25	50	10
H_2O (ml)	15.5	23.1	11.9	17.1	8.2	12.3	6
20% APS (μl)	100	200	100	200	100	200	75

7. Replace this solution with 50 ml 50 mM Tris-HCl, pH 6.8, 1% (v/v) 2-mercaptoethanol. Incubate for 15 min.

8. Remove clips and bottom spacer from the SDS gel. Remove butanol from gel surface. Lay the gel on the bench with the notched plate uppermost and raise the top end to an angle of around 30° (resting it on the bottom buffer reservoir is fine).

9. Fill the trough above the stacking gel with 50 mM Tris-HCl, pH 6.8.

10. Taking care not to distort or tear the one-dimensional gel strip, insert it horizontally into the trough, and gently tamp it down until it sits on top of the stacking gel. Make sure the two surfaces are completely in contact, with no air bubbles.

11. Clamp bottom and top reservoirs to gel plates, and fill them with 1× electrode buffer.

12. Connect the power supply and electrophorese at 3 mA for 30–60 min.

13. Increase current to 25 mA for 15% gels, 20 mA for 12% gels, and continue electrophoresis for 18–20 hr.

14. Switch off power, discard electrode buffer, and pry apart the plates with a knife or spatulas.

15. Discard the top two-thirds of the gel and place the bottom one-third in stain. Replace stain after approximately 30 min.

16. Destain after 1–4 hr in 5:16:1; then photograph, prepare for fluorography,[21] autoradiography, or silver stain.

Notes

1. After destaining, the one-dimensional gel swells so that it fits snugly into a 0.75-mm-wide slot. Staining the one-dimensional gel makes it possi-

[21] R. A. Laskey and A. D. Mills, *Eur. J. Biochem.* **56**, 335 (1975).

ble to cut out lanes and select segments to be run in the second dimension.

4. The width of the lanes in the one-dimensional gel is about 1 cm; thus the strips fit into the loading trough with no gel or protein jutting above the edge of the notched plates.

5–7. The rationale for not equilibrating the gel strips with solutions containing SDS is as follows: when the subsequent electrophoresis is begun, free histone H1 molecules, being cationic, migrate upward, while H1–SDS complexes, which form at the downward-moving front of SDS, migrate downward, concentrating the zone of H1 molecules to produce more well-defined spots. It is frequently convenient to equilibrate a whole one-dimensional gel in 500 ml of the equilibration solutions and to cut out individual lanes just prior to loading them. We have used one-dimensional gels, a few weeks old, with no obvious difference in resolution, as long as they have not been extensively destained or examined on a light box for more than a few minutes.

10. When excising one-dimensional strips, about 1.5–2.5 cm of gel above the slowest and below the fastest migrating band should be included. If not, these bands may give rise to distorted spots in the second dimension. Routinely, two one-dimensional samples can be loaded onto one two-dimensional gel. A flexible plastic ruler is useful for handling gel strips and for tamping them down.

Examination and Analysis of Gels

Most samples of H1 from nongrowing cells or tissues contain little or no phosphorylated H1 detectable with Coomassie blue or silver.[22,23] In such cases, H1 may be resolved into four to seven spots, depending on the age of the animal from which the tissues were removed or how long tissue culture cells have been in a nondividing state. These spots represent H1a, b, c, d, and e and two forms of H1°.[22–24] An H1 subtype restricted to the testis, H1t, has also been identified.[24–27]

A substantial portion of the H1 from growing cells or tissues is phosphorylated, with the number and amount of phosphorylated forms depending on subtype.[5,28,29] Most normal growing somatic cells contain H1a,

[22] R. W. Lennox and L. H. Cohen, *J. Biol. Chem.* **258**, 262 (1983).

[23] R. W. Lennox, *J. Biol. Chem.* **259**, 669 (1984).

[24] W. S. Kistler and M. E. Geroch, *Biochem. Biophys. Res. Commun.* **63**, 378 (1975).

[25] S. M. Seyedin and W. S. Kistler, *J. Biol. Chem.* **255**, 5949 (1980).

[26] M. L. Meistrich, L. R. Bucci, and W. A. Brock, *Exp. Cell Res.* **140**, 111 (1982).

[27] R. W. Lennox and L. H. Cohen, *Dev. Biol.* **103**, 80 (1984).

[28] L. R. Gurley, J. A. D'Anna, S. S. Barham, L. L. Deaver, and R. A. Tobey, *Eur. J. Biochem.* **20**, 1445 (1981).

[29] K. Ajiro, T. W. Borun, and L. H. Cohen, *Biochemistry* **20**, 1445 (1981).

b, c, d, and e and at least ten phosphorylated forms in interphase[7] and more during mitosis.[4] Each phosphate group slows migration in the first dimension by about 1%. In some subtypes, moreover, there is at least one site at which a phosphate group slows migration slightly in the second dimension, resulting in the resolution of isomers differing only in the position, but not the number, of phosphate groups.[7] This is most evident in the case of H1b, which undergoes several phosphorylations, one of which affects mobility in both dimensions.

Phosphorylated forms of H1 can be identified by comparing ^{32}P- and ^{3}H-labeled samples or by determining which spots disappear when alkaline phosphatase-treated samples are electrophoresed.[7] In addition, incubating cells with ^{3}H-labeled amino acids for short periods of time, that is, 15 min or less, results in only unphosphorylated components being labeled since phosphorylation of newly synthesized H1 does not appear to occur until at least around 30 min after synthesis.[30] Figure 1 shows the resolution of H1 possible using the methods described here. Inspection of Fig. 1 shows that electrophoresis only in acid–urea gels or in SDS-containing gels is not sufficient to resolve all subtypes or all phosphorylated forms of H1. Twelve components are visible in the samples labeled with [^{3}H]lysine for 16 hr. Additional components can be detected when ^{32}P-labeled H1 is analyzed. This method offers several advantages over other methods, such as ion-exchange chromatography, for studying H1 subtypes and their metabolism. It is considerably quicker and requires only small quantities of material. It provides information about the relative size of the various subtypes and their charge to mass ratios. The amount of phosphorylated H1 and the number of phosphorylated forms of each subtype can be easily determined. Structural variations of subtypes among species are readily detected.

[30] R. W. Lennox, unpublished observations, 1987.

Fig. 1. Two-dimensional gel electrophoresis of H1 from rat and mouse. Rat cells (RBL-1) and mouse cells (EL-4) were incubated in Dulbecco's modified Eagle's medium containing 10% (for short-term labeling) or 100% (for long-term labeling) the normal concentration of lysine, 100 μCi/ml [^{3}H]lysine (short-term) or 10 μCi/ml [^{3}H]lysine (long-term), 10% (v/v) fetal calf serum, 100 units/ml penicillin, and 100 μg/ml streptomycin for 5 min, 2 hr, or 16 hr. Cells were then harvested, and H1 was isolated and separated by two-dimensional gel electrophoresis as described here. Fluorograms[21] were exposed to preflashed Kodak X-Omat X-ray film at $-70°$. (A) RBL-1, 5 min; (C) RBL-1, 2 hr; (E) RBL-1, 16 hr; (B) EL-4, 5 min; (D) EL-4, 2 hr; (F) EL-4, 16 hr. Unphosphorylated forms of the various subtypes are identified by the letters above or below the appropriate spot. Note that the rat H1 subtype we refer to as H1e has been called H1d by Kistler;[24] similarly, the rat H1 subtype we refer to as H1d has been called H1e.

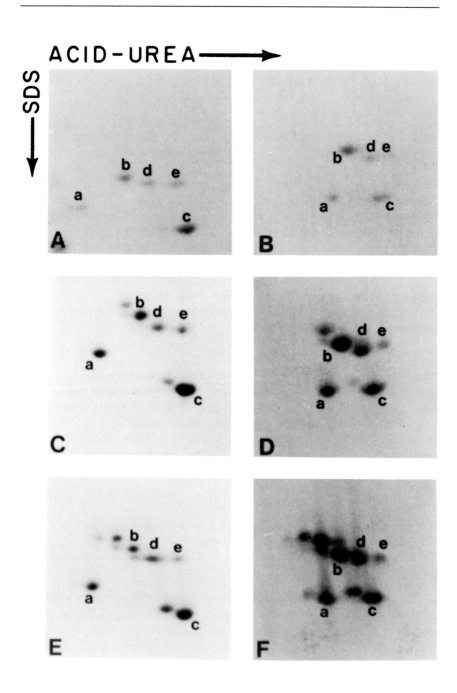

Electrophoresis of H1 in SDS-Containing Minigels

Since a number of manufacturers supply equipment for pouring and electrophoresing minigels cast between lantern slide cover glasses, each using slightly different methods, we do not describe pouring or assembling such gels in detail.

1. Gels may be poured singly between pairs of lantern slide glasses, assembled as described above, but with the comb in place such that the stacking gel has slots. If a casting box is used, as many as 12 may be poured at once.

2. The polymerized gel is assembled into the electrophoresis apparatus. Where required, agarose is used to seal the gel in place.

3. Electrophoresis buffer is added to the buffer chambers. The comb is removed, the wells rinsed with electrophoresis buffer using a syringe equipped with a 25-gauge needle, and samples $(0.1 - 7.0 \mu l)$ applied with a 5- or 10-μl syringe.

4. Samples are electrophoresed at 7.5 mA for 5 min and at 25 mA for 1 dye time plus 5–15 min for a total time of 45–55 min.

5. Gels are stained and destained and then processed for further treatment as above.

Notes

1. It is convenient to pour a number of gels at one time in a gel casting box. Gels may be stored for about 1 week, with no great loss in quality, in a box with a tight-fitting lid containing paper towels dampened with H_2O.

5. Gels may be soaked in methanol–H_2O–acetic acid (5:16:1) for around 15 min then transferred to 50% (w/v) methanol for about 30 min and immediately silver stained, if desired. This procedure is useful, for example, when analyzing aliquots of fractions separated by column chromatography.

Electrophoresis of Core Histones in Acetic Acid – Urea and
  Triton X-100 – Acetic Acid – Urea Gels

Method

The method of Panyim and Chalkley[6] for electrophoresis in gels containing acetic acid and urea separates histones on the basis of differences in net charge and mass. It resolves H3 and H4 and their modified forms but does not completely resolve H2A and H2B. The incorporation of a nonionic detergent, such as Triton X-100,[8] not only resulted in resolution of all the histone classes, but also revealed the existence of subtypes, generally differing in a few amino acid residues, of histones H3, H2B, and H2A.[9,10] Since these differences occur in hydrophobic regions of the molecule, they affect detergent binding and, hence, electrophoretic mobility. The high degree of discrimination afforded by this method is illustrated by the fact that isomers of chemically oxidized H2B, differing only in the position of a single oxygen atom (i.e., methionine sulfoxide at residue 59 or 62), can be resolved.[8] The presence of Triton neither improves nor diminishes the resolution of phosphorylated or acetylated forms, in which the modifications are generally in hydrophilic regions of the protein.

Among histones, Triton affinity is usually in the order H2A > H3 > H2B > H4 > H1 (but in sea urchin, H4 > H2B). Optimal resolution of the subtypes of any particular histone is usually obtained under subsaturating conditions for that histone, where differences in affinity for Triton, as distinct from capacity, can be exploited. These conditions are obtained by varying the concentration of urea (which lowers Triton–protein affinity) rather than the concentration of Triton, which must be kept high enough to minimize the effects of changes in free Triton concentration caused by the binding of Triton to the proteins. Table II gives recipes and running conditions for gels containing different urea concentrations, some described by Zweidler[10] and some used in this laboratory. No one condition is satisfactory for resolving histones from all species, and two conditions may be required even with histones from a single species. For mammalian systems, 7.5 M urea resolves most of the histones and their subtypes. Maximal resolution of H2A is obtained with 8.8 M urea, particularly useful for separating the large number of H2A subtypes present in sea urchins. One H2B subtype is resolved in 8 M urea that is not resolved in 8.8 M urea, and at 8.8 M urea one is resolved that is not resolved in 8 M urea. H3 subtypes, as well as some H2B subtypes, are best resolved in 7.5 or 6 M urea, depending on the organism. The mobility of H1 is generally not affected by Triton, except that at very low urea concentrations (0.5 M) some subtypes of sea urchin and wheat H1 are retarded.

1. Assemble glass plates 35 cm by 25 cm by 0.5 cm (one notched) with Teflon spacers 0.5 mm thick, and seal the sides and bottom with 1% (w/v) agarose, as above.

2. Prepare the gel solution according to the recipes given in Table II and degas it before adding APS. Pour the gel to within 2.0–2.5 cm of the top of the notched plate, overlay with butanol, and allow to polymerize for 60–90 min.

3. Remove the butanol overlay, prepare the loading gel (see Table II), degas it, and pour it to a height of 1 cm. Overlay with butanol and allow to polymerize for 60–90 min.

TABLE II

PREPARATION OF TRITON–ACID–UREA GELS FOR ELECTROPHORESIS OF CORE HISTONES[a]

Urea concentration (M):	3		6		7.5		8.0		8.8		2.5	
	S	L	S	L	S	L	S	L	S	L	S	L
Urea (g)	7.2	1.8	14.4	3.6	18.0	4.5	19.2	4.8	21.1	5.3	6.0	1.5
60% acrylamide (ml)	8	1	8	1	8	1	8	1	8	1	10	1
2% MBA (ml)	1.6	0.2	1.6	0.2	1.6	0.2	1.6	0.2	1.6	0.2	2	0.2
10% Triton X-100 (ml)	1.5	0.37	1.5	0.37	1.5	0.37	1.5	0.37	1.5	0.37	0	0
Acetic acid (ml)	2	0.5	2.0	0.5	2	0.5	2	0.5	2	0.5	2	0.5
TEMED (μl)	200	50	200	50	200	50	200	50	200	50	200	50
20% APS (μl)	200	75	160	75	130	60	120	60	100	50	400	100
Final volume (ml)	40	10	40	10	40	10	40	10	40	10	40	10
Electrophorese at 400 V for time (hr)	16		18		18		22		25		13	

[a] The columns headed S are for separating gels, the columns headed L for loading gels. The final concentration of Triton X-100 is generally given as 6 mM. Recipes for stock solutions are given in the text.

4. Prerun and scavenge the gels and load the samples, as described above for acid–urea gels for separating H1 proteins.

5. Electrophorese at the voltages for the times listed in Table II.

6. Stain and destain the whole gel as described above.

Notes

1, 5. Shorter gels are sufficient for some purposes. As an approximation, adjust *both* the voltage and the running time in direct proportion to the gel length, or adjust only the running time in proportion to the square of the gel length. Cytochrome c (40–50 μg) can be run as a visible marker. In gels containing 6 M urea or greater, cytochrome c runs with H2B, and electrophoresis should be stopped when the marker is about 4 cm from the bottom to ensure that H4 is not lost. In gels containing 3 M urea, cytochrome c can be run to the end of the gel.

4. The scavenging step is especially important in Triton-containing gels in order to prevent the oxidation of methionine-containing proteins during electrophoresis. Since oxidation of methionyl residues diminishes Triton binding, protein molecules accelerate as they become oxidized, resulting in smeared bands. Obviously, only proteins that have methionine and that bind Triton under the conditions of the run are affected in this way. If oxidation of the proteins has occurred prior to electrophoresis, artifactual bands will be obtained. Methionine oxidation occurs fairly rapidly in denatured proteins unless prevented by the addition of a protective thioether. It is possible, incidentally, to determine the minimum number of methionyl residues by partially oxidizing the sample with H_2O_2 before running.

Electrophoresis of Core Histones in Gels Containing SDS

Two-Dimensional Gels for Core Histones

The methods for preparing the gels and equilibrating samples are essentially the same as those described above for histone H1. Core histones are resolved best in gels containing SDS and 18% acrylamide. The recipes for preparing gels are given in Table I and the methods outlined above. Note, however, that SDS gels containing 18% acrylamide can only be 20 cm long (requiring 40 ml gel solution) and are electrophoresed at 16 mA for approximately 19 hr. With 35-cm-long gels, SDS accumulation at the chloride boundary increases resistance, with consequent voltage increases and severe overheating before reasonable separation is obtained.

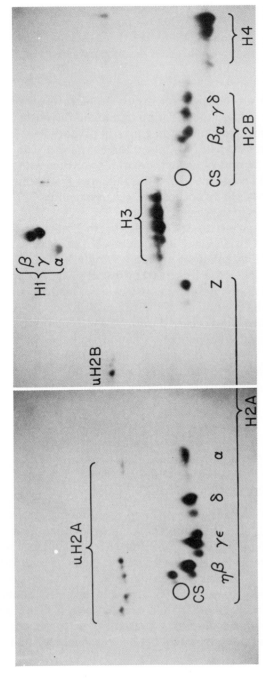

FIG. 2. Fluorogram of a two-dimensional gel of histones from sea urchin (*Strongylocentrotus purpuratus*) embryos labeled for 6 hr with 2 μCi/ml [³H]leucine at the mesenchyme blastula (22 hr) stage. The one-dimensional gel contained 6 m*M* Triton X-100 and 8 *M* urea. The lane was cut out and divided into two lengths, which were then run in two identical SDS-containing gels as described in the text. The circles show the positions of two histone subtypes, H2A$_{cs}$ and H2B$_{cs}$, which are present but not synthesized at this stage of development. The complexity in the H3 region is due to the presence of both acetylated and phosphorylated forms, as described in the text.

Interpretation of Gels

Figure 2 illustrates the resolution of sea urchin histones by the two-dimensional method. Under the conditions employed, eight subtypes of H2A and five of H2B are resolved, as well as most modified forms of H2A, H3, and H4. Generally, each acetylation of these proteins retards their migration in the TAU dimension (in inverse relation to the number of basic residues in the protein) and accelerates it very slightly in the SDS dimension. In the case of H3, a small amount of each form is also phosphorylated, which retards migration in the first dimension only. Thus, two overlapping, slightly tilted tiers of H3 spots are seen, the darker tier comprising the parent protein and its variously acetylated forms, and the paler tier at its left comprising their phosphorylated forms. A few other modified histones are incompletely resolved as well: acetylated H2Aϵ overlaps unacetylated H2Aγ, and the H2B subtypes are not separated well enough to detect most of their acetylated or phosphorylated forms. Ubiquitination retards migration considerably in both dimensions, with the ubiquitinated set of H2A subtypes overlapping the parent set in the first dimension.

Electrophoresis of Core Histones in Minigels

The methods for preparing minigels are given above, and the recipes for 18% gels are provided in Table I. Electrophoresis is usually carried out at 7 mA for 5 min, then at 20–25 mA for 1 dye time plus 10–15 min, for a total time of 50–60 min.

Acknowledgments

Work from this laboratory was supported by Grants GM-24019, CA-06927, and RR-05539 from the National Institutes of Health and an appropriation from the Commonwealth of Pennsylvania.

[26] Chemical Cross-Linking of Histones

By JEAN O. THOMAS

Introduction

Bifunctional reagents, particularly those that react with amino groups, which are generally accessible on protein surfaces, have proved useful in the analysis of several features of chromatin structure. They are powerful

Copyright © 1989 by Academic Press, Inc.
All rights of reproduction in any form reserved.

tools for the characterization of histone oligomers and chromatin from various sources and prepared by various procedures, including reconstitution, and for the analysis of interactions of histones with each other and with nonhistone proteins in solution and in chromatin.

Broadly, information may be gained at three levels. First, and most simply, the oligomeric state of the histones and the stability of histone oligomers as a function of protein concentration and ionic strength may be determined from the relative molecular masses of the cross-linked oligomers, which are most easily estimated by polyacrylamide gel electrophoresis in the presence of sodium dodecyl sulfate (SDS). Second, the nearest neighbors within a protein oligomer may be determined. In this case, cleavable cross-linkers are needed so that fractionated cross-linked products may be cleaved at the cross-links to regenerate their component histones for identification. The whole procedure is most simply carried out by SDS–polyacrylamide gel electrophoresis in two dimensions at right angles to each other, with cleavage of the cross-links between the two dimensions. Third, particular regions of a cross-linked histone pair may be identified; this is achieved by judicious enzymatic or chemical cleavage followed by fractionation of the cross-linked fragments and identification of the component peptides after cleavage of the cross-links. Examples of each type of procedure are described.

Reagents

Discussion is confined to the use of bisimidoesters and bis(N-hydroxysuccinimide esters), which cross-link the nucleophilic, unprotonated ε-amino groups of lysine side chains in proteins under appropriate conditions, and which have proved particularly useful in studies of histones and chromatin.[1] Disulfide-containing members of both classes of reagent are available and permit cleavage of cross-links for identification of cross-linked components. For a comprehensive survey of a wide range of bifunctional reagents, see the article by Ji in this series,[2] and references therein to earlier reviews. Some early studies of histone cross-linking with other types of reagents have been mentioned previously.[1]

Imidoesters are readily soluble in aqueous solution and react specifically with amino groups in the pH range 7–10.[3,4] However, they are rapidly hydrolyzed at the lower pH values (for methyl and ethyl acetimi-

[1] J. O. Thomas and R. D. Kornberg, *Methods Cell Biol.* **18**, 429 (1978).

[2] T. H. Ji, this series, Vol. 91, p. 580.

[3] M. J. Hunter and M. L. Ludwig, this series, Vol. 25, p. 585.

[4] J. K. Inman, R. N. Perham, G. C. DuBois, and E. Appella, this series, Vol. 91, p. 559.

date the half-life[5] is ~2–5 min at pH <8.5, ~12 min at pH 9, and ~25 min at pH 10), and amidination of proteins is therefore usually carried out at alkaline pH to reduce competition from reagent hydrolysis. Provided that there is no risk of structural distortion at high pH, this has the advantage of suppressing an undesirable side reaction with imidoesters which leads to reaction of two protein amino groups at one imidoester group;[1,5] however, with disulfide-containing bifunctional reagents, high pH is likely to promote undesirable disulfide interchange.

N-Hydroxysuccinimide esters are poorly soluble in aqueous solutions, and the reagent is therefore dissolved in a suitable organic solvent (see below); its presence during the reaction (usually at a concentration of less than 0.5%, v/v) has no obvious ill effect. N-Hydroxysuccinimide esters react preferentially with amino groups in proteins; the product of reaction with the imidazole groups of the histidine side chain is unstable and easily hydrolyzed. Reagent hydrolysis again competes with protein modification, and since N-hydroxysuccinimide esters are hydrolyzed faster the higher the pH (half-life[6] 4–5 hr at pH 7.5 <10 min at pH 8.6), modification of proteins is usually carried out at pH 7–8; under these conditions reaction with amino groups is rapid at low reagent concentration.

Imidoesters have the desirable property that the charge on the amino group is conserved after amidination (although shifted one atom further from the peptide backbone), whereas a positive charge is lost at each amino group converted to an amide by acylation with N-hydroxysuccinimide esters. However, against this very real advantage of imidoesters has to be weighed the alkaline conditions necessary for reasonably rapid reaction with amino groups without unduly large excesses of reagent. In summary, both types of reagent are powerful tools, but both should be used with careful consideration of the pros and cons, and ideally in parallel.

The hydrolysis that occurs with both types of bifunctional reagent may occur at both ends or only at one end, and this may occur before or after reaction with the protein. Increasing the reagent concentration to compensate for complete hydrolysis may not significantly increase the degree of cross-linking, because it will also increase the rate of monofunctional reaction with the protein and reduce the concentration of amino groups left free to participate in bifunctional reactions. A better ploy in cases where more extensive cross-linking is required is to make multiple additions of reagent at low concentration.

Three useful and readily available reagents for protein cross-linking (Fig. 1) are the two bisimidoesters dimethyl suberimidate (I), first used by

[5] D. T. Browne and S. B. H. Kent, *Biochem. Biophys. Res. Commun.* **67,** 126, 133 1975).
[6] A. J. Lomant and G. Fairbanks, *J. Mol. Biol.* **104,** 243 (1976).

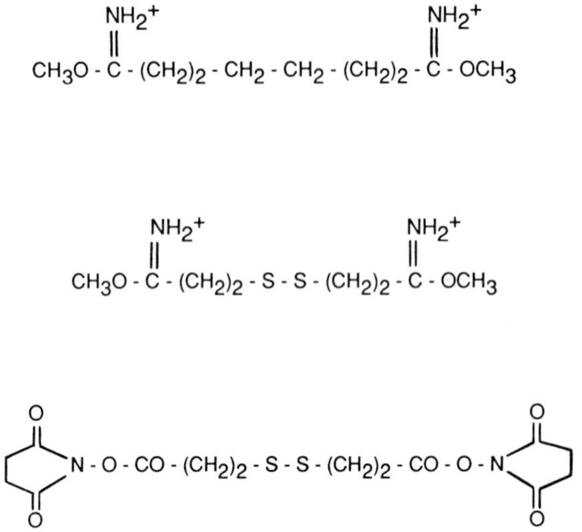

$$\overset{\overset{\displaystyle NH_2^+}{\|}}{CH_3O - C} - (CH_2)_2 - CH_2 - CH_2 - (CH_2)_2 - \overset{\overset{\displaystyle NH_2^+}{\|}}{C} - OCH_3 \qquad (I)$$

$$\overset{\overset{\displaystyle NH_2^+}{\|}}{CH_3O - C} - (CH_2)_2 - S - S - (CH_2)_2 - \overset{\overset{\displaystyle NH_2^+}{\|}}{C} - OCH_3 \qquad (II)$$

N - O - CO - (CH₂)₂ - S - S - (CH₂)₂ - CO - O - N   (III)

FIG. 1. Some reagents used for cross-linking: I, dimethyl suberimidate; II, dimethyl bisdithiopropionimidate; III, dithiobis(succinimidyl propionate).

Davies and Stark[7] to determine the subunit structure of proteins, and dimethyl bisdithiopropionimidate (II), described by Wang and Richards,[8] and the bis(N-hydroxysuccinimide ester) dithiobis(succinimidyl propionate) (III), introduced by Lomant and Fairbanks.[6] They have a maximum span of about 11–12 Å in cross-linking reactions; the protein–protein cross-links introduced with II and III may be cleaved by thiolysis. All three reagents are supplied by Pierce Chemical Company (Rockford, IL). They should be stored under anhydrous conditions at −20° and are stable for at least 1 year. Bisimidoesters with a shorter span than dimethyl suberimidate are also commercially available (Pierce). Dimethyl malonimidate and succinimidate (one and two methylene groups, respectively, replacing the six in I form cross-links of maximum length 5 and 6 Å between amino groups. These give little cross-linking of histones in chromatin beyond dimers, reflecting the restricted probability of finding two lysine amino groups suitably disposed for participation in a cross-link; therefore, they are unsuitable for use in determining the largest oligomer size by cross-linking, for which purpose reagents the length of dimethyl suberimidate are well suited. There are no cleavable analogs, so these reagents have not been used for analysis of histone contacts in chromatin.

[7] G. E. Davies and G. R. Stark, Proc. Natl. Acad. Sci. U.S.A. 66, 651 (1970).
[8] K. Wang and F. M. Richards, Isr. J. Chem. 12, 375 (1974).

Dimethyl Suberimidate and Dimethyl Bisdithiopropionimidate. The compounds are supplied as the dihydrochlorides. We dissolve them immediately before use, usually at a concentration of approximately 11 mg/ml, in the buffer to be used for cross-linking (see below), but at a pH high enough to neutralize the HCl (and prevent hydrolysis of the reagent; see above). For example, if sodium borate, pH 9, is the cross-linking buffer, the reagent is dissolved in the borate buffer at pH 10.5–11. If a triethanolamine (TEA) buffer (e.g., 100 mM triethanolamine-HCl, pH 8.5) is to be used, the stock solution of reagent may be prepared in free base at the buffer concentration (i.e., 100 mM triethanolamine). Reagent solution prepared as described and diluted 10-fold into the protein solution barely alters its pH.

Dithiobis(succinimidyl Propionate). Because dithiobis(succinimidyl propionate) is relatively insoluble in water, a stock solution is prepared immediately before use at 50 mg/ml in dry dimethylformamide (DMF); when diluted around 250-fold into the buffered solution of protein to be cross-linked, it remains soluble [final DMF concentration 0.4% (v/v)]. With ice-cold solutions, a transient opalescence appears on addition of the reagent, but this disappears instantly, presumably accelerated by rapid reaction of the reagent with the protein. In principle, other water-miscible, nonreactive organic solvents, such as dimethyl sulfoxide, dioxane, or acetonitrile, should serve the same purpose.

Buffers and Protein Concentration

Histones or chromatin to be cross-linked should be in an amine-free buffer at pH 7.5–9.0, the pH depending on the reagent: phosphate and HEPES for around pH 7–7.5; triethanolamine for pH 7.5–8.5; and borate for pH 8.5–9.5. Related buffers are also suitable. Low protein concentrations are used to favor intramolecular reaction (within a protein oligomer) over intermolecular reaction. A concentration of around 0.5–1.0 mg/ml is a good starting point for free histones; for nucleosome core particles or chromatin, a concentration corresponding to $A_{260} = 1-10$ (histone concentration ~ 50–500 μg/ml).

Cross-Linking

The basic cross-linking procedure is the same with both types of reagent. To a solution of chromatin or histones at the appropriate concentration, add 1/10 volume of an 11 mg/ml stock solution of dimethyl suberimidate or dimethyl bisdithiopropionimidate to give a final concentration of 1 mg/ml (~ 3.7 and 3.2 mM, respectively). For dithiobis(succinimidyl propionate), add 1/250 volume of a 50 mg/ml stock solution in DMF to give a final reagent concentration of approximately 0.2 mg/ml (~ 0.5 mM). To determine the time course of cross-linking, at various times [e.g., up to 2 hr with dimethyl suberimidate or 30 min with dithiobis(succinimidyl propionate)] take samples sufficient for analysis in SDS–polyacrylamide gels into an equal volume of 50% (w/v) trichloroacetic acid (TCA) on ice. After about 30 min, collect the precipitated histones (or histones plus DNA in the case of chromatin) by centrifugation, wash the pellet with cold acetone–10 mM HCl, then with cold acetone, dry *in vacuo*, and analyze the cross-linked products by SDS–polyacrylamide gel electrophoresis (see below). Note that removal of the DNA from cross-linked chromatin before electrophoresis is not necessary. This is easily achieved if required, however, by dispensing the chromatin after cross-linking into H$_2$SO$_4$ (final concentration, 0.2 M) instead of into TCA. Allow 30 min at 4° to extract the histones, and then collect them by TCA precipitation as above. Alternatively, the TCA-precipitated histone–DNA mixture may be extracted with H$_2$SO$_4$, etc.

When cross-linked samples are required for purposes other than immediate analysis in SDS gels, the cross-linking section may be quenched with 1/20 volume 1 M glycine, or some other amino compound, and then dialyzed and freeze-dried, or collected by TCA precipitation as above. Cross-linking with bisimidoesters may also be stopped by lowering the pH with 1/4 volume 3 M sodium acetate, pH 5, to hydrolyze the reagent.

Analysis of Cross-Linked Products by SDS–Polyacrylamide Gel Electrophoresis

Dissolve the TCA-precipitated or freeze-dried sample at around 0.5–1 mg/ml in the appropriate "sample buffer" (see below), omitting 2-mercaptoethanol for samples cross-linked with dimethyl bisdithiopropionimidate or dithiobis(succinimidyl propionate) in order to preserve the S–S-containing cross-links. Preheat the sample buffer in a boiling water bath and add 0.01 volume phenylmethylsulfonyl fluoride (PMSF, 50 mM in propan-2-ol) immediately before addition to the sample to ensure rapid denaturation and inhibition, respectively, of any contaminating proteases.[9,10] Immediately reheat the sample solution in a stoppered microfuge tube for 1 min in boiling water; allow the solution to cool, and centrifuge briefly to collect the sample into the bottom of the tube.

[9] K. Weber, J. R. Pringle, and M. Osborn, this series, Vol. 26, p. 3.
[10] J. O. Thomas, *in* "Techniques in Protein and Enzyme Biochemistry" (H. L. Kornberg, J. C. Metcalfe, D. H. Northcote, and K. F. Tipton, eds.), B. 106, p. 1. Elsevier/North-Holland, Amsterdam, 1978.

The type of gel chosen for electrophoresis will vary with the task at hand. A "low-resolution" gel would be sufficient if the aim were merely to resolve classes of oligomers (e.g., dimers, trimers, and tetramers) when resolution of individual dimers, for example, is not required and might even be an undesirable complication. "High-resolution" gels are essential when separation of individual dimers is required for subsequent determination of their composition, or when cross-linked fragments of histones must be resolved for fine mapping of cross-links. The use of both gel systems for analysis of histone cross-linking has been described previously.[10] Details are given again here for convenience.

Low-Resolution Phosphate Gels

SDS 5% or 7.5% polyacrylamide gels are suitable for the separation of cross-linked histone oligomers. We find that tube gels (as in the original procedures[9,11,12]) consistently give sharper bands than gels of the same composition cast as slabs; examples of both are given below.

Pour the tube gels to a height of 7 cm in chromic acid-cleaned Pyrex glass tubes (9 × 0.45 cm i.d.), standing vertically and sealed at the bottom with Parafilm. Layer 0.1% (w/v) SDS carefully onto the surface to give a flat meniscus. The gels set within an hour.

5 or 7.5% polyacrylamide–SDS gels: These quantities are for about eight tube gels; for a slab gel, 12 × 15 × 0.15 cm, multiply by 2.5.

	5%	7.5%
Acrylamide–N,N'-methylenebisacrylamide (30%/0.8%, w/v)	3.33 ml	5.00 ml
Sodium phosphate (1 M, pH 6.6)	2.00 ml	2.00 ml
Water	14.26 ml	12.59 ml
N,N,N',N'-tetramethylethylenediamine (TEMED)	0.01 ml	0.01 ml
SDS (10%, w/v)	0.2 ml	0.2 ml
Ammonium persulfate (10%, w/v; fresh)	0.2 ml	0.2 ml

Before addition of TEMED, SDS, and ammonium persulfate, degas on a water pump for 3–4 min (O_2 inhibits polymerization).

Running buffer: sodium phosphate (1 M, pH 6.6) 100 ml, SDS (10%, w/v) 10 ml, water 890 ml

[11] A. L. Shapiro, E. Viñuela, and J. V. Maizel, *Biochem. Biophys. Res. Commun.* **28,** 815 (1967).
[12] K. Weber and M. Osborn, *J. Biol. Chem.* **244,** 4406 (1969).

> Sample buffer: sodium phosphate (1 M, pH 6.6) 0.05 ml, SDS (10%, w/v) 0.5 ml, 2-mercaptoethanol 0.05 ml, glycerol 0.5 ml, bromphenol blue (1%, w/v) 0.05 ml, water 3.9 ml

Layer the samples onto the gels, already supported in the electrophoresis apparatus with buffer in the reservoirs. A sample loading of 10–20 μg is necessary for a mixture of cross-linked proteins to be stained with Coomassie Brilliant Blue R, aiming at 2–5 μg per band. Conventional molecular weight markers are not useful because of the anomalously low migration of histones in SDS gels resulting from their high intrinsic positive charge.

After electrophoresis at about 7 mA per gel for about 3 hr, the bottom of the bromphenol blue band is about 1 cm from the bottom of the gel; the fastest migrating histone, H4, travels with the dye in this gel system. Loosen the gels from their tubes by squirting 0.1% SDS through an 18-gauge syringe needle rotated around the inside of the tube, and allow the gel to drop directly into fixer (methanol–acetic acid–H_2O, 5:1:5, v/v) in glass test tubes. After about 1 hr at room temperature, stain the gels with 0.1% (w/v) Coomassie Brilliant Blue R-250 in the same solution for 30 min at 37° and then destain with 7.5% acetic acid–5% methanol (v/v) at 37°. Bands are visible within 15 min, and the gels may be completely destained in a few hours with several changes of destaining solution.

High-Resolution Discontinuous Tris Slab Gels

The following high-resolution system is based on that of Laemmli[13] with the following modifications:[1,14] lowering of the N,N'-methylenebisacrylamide to acrylamide ratio in the separating gel to 0.15:30; increase of this ratio in the stacking gel to 1.5:30; increase of the Tris-HCl concentration in the separating gel to 0.75 M; and the use of an electrode buffer (pH 8.4) of composition 50 mM Tris–0.38 M glycine–0.1% (w/v) SDS. The details given below are for a resolving gel 12 cm long by 15 cm wide by 0.15 cm thick.

18% Polyacrylamide resolving gel, pH 8.8:

Acrylamide–N,N'-methylenebisacrylamide (30%/0.15%, w/v)	30 ml
Tris-HCl (3 M, pH 8.8)	12.5 ml
Water	6.25 ml
N,N,N',N'-Tetramethylethylenediamine	0.01 ml
SDS (10%, w/v)	0.5 ml
Ammonium persulfate (10%, w/v; fresh)	0.5 ml

[13] U. K. Laemmli, *Nature (London)* **227,** 680 (1970).
[14] J. O. Thomas and R. D. Kornberg, *Proc. Natl. Acad. Sci. U.S.A.* **72,** 2626 (1975).

Pour the mixture to a depth of 15 cm in a vertically supported gel mold, overlayer the solution with 0.1% SDS, and allow the gel to set for about 1 hr. Pour off the SDS and replace with a stacking gel.

3% Polacrylamide stacking gel, pH 6.8:

Acrylamide–N,N'-methylenebisacrylamide (10%/0.5%, w/v)	6.0 ml
Tris-HCl (0.5 M, pH 6.8)	4.8 ml
Water	8.8 ml
N,N,N',N'-Tetramethylethylenediamine	0.01 ml
SDS (10%, w/v)	0.2 ml
Ammonium persulfate (10%, w/v; fresh)	0.2 ml

Sample buffer: Tris-HCl (0.5 M, pH 6.8) 0.5 ml, SDS (10% w/v) 0.5 ml, 2-mercaptoethanol 0.05 ml, glycerol 0.5 ml, bromphenol blue (1%, w/v) 0.05 ml, water 3.4 ml

Running buffer: The electrode buffer is prepared by 5-fold dilution of a solution containing (in 1 liter) 30 g Tris base, 140 g glycine, and 50 ml SDS (10%, w/v). The pH of the diluted solution is 8.4 and should not be adjusted.

Carry out electrophoresis at a constant current of 30 mA at room temperature; the dye reaches the end of the gel in about 6 hr for 12-cm gels. Fix, stain, and destain the slabs as described above for tube gels.

Some Examples

Some representative examples taken from work in this laboratory are now given of the various ways in which chemical cross-linking has provided information about histones and chromatin, from gross features to finer detail, as outlined in the Introduction. The basic cross-linking procedure is essentially as described above; particular details are given in the text or in the figure legends.

Determination of the Oligomeric State of Histones in Solution and in Chromatin

Cross-linking provided the first demonstration of the tetrameric nature of the complex of the arginine-rich histones H3 and H4 extracted from chromatin by mild methods[15] and of the octameric nature of the protein core of the nucleosome in chromatin and free in solution.[14,16] It continues

---

[15] R. D. Kornberg and J. O. Thomas, *Science* **184,** 865 (1974).
[16] J. O. Thomas and P. J. G. Butler, *J. Mol. Biol.* **116,** 769 (1977).

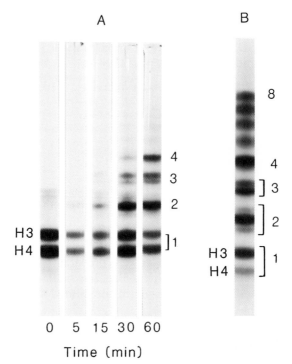

Fig. 2. (A) Cross-linking of histones H3 and H4 with dimethyl suberimidate shows that they exist as a tetramer in solution. Histones H1, H2A, and H2B were removed from chromatin by gel-exclusion chromatography in 1 M NaCl; the remaining "H3/H4 chromatin" was dialyzed into buffer at pH 9 and ionic strength 2 M (1.95 M NaCl, 137 mM sodium borate), cross-linked at 4° with dimethyl suberimidate (1 mg/ml) for the times indicated, and analyzed in SDS–5% polyacrylamide tube gels in phosphate buffer. [From P. G. Stockley and J. O. Thomas, *FEBS Lett.* **99,** 129 (1979).] (B) Cross-linking of "H3/H4 chromatin" with dimethyl suberimidate. H3/H4 chromatin, prepared as in A, was cross-linked in 100 mM sodium borate, pH 9.5, with 1 mg/ml dimethyl suberimidate and analyzed in a 5% gel as above (but run for longer than in A). The cross-linking pattern suggests a mixture of tetramers of H3 and H4 (note the unusually strong tetramer band) and octamers (see text). The three dimers [strong H3–H4 and weaker flanking (H3)₂H4 and (H4)₂H3] are characteristic of the arginine-rich tetramer.

to serve as a useful and simple way of checking the oligomeric state of histones and histone complexes isolated in various ways, and from various sources. Figure 2A shows that H3 and H4, left after salt-stripping of H1, H2A, and H2B from chromatin, are displaced from the DNA as tetramers when the ionic strength is raised to 2 M.[17] However, a significant proportion of the H3 and H4 that remains bound to DNA exists as octamers (Fig.

[17] P. G. Stockley and J. O. Thomas, *FEBS Lett.* **99,** 129 (1979).

2B), presumably resulting from the sliding of tetramers in the 1 M NaCl used to dissociate H1, H2A, and H2B. Note that a graph of the relative molecular masses of cross-linked products based on known monomer M_r values for the constituent proteins is essentially linear; such a plot thus serves as the first check on the identity of cross-linked products since, as already noted, standard molecular weight markers are inappropriate for histones.

A directly related use of cross-linking is to examine the stability of oligomers, for example, as a function of dilution or ionic strength. Figure 3A illustrates the dependence of the stability of the histone octamer as a

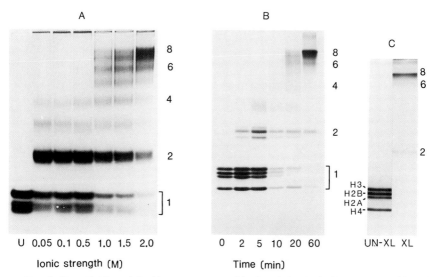

FIG. 3. Cross-linking of the histone octamer with dimethyl suberimidate. (A) Stability of the histone octamer (extracted with 2 M NaCl) as a function of ionic strength at pH 8, monitored by cross-linking with dimethyl suberimidate. The octamer was extracted from H1-depleted chromatin in 2 M NaCl at pH 9 [J. O. Thomas and P. J. G. Butler, *J. Mol. Biol.* **116**, 769 (1977)]. Samples were dialyzed into 274 mM triethanolamine-HCl, pH 8 (ionic strength 0.1 M), containing various concentrations of NaCl to give the final ionic strengths shown, and then cross-linked with dimethyl suberimidate (1 mg/ml) for 90 min at 25°. They were analyzed in an SDS–7.5% polyacrylamide slab gel in phosphate buffer. (B) Time course of cross-linking of the histone octamer at pH 8.5, ionic strength 2 M (1.95 M NaCl, 137 mM triethanolamine-Cl) with 1 mg/ml dimethyl suberimidate at 21° Samples removed at various times were analyzed in an SDS–18% polyacrylamide slab gel in the discontinuous Tris buffer system. (C) Histone octamer at 1 mg/ml prepared as in A and cross-linked at pH 9 in 1.95 M NaCl, 137 mM sodium borate, with 1 mg/ml dimethyl suberimidate at 22° for 45 min. UN-XL and XL indicate uncross-linked and cross-linked octamer; gel as in B. The small amounts of hexamer and dimer arise from slight dissociation of the octamer (see text). Slight aggregation of octamers giving 16-mers, etc., is also apparent. [Adapted from J. O. Thomas and P. J. G. Butler, *J. Mol. Biol.* **116**, 769 (1977).]

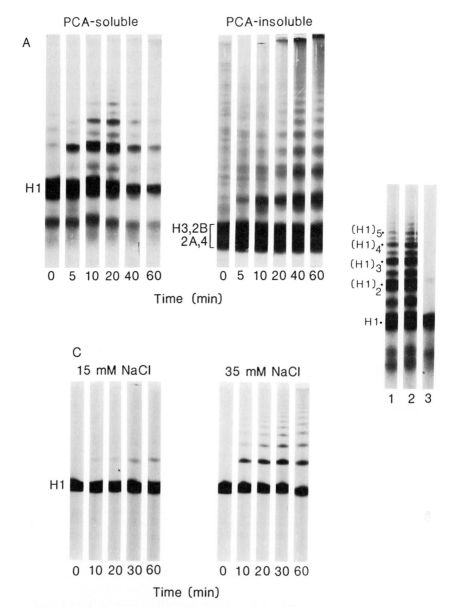

FIG. 4. Analysis of H1 cross-linking in chromatin and H1–DNA complexes. (A) Cross-linking of rat liver chromatin with dimethyl suberimidate: analysis of the cross-linked products of H1 and of the core histones. Cross-linking was carried out in 100 mM triethanolamine-HCl, pH 8.5, at 21° with 1 mg/ml dimethyl suberimidate for the times indicated; the products were fractionated with 5% (v/v) perchloric acid (PCA), and the proteins in the soluble and insoluble fractions were analyzed in SDS–5% polyacrylamide phosphate tube gels. The PCA-soluble products are H1 polymers, $(H1)_n$ and $(H1)_n$–X (see text). Material migrating ahead of H1 in the PCA-soluble fraction is partly HMG proteins and partly

function of ionic strength. The octamer is stable in 2 M NaCl at pH 8, and the main cross-linked band is an octamer, with some hexamer and dimer arising from dissociation of the octamer. As the ionic strength is lowered, hexamers dissociate into tetramers and dimers,[14,18] and the cross-linked dimer band becomes more prominent. The octamer stability falls off sharply between ionic strength 1 and 0.05 M, where the largest oligomer is a tetramer. The resistance of the tetramer to dissociation on further lowering of the ionic strength down to 0.05 M is evident from the persistence of a cross-linked tetramer as the largest oligomer. Note that the conditions are such that cross-linking is incomplete and a ladder of bands corresponding to intermediates in cross-linking is observed up to the largest oligomer size, which is thereby easily identified. The intermediates can be abolished by either increasing the pH (which increases the concentration of unprotonated amino groups and decreases the loss of reagent by hydrolysis; see above) or by multiple additions of reagent to offset its destruction by hydrolysis. Figure 3C shows such a case where the octamer at high ionic strength is cross-linked at pH 9. The main species is the fully cross-linked octamer; the faint cross-linked hexamer and dimer bands arise from true hexamers and dimers resulting from slight dissociation of the octamer at this protein concentration.

The suitability of the low-resolution phosphate gel (Fig. 3A) for analysis of different classes of cross-linked oligomer is evident from comparison with Fig. 3B, which shows an analysis of octamer cross-linking in a high-

---

[18] J. O. Thomas and R. D. Kornberg, *FEBS Lett.* **58**, 353 (1975).

---

contaminating H2B. The background of bands throughout the gel in the absence of cross-linking in the PCA-insoluble fraction, containing core histones, is present because the total PCA-insoluble fraction was analyzed; it largely disappears if a 0.2 M H$_2$SO$_4$ extract of this fraction is analyzed instead. (B) Assessment by cross-linking of the fidelity of H1 binding in chromatin after dissociation and reassociation. Analysis of the H1-containing (5% PCA-soluble) fractions from (lane 1) chromatin treated with 0.5 M NaCl for 12 hr at 0° before dialysis into 100 mM triethanolamine-HCl, pH 8.5, and cross-linking with 1 mg/ml dimethyl suberimidate for 20 min at 21°; (lane 2) control chromatin not exposed to salt; and (lane 3) chromatin cross-linked in the presence of 0.5 M NaCl. The gels were SDS–5% polyacrylamide phosphate tube gels. [A and B from J. O. Thomas and A. J. A. Khabaza, *Eur. J. Biochem.* **112**, 501 (1980).] (C) Analysis of the salt-dependent cooperativity in H1–DNA binding by cross-linking. Cross-linking of H1–DNA complexes (10% H1:DNA, w/w) was carried out in 1 mM sodium phosphate, 0.2 mM Na$_2$EDTA, 0.25 mM PMSF, pH 7.4, containing 15 or 35 mM NaCl, with 0.2 mg/ml dithiobis(succinimidyl propionate) at 23° for the times indicated. The samples were analyzed in SDS–5% polyacrylamide tube gels in phosphate buffer. The H1 molecules are in close proximity in 35 mM NaCl, but not in 15 mM NaCl. [From D. J. Clark and J. O. Thomas, *J. Mol. Biol.* **187**, 569 (1986).]

resolution, 18% polyacrylamide slab gel. The dimers are well resolved, which is necessary if their composition is to be analyzed (see next section), but the higher oligomers are smeared and hard to distinguish clearly at intermediate times of cross-linking, although the octamer is still clearly recognizable as the end product of cross-linking.

Cross-linking has revealed a number of interesting aspects of the behavior of histone H1, free in solution, in chromatin, and in H1–DNA complexes. Figure 4A illustrates, first, that H1 molecules on successive nucleosomes in rat liver chromatin are sufficiently close to be cross-linked by reagents of span 11–12 Å;[19] they are also cross-linked by much shorter reagents.[20] The H1-containing cross-linked products, extracted with 5% (v/v) perchloric acid,[21] appear as an alternating series of strong $(H1)_n$ bands and weaker $(H1)_n – X$ bands. (The composition of the two sets of bands was identified using cleavable cross-linking reagents, as described in the next section, and X was shown to be any of the core histones.[19]) The value of following the progress of cross-linking is clear from Fig. 4A: the number of cross-linked bands increases with time, up to (in this case) 20 min, and then decreases. The decrease is accompanied by a reduction in the total amount of perchloric acid-soluble material extracted and is due to increased cross-linking of the H1 cross-linked polymer to the core histones. The resulting PCA-insoluble product is large and does not enter the gel (Fig. 4A). The core histones (5% PCA-insoluble) are cross-linked much more slowly than H1, presumably because they are both less lysine-rich and "inside" the nucleosome. The bands in the gel do not stop at octamer, indicating that nucleosomes are in close proximity in the nucleosome filament at this pH and ionic strength.

Figure 4B shows how cross-linking can be generally useful in checking (albeit at a gross level) the fidelity of chromatin reconstitution procedures. In this case, H1 was merely dissociated from chromatin by raising the ionic strength to 0.5 M and then allowed to reassociate by lowering the ionic strength.[19] The H1 cross-linking pattern was identical with that of untreated chromatin. However, when chromatin was treated with cross-linking reagent in 0.5 M NaCl, no H1 cross-linked polymers were formed because H1, free in solution, exists as a monomer (Fig. 4B).

Cross-linking has also been useful in determining whether the binding of H1 to DNA is cooperative or noncooperative.[22] The results in Fig. 4C, in conjunction with hydrodynamic analysis, show that when limited amounts

[19] J. O. Thomas and A. J. A. Khabaza, *Eur. J. Biochem.* **112**, 501 (1980).
[20] R. Ring and R. D. Cole, *J. Biol. Chem.* **258**, 15361 (1983).
[21] E. W. Johns, *Biochem J.* **92**, 55 (1964).
[22] D. J. Clark and J. O. Thomas, *J. Mol. Biol.* **187**, 569 (1986).

of H1 (10% H1 : DNA, w/w) are added to DNA, the binding is noncooperative in 15 mM NaCl at pH 7, but cooperative in 35 mM NaCl. This leads to essentially no cross-linking in one case but to an array of cross-linked oligomers in the other. Further details are given in the legend for Fig. 4.

In summary, insight may be gained into a wide range of aspects of histone behavior simply by cross-linking with readily available reagents and analyzing the products in a one-dimensional, low-resolution gel.

Determination of Histone Neighbors in Chromatin Using Cleavable Cross-Linkers and Two-Dimensional ("Diagonal") Gel Electrophoresis

In early studies of histone neighbors within the nucleosome,[18] we used a two-stage cross-linking procedure in which lysine amino groups were first amidinated with a thiol-containing imidoester (methyl 3-mercaptopropionimidate[23]) and S–S-containing cross-links then generated by oxidation with H_2O_2; the procedure has been described in detail elsewhere.[1] Iminothiolane, the cyclized form of the higher homolog methyl 4-mercaptobutyrimidate, has similarly been extensively used in ribosome cross-linking studies[24] as well as in other studies of chromatin,[25,26] including the recent demonstration of cross-linking of HMG 17 to H2A in a reconstituted nucleosome core particle–HMG 17 complex.[27] The two-stage procedure and exposure to hydrogen peroxide can be avoided (especially important for proteins that contain tryptophan; histones do not) by the use of a disulfide-containing bisimidoester. For example, dimethyl bisdithiopropionimidate (Fig. 1, **II**) has been used in H1 cross-linking studies[19,20] and recently in an investigation of the interaction of HMG 1 and 2 with nucleosome core histones.[28]

Instead of imidoesters, we now usually use dithiobis(succinimidyl propionate), the main advantage being that cross-linking can be carried out at neutral rather than alkaline pH to minimize the risk of disulfide interchange between the cross-links (see above). [Fortunately, histones themselves contain hardly any cysteine (e.g., only four per nucleosome in bovine chromatin—two each in the two H3 molecules; two per nucleo-

[23] R. N. Perham and J. O. Thomas, *J. Mol. Biol.* **62**, 415 (1971).

[24] J. W. Kenny, J. M. Lambert, and R. R. Traut, this series, Vol. 59, p. 534.

[25] V. A. Pospelov, A. M. Jerkin, and A. T. Khachatrian, *FEBS Lett.* **128**, 315 (1981).

[26] L. G. Nikolaev, B. O. Glotov, V. K. Dashkevich, S. F. Barbashov, and E. S. Severin, *FEBS Lett.* **163**, 66 (1983).

[27] G. R. Cook, P. Yau, H. Yasuda, R. R. Traut, and E. M. Bradbury, *J. Biol. Chem.* **261**, 16185 (1986).

[28] M. Štros and A. Kolíbalová, *Eur. J. Biochem.* **162**, 111 (1987).

some in many other species) and no cystine.] A suitable time of cross-linking for analysis of cross-linked dimers should be selected on the basis of a time course of cross-linking on an analytical scale followed by SDS–polyacrylamide gel electrophoresis. Both "early" (little cross-linking beyond dimer) and "late" (many higher oligomers) times of cross-linking should be examined since some dimers may, in principle, be more readily cross-linked further than others and, thus, be underrepresented in the dimer band at later times.

For two-dimensional SDS–polyacrylamide gel electrophoresis[1,18] the cross-linked sample is dissolved at a concentration of about 1 mg/ml in the appropriate hot sample buffer (see above), but without 2-mercaptoethanol. A gel system is used for the first dimension which gives optimal resolution of the cross-linked oligomer bands. Cross-linked dimers of the core histones may be resolved in a 12-cm-long, 18% polyacrylamide–SDS slab gel with the discontinuous Tris buffer system (cf. Fig. 3B).[18] For the homo- and heterodimers of H1 and H5 in chicken erythrocyte chromatin, a 9% polyacrylamide–SDS slab gel in the same buffer system is suitable for the first dimension.[29] We use a 0.9-mm-thick gel with wide (2 cm) sample wells to ensure a sufficiently high "loading" for the second dimension without overloading the first. Electrophoresis is carried out as described above. The gel is not usually fixed or stained (although light staining, without prefixation, seems to have no adverse effect); duplicate first-dimension samples are run for this purpose. A gel strip containing the sample track is soaked successively for 20-min periods in 0.12 M Tris-HCl–0.1% (w/v) SDS–1.4 M 2-mercaptoethanol at pH 8.8, and then pH 6.8 (twice); this is conveniently done in a sealed plastic box with gentle agitation (use a fume cupboard!). The strip, which by now has swollen somewhat, is then placed on top of a 1.5-mm-thick slab for electrophoresis in the second dimension at right angles to the first in the high-resolution gel system (see above). From bottom to top, the slab consists of 18% polyacrylamide–SDS separating gel (12 cm), 3% polyacrylamide–SDS stacking gel (1.5 cm), and 1% (w/v) agarose (0.5 cm) containing 0.12 M Tris-HCl (pH 6.8)–0.1% SDS–1.4 M 2-mercaptoethanol.[1,18] The first-dimension strip is sealed onto the second gel with the same agarose mixture, but lacking mercaptoethanol and containing bromphenol blue as tracking dye. Electrophoresis is carried out as above.

Figure 5A shows the analysis of cross-linked dimers of H5 (the tissue-specific H1 variant) and H1 from chicken erythrocyte nuclei using dithiobis(succinimidyl propionate) as cross-linking agent.[29] The presence of the cross-linked dimers (H5)$_2$, H5–H1B, H5–H1A, (H1B)$_2$, H1A–H1B, and

[29] A. C. Lennard and J. O. Thomas, *EMBO J.* **4**, 3455 (1985).

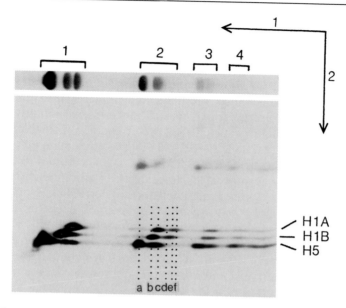

Fig. 5. Analysis of histone neighbors in chromatin using cleavable cross-linkers and two-dimensional SDS–polyacrylamide gel electrophoresis: analysis of cross-linked dimers of H5 and H1 from chicken erythrocyte nuclei. Nuclei were cross-linked at 23° with dithiobis(succinimidyl propionate) (0.2 mg/ml) for 10 min at pH 7.5. The H5 and H1 were extracted, and a sample of 15 μg was subjected to two-dimensional SDS–polyacrylamide gel analysis. First dimension (no 2-mercaptoethanol), 9% polyacrylamide in the discontinuous Tris buffer system; second dimension, 18% polyacrylamide in the same buffer; the gel was silver-stained. Top, horizontal: a Coomassie Blue-stained counterpart of the first-dimension gel. Spots lying on the vertical lines arise from components that were initially cross-linked to each other; lines a, b, c, d, e, and f, therefore, indicate the cross-linked dimers (H5)$_2$, H5–H1B, H5–H1A, (H1B)$_2$, H1A–H1B, and (H1A)$_2$, respectively. [From A. C. Lennard and J. O. Thomas, *EMBO J.* **4**, 3455 (1985).]

(H1A)$_2$, which are also obtained from extended chromatin and dinucleosomes, shows that H5 and both H1 subtypes, H1A and H1B, are interspersed along the nucleosome filament rather than segregated into blocks.

Assignment of H5–H5 Cross-Links to Particular H5 Domains

H1 molecules in the relaxed nucleosome filament are cross-linked into polymers, suggesting a linear array of H1 molecules in close proximity (see above). H1 has a tripartite structure with a globular middle (G) flanked by an amino-terminal (N) and a longer, basic carboxy-terminal (C) tail, both of which are random coil when H1 is free in solution. To determine whether the arrangement is polar (head-to-tail), or whether the molecules are randomly oriented, a combination of cross-linking with cleavable reagents, selective proteolytic excision of cross-linked fragments, and "diagonal" two-dimensional gel electrophoresis for identification of the fragments has been used. The strategy is outlined in Fig. 6.

For illustration, the procedure is described in detail for H5 and H1 molecules in nuclei, extended chromatin, and dinucleosomes from chicken erythrocytes, cross-linked at pH 7.5 with dithiobis(succinimidyl propionate).[29] Nuclei are resuspended in 0.34 M sucrose, 5 mM triethanolamine-HCl, 3 mM MgCl$_2$, 0.2 mM PMSF, pH 7.5, at approximately $A_{260} = 6$ for cross-linking; long extended chromatin and dinucleosomes obtained by fractionation of a micrococcal nuclease digest of chromatin[30] in sucrose gradients are dialyzed into 5 mM triethanolamine-HCl, 0.25 mM PMSF, 0.2 mM Na$_2$EDTA at about $A_{260} = 1$. To cross-link nuclei or long chromatin, dithiobis(succinimidyl propionate) (50 mg/ml in DMF) is added to a final concentration of 0.2 or 0.4 mg/ml, respectively, and the reaction is allowed to proceed for 10 min at 23°. To achieve comparable levels of cross-linking of short oligomers (e.g., dinucleosomes) three additions of reagent (0.2 mg/ml) are made at 10-min intervals. The reaction is quenched with 50 mM glycine. All H1- and H5-related species, cross-linked or not, are extracted with 5% (v/v) perchloric acid[21] on ice, precipitated with 25% (w/v) trichloroacetic acid, and the pellet is washed with acetone–10 mM HCl and then acetone. Ideally, cross-linked dimers and larger oligomers are separated from residual monomers by gel filtration (e.g., through a column of Sephadex G-100 in 10 mM HCl), and the eluate is monitored by absorbance at 230 nm and by SDS–18% polyacrylamide gel electrophoresis. With small amounts of material (e.g., from dinucleosomes) the gel filtration step may be omitted and the unfractionated mixture of residual monomeric and cross-linked products analyzed. The gels are aesthetically less pleasing (compare Fig. 8 with Fig. 7, below) but present no difficulties in interpretation (although the contribution of intramolecular cross-linking in the monomer to the result has to be evaluated; see below).

The cross-linked H5–H1 mixture is dissolved in 50 mM Tris-HCl, pH 6.5, at 0.2 mg/ml (assuming $A_{230} = 1.85$ for 1 mg/ml[31]) and digested at 37° with α-chymotrypsin (Sigma, St. Louis, MO) at an enzyme to histone ratio of 1:500 (w/w). Cleavage occurs preferentially at the conserved phenylalanine (Phe-93 in H5; Phe-105 in chicken H1), roughly in the middle of the amino acid sequence, just within the globular region. A suitable digestion time is determined on an analytical scale by SDS–18%

[30] M. Noll, J. O. Thomas, and R. D. Kornberg, *Science* **187,** 1203 (1975).
[31] R. D. Camerini-Otero, B. Sollner-Webb, and G. Felsenfeld, *Cell* **8,** 333 (1976).

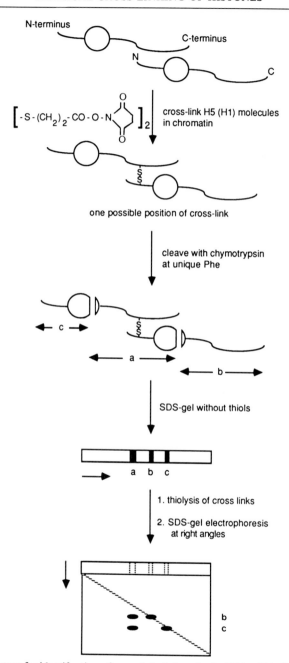

FIG. 6. Strategy for identification of cross-linked domains in H5 (or H1) dimers (see text). [From A. C. Lennard and J. O. Thomas, *EMBO J.* **4**, 3455 (1985), with slight modification.]

polyacrylamide gel electrophoresis of samples removed at various times and precipitated with an equal volume of 50% TCA. A digestion time should be chosen such that the amino-terminal fragment (NG), comprising the amino-terminal "half" of the H5(H1), is not appreciably degraded further; this is about 2 min under the conditions given above.

The sample is subjected to two-dimensional gel electrophoresis with thiolysis between the first and second dimensions, as described above for cross-linked histone oligomers, to determine the composition of the cross-linked proteolytic fragments. Suitable loadings for the first-dimension gel are about 15 μg of digested cross-linked H5(H1), or 40 μg of a digest of unfractionated monomeric and cross-linked H5(H1). A marker for the products generated in the second dimension is run in a slot formed in the agarose layer above the stacking gel with a strip ($\sim$ 1 cm) of perspex held vertically. The marker can be either a chymotryptic digest of H5(H1) or a sample of the cross-linked digest that was loaded onto the first dimension gel, but reduced with 2-mercaptoethanol. The gel is silver-stained.[32] The slow migration of the very basic carboxy-terminal fragment, relative to that of the NG fragment which has similar molecular mass (Fig. 7), is presumably due to its much higher intrinsic positive charge; differences in basic character probably also account for differences in the molar staining ratios of the two fragments with Coomassie Blue (not shown) and for distinctive colors on silver staining.

Figure 7 illustrates the result for H5, H1 cross-linked in nuclei and extended chromatin, and Fig. 8 the result for dinucleosomes.[29] In the former the cross-linked material was isolated by gel filtration; in the case of dinucleosomes this step was omitted, accounting for the much larger amount of material on the "diagonal." Cross-linked products which migrate close to H5,H1 in the first dimension, but which give off-diagonal peptides in the second dimension, represent two half-molecules cross-linked together. The components of such cross-linked products from H5 lie on the vertical lines labeled a, b, and c in Fig. 7A and 7B (for clarity, the corresponding weaker H1 peptides are ignored). They have the same electrophoretic mobility as purified, characterized NG and C fragments of H5 (designated NG.H5 and C.H5) and arise from NG–NG, NG–C, and C–C cross-linking, respectively. The other major off-diagonal peptides (lines d–f) arise, respectively (see Fig. 7C), from H5–NG.H5, H5–C.H5, and (H5)$_2$, all products of incomplete cleavage by chymotrypsin. Therefore, in the condensed chromatin present within nuclei (Fig. 7A), cross-linking can occur between the NG and C domains of H5 in all combinations (Fig. 7A). Least favored (as judged from the relative intensities of the stained spots) is

[32] W. Wray, T. Boulikas, V. P. Wray, and R. Hancock, *Anal. Biochem.* **118**, 197 (1981).

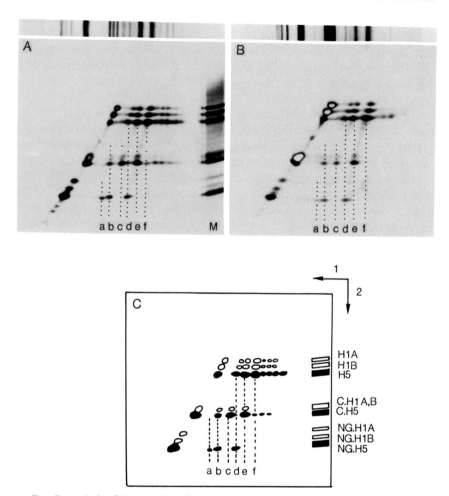

FIG. 7. Analysis of the domains of neighboring H5 molecules cross-linked in nuclei and extended chromatin. Cross-linked H5(H1) polymers generated by treatment of nuclei or extended chromatin with dithiobis(succinimidyl propionate) at pH 7.5 were isolated by gel filtration, and 14 μg were digested with chymotrypsin for 2 min and analyzed by two-dimensional SDS–gel electrophoresis (18% polyacrylamide in both dimensions) as in Fig. 6; (A) nuclei, (B) extended chromatin. The gels were silver-stained. Top, horizontal: Coomassie blue-stained counterparts of the first-dimension gels. M (applied to the second-dimension gel) is the initial mixture of fragments reduced with 2-mercaptoethanol. The identification of the cross-linked peptides, giving rise to spots on vertical lines a–f, is given in the text. (C) Schematic representation and identification of the peptides in A. The solid spots represent peptides that derive from H5; the corresponding H1 peptides migrate just behind them and are particularly visible in the gels for the carboxy-terminal fragments, designated C.H5 and C.H1A,B. [From A. C. Lennard and J. O. Thomas, *EMBO J.* **4**, 3455 (1985).]

NG–NG cross-linking; the amounts of NG–C and C–C cross-linking are roughly equal. In the nucleosome filament at low ionic strength there is virtually no NG–NG cross-linking, and C–C cross-linking is now less than NG–C cross-linking (Fig. 7B). (Note that the intensity of the C spot will be twice as great for C–C cross-links as for the same number of NG–C cross-links.)

In purified dinucleosomes (Fig. 8) the cross-linking is almost entirely NG–C. Since, in this case, a mixture of unfractionated residual monomer and uncross-linked dimer was analyzed, the cross-linked chymotryptic fragments with M_r close to those of H5 and H1 might, in principle, have arisen from cross-linked dimers as required or from intramolecularly cross-linked monomers. Intramolecular cross-linking was shown to be slight.[29] The strong NG–C cross-linking in dinucleosomes, together with the predominance of NG–C cross-linking in extended chromatin, therefore suggests that H5 molecules are arranged in a polar, head-to-tail man-

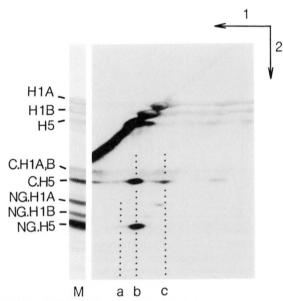

FIG. 8. Analysis of the domains of H5(H1) cross-linked in dinucleosomes. H5 and H1 were extracted from dinucleosomes after cross-linking at approximately $A_{260} = 2$ with dithiobis(succinimidyl propionate) (three treatments of 0.2 mg/ml for 10 min each at 23°), under the low ionic strength conditions for extended chromatin, and analyzed without gel filtration. The procedure was otherwise as in the legend to Fig. 7, except that a total of 40 μg of material was analyzed. Only the region of interest of the gel is shown for clarity. Marker (M) and vertical lines a, b, and c are as in Fig. 7. [From A. C. Lennard and J. O. Thomas, *EMBO J.* **4**, 3455 (1985).]

ner on successive nucleosomes. The C–C cross-linking in extended long chromatin may be due to random collisions between regions of the inherently flexible nucleosome filament, or it may possibly be due to some folding into prototype higher order structures even at low ionic strength.[33]

Acknowledgment

Work in this laboratory was supported by the Science and Engineering Research Council of the U.K.

[33] J. Bordas, L. Perez-Grau, M. H. J. Koch, M. C. Vega, and C. Nave, *Eur. Biophys. J.* **13,** 157 (1986).

Section IV

Assembly and Reconstitution of Chromatin

[27] Assembly of Nucleosomes and Chromatin *in Vitro*

By Daniela Rhodes *and* Ronald A. Laskey

Guide to Selection of Methods for Assembling Chromatin or Nucleosome Cores

A wide range of methodological approaches is available for assembling purified DNA into nucleosome cores or more complex forms of chromatin. This chapter considers their relative merits and advises on the selection of methods. In addition, it describes selected protocols in detail.

Even before the discovery of the nucleosome, material with the physical properties of chromatin had been reconstituted from mixtures of histones and DNA by dialyzing from 2 M NaCl to lower ionic strength.[1,2] In spite of advances in knowledge of nucleosome structure and assembly, we shall argue that variants of salt dialysis protocols remain the most useful methods for reconstituting nucleosome core particles or for accurately positioning individual core particles on purified DNA. It is clear, however, that salt gradient dialysis fails to reflect the complexity of the cellular assembly pathways and that it also fails to space nucleosomes at regular intervals of 200 base pairs (bp) from each other. Regular spacing of nucleosomes at the physiological interval of 200 bp has been achieved in crude cell-free extracts either from animal eggs that contain stored histone pools[3-5] or from cultured cells.[6]

Fractionation of crude cell-free assembly extracts has resulted in the identification of proteins that are involved in the cellular process of nucleosome assembly.[7,8] One of these, nucleoplasmin, was identified by its ability to assemble nucleosome cores from purified histones and DNA *in vitro*.[7] We describe an assembly protocol using nucleoplasmin as an assembly factor (see below), but we would not recommend this protocol for routine use for the following reasons. First, the optimum concentration ratios of the reagents are extremely sharp and have to be titrated carefully

[1] R. D. Kornberg, *Science* **184**, 686 (1974).

[2] P. Oudet, M. Gross-Bellard, and P. Chambon, *Cell* **4**, 281 (1975).

[3] R. A. Laskey, A. D. Mills, and N. R. Morris, *Cell* **10**, 237 (1977).

[4] T. Nelson, T.-S. Hsieh, and D. Brutlag, *Proc. Natl. Acad. Sci. U.S.A.* **76**, 5510 (1979).

[5] G. C. Glikin, I. Ruberti, and A. Worcel, *Cell* **37**, 33 (1984).

[6] B. Stillman, *Cell* **45**, 555 (1986).

[7] R. A. Laskey, B. M. Honda, A. D. Mills, and J. T. Finch, *Nature (London)* **275**, 416 (1978).

[8] J. A. Kleinschmidt, E. Fortkamp, G. Krohne, H. Zentgraf, and W. W. Franke, *J. Biol. Chem.* **260**, 1166 (1984).

Copyright © 1989 by Academic Press, Inc.
All rights of reproduction in any form reserved.

with each reagent preparation. Second, it is difficult to achieve complete assembly of the full nucleosome complement on the template by this method. Third, it is now clear that the role of nucleoplasmin in the *Xenopus* egg involves transfer of only histones H2A and H2B, and not H3 or H4, so its ability to assemble entire nucleosomes cores *in vitro* does not reflect the full cellular process.[8,9]

Preliminary experiments with the combination of nucleoplasmin and the acidic nuclear protein N1, which binds and transfers histones H3 and H4, have appeared promising, but so far it has not been possible to purify enough N1 to assess this more physiological combination as a practical assembly method. In the meantime, other methods using acidic assembly factors can be considered as alternatives. For example, improved assembly using nucleoplasmin has been reported when highly acetylated histones are used[10] (see Sealy *et al., this volume [30]). Alternatively, other acidic polymers such as poly(glutamic acid)[11] (see Stein, this volume [28]) or even RNA[12] can assemble nucleosome cores at physiological ionic strength. Similar results have also been reported for HMG 1.[13] While crude, cell-free extracts, which use the cellular assembly mechanisms, can obviously offer specific advantages, salt gradient dialysis and salt-mediated histone exchange deserve consideration as practical alternatives to methods using purified assembly factors.

Finally, purified eukaryotic transcription factors have been successfully incorporated into reconstituted chromatin using either salt gradient dialysis or crude cell-free extracts.[14-17] Selected examples are described.

Reconstitution of Nucleosome Cores by the Salt Gradient Method

The simplest and hence the most commonly used method for assembly of nucleosome cores *in vitro* is the salt dialysis or exchange method in which a DNA fragment of suitable length is mixed with the histone octamer in high salt. As the salt concentration is reduced stepwise or continuously, nucleosome cores are formed. This method produces nucleosome core particles that are authentic by various criteria (histone content, elec-

[9] S. M. Dilworth, S. J. Black, and R. A. Laskey, *Cell* **51**, 1009 (1987).
[10] L. Sealy, M. Cotten, and R. Chalkley, *Biochemistry* **25**, 3064 (1986).
[11] A. Stein, J. P. Whitlock, and M. Bina, *Proc. Natl. Acad. Sci. U.S.A.* **76**, 5000 (1979).
[12] T. Nelson, R. Wiegand, and D. Brutlag, *Biochemistry* **20**, 2594 (1981).
[13] C. Bonne-Andrea, F. Harper, J. Sobczak, and A.-M. De Recondo, *EMBO J.* **3**, 1193 (1984).
[14] D. Rhodes, *EMBO J.* **4**, 3473 (1985).
[15] B. M. Emerson, C. D. Lewis, and G. Felsenfeld, *Cell* **41**, 21 (1985).
[16] M. S. Schlissel and D. D. Brown, *Cell* **37**, 903 (1984).
[17] J. M. Gottesfeld and L. S. Bloomer, *Cell* **28**, 781 (1982).

tron microscopy, DNase I digestion studies, and physicochemical methods). Depending on whether small or large amounts of nucleosome cores are required, the reconstitution is carried out by one or the other of two variants of the method.

Note About Length of DNA Used in Reconstitution

Because histone octamers can position themselves on the DNA in a sequence-specific manner (see Ref. 18 and references therein), when the formation of one nucleosome core is required, it is essential to use a DNA fragment that is long enough to allow positioning of one histone octamer but not two. A suitable DNA length is probably 200–270 bp.[14,19] Several histone octamers can be reconstituted on a long DNA fragment,[20] but it should be noted that it is not possible by the salt reconstitution method to obtain the 200-bp spacing characteristic of native chromatin (see below).

Bulk Reconstitution

Pure histone octamer, containing equimolar amounts of each of the four histones, can be prepared from any of several sources, such as chicken blood and rat liver.[21,22] It is added to a DNA fragment in a 0.6–0.9 molar ratio and diluted to a maximum DNA concentration of 1 mg/ml in a final concentration of $2 M$ NaCl (or KCl), 20 mM potassium cacodylate (pH 6.0), 1 mM EDTA, and 0.5 mM benzamidine. The ratio of DNA to protein is kept below 1 in order to obtain a reconstituted sample that is soluble. The salt concentration is reduced slowly either by stepwise dialysis at 4°, 2.0 M NaCl (4 hr), 0.85 M (6 hr), 0.65 M (6 hr), 0.5 M (8 hr), and finally 0.1 M NaCl (8 hr),[2,23] or by a decreasing salt gradient procedure. Any precipitate present at the end of reconstitution is removed by centrifugation at 10,000 g for 15 min at 4°.

Purification of reconstituted nucleosome core particles from unreconstituted DNA and histones is carried out on sucrose gradients[24] or, for single nucleosomes on short DNA fragments, preferably by high-performance liquid chromatography (HPLC) using a TSK DEAE/5PW column (a 2.15×15 cm column is suitable for about 9 mg of reconstituted sample),

[18] S. C. Satchwell, H. R. Drew, and A. A. Travers, *J. Mol. Biol.* **191**, 659 (1986).
[19] R. T. Simpson and D. W. Stafford, *Proc. Natl. Acad. Sci. U.S.A.* **80**, 51 (1983).
[20] H. R. Drew and C. R. Calladine, *J. Mol. Biol.* **195**, 143 (1987).
[21] J. O. Thomas and P. J. G. Butler, *J. Mol. Biol.* **116**, 769 (1977).
[22] R. Simon and G. Felsenfeld, *Nucleic Acids Res.* **6**, 689 (1979).
[23] D. Rhodes, *Nucleic Acids Res.* **6**, 1805 (1979).
[24] J. T. Finch, M. Noll, and R. D. Kornberg, *Proc. Natl. Acad. Sci. U.S.A.* **72**, 3320 (1975).

as done by Richmond *et al.*,[25] to obtain and crystallize milligram quantities of nucleosome cores containing a 146-bp specific-sequence DNA fragment. The fact that such reconstituted nucleosome cores give X-ray diffraction patterns that are the same as those from native cores is the most definitive proof that salt reconstitution produces authentic nucleosome cores. This reconstitution method can also be used for the preparation of microgram or nanogram quantities of nucleosome cores, but on a small scale is less reproducible than the method described below.

Histone Transfer Method

The histone transfer method is the method of choice for the production of small (nanogram to microgram) quantities of reconstituted nucleosome cores for study of, for instance, nucleosome positioning.[14,26-28] In this procedure, a suitable amount of radioactively labeled DNA fragment is incubated with a large molar excess of nucleosome cores or long H1-stripped chromatin. (The presence of a radioactive label on the DNA permits visualization and subsequent analysis of the reconstituted sample.) Because the efficiency of transfer of histone octamer (or octamers) to DNA fragments of different length and sequence varies, it is advisable to investigate different molar ratios of histone source to DNA. Typically, 1 pmol labeled DNA is mixed with about 100-fold molar excess (100 pmol or 1 mg/ml) of nucleosome cores (prepared most easily from chicken blood, but also from any other suitable or convenient source; see Lutter, this volume [13]) in 20 μl of a solution containing 1 M NaCl, 20 mM Tris-HCl (pH 7.4), 0.2 mM phenylmethylsulfonyl fluoride, and 0.2 M EDTA. Incubation at 37° for 20 min allows the histone octamer to become detached from the DNA, and nucleosome cores are reformed as the salt concentration is lowered to 0.1 M by the stepwise addition of 10-μl aliquots of 10 mM Tris-HCl (pH 7.4), 0.2 mM EDTA every 15 min at 20°. Transfer of histones from long to short DNA fragments is disfavored.[29] Therefore, if long H1-stripped chromatin is to be used as histone source, it is necessary to maintain the concentration of chromatin high and constant (1.5–2 mg/ml) in order to obtain efficient transfer. In this case the salt concentration is reduced by dialysis as described above. By fine adjustment of histone source to DNA ratio, reconstitution using this method can be very close to 100%.

[25] T. J. Richmond, M. A. Searles, and R. T. Simpson, *J. Mol. Biol.* **199**, 161 (1988).
[26] J. E. Germond, M. Bellard, P. Oudet, and P. Chambon, *Nucleic Acids Res.* **3**, 3173 (1976).
[27] W. Linxweiler and W. Hörz, *Cell* **42**, 281 (1985).
[28] H. R. Drew and A. A. Travers, *J. Mol. Biol.* **186**, 773 (1985).
[29] S. Caron and J. O. Thomas, *J. Mol. Biol.* **146**, 513 (1981).

Reconstitution of Nucleosome Cores Using Acidic Assembly Factors

Nucleosome core assembly using poly(glutamic acid) as an assembly factor is described elsewhere in this volume (Stein [28]). Here we describe a procedure[30] for the assembly of nucleosome cores by nucleoplasmin, which is an acidic thermostable protein from *Xenopus* eggs.[7,9,30] It should be noted that nucleoplasmin is only one half of the cellular assembly pathway in *Xenopus* eggs[9] and that this method has the drawbacks described above in the method selection guide.

Assembly can be achieved over a range of histone to DNA weight ratios (1.6–4.8:1), but the optimal nucleoplasmin to histone ratio also varies (0.4–2.6:1). A ratio that has been used successfully is 4.5 μg nucleoplasmin, 1.5 μg core histones, and 0.5 μg DNA in 50 μl of 50 mM Tris-HCl (pH 8.0), 1 mM EDTA. Higher salt concentrations up to 150 mM NaCl or KCl have also been used. It is important that nucleoplasmin be present before the DNA and histones are mixed, or precipitation will result. Nucleoplasmin can be purified from unfertilized eggs of *Xenopus laevis* by a combination of ammonium sulfate precipitation, hydrophobic interaction chromatography, and ion-exchange chromatography, as described elsewhere.[31]

Nondenaturing Agarose Gels

The efficiency of reconstitution and the number of histone octamers bound to the labeled DNA fragment can be monitored very simply by analysis in agarose[14] (or nondenaturing polyacrylamide gels[27]) in which nucleoprotein complexes migrate more slowly than the naked DNA. Gels made from 0.7% agarose are suitable for analysis of one and up to three or four nucleosome cores. Samples are loaded in 10% glycerol, and electrophoresis carried out in flatbed minigels (10 × 10 × 0.5 cm) at 20–30 mA. The buffer in the gel and for electrophoresis contains 40 mM Tris–borate (pH 8.3) and 1.25 mM EDTA. In order to visualize bands rapidly, the gel is laid on a piece of DEAE paper, dried under vacuum at 40–50°, and subjected to autoradiography with or without an intensifying screen.[32] If the DNA (or protein) in a complex is to be extracted after electrophoresis (see below), the gel can be placed in a plastic bag and subjected to autoradiography at 4° for 2–4 hr. In this case, it is necessary to increase the amount of radioactivity loaded on the gel.

[30] W. C. Earnshaw, B. M. Honda, R. A. Laskey, and J. O. Thomas, *Cell* **21,** 373 (1980).
[31] C. Dingwall, S. V. Sharnick, and R. A. Laskey, *Cell* **30,** 449 (1982).
[32] R. A. Laskey, this series, Vol. 65, p. 363.

DNase I Footprinting Analysis

The authenticity of a reconstituted nucleosome core particle is most easily tested by carrying out DNase I digestion studies. The resulting digestion pattern should show a characteristic pattern of cuts occurring every 10 nucleotides, as seen for native nucleosome cores.[33,34] Such a pattern, when compared to that of the DNase I-treated, naked DNA fragment, gives information on the location of the histone octamer on the DNA. The quality of the pattern indicates how well positioned the histone octamer is. Typically, 100 μl of a reconstituted nucleosome core preparation is supplemented with 10 μl 20 mM MgCl$_2$ and incubated with 5 μl DNase I at 16 units/ml at 20°. In a parallel experiment, a DNA sample that has been taken through all the steps of reconstitution, but has not been exposed to histones, is also digested with DNase I. Aliquots of 20 μl are taken out at times between 0 and 2 min, and digestion is stopped by addition into 20 μl of a solution containing 5 mM EDTA, 1% sodium dodecyl sulfate (SDS), followed by treatment with proteinase K (0.5 mg/ml) at 37° for 30 min. The samples are then extracted twice with phenol–chloroform, precipitated with ethanol, and the DNA is fractionated in 6 or 8% denaturing polyacrylamide gels.[35] The gels are fixed in 10% acetic acid (v/v) and 10% methanol (v/v), dried under vacuum at 70°, and subjected to autoradiography against an intensifying screen at −70°.[33]

Assembly of Regularly Spaced Nucleosomes Using Crude Cell Extracts

Cellular mechanisms of nucleosome assembly can be exploited to assemble regularly spaced chains of nucleosomes onto purified DNA. We describe here a method using extracts of unfertilized eggs of *Xenopus laevis*,[7,9] which contain a stored histone pool sufficient for at least 6000 diploid nuclei each. The histones are stored as complexes with two acidic assembly factors, nucleoplasmin and N1.[8,9] It should be noted that although this method yields nucleosomes spaced at intervals of 200 bp,[7] histone H1 appears to be enigmatically absent from the assembled product.[9] Alternative methods have been described using ovarian oocytes of *Xenopus laevis*[5] or eggs of *Drosophila melanogaster*.[4] In addition, Stillman has described a method that allows assembly of regularly spaced nucleosomes using extracts of cultured mammalian cells.[6]

Unfertilized eggs are obtained from female *Xenopus laevis* by injection

[33] L. C. Lutter, *J. Mol. Biol.* **124**, 391 (1978).
[34] A. Prunell, R. D. Kornberg, L. Lutter, A. Klug, M. Levitt and F. Crick, *Science* **204**, 855 (1979).
[35] F. Sanger and A. R. Coulson, *FEBS Lett.* **87**, 107 (1978).

of chorionic gonadotropin[36] and collected in high-salt Barth's solution (110 mM NaCl, 2 mM KCl, 1 mM MgSO$_4$, 0.5 mM Na$_2$HPO$_4$, 2 mM NaHCO$_3$, 15 mM Tris-HCl, pH 7.4). After washing, eggs are dejellied in 2% cysteine-HCl in H$_2$O, adjusted to pH 7.8 with NaOH, and rinsed repeatedly in extraction buffer (50 mM HEPES–KOH, pH 7.4, 50 mM KCl, 5 mM MgCl$_2$, 2 mM 2-mercaptoethanol). It is essential that all abnormal or damaged eggs be removed at this stage, particularly any that are mottled. Eggs are allowed to settle in a centrifuge tube at 4°, and excess buffer is removed, leaving only interstitial buffer between the eggs. The eggs are broken by centrifugation at 9000 g for 15 min and the central, clear cytoplasmic layer is isolated and recentrifuged at 145,000 g_{max} for 20 min at 4°.

For optimal assembly of spaced nucleosomes, extracts are supplemented with 1 mM ATP and 1.5 mM MgCl$_2$ and incubated at 21°–25° with 0.5–1 μg DNA in 30 μl extract. When nick-translated DNA is used, it should be rapidly ligated by the extract. Extracts that contain endogenous nucleases should be discarded. This problem can be ovecome by attention to egg quality.

An alternative type of cell-free extract has been reported to assemble and space nucleosomes correctly *in vitro*.[6] Derived from cultured mammalian cells, it has an interesting property in that nucleosome assembly is dependent on DNA replication *in vitro*. Only SV40 viral templates replicate in the cell-free system. Therefore, only molecules containing an SV40 replication origin will replicate and hence assemble into nucleosomes. Further details of this procedure are described by Stillman.[6]

Introduction of Transcription Factors into Chromatin

Introduction of transcription factors into chromatin has so far been concerned with answering two different types of questions: (1) what is the reconstitution of a chromatin template that allows transcription; and (2) what is the location of a nucleosome core with respect to the DNA binding site of a specific transcription factor essential for transcription?

Reconstitution of a Chromatin Template That Allows Transcription

The picture emerging from studies of chromatin template reconstitution for both *Pol*II and *Pol*III genes is that preincubation with transcription factors prior to or simultaneously with *in vitro* nucleosome assembly results in a chromatin fiber that can be transcribed (5 S RNA,[16,17] chicken

---

[36] J. B. Gurdon and L. Wakefield, *in* "Microinjection and Organelle Transplantation Techniques: Methods and Applications" (J. E. Celis, A. Graessmann, and A. Loyter, eds.), p. 269. Academic Press, London, 1986.

adult β-globin gene,[15] and adenovirus major late promoter[37]). These observations suggest that binding of at least one essential transcription factor to its recognition sequence results in a chromatin structure in which the promoter region of the gene is rendered accessible to the other components of the transcriptional machinery. By contrast, addition of transcriptional components after nucleosome assembly fails to activate transcription. It should be noted that for this type of study, because transcription factors are available in very small quantities only, there has been no characterization of the proteins present on the chromatin fiber or nucleosome position in respect to regulatory sequences (but see Ref. 38).

Because binding conditions for the various factors involved in transcription vary with each gene system studied, the reader should refer to the references provided for more detailed information. Here, we simply aim to outline a general approach:

Step 1. Preincubation of DNA with purified or partially purified specific transcription factor(s)[16,17,31] or nuclear extracts.[15] Note that in order to facilitate subsequent analysis, the DNA from the gene of interest forms part of a plasmid.

Step 2. Assembly of nucleosomes using either crude cell-free extracts[3-6] that allow reconstitution at physiological salt conditions[17,37] or the salt method,[17] both of which are described in this chapter.

Step 3. Separation of reconstituted chromatin template from unreconstituted DNA either by sedimentation through a sucrose gradient[17,37] or by gel filtration.[37]

Step 4. Analysis of nucleosome assembly, generally involving digestion with DNase I and micrococcal nuclease together with analysis of DNA topology. The quality of the digestion patterns, together with measurements of the number of superhelical turns introduced into a plasmid on nucleosome formation (one nucleosome results in one negative superhelical turn), gives some indication of the efficiency of nucleosome assembly.[17,37,39]

Step 5. Assay of transcriptional activity. Supplementation with additional factors involved in transcription, as well as polymerase, may be necessary. The number of components to be added is dependent both on the number of essential factors present in the preincubation mixture prior to nucleosome assembly and on the stability of the assembled transcriptional complex to subsequent fractionation (Step 3). There is, however,

[37] J. L. Workman and R. G. Roeder, *Cell* **51**, 613 (1978).
[38] Y. Lorch, J. W. Le Pointe, and R. D. Kornberg, *Cell* **49**, 203 (1987).
[39] J. E. Germond, B. Hirt, P. Oudet, M. Gross-Bellard, and P. Chambon, *Proc. Natl. Acad. Sci. U.S.A.* **72**, 1843 (1975).

evidence that some transcriptional or preinitiation complexes containing several factors,[17] and even the polymerase,[37] are stable to both subsequent nucleosome assembly and fractionation.

Reconstitution of Complex Between Nucleosome Core and Transcription Factor

Because transcription factors are mostly very rare proteins, studies aimed at reconstructing the transcriptional complexes present in chromatin from purified components have so far been limited to the 5 S RNA gene system mentioned above. The *Xenopus* transcription factor IIIA (TFIIIA), specific for 5 S RNA genes, can be purified in milligram quantities,[40,41] and this has allowed the study of the structural relationship between the histone octamer, TFIIIA, and a DNA fragment containing the 5 S RNA gene in a triple complex reconstituted *in vitro*. Reconstitution results in a precisely positioned nucleosome core in which the binding site of TFIIIA is located partly inside and partly outside the nucleosome core.[14] This nucleosome position permits the transcription factor to bind through the displacement but does not allow removal of histones from the region of the TFIIIA-binding site. (*Note:* Because TFIIIA requires zinc for sequence-specific binding to DNA, EDTA must be omitted from all reactions and fractionation systems in which TFIIIA is required to bind to DNA.)

Step 1. Reconstitution of a nucleosome core on the 5 S RNA gene. A 236-bp radioactively labeled DNA fragment containing the 5 S RNA gene (120 bp) and flanking sequences is reconstituted into a nucleosome core using the histone transfer method described above.

Step 2. Analysis of nucleosome core formation. Analysis is carried out on nondenaturing agarose gels, and the localization of the histone octamer on the DNA fragment is done by DNase I footprinting. Both of these methods have been described earlier in this chapter.

Step 3. Determination of amount of transcription factor. In order to find the concentration of transcription factor required for formation of the triple complex, reconstituted nucleosome cores are supplemented with 10 μl of 10 $\times$ binding buffer [200 mM Tris-HCl (pH 7.4), 20 mM MgCl$_2$, 200 μM Zn(OAc)$_2$, 10 mM dithiothreitol, 60% glycerol, 1% (v/v) Nonidet P-40], and 10-μl samples are incubated with increasing amounts of TFIIIA for 20 min at 20°. TFIIIA can be added in the form of RNase-treated 7 S particles.[40] As a control, and also for comparison, aliquots of unreconstituted DNA fragments are also incubated with transcription factor under

[40] H. R. B. Pelham and D. D. Brown, *Proc. Natl. Acad. Sci. U.S.A.* 77, 4170 (1980).
[41] J. Miller, A. D. McLachlan, and A. Klug, *EMBO J.* 4, 1609 (1985).

the same experimental conditions. The complexes thus formed are then analyzed by electrophoresis in the nondenaturing agarose gel system described above. The triple complex migrates as a distinct band more slowly than the nucleosome core. Under these reaction conditions, 3- to 10-fold molar excess (3–10 nM TFIIIA over nucleosome cores containing the 5 S RNA gene (1 nM) is required to saturate binding. The excess of a transcription factor required to saturate binding is dependent on both the molarity of the binding site in the reaction mixture and the binding constant of the factor of interest. (Measured binding constants vary enormously, from 10^9 to 10^{14} M^{-1}.)

Step 4. DNase I footprinting. Analysis of the triple complex is carried out to determine how binding of TFIIIA is accommodated. Although DNase I digestions can be carried out on samples that by agarose gel analysis appear to contain entirely triple complex, it is more prudent to observe the following procedure. Conditions for DNase I digestion are as described for nucleosome cores, but the concentration of DNase I is decreased to one-fifth. Several time points are taken, but as EDTA would cause TFIIIA to become detached from DNA, the digestion is slowed down by chilling on ice and rapidly applying aliquots to the gel at 4°. To improve the resolution between nucleosome core and triple complex, it is advisable to use large (20 × 20 × 0.5 cm) flatbed agarose gels. After electrophoresis at 4° and autoradiography of the wet gel at 4°, the band identified to be that of the triple complex is cut out, and the DNA is extracted and analyzed in a denaturing polyacrylamide gel, as described above. This procedure ensures that the DNase I digestion pattern is characteristic of the triple complex and not contaminated by fragments arising from the two possible dimer complexes, namely, the nucleosome core or DNA–TFIIIA complex. Quantitative analysis of complex DNase I digestion patterns can often be very informative (for this type of analysis, see Refs. 14 and 18).

Step 5. Analysis of protein content. The use of nondenaturing gels to fractionate a protein—DNA complex permits the analysis of the proteins present in the complex.[14]

Future Developments

At present there is still a gap between salt reconstitution methods for nucleosome cores and crude, unfractionated cell-free systems that exploit cellular assembly mechanisms. Although we now know the assembly pathway and the identity of assembly factors for the *Xenopus* egg, it has not been possible to obtain enough of one of them (N1) to make practical use of the physiological pathway as an assembly method *in vitro*. However, there is obvious room for progress there. Another area where progress appears likely is Stillman's cultured cell extract.[6] Not only might this elucidate the relationship between nucleosome assembly and DNA replication, but it may also become more generally applicable to nonviral templates as nonviral DNA replication systems emerge.[42,43]

---

[42] J. J. Blow and R. A. Laskey, *Cell* **47**, 557 (1986).
[43] J. Newport, *Cell* **48**, 205 (1987).

[28] Reconstitution of Chromatin from Purified Components

By ARNOLD STEIN

Several useful methods have resulted from attempts to reconstitute purified histones and DNA into chromatin *in vitro*. For example, core histones can be annealed with DNA to form nucleosomes using high salt concentrations ($\geq 0.80\ M$ NaCl).[1] Alternatively, core histones can be added very slowly to DNA at physiological salt concentrations to form nucleosomes.[2] Both of these methods have been devised to circumvent the aggregation pathways that are encountered on directly mixing the components at physiological salt concentrations. The aggregation pathways can also be avoided by using a negatively charged third component as an assembly factor.[3-5] The use of poly(glutamic acid) as a nucleosome and chromatin assembly factor is described here.

Materials and Handling Procedures

Chicken whole blood with sodium citrate is purchased from Pel-Freeze (Rogers, AK); it is maintained at 4° until use. Micrococcal nuclease (Cooper, Malvern, PA) is dissolved in water at 10 units/μl and stored frozen in small portions. A thawed portion can be stored at 4° for more than 1 week without loss of activity. DNase I (Miles, Elkhart, IN) is dissolved in 0.01 N

---

[1] P. Oudet, M. Gross-Bellard, and P. Chambon, *Cell* **4**, 281 (1975).
[2] A. Ruiz-Carrillo, J. L. Jorcano, G. Eder, and R. Lurz, *Proc. Natl. Acad. Sci. U.S.A.* **76**, 3284 (1979).
[3] A. Stein, J. P. Whitlock, and M. Bina, *Proc. Natl. Acad. Sci. U.S.A.* **76**, 5000 (1979).
[4] W. C. Earnshaw, B. M. Honda, R. A. Laskey, and J. O. Thomas, *Cell* **21**, 373 (1980).
[5] T. Nelson, R. Wiegand, and D. Brutlag, *Biochemistry* **20**, 2594 (1981).

Copyright © 1989 by Academic Press, Inc.
All rights of reproduction in any form reserved.

HCl at 2.0 units/μl and stored frozen in 50-μl portions; a portion is thawed, added to 450 μl of 10 mM Tris-HCl, pH 7.5, 5 mM MgCl$_2$, 1 mg bovine serum albumin/ml, and incubated 1 hr at 0° just before use;[6] the remainder of this sample is discarded. The large fragment of *Escherichia coli* DNA polymerase I (Klenow fragment) is purchased from Bethesda Research Laboratories (BRL, Bethesda, MD). Proteinase K (Boehringer-Mannheim, Indianapolis, IN) is dissolved in water just before use. A 10 mg/ml stock solution of poly(glutamic acid) of average molecular weight 64,000 (Miles) in 20 mM Tris-HCl (pH 7.2), 0.2 mM Na$_2$EDTA is prepared by neutralization with sodium hydroxide; this poly(glutamic acid) is insoluble at acidic pH. This stock solution is stored at 4°. Hydroxyapatite (DNA grade) is purchased from Bio-Rad (Richmond, CA). Sephacryl S-200 (Superfine) is purchased from Pharmacia (Piscataway, NJ). 2'-Deoxyadenosine 5'-triphosphate and thymidine 5'-triphosphate are purchased from Boehringer-Mannheim. Poly[d(A-T)] for use as a primer can be purchased from Miles Laboratories. Urea for gel electrophoresis is purchased from Bethesda Research Laboratories.

It should be emphasized that low molecular weight (<60,000) poly(glutamic acid) is not suitable for chromatin reconstitution experiments, as initially reported.[3] It has also been found that high molecular weight sodium polyglutamate (from a different manufacturer) is not suitable for the procedures described here.

Routine Procedures

Nucleosome Reconstitution by Salt-Gradient Dialysis. DNA or poly[d(A-T)] duplex in a buffer (pH 8.0) containing 1 mM Na$_2$EDTA is adjusted to about 1.0 M NaCl and mixed with the required amount of core histones (in the 2.5 M NaCl stock solution buffer) for a final nucleic acid concentration of about 5 A_{260} units/ml. This mixture (generally 0.4–1.0 ml) is dialyzed (in Spectrapore tubing, 1.0 cm dry width, 12,000–14,000 M/W cutoff) for 1.5 hr at room temperature against 1.0 liter of 0.80 M NaCl, 20 mM Tris-HCl (pH 7.2), 0.2 mM Na$_2$EDTA, 1 mM 2-mercaptoethanol. The sample is next dialyzed for 1.5 hr at room temperature (or overnight at 4°) against 1 liter of 0.15 M NaCl, 20 mM Tris-HCl (pH 7.2), 0.2 mM Na$_2$EDTA, and finally against 1 liter of 20 mM Tris-HCl (pH 7.2), 0.2 mM Na$_2$EDTA for 1.5 hr at room temperature (or overnight at 4°). Usually, greater than 90% of the A_{260} is recovered.

Preparation of Samples for Electrophoresis. Generally, a portion containing 5 μg of nucleic acid from a micrococcal nuclease digest is removed

---

[6] P. W. J. Rigby, M. Dieckmann, C. Rhodes, and P. Berg, *J. Mol. Biol.* **113**, 237 (1977).

for a time point and added to an equal volume of 0.2% sodium dodecyl sulfate (SDS), 20 mM Na$_2$EDTA solution to stop the reaction. Freshly dissolved proteinase K is added to a concentration of approximately 0.2 mg/ml, and the samples are incubated 2 hr or more at 37°. Samples are next adjusted to 1% SDS, 0.1 M Tris-HCl (pH 8.0), 10 mM Na$_2$EDTA, and extracted twice with phenol (distilled and saturated with the same buffer; stored frozen in portions) and once with chloroform – isoamyl alcohol (24:1, v/v). Some of the remaining polyglutamic acid coextracts with DNA and does not (in these quantities) interfere with electrophoresis or stain with ethidium bromide. Samples are finally supplemented with an additional 0.1 M NaCl and 50 μg of polyglutamic acid (as a carrier), precipitated with 2.5 volumes of ethanol (for 10 min using dry ice), dried under vacuum, and dissolved in sample buffer.

Agarose Gel Electrophoresis. For analysis of micrococcal nuclease DNA ladders, a 2.5% gel in 40 mM Tris–acetate, 0.5 mM Na$_2$EDTA, pH 8.0, is used in a 4-mm-thick, 130-mm-long horizontal apparatus (Savant). Electrophoresis is for 3 hr at 100 V. For analysis of DNA topoisomers, a 1% gel in 0.36 M Tris, 30 mM NaH$_2$PO$_4$, 1 mM EDTA, pH 7.8, is used[7] in the same apparatus. In this case, the sample buffer contains 10 mM Tris-HCl (pH 8.0), 1 mM Na$_2$EDTA, 1% SDS, 10% (w/v) sucrose, and 0.05% bromphenol blue. Electrohoresis is for 15 hr at 30 V. Gels are stained with 2 μg of ethidium bromide/ml for 15 min, soaked in deionized water for several hours, and photographed under ultraviolet illumination through a Kodak No. 9 Wratten filter.

Preparative Procedures

Preparation of Core Histones. The following procedure for rapidly preparing core histones from nuclei is optimized for chicken erythrocyte histones but is easily adapted to any chromatin from which H1 (or H5) can be selectively removed by elution with a particular NaCl-containing solution. Nuclei [prepared by washing cells repeatedly with 1% Triton X-100,[8] 0.25 M sucrose, 10 mM Tris-HCl (pH 8.0), 3 mM CaCl$_2$ and then removing the detergent before using or freezing] containing about 6 mg DNA/ tube are suspended in 10 ml of 0.7 M NaCl, 50 mM sodium phosphate (pH 6.8) and lysed by occasional gentle stirring for 30 min at 0°. To each tube, 4.0 g hydroxyapatite is added, making a paste. Volumes are then increased to 40 ml with the same buffer, the protease inhibitor phenyl-

[7] J. E. Germond, B. Hirt, P. Oudet, M. Gross-Bellard, and P. Chambon, *Proc. Natl. Acad. Sci. U.S.A.* **72,** 1843 (1975).

[8] C. W. Hymer and E. L. Kuff, *J. Histochem. Cytochem.* **12,** 359 (1964).

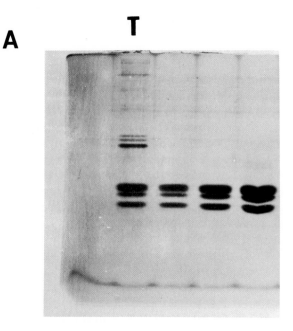

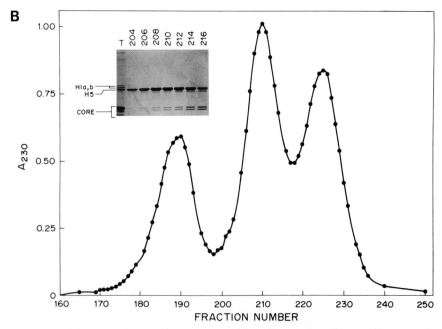

FIG. 1. Chicken erythrocyte histones prepared by salt extraction of immobilized nuclei. (A) SDS–polyacrylamide gel of total SDS-extracted nuclear proteins (T) and salt-extracted core histones at increasing loads. Samples were mixed directly with sample buffer, heated, and loaded on the gel. (B) Gel filtration column elution profile and analysis of selected

methylsulfonyl fluoride (PMSF) is added to a concentration of 0.2 mM, and the suspension is stirred for 10 min at 0°. The hydroxyapatite-immobilized chromatin is collected by low-speed centrifugation (2000 g for 5 min at 4°) and subsequently washed 6 times by the above procedure to remove histones H1 and H5. The pellets are then suspended in 40 ml each of 2.5 M NaCl, 50 mM sodium phosphate (pH 6.8) by gentle stirring for 15 min at 0° to elute the core histones from the immobilized DNA. After a low-speed spin, as described above, the pellets are resuspended in 40 ml each of 2.5 M NaCl-containing buffer, and the supernatants are combined with those of the first extraction. Core histones are concentrated to 1–5 mg/ml by ultrafiltration in an Amicon cell using a YM10 membrane.

The core histone concentration can be measured spectrophotometrically using an extinction coefficient ($A_{280}^{1\%}$ of 4.4.[9] To check that the A_{280} is not in part due to the presence of UV-absorbing contaminants, a possible problem with metabolically active cells, the bicinchonic acid protein assay[10] can be performed in 0.5% SDS at 60°, using the protein component of nucleosome core particles as a protein standard; the amount of standard core histone protein added is calculated from the A_{260} value and the molecular weights of protein and DNA in the complex. These two methods give the same concentration value for chicken erythrocyte core histones.

Possible problems with this procedure are the incomplete removal of histones H1 and H5 or the depletion of H2A and H2B histones. The use of more than 6 mg DNA per 40 ml of batch extraction buffer or lower NaCl concentrations can lead to inefficient washing and H1/H5 contamination. Conversely, more dilute extraction conditions than stated above can remove too much H2A and H2B. Depletion of a small amount ($<$ 10% here) of H2A plus H2B generally occurs when preparing core histones by most salt extraction methods. A typical preparation of core histones is shown in Fig. 1A compared with total SDS-extracted nuclear proteins from chicken erythrocyte nuclei. If it is desired to rigorously maintain the native core histone stoichiometry, core histones can be batch-extracted with hydroxyapatite from repurified H1- and H5-stripped (in a 0.6 M NaCl-containing

[9] A. Stein and D. Page, *J. Biol. Chem.* **255**, 3629 (1980).
[10] P. K. Smith, R. I. Krohn, G. T. Hermanson, A. K. Mallia, F. H. Gartner, M. D. Provenzano, E. K. Fujimoto, N. M. Goeke, B. J. Olson, and D. C. Klenk, *Anal. Biochem.* **150**, 76 (1985).

fractions. A 1.6-ml sample was loaded onto a 1.5 × 75 cm column, and 0.50-ml fractions were collected. For gel analysis, 50-μl portions of fractions were precipitated with trichloroacetic acid (20%), washed with HCl-acidified acetone, twice with acetone, then dried under vacuum and dissolved in sample buffer. Lanes labeled T are total histone markers.

buffer) mononucleosomes as previously described.[9] No differences in chromatin reconstitution have been observed by us when such core histones were compared with core histones prepared as described above. Core histones are stored frozen in the 2.5 M NaCl buffer in small portions.

Preparation of Histones H1 and H5. There is evidence that irreversible changes in purified histone H1 molecules in solution may occur merely by decreasing the pH below about 6.0 or by decreasing the ionic strength below about 0.5 M.[11] Since H1 or H5 is the active component of the chromatin reconstitution system described here, fully functional molecules are essential. Thus, procedures that denature or even unfold these histones should be avoided.

The procedure is illustrated here using chicken erythrocyte nuclei. However, it works well for nuclei from all of the species and tissues that have been examined. Nuclei containing about 25 mg of DNA/tube are washed twice with 40 ml of 0.35 M NaCl, 10 mM Tris-HCl (pH 8.0), 1 mM Na$_2$EDTA to remove most nonhistone proteins. The washed nuclei are then resuspended in 10 ml of 0.7 M NaCl, 50 mM sodium phosphate (pH 6.8) and equilibrated 30 min at 0° with occasional stirring to lyse; then 4.0 g hydroxyapatite is added as described above. The volume is then increased to 40 ml with the same buffer, containing in addition 0.2 mM PMSF, and samples are stirred 15 min at 0° to elute H1 and H5 from the immobilized chromatin. After a low-speed spin, the pellet is extracted with an additional 40 ml of buffer, and the protein in the combined supernatants is concentrated to 1–2 mg/ml by ultrafiltration in an Amicon cell using a YM10 membrane. The extinction coefficient ($A_{280}^{1\%}$) used for H1 or H5 is 2.0.[12]

Histones H1 and H5 can be easily purified by gel filtration in 0.6 M NaCl, 50 mM sodium phosphate (pH 6.8) using Sephacryl S-200 (Superfine).[13] Figure 1B shows a typical column elution profile and an analysis of the proteins in selected fractions. Histone H1 is fairly well resolved, whereas histones H2A plus H2B generally appear as a shoulder on the late-eluting side of H5 and can be removed completely by refractionation. Appropriate fractions are combined and concentrated using Centricon microconcentrators (Amicon). Samples are stored frozen in the same buffer.

Acid-extracted H5, prepared as described by Johns and Diggle,[14] generally was not as effective in aligning nucleosomes on poly[d(A-T)]·

[11] S. H. Brand, N. M. Kumar, and I. O. Walker, *FEBS Lett.* **133**, 63 (1981).
[12] E. W. Johns, *in* "Histones and Nucleohistones" (D. M. P. Phillips, ed.), p. 37. Plenum, London, 1971.
[13] A. Stein and J. Lauderdale, unpublished results, 1988.
[14] E. W. Johns and J. H. Diggle, *Eur. J. Biochem.* **11**, 495 (1969).

poly[d(A-T)] (see below) as salt-extracted H5, prepared as described above. Occasionally, a good reaction was obtained when acid-extracted H5 was used in a greater excess than the salt-extracted material and "renatured" in the 0.15 M NaCl-containing buffer with 2 mg poly(glutamic acid)/ml. These observations suggest that all of the acid-denatured H5 molecules do not fully renature.

Preparation of Suitable Poly[d(A-T)]·Poly[d(A-T)]. Commercial preparations of poly[d(A-T)] are generally of low molecular weight, and the duplex form obtained from the lyophilized product appears to be highly branched when examined by electron microscopy.[15] It is thus not surprising that these preparations work poorly in the reconstitution of chromatin. A more suitable form of the polynucleotide duplex can be made using a modification of the procedure of Setlow *et al.*[16] The reaction mixture contains the following: 66 mM potassium phosphate (pH 7.4), 6.6 mM MgCl$_2$, 1.5 mM dATP, 1.5 mM dTTP, 1 mM 2-mercaptoethanol, 15 μg primer/ml, and 125 units *E. coli* DNA polymerase I Klenow fragment/ml. This mixture is incubated at 37° for 20–30 hr. NaCl is then added to a concentration of 1 M, Na$_2$EDTA (from a 0.5 M, pH 8.0, stock solution) is added to a concentration of 10 mM, and the clear solution is dialyzed for at least 20 hr against 1.0 M NaCl, 10 mM Tris-HCl, pH 8.0, 1 mM Na$_2$EDTA at room temperature. This procedure avoids phenol extraction, ethanol precipitation, or low ionic strength, conditions that could lead to chain branching by destabilization of base pairing. After extensive dialysis, the sample concentration is determined in a 1.0 M NaCl-containing buffer from the absorbance at 260 nm, using a value of 20 A_{260} units (10-mm-path length cell) per milligram of polynucleotide. The same concentration value, typically around 0.65 mg/ml, is obtained when based on ethanol-precipitable material, indicating that unreacted deoxynucleotides are effectively removed by dialysis under the above conditions.

The molecular weight of the poly[d(A-T)] product is strongly dependent on the size of the primer, and the yield of the reaction is dependent on the concentration of primer 3'-hydroxyl ends. To prepare the primer, poly[d(A-T)] is digested with DNase I for 8 min at room temperature under the following conditions: 1 mg poly[d(A-T)]/ml, 10 mM Tris-HCl, pH 8.0, 10 mM MgCl$_2$, 2 units DNase I/ml. The reaction is stopped by adjusting the solution to 20 mM Na$_2$EDTA, 1% SDS, and the polyd(A-T) is purified by phenol extraction and dissolved in 0.2 mM Na$_2$EDTA. Figure 2 shows that freshly renatured primer and product migrate on a 1% agarose gel with average mobilities corresponding to *DNA* fragments that

[15] M. Bina and A. Stein, unpublished observations, 1983.
[16] P. Setlow, D. Brutlag, and A. Kornberg, *J. Biol. Chem.* **247**, 224 (1972).

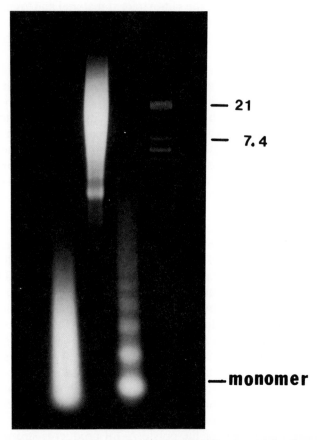

Fig. 2. Agarose gel (1%) of poly[d(A-T)]·poly[d(A-T)], prepared for electrophoresis as described in the text, along with DNA size markers. Lanes 1 and 2 are 5 μg each of primer and product, respectively. Lane CE shows DNA fragments that are multiples of about 200 base pairs, obtained from micrococcal nuclease digestion of chicken erythrocyte chromatin. Lane M is an *Eco*RI limit digest of bacteriophage λ DNA; DNA fragment sizes in kilobase pairs are indicated.

are about 300 and 12,000 base pairs (bp) long, respectively. Samples that are not heated just before loading streak upward on the gel owing to aggregation induced by ethanol precipitation of the samples in preparation for electrophoresis. This gel serves merely as a convenient method for sample characterization, rather than a measure of actual duplex length.

Actual duplex lengths are probably greater than apparent lengths because of secondary structure formation during electrophoresis (and preparations of samples for electrophoresis). Finally, the poly[d(A-T)], prepared as described here, is more homogeneous in size than the material used in initial studies,[17] which was prepared by brief sonication of very high molecular weight polynucleotide.

Preparation of Topoisomerase I Extract. To prepare the crude topoisomerase I extract, nuclei from 3 liters of HeLa cells ($\sim 500,000$ cells/ml) are suspended in 10 ml of 0.35 M NaCl, 10 mM Tris-HCl (pH 8.0), 1 mM Na$_2$EDT, and stirred for 15 min at 0°. This supernatant is stored frozen in portions. Generally, 1 μl can relax 1 μg of supercoiled DNA in 10 min in 0.1 M NaCl, 10 mM Tris-HCl (pH 8.0), 1 mM Na$_2$EDTA, 200 μg/ml polyglutamic acid at 37°.

Reconstitution of Nucleosomes at Physiological Ionic Strength

It is desirable to be able to form nucleosomes efficiently on DNA in a small-scale reaction. That is, a large fraction of the histones added should be incorporated into nucleosomes. The reaction efficiency can be measured by the supercoiling assay, whereby a linking number change (ΔL) equal to -1.0 is obtained for each nucleosome formed on a closed circular plasmid.[7,18] For example, association of 10 histone octamers (octamer MW 1.10×10^5) with a molecule of SV40 DNA (MV 3.46×10^6) corresponds to a histone to DNA ratio of 0.32. Thus, reconstitution of 1.0 μg of SV40 DNA with 0.32 μg of core histones should produce an average ΔL of -10 if all of the histones added form nucleosomes. Similarly, at a histone to DNA ratio of 0.83, DNA molecules should be produced as supercoiled as is native SV40 DNA ($\Delta L = -26$).[19]

Examination of the literature indicates that few methods achieve this efficiency when microgram amounts of histone are used. Unaccounted for histones could complicate the analysis of the experiment being performed. Up to one-half of the histones added to the reaction mixture can apparently be lost during dialysis[7] or can apparently form nonproductive complexes with the assembly factor or DNA.[3–5,20] In contrast, a salt-step procedure, whereby a small volume of sample is diluted stepwise from a high-salt

[17] A. Stein and M. Bina, *J. Mol. Biol.* **178**, 341 (1984).
[18] R. T. Simpson, F. Thoma, and J. M. Brubaker, *Cell* **42**, 799 (1985).
[19] C. Ambrose, R. McLaughlin, and M. Bina, *Nucleic Acids Res.* **15**, 3703 (1987).
[20] I. Oohara, A. Suyama, and A. Wada, *Biochim. Biophys. Acta* **741**, 322 (1983).

solution,[21] does approach the calculated efficiency[22] and, thus, can serve as a positive control.

The following procedure is designed to efficiently reconstitute nucleosomes in a 0.1 M NaCl-containing buffer. First, a core histone plus poly(glutamic acid) stock solution (H + PGA) is prepared by mixing 1.0 mg of core histones (2–4 mg/ml in 2.5 M NaCl buffer) with 5.0 mg of poly(glutamic acid) (from the 10 mg/ml stock solution). The mixture is then dialyzed 16 hr at 0° against 1 liter of 0.10 M NaCl, 10 mM Tris-HCl (pH 8.0), 1 mM Na$_2$EDTA. This dialyzed sample is centrifuged for 2 min at 13,000 g in a microfuge to remove the high molecular weight aggregates that form. The histone concentration is measured from the absorbance at 280 nm [1.0 mg of core histones/ml has an absorbance of 0.44; 5.0 mg of poly(glutamic acid)/ml has an absorbance of approximately 0.020]. Usually, more than 70% of the histones are recovered in the cleared solution. This solution may be stored frozen. Also, it can be diluted to an appropriate concentration using the 0.1 M NaCl-containing buffer without deleterious effects.

An appropriate amount (in a volume up to 10 μl) of H + PGA is then added to a 200-μl reaction mixture containing 1.0 μg DNA, 200 μg poly(glutamic acid)/ml, 10 mM Tris-HCl (pH 8.0), 0.10 M NaCl, 1 mM Na$_2$EDTA, 1 mM 2-mercaptoethanol, and 1 μl topoisomerase I extract (about 6 units). The mixture is incubated 2 hr at 37°.

Figure 3 shows the extents of supercoiling induced in initially relaxed, closed circular DNA at histone to DNA ratios of 0.30–1.0. At 0.30 μg of core histones per microgram of DNA, the average ΔL value is approximately -10 (deduced by counting the topoisomers), the value expected if all of the histones that were added formed nucleosomes. Similarly, at a ratio of 0.70, the DNA is highly supercoiled and indistinguishable (on this gel system) from native SV40 DNA. At a ratio of 1.0, topoisomerase I activity was apparently inhibited by high levels of histones that rapidly associated with the relaxed DNA, making a large fraction of the molecules inaccessible.

Alignment of Nucleosomes on DNA

Nucleosomes in chromatin are arranged in a regular fashion and possess a measurable spacing periodicity (repeat length), which in most species and cell types is close to 200 bp.[23] In contrast, nucleosomes reconstituted from DNA and purified histones, as described above, or using other

[21] P. Oudet, J. E. Germond, M. Sures, D. Gallivitz, M. Bellard, and P..Chambon, *Cold Spring Harbor Symp. Quant. Biol.* **42**, 287 (1978).

[22] A. Stein, *Nucleic Acids Res.* **8**, 4803 1980).

[23] R. D. Kornberg, *Annu. Rev. Biochem.* **46**, 931 (1977).

I Ir .3 .5 .7 1

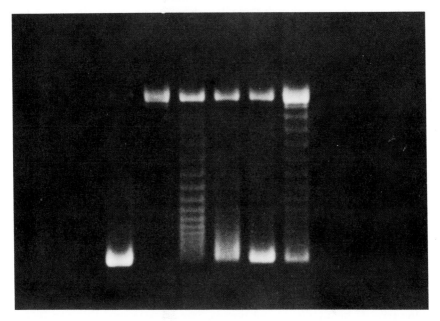

Fig. 3. Supercoiling assay for nucleosome reconstitution. Lane I, native, supercoiled SV40 DNA (BRL); lane Ir, relaxed (and nicked) SV40 DNA, the starting material used for reconstitution; lanes .3, .5, .7, and 1.0 show the extents of supercoiling induced in 1.0-μg DNA samples by 0.30, 0.50, 0.70, and 1.0 μg core histones, respectively, for the reaction described in the text. The more slowly migrating band in lane .7 is nicked DNA.

methods, are irregularly spaced and tend to pack closely together. Moreover, similar results are obtained when the "linker histone," H1 (or H5), is included in the reconstitution protocol.[24,25] The failure to be able to reconstitute a native like chromatin structure in a simple *in vitro* system has led to the view that additional factors are required and that the cellular assembly mechanism may be complex.[26,27]

There are two obvious problems that one encounters in attempting to reconstitute a chromatin-like structure from purified histones and DNA. First, the addition of histone H1/H5 to DNA, or to DNA reconstituted with core histones, at physiological salt concentrations generally leads to extensive aggregation. It is probably the interactions between H1/H5 and

[24] M. Yaneva, B. Tasheva, and G. Dessev, *FEBS Lett.* **70,** 67 (1976).

[25] M. Steinmetz, R. E. Streeck, and H. G. Zachau, *Eur. J. Biochem.* **83,** 615 (1978).

[26] R. A. Laskey and W. C. Earnshaw, *Nature (London)* **286,** 763 (1980).

[27] G. C. Glikin, I. Ruberti, and A. Worcel, *Cell* **37,** 33 (1984).

stretches of naked DNA that lead to aggregation. Second, it is clear that DNA can accommodate approximately 50% more nucleosomes than the number that occurs per unit DNA length in native chromatin,[28] and that reconstitution using core histones alone leads to a broad distribution of histone to DNA ratios (see Fig. 3). Thus, it would not seem possible to be able to convert DNA molecules that are already covered with closely packed nucleosomes into a chromatin-like arrangement possessing nucleosomes that are spaced 50 bp apart. Also, at lower than physiological ratios, nucleosome-free gaps of irregular size might be formed.

The first problem can be solved by using poly(glutamic acid) as an H1/H5 assembly factor. The results of these studies, which use poly(glutamic acid) to prevent nucleohistone aggregation, indicate that some correctly spaced nucleosomes can be generated when histone H5 is added to DNA that had been reconstituted with core histones at a low histone to DNA ratio (0.6, w/w).[29] Attempts to reconstitute chromatin-like structures at higher ratios, by adding H5 and core histones simultaneously or by adding H5 before all of the core histones were added, were unsuccessful. Thus, the second problem remains unsolved in general (see below).

Although it was not possible to reconstitute long arrays of correctly spaced nucleosomes, these experiments revealed that histone H5 can correctly align at least some types of disordered nucleosome arrangements. One type of highly disordered nucleosome arrangement that can be correctly aligned is the product of perturbing H1/H5-stripped chicken erythrocyte chromatin under conditions where histone octamers slide but do not dissociate from the DNA to an appreciable extent.[30] For example, Fig. 4a shows the DNA fragments resulting from micrococcal nuclease digestion of H1/H5-stripped chromatin thoroughly perturbed (by sliding) by incubation in a 0.7 M NaCl-containing buffer for 2 hr at 37°, with subsequent dialysis to remove the NaCl. No trace of the 200-bp ladder reflecting the native nucleosome arrangement remains. Such digests exhibit only 150-bp periodicities superimposed on very high backgrounds. When this sample was incubated with H5 and poly(glutamic acid) in a 0.1 M NaCl-containing buffer, an ordered, native-like nucleosome arrangement was restored to a large extent, as evidenced by the 200-bp periodicity generated (Fig. 4b). On incubation with poly(glutamic acid) alone, no native-like bands appear on the gel (Fig. 4c). Additionally, for chromatin perturbed less extensively, an essentially complete realignment is obtained,[30] whereas when core histones are reassociated with size fractionated sonicated erythrocyte DNA a significantly poorer realignment than in Fig. 4b occurs.

[28] M. Noll, S. Zimmer, A. Engel, and J. Dubochet, *Nucleic Acids Res.* **8**, 21 (1980).
[29] P. Künzler and A. Stein, *Biochemistry* **22**, 1783 (1983).
[30] A. Stein and P. Künzler, *Nature (London)* **302**, 549 (1983).

The nucleosome alignment reaction can be performed as follows.[30] Perturbed chromatin (stripped of H1/H5) is adjusted to 2 A_{260} units/ml, 2.0 mg poly(glutamic acid)/ml, and dialyzed against 0.1 M NaCl, 20 mM Tris-HCl (pH 7.2), 0.2 mM Na$_2$EDTA. An amount corresponding to a 10-fold molar excess of H5 in a solution containing 2.0 mg poly(glutamic acid)/ml, 0.1 M NaCl, 20 mM Tris-HCl (pH 7.2), 0.2 mM Na$_2$EDTA is mixed with the perturbed chromatin (generally <0.2 volumes H5 solution to 1 volume of perturbed chromatin solution). The mixture (a clear solution) is then incubated 3 hr at 37° and dialyzed against 20 mM Tris-HCl (pH 7.2), 0.2 mM Na$_2$EDTA for several hours at 5°.

An improved H5-addition procedure is given below (complete recon-

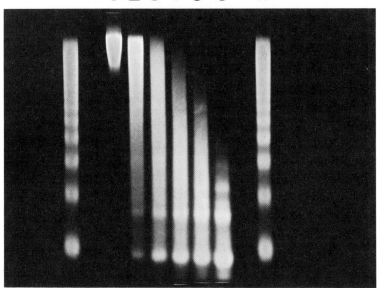

FIG. 4. Agarose gel electrophoresis of DNA fragments, produced by micrococcal nuclease digestion of an extensively rearranged H1/H5-stripped chromatin sample; (a) untreated, (b, p. 598) incubated with H5 and poly(glutamic acid), or (c, p. 598) incubated with poly(glutamic acid) alone. Sample a was digested with 0.04 units of enzyme/μg DNA at 37° for 0, 0.5, 1, 1.5, 2, and 4 min (lanes 1–6, respectively). Lanes n show DNA fragments from a digest of native chromatin. Samples b and c were each digested with 2.0 units of enzyme/μg DNA at 37° for 1, 2, and 4 min. [The presence of poly(glutamic acid) inhibits micrococcal nuclease.] These three digests are displayed in order of decreasing extent of digestion from top to bottom; the bottommost tracing for each was produced from a native chromatin sample. All DNA samples in b and c were run on a single gel together with a $Hinc$II digest of RF ϕX174 DNA for calibration. The photographic negative of the gel was scanned using a densitometer (E-C Apparatus). (Reproduced, with permission, from Ref. 30.)

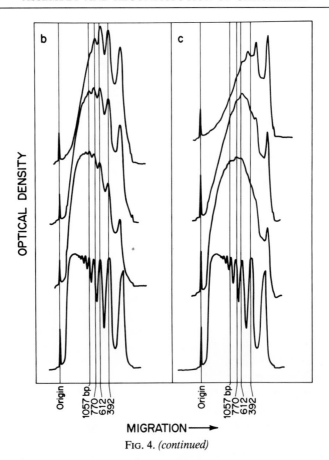

FIG. 4. *(continued)*

stitution of poly[d(A-T)] chromatin) that requires only a 2-fold molar excess of H5 for nucleosome alignment. This method should also work here, but it has not been tested with perturbed chromatin.

Complete Reconstitution of Chromatin-Like Structures

Interestingly, a highly ordered and physiologically spaced nucleosome arrangement can be generated on the synthetic polynucleotide poly[d(A-T)]·poly[d(A-T)] using essentially the same method that was given in the last section.[17] Poly[d(A-T)] duplex differs from DNA in a number of ways that might influence the nucleosome alignment reaction. First, the use of this synthetic polynucleotide does not allow the possibility for base sequence-specific interactions with histone octamers to occur.

Also, the poly[d(A-T)] duplex differs slightly in structure from DNA,[31-33] it could be more flexible than DNA, and it appears to have a higher affinity for histone H1 than DNA.[34] It is not presently known which of these factors (if any) is most important.

In the first step of the procedure, poly[d(A-T)]·poly[d(A-T)] is simply reconstituted with core histones to form randomly arranged nucleosomes. The amount of core histones used should be slightly less than what would correspond to a physiological ratio (85–95%); otherwise, some molecules will possess too many nucleosomes for spacing and alignment to be possible. Conventional salt-gradient dialysis (described under Routine Procedures) has usually been used instead of the poly(glutamic acid)-mediated reconstitution method (described above) in order to maintain two distinct phases in the reaction: nucleosome reconstitution and nucleosome alignment.

In the second step of the procedure, nucleosomes are aligned and regularly spaced by histone H1 or H5 concomitantly with the formation of so-called 300 Å fibers in regions of polynucleotide molecules.[17] The final reaction mixture contains TE buffer [20 mM Tris-HCl (pH 7.2), 0.2 mM Na$_2$EDTA], 0.15 M NaCl (adjusted using a 2.0 M NaCl stock in TE buffer), 2.0 mg polyglutamic acid/ml (added from a 10 mg/ml stock in TE buffer), 2.0 A_{260} units core histone-reconstituted poly[d(A-T)]/ml (added from a solution of ~4.0 A_{260} units/ml in TE buffer), 30 μg histone H5/ml [added from a concentrated (1–2 mg/ml) solution in 0.6 M NaCl, 50 mM sodium phosphate, pH 6.8]. Next, the sample is mixed well and incubated at 37° for 4 to 12 hr. Generally, a 200-μl reaction is incubated, without mixing, in a 400-μl capped polypropylene tube in a thermostated water bath.

Reaction components should be added in the order listed above for the best results, and a clear solution should be obtained. The amount of histone H5 in the above reaction corresponds to a 2-fold molar excess. This relatively small excess of added H5 is a substantial improvement over the 6- to 10-fold molar excess required in initial experiments[17] and is a consequence of the improved procedures given here. Some H5 apparently interacts with the poly(glutamic acid), and some probably binds nonspecifically. Also, a small amount of H5 proteolysis occurs during its preparation and the long 37° incubation. The poly(glutamic acid) component prevents

[31] A. Klug, A. Jack, A. Viswamitra, O. Kennard, Z. Shakked, and T. A. Steitz, *J. Mol. Biol.* **131,** 669 (1979).
[32] H. Shindo, R. T. Simpson, and J. S. Cohen, *J. Biol. Chem.* **254,** 8125 (1979).
[33] S. Arnott, R. Chandrasekaran, L. C. Puigjaner, J. K. Walker, I. H. Hall, D. B. Birdsall, and R. L. Ratliff, *Nucleic Acids Res.* **11,** 1457 (1983).
[34] J. Sponar and Z. Sormova, *Eur. J. Biochem.* **29,** 99 (1972).

aggregation of H1/H5-containing nucleoprotein complexes, possibly by decreasing the affinity of H1/H5 for histone-free regions of the polynucleotide relative to that of nucleosome-containing regions.[17]

Gel Electrophoresis Conditions and Analysis of the Nucleosome Spacing Periodicity for Poly[d(A-T)] Chromatin. It is necessary to run poly[d(A-T)] fragments under denaturing conditions to prevent substantial hairpin formation, which interferes with the analysis.[35] Generally, as a reference, DNA fragments from a micrococcal nuclease digestion of chicken erythrocyte chromatin are run on the same gel. Thus, DNA and poly[d(A-T)] samples are heated for 2 min at 100° and loaded on a 4% (w/v) polyacrylamide gel containing 7 M urea, using the Tris–borate buffer system, as described by Maniatis.[36] To obtain even and undistorted gel slots, a polymerization time of 20 min is used and samples are loaded immediately after removing the comb from the electrode buffer-containing apparatus and gently rinsing out the slots. Samples are run into the 15-cm-long, 3-mm-thick gel for 15 min at 100 V and then electrophoresed, 4.5 hr at 150 V. Gels are stained with 3 μg ethidium bromide/ml for 15 min, soaked in water overnight, and photographed under UV illumination using a Kodak No. 9 Wratten filter.

A photograph of a typical gel is shown in Fig. 5. This gel system is capable of resolving up to seven multiples of the unit "200" nucleotide repeat. One finds that the band positions and widths are very similar to the reconstituted and native chromatin samples, although the background fluorescence is slightly higher in the digests of the reconstituted sample. The existence of seven multiples of 200 nucleotides with only a small amount of closely packed dimer (arrow) in the rather extensively digested sample at far left suggests that nearly all of the nucleosomes were correctly aligned.

In order to precisely size the poly[d(A-T)] fragments produced from a digest, DNA restriction-fragment markers must be used to calibrate the gel. Unfortunately, most DNA restriction fragments run anomalously on ureacontaining polyacrylamide gels at DNA concentrations that are high enough for ethidium bromide staining to be used;[37] DNA strand renaturation probably occurs during electrophoresis. A 4% polyacrylamide gel containing 98% formamide[36] is a suitable gel system for this purpose.[37] For example, Fig. 6A shows micrococcal nuclease ladders, produced from native HeLa chromatin, along with a 123-nucleotide DNA ladder (BRL) run on this denaturing gel system as a test case. The nucleosome spacing periodicity is best obtained from such data by plotting nucleosome oligomer size (in nucleotides) versus oligomer number (monomer excluded)

---

[35] R. T. Simpson and P. Kunzler, *Nucleic Acids Res.* **6,** 1387 (1979).
[36] T. Maniatis and A. Efstratiadis, this series, Vol. 65, p. 299.
[37] R. Rolfes and A. Stein, unpublished results, 1985.

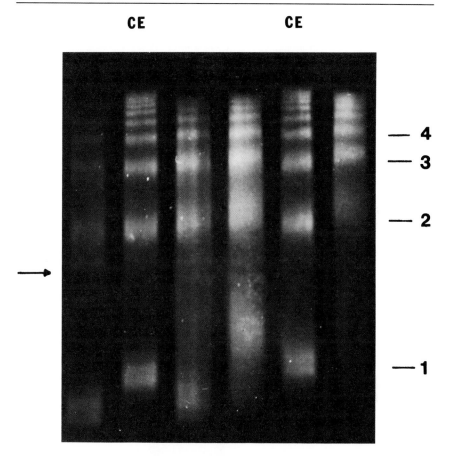

Fɪɢ. 5. Urea-containing (4%) polyacrylamide gel electrophoresis of poly[d(A-T)] fragments produced by micrococcal nuclease digestion of chromatin reconstituted *in vitro*. A sample containing 20 μg poly[d(A-T)] was digested with 4 units of enzyme for 1, 1.5, 2, or 2.5 min (unlabeled lanes, from right to left). Lanes CE are single-stranded DNA fragments from a micrococcal nuclease digestion of chicken erythrocyte chromatin. Bands numbered 1–4 denote the mono-, di-, tri-, and tetranucleosome fragments, respectively.

and taking the slope of the best straight line through the points.[38] Using this method, a repeat length value of 187 ± 3 nucleotides is obtained by averaging the values measured from lanes 2–5, which showed no systematic trend. This value of the average repeat length is in excellent agreement with literature values for HeLa chromatin.[39] Figure 6B shows results obtained for chromatin reconstituted from poly[d(A-T)] and chicken erythrocyte

[38] J. O. Thomas and R. J. Thompson, *Cell* **10,** 633 (1977).
[39] J. L. Compton, M. Bellard, and P. Chambon, *Proc. Natl. Acad. Sci. U.S.A.* **73,** 4382 (1976).

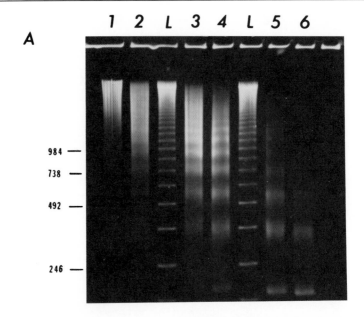

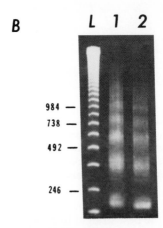

FIG. 6. Formamide-containing (4%) polyacrylamide gel electrophoresis with single-stranded DNA size markers. (A) Single-stranded DNA fragments produced from HeLa chromatin digested with micrococcal nuclease for 0.5, 1, 2, 4, 8, or 16 (lanes 1–6, respectively). (B) Poly[d(A-T)] fragments produced from chromatin reconstituted *in vitro,* using chicken erythrocyte histones, and digested with micrococcal nuclease for 30 sec (lane 1) or 1 min (lane 2). Lanes L are 123-nucleotide ladders; several marker DNA fragment sizes in nucleotides are indicated.

histones. All components were prepared independently from those used in the experiment shown in Fig. 5. By the method described above, the repeat length is calculated to be 209 ± 3 nucleotides, in excellent agreement with literature values for chicken erythrocyte chromatin.[40] Additionally, the poly[d(A-T)] fragment derived from the nucleosome monomer in Fig. 6B, lane 2, is measured to be 148 ± 3 nucleotides, consistent with core particle length DNA.[41]

Comments on Poly(glutamic Acid)-Mediated Nucleosome Alignment Reaction. It has been possible to generate spaced nucleosome arrays using histones from all species and cell types that have been examined thus far in our laboratory. However, reaction conditions must be modified in some cases, suggesting that different types of chromatin may have rather different stabilities. The following are variables that have been found to require adjustment for some alignment reactions to proceed: H1 to nucleosome ratio, core histone to polynucleotide ratio, NaCl concentration, and incubation time. Interestingly, when H5 and chicken erythrocyte core histones are used, the reaction is largely insensitive to changes in these parameters.[42]

[40] N. R. Morris, *Cell* **9,** 627 (1976).
[41] A. Prunell, R. D. Kornberg, L. C. Lutter, A. Klug, M. Levitt, and F. H. C. Crick, *Science* **204,** 855 (1979).
[42] A. Stein and M. Mitchell, *J. Mol. Biol.* **203,** 1029 (1988).

[29] Assembly of Chromatin with Oocyte Extracts

By Akiko Shimamura, Bret Jessee, and Abraham Worcel

Introduction

Chromatin can be assembled on circular DNA molecules *in vitro* using a high-speed supernatant of *Xenopus* oocytes (oocyte S-150[1]). The oocyte S-150 provides both the histone proteins and the enzymes used in this complex process, which closely mimics the one observed *in vivo* when circular DNA molecules are injected into the germinal vesicle.[2] When the procedures described below are followed, the circular DNA is quantitatively assembled into a minichromosome displaying a full complement of regularly spaced nucleosomes.

[1] G. C. Glikin, I. Ruberti, and A. Worcel, *Cell* **37,** 33 (1984).
[2] M. Ryoji and A. Worcel, *Cell* **37,** 21 (1984).

METHODS IN ENZYMOLOGY, VOL. 170

Copyright © 1989 by Academic Press, Inc.
All rights of reproduction in any form reserved.

The assembly reaction requires ATP and Mg^{2+},[1,3,4] and thus an ATP-regenerating system should be used to ensure that optimal ATP levels are maintained throughout the incubation. The reaction is very sensitive to dilution of the S-150 and to ionic strength. The procedures detailed below must be followed if native chromatin is to be produced; any changes introduced in this experimental protocol should be carefully evaluated, and their effect should be ascertained by appropriate MNase digestions of the product. Limited amounts of DNA must be used, so as not to exceed the assembly capacity of the oocyte S-150. In order to facilitate analysis of small amounts of chromatin, one may use DNA circles labeled with high specific activity ^{32}P, as described below.

Single-Site Labeling of Circular DNA

Circular DNA molecules tagged with ^{32}P are convenient templates for chromatin assembly because they greatly facilitate structural analysis of the overall chromatin organization and the localized nucleoprotein at a given site. The ^{32}P label can be used to (1) follow the changes in DNA topology during incubation in the S-150 after appropriate gel electrophoresis and radioautography (see Figs. 1, 2, and 3 in Ref. 1); (2) display the nucleosomal ladder and the nucleosome or subnucleosome limit digest after MNase digestions (see Fig. 4 in Ref. 1; see also Figs. 1 and 9 in Ref. 3); (3) generate end-labeled, *double-stranded* DNA fragments after digestion with a restriction enzyme that cleaves next to the labeled site (such a direct end-label technique allows precise and fast mapping of the nuclease hypersensitive sites and the nucleosome phasing on the assembled chromatin[5,6]); (4) generate end-labeled, *single-stranded* DNA fragments after digestion with two restriction enzyme isoschizomers that generate either 5'- or 3'-labeled DNA (such a direct chromatin footprint technique permits detailed analysis of the nucleoprotein structure assembled around the ^{32}P label[5,7]); (5) footprint the nucleoprotein structure within the gene, at the 5' or 3' boundary, or removed from the gene, within the vector DNA sequences of the plasmid.[8] The following is our protocol for making tagged DNA circles.

1. Cleave 10 μg supercoiled DNA in a 50-μl reaction volume using a restriction enzyme that cuts only once within the circle. Preferably,

[3] I. Ruberti and A. Worcel, *J. Mol. Biol.* **189,** 457 (1986).
[4] J. A. Knezetic and D. S. Luse, *Cell* **45,** 95 (1986).
[5] G. Gargiulo and A. Worcel, *J. Mol. Biol.* **170,** 699 (1983).
[6] G. Gargiulo, F. Razvi, I. Ruberti, I. Mohr, and A. Worcel, *J. Mol. Biol.* **181,** 333 (1985).
[7] F. Razvi, G. Gargiulo, and A. Worcel, *Gene* **23,** 175 (1983).
[8] M. Ryoji and A. Worcel, *Cell* **40,** 923 (1985).

the enzyme should generate 5′ protruding ends to permit efficient kinase and ligase reactions.

2. For the phosphatase reaction, add 10 units (0.5 μl) calf intestine alkaline phosphatase (Boehringer-Mannheim, Indianapolis, IN). Incubate 30 min at 37°.

3. Add sodium dodecyl sulfate (SDS) to 0.1% final concentration, EDTA to 8 mM final concentration, and 30 μg proteinase K. Incubate 1 hr at 37°.

4. Extract with 1 volume of phenol–chloroform (2 times). Extract with 1 volume of chloroform (2 times). Add 1/10 volume of 3 M sodium acetate, pH 7.0. Add 2 volumes ethanol. Precipitate DNA at −70° for 1 hr.

5. Centrifuge DNA in microfuge 15 min at 4°. Wash pellet in 80% ethanol. Resuspend pellet in 25 μl H_2O.

6. For the kinase reaction, add to DNA 1/10 volume 10 × kinase buffer [500 mM Tris-HCl, pH 7.6, 100 mM $MgCl_2$, 50 mM dithiothreitol (DTT), 1 mM spermidine, 1 mM EDTA; filter, sterilize, and store in 10-μl aliquots at −20°], 500 μCi [γ-^{32}P]ATP (6000 Ci/mmol), and 30 units T4 kinase in a final reaction volume of 40 μl. Incubate 40 min at 37°. Inactivate kinase 10 min at 65°.

7. For the ligation reaction, transfer the DNA to 20 ml ligation buffer [2 × ligation buffer: 40 mM HEPES, pH 7.4, 20 mM $MgCl_2$; add fresh DTT to 1 mM and fresh ATP to 100 μM (in 1 × buffer) just before the ligase] in a 50-ml centrifuge tube (0.5 μg DNA/ml to minimize intermolecular ligation). Add 10–20 units of T4 ligase. Incubate overnight at 4°.

8. Concentrate DNA by 1-butanol extractions as follows: Add 30 ml 1-butanol. Mix. Spin on a table-top centrifuge to allow phases to separate. Discard upper butanol phase. Repeat until aqueous phase volume is about 400 μl (~ 5 butanol extractions). Smaller volumes of butanol should be used as lower volumes are reached since overconcentration will cause the salts to precipitate. Transfer DNA to a 1.5-ml Eppendorf tube.

9. Extract DNA with 1 volume of phenol–chloroform (2 times). Extract with 1 volume of chloroform (2 times). Extract with 1 volume of ether (2 times). Remove traces of ether in Speedvac (Savant) for 30 min.

10. Add NaCl (from stock of 5 M NaCl) to 0.5 M, and polyethylene glycol (from stock of 50% PEG in 0.5 M NaCl) to 10% final concentration. Precipitate 3 hr on ice (this precipitates the DNA, leaving free ATP in solution). Spin in microfuge 30 min at 4°. Discard supernatant.

11. Remove traces of PEG by washing DNA in 80% ethanol. Spin in microfuge 5 min at 4°.
12. Repeat step 11.
13. Resuspend DNA in 200 μl 10 mM HEPES, pH 7.5. Extract with 1 volume of chloroform (2 times). Extract with 1 volume of ether (2 times).
14. Remove traces of ether in Speedvac for 30 min. Precipitate DNA with ethanol as described. Wash pellet with 80% ethanol. Resuspend DNA in desired buffer.

To minimize radiolysis, the ^{32}P-labeled DNA should be diluted to 5–20 μg/ml and stored at $-20°$. The DNA concentration should be checked by linearizing an aliquot of the labeled DNA with a restriction enzyme and comparing it in an ethidium bromide-stained gel with markers of the same nonradioactive linear DNA of known concentrations. The topology of the DNA should be checked periodically by running the DNA out in a gel and either staining with ethidium bromide or performing radioautography. The DNA should be discarded when linear and nicked circular DNA molecules become prominent.

Preparation of Oocyte S-150

The oocyte S-150 is prepared by high-speed centrifugation of dispersed whole oocytes. It is essential to use ovaries containing predominantly large, stage 6 oocytes because the presence of small oocytes within the cell population results in defective extracts. The oocytes are dispersed by digestion with 0.15% collagenase type II (Sigma, St. Louis, MO) for 2–3 hr. We have noticed that the specific activity of the Sigma type II collagenase can vary up to 5-fold between different batches. It is important to use a collagenase of high specific activity because impure collagenase preparations yield extracts with high nuclease and protease activities.

The total protein content of the oocyte S-150, prepared as described below, varies between 2 and 4 mg/ml. The amount of endogenous histones available to make chromatin must be determined for each oocyte S-150 by analyzing the DNA topology and MNase ladder after a 3-hr incubation with increasing amounts of DNA (see Fig. 2 in Ref. 1). Three microliters oocyte S-150 will usually assemble between 15 and 50 ng of DNA into chromatin. Higher levels of DNA will result in partially supercoiled DNA or relaxed DNA molecules; under conditions of DNA excess, the circular DNA molecules will not be fully covered with protein, and thus partially or totally protein-free DNA molecules will be recovered after the 3-hr incubation.

To ensure that all of the DNA is fully packaged into chromatin, the amounts of DNA used should never exceed the capacity of the oocyte S-150. We have attempted to supplement the oocyte extract with either calf thymus or *Xenopus* histones, but we have been unable to produce native chromatin under these conditions. Conventional histones tend to precipitate at the physiological ionic strength and neutral pH used in our assembly system. Addition of conventional nucleosomal histones, with or without histone H1 (from a concentrated solution of histones in very dilute acid), sometimes increases the capacity of the oocyte S-150 to supercoil DNA into nucleosomes; however, even in these successful cases, the nucleosomes formed do not display the 170–200 bp periodicity characteristic of native chromatin.[9]

Two ovaries yield about 15–30 ml of oocyte S-150, which is sufficient for 5000 assays when using ^{32}P-labeled DNA as template for assembly. The oocyte S-150 remains fully active for many months when stored at $-70°$. The ease of preparation, the large amount of extract produced, and the stability of the extract more than compensate for the minor inconveniences of the required DNA titration and for occasional occurrence of defective extracts. In the procedure presented below, the oocyte S-150 is prepared with low salt in order to maintain isotonicity during the chromatin assembly reaction with the ATP-regenerating system. A modification of this procedure that allows visualization of the proteins assembled onto the DNA is described.[10]

1. Anesthetize a large, adult female frog [*Xenopus laevis* (Nasco, Fort Atkinson, WI)] in ice for 45–60 min.
2. Remove ovaries and rinse in 3 changes of OR-2 buffer[11] to remove any blood [10× OR2-A: 50 mM HEPES, 10 mM Na$_2$HPO$_4$, 825 mM NaCl, 25 mM KCl (final pH 7.6); 100× OR2-B: 100 mM CaCl$_2$, 100 mM MgCl$_2$; dilute and mix OR2-A and OR2-B just before use].
3. Add ovaries to 150 ml of OR-2 buffer containing 0.15% type II collagenase (Sigma) in a 500-ml Erlenmeyer flask.
4. Shake gently in a rotary shaker (60 rpm) at room temperature for 2–3 hr until the oocytes are dispersed.
5. Decant the collagenase buffer and wash the oocytes 10 times in 150-ml volumes of OR-2 buffer. For each wash, the buffer should be decanted after the large stage-6 oocytes have been allowed to settle but while the small white oocytes are still floating. These

[9] I. Ruberti and A. Worcel, unpublished observations.
[10] A. Shimamura, D. Tremethick, and A. Worcel, *Mol. Cell Biol.* **8,** 4257–4269 (1988).
[11] R. A. Wallace, D. W. Jared, J. N. Dumont, and M. W. Sega, *J. Exp. Zool.* **184,** 321 (1973).

washes eliminate the immature oocytes, connective tissue, and any remaining traces of collagenase.

6. Transfer the oocytes to a petri dish and remove any remaining clusters of oocytes or connective tissue. Gently load the dispersed oocytes into ultracentrifuge tubes with the wide end of a Pasteur pipette. The settled oocytes should fill slightly more than half the volume of the tube. Remove excess buffer.

7. Wash the oocytes quickly with 3 changes of cold extraction buffer (20 mM HEPES, pH 7.5, 5 mM KCl, 1.5 mM MgCl$_2$, 1 mM EGTA, 10% glycerol, 10 mM β-glycerophosphate, 0.5 mM DTT). This is best done by squirting buffer to disperse the oocytes, though care must be taken to avoid lysis. The buffer is then removed after the oocytes have settled.

8. Add extraction buffer to fill the remaining volume of the tube. The volume of the settled oocytes should be slightly larger than the buffer volume. The buffer should remain clear; cloudiness is indicative of lysis.

9. Spin the oocytes in a Beckman SW50.1 rotor for 30 min at 40,000 rpm at 4°.

10. Remove the clear supernatant that forms between the upper phase (which contains lipid) and the pellet (which contains yolk platelets, pigment, and cell debris), by inserting a 5-ml syringe carrying an 18-gauge needle through the side of the tube just above the pellet. Gently transfer the extract to a sterile tube on ice. After all the extract has been pooled, mix gently to ensure homogeneity.

11. Aliquot the extract and store at −70°. Once a tube of extract has been thawed, it should be used immediately, and any remaining extract should be discarded.

Assembly of Chromatin

The assembly reaction works best in small volumes at either 22° or 37°, although it is possible to scale up the reaction to larger volumes at 37°. We present below the protocol for the assembly reaction with ^{32}P-labeled DNA in a 10-μl final volume and the protocol for the assembly reaction with nonradioactive DNA in a 500-μl final volume.

Assembly on ^{32}P-Labeled DNA

1. To a 0.5-ml microfuge tube, add the following: 1 μl 10× chromatin assembly mix (see below), 1 μl ^{32}P-labeled DNA, 1 μl nonradioactive DNA (if desired) or extraction buffer, 1 μl 10 ng/μl creatine phos-

phokinase, and 6 μl S-150 extract. All reagents and DNA stock solutions are made up in extraction buffer to maintain the ionic strength of the reaction. The creatine-phosphokinase solution is stable on ice for at least 1 month. (The 10× chromatin assembly mix contains 30 mM ATP, 50 mM MgCl$_2$, and 400 mM creatine phosphate. ATP and creatine phosphate stocks are made on ice and adjusted to pH 7–8 with concentrated Tris base using pH indicator paper. Reagents are stored individually at −70° to inhibit hydrolysis, which is enhanced at 4° and by Mg^{2+}.)

2. Incubate the reaction for 6 hr at room temperature or 2 hr at 37°.
3. Stop the reaction with 3 μl of stop solution (1% SDS, 80 mM EDTA).
4. Add 2 μl of a 10 mg/ml proteinase K solution and incubate at 37° for at least 1 hr.
5. Add 1/5 volume sample dye (50% glycerol, 0.5% bromphenol blue, 0.5% xylene cyanol, 5 mM EDTA), and run the DNA on a 1% agarose gel in Tris–glycine buffer (120 g Tris–576 g glycine in 20 liters H$_2$O final volume). Dry gel and visualize DNA by autoradiography.

Assembly on Nonradioactive DNA

Set up the chromatin assembly reactions keeping the ratio of the extract volume to the total reaction volume the same as for the ^{32}P-labeled DNA reactions. The reaction volumes may vary from 50 μl to 1 ml in a 1.5-ml microfuge tube.

1. For the sample reaction (500-μl final volume), add the following to a 1.5-ml microfuge tube: 50 μl 10× chromatin assembly mix, 5 μl 100 ng/μl creatine phosphokinase, 145 μl 10 μg/ml plasmid DNA in extraction buffer, and 300 μl S-150 extract.
2. Incubate at 37° for 2 hr.
3. Stop the reaction with 1/4 volume of a 2.5% sarkosyl–100 mM EDTA solution.
4. Add 1 μl of a 10 mg/ml RNase A solution (DNase-free) and incubate at 37° for 30 min.
5. Add SDS to a final concentration of 0.2%, add proteinase K to a final concentration of 1 mg/ml, and incubate at least 1 hr at 37°.
6. Add ammonium acetate to a final concentration of 3 M. Add 2 volumes of ice-cold ethanol and place on ice for 5 min. Spin in microfuge for 30 min at 4°.
7. Wash pellet with 80% ethanol. Dry pellet in Speedvac.
8. Dissolve samples in TE buffer. Add 1/5 volume sample dye and run

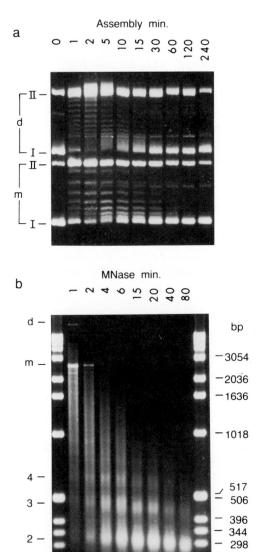

samples on a 1% agarose gel. Visualize DNA with ethidium bromide (see Fig. 1a).

Micrococcal Nuclease Digestion

For ^{32}P-labeled DNA

1. Set up desired number of individual chromatin assembly reactions as described in Assembly on ^{32}P-Labeled DNA, steps 1 and 2.
2. Pool all the individual reactions into a 1.5-ml microfuge tube and bring to room temperature. Add CaCl$_2$ to a final concentration of 3 mM. Add micrococcal nuclease (Boehringer Mannheim) to a final concentration of 0.3 units/μl.
3. Remove aliquots at the desired time intervals, and transfer them to microfuge tubes containing 1/3 volume of stop solution.
4. Process as in Assembly on ^{32}P-Labeled DNA, step 4. Run DNA samples on a 1.5% agarose gel until the bromphenol blue has run two-thirds the length of the gel. Place gel on DEAE chromatography paper (to trap any small fragments of DNA) and dry for autoradiography.

For Nonradioactive DNA

1. Set up chromatin asssembly reactions as described in Assembly on Nonradioactive DNA, steps 1 and 2.
2. Bring the reactions to room temperature. Add CaCl$_2$ to a final concentration of 3 mM. Add micrococcal nuclease to a final concentration of 0.3 units/μl.
3. Remove aliquots at desired times and transfer to microfuge tubes containing 1/4 volume of a 2.5% sarkosyl – 100 mM EDTA solution.

Fig. 1. Chromatin assembly on nonradioactive DNA. (a) Time course of DNA supercoiling. Supercoiled pUC9 DNA (1.5 μg) was assembled into chromatin at 37° in a standard 500-μl reaction (see text). Aliquots of 50 μl were removed from the reaction at the times indicated and processed as described (see also Fig. 3 in Ref. 1). I, Supercoiled DNA; II, closed circular relaxed DNA and nicked relaxed DNA; m, monomer pUC DNA; d, dimer pUC DNA. (b) MNase digestion of the assembled chromatin. Supercoiled pUC9 DNA (1.5 μg) was assembled into chromatin in the standard 500-μl reaction for 4 hr at 37°, and the chromatin was then digested with MNase as described. Aliquots of 50 μl were removed from the MNase digestion mixture at the indicated times and processed as described. The outside lanes contain molecular size markers. The DNA samples were applied to the wells (made with a 30-well comb) and electrophoresed for 2 hr at 70 V in a 15 × 10 cm agarose gel. m, Monomer linear pUC DNA; d, dimer linear pUC DNA; numerals 1 to 4 indicate the positions of the nucleosome oligomers.

4. Process as in Assembly on Nonradioactive DNA, steps 4–7. Dissolve samples in TE buffer. Add 1/5 volume of orange G dye solution (50% glycerol, 0.1% orange G, 5 mM EDTA). Run samples on a 1.5% agarose gel until the dye has reached the bottom. Visualize DNA with ethidium bromide (see Fig. 1b).

Acknowledgments

This research was supported by grants from the National Institutes of Health to A.W. and by an NIH MSTP fellowship to A.S.

[30] Purification of *Xenopus* Egg Nucleoplasmin and Its Use in Chromatin Assembly *in Vitro*

By Linda Sealy, Richard R. Burgess, Matt Cotten, and Roger Chalkley

The purification of *trans*-acting regulatory proteins has been studied extensively. These agents modulate the transcriptional activity of genes, though their mechanism of action is not understood. As a consequence there has been renewed interest in reassembling chromatin containing these factors to assay the interrelationship between structure and function *in vitro.* This has led to an increased interest in systems for assembling histones onto DNA under physiological conditions. It is generally thought that chromatin assembly in the cell is mediated through the action of a histone-binding assembly agent. Unfortunately, such a protein has not been unequivocally identified in somatic cells, primarily because it is likely to be present in small amounts. However, the *Xenopus laevis* frog oocyte stores large quantities of histone for assembly after fertilization. These histones are associated with two karyophilic proteins, N1/N2 and nucleoplasmin.[1] Although whole oocyte homogenates are capable of directing chromatin assembly, the many uncontrolled variables so introduced have encouraged the purification of individual factors such as nucleoplasmin.[2] We have shown that nucleoplasmin obtained from unfertilized eggs, rather than oocytes, is far superior in promoting the formation of nucleosomes onto DNA *in vitro,* owing to massive additional phosphorylation of the

---

[1] J. A. Kleinschmidt, E. Fortkamp, G. Krohne, H. Zentgraf, and W. W. Franke, *J. Biol. Chem.* **260**, 1166 (1985).
[2] R. A. Laskey, B. A. Honda, A. D. Mills, and J. T. Finch, *Nature (London)* **275**, 416 (1978).

Copyright © 1989 by Academic Press, Inc.
All rights of reproduction in any form reserved.

protein.[3,4] Details of the purification of nucleoplasmin from *Xenopus laevis* frog eggs and the use of this protein as an assembly agent are presented below.

Care and Maintenance of Frogs

The primary key to successful purification of the chromatin assembly agent, nucleoplasmin, is to obtain a sufficient quantity of unfertilized eggs of high quality. The number of eggs laid by any particular female frog after injection with human chorionic gonadotropin (HCG) is intrinsically variable. However, we have observed that appropriate care and feeding of the frogs can greatly improve the average yield of quality eggs obtained. We have obtained our best yields of high-quality eggs from female frogs that were purchased approximately 6 months before reaching maturity and maintained during that interval on a rich diet. However, if this is not feasible, it would probably be wise to wait at least 2 weeks to 1 month after receiving a fully mature female frog before injecting the animal, since frogs subjected to the shock of shipment as well as new living conditions will probably not perform well.

During the course of our studies with nucleoplasmin, we have maintained the frogs in glass aquariums or plastic tubs in tap water (allowed to stand overnight to permit dechlorination) at room temperature (22°). The frogs were maintained on a diet of fresh calf liver (frog brittle is also available; Nasco, Fort Atkinson, WI) and fed every other day; the tank water was changed approximately 6–8 hr after feeding. Maintaining clean water in the tanks is the most effective way we have found to combat diseases. Nonetheless, *Xenopus* frogs still occasionally develop a skin disease known as "red leg," which can be recognized by the appearance of blotches on the skin with loss of pigment. In advanced stages the frogs' skin also becomes extremely dry. We have had no success in treating this problem, and the best course of action is probably to remove the infected animal from the tank as soon as possible to prevent the disease from spreading to other frogs. Further advice on care and treatment of *Xenopus* frogs can be obtained from the suppliers (Nasco or Xenopus I, Ann Arbor, MI) or from appropriate texts.[5]

Egg Collection

In the course of injecting some 150 different frogs over a 2-year period, we have found that the yield of packed, jellied eggs per frog can vary over a

[3] L. Sealy, M. Cotten, and R. Chalkley, *Biochemistry* 25, 3064 (1986).

[4] M. Cotten, L. Sealy, and R. Chalkley, *Biochemistry* 25, 5063 (1986).

[5] E. Deuchar, *in* "*Xenopus,* The South African Clawed Frog," p. 1. Wiley, London, 1975.

wide range from as little as 0–5 ml to as great as 80–90 ml (with no direct relationship to size of the frog). The median value is 30–35 ml/frog. The quality of the eggs can also vary. High-quality eggs (those that give excellent yields of nucleoplasmin) have a buff-colored hemisphere and an olive green-colored hemisphere with an extremely sharp division at the equator (often appearing as a thin, distinct dark line). As the eggs deteriorate, the buff and green colors become mottled; finally the egg becomes light gray or white and disintegrates. We have observed that some females lay deteriorated eggs; however, even eggs of high quality will deteriorate over time. This is most rapid when the eggs are kept in tap water at room temperature; the problem can be substantially reduced by frequent egg collection and placing the eggs in a saline solution at 4°. Given that the average frog will usually commence laying approximately 6–8 hr after injection and continue to lay for a period of 12–24 hr, we have developed the following protocol for egg collection.

Each female frog is injected with 0.5 ml (small frogs) or 1.0 ml (large frogs) of HCG (1000 units/ml, rehydrated as directed by the manufacturer) in the evening (5–8 p.m.). It is preferable not to feed the frogs on the day of injection. Each frog is then placed in a separate collection bin in Barth's modified saline solution (15 mM Tris, pH 7.5, 1 mM MgSO$_4$, 1 mM CaCl$_2$, 0.5 mM Na$_2$HPO$_4$, 110 mM NaCl, 2 mM KCl, 2 mM NaHCO$_3$, prepared using single-distilled H$_2$O). As the eggs have a half-life of approximately 4–6 hr in this saline solution at room temperature, it is necessary to collect the eggs at 2- to 3-hr intervals and place them in ice-cold Barth's saline in a refrigerator to minimize deterioration. We find the eggs are conveniently collected using a wide-mouth 10- or 25-ml pipette. The frogs will shed slime into the water, which one should avoid collecting. However, the frogs appear to suffer no ill effects from being maintained in the saline solution for as long as 36 hr. Since the frogs should begin to lay in the early morning hours, a collection at 6–7 a.m. will allow these first eggs to be stored approximately 4–6 hr after laying. Subsequent collections can be continued throughout the day, and most frogs will have nearly finished production by approximately 3 p.m., at which time the protein purification procedure can begin.

If difficulties in obtaining a sufficient quantity of high-quality eggs are encountered with this protocol, it is possible to prime the frogs prior to HCG injection with pregnant mare serum. We find this improves primarily the quality rather than the quantity of eggs produced and may be most useful for mature frogs that were only just obtained or have not received premium care and feeding. An injection of 200 units (Sigma, St. Louis, MO; hydrated in physiological saline) is given 2 days prior to the HCG (1000 units/frog) injection. Primed frogs generally begin to lay sooner (in

3–5 hr) than unprimed frogs and may produce their eggs over a shorter period. However, the eggs are still subject to deterioration and should be collected and placed at 4° at frequent intervals.

Female frogs that have laid eggs can be "reutilized" for egg production after a suitable interval. Since oocyte maturation requires approximately 100 days in *Xenopus* frogs, maximum production would require a waiting interval of at least 3 months between hormone injections. We have cycled frogs at 4- to 6-month intervals, although frogs that lay very poorly are generally not maintained.

Purification of Nucleoplasmin from Unfertilized Eggs

The following purification scheme has been adapted and extended from those previously published.[2,6,7] It is designed for processing approximately 250–300 ml of packed, jellied eggs. Given an average yield of 30–35 ml/frog, injection of 8–10 frogs should yield this quantity of unfertilized eggs. The yield of purified nucleoplasmin is roughly equivalent in milligrams to the milliliters of packed, jellied eggs used in the procedure. Thus, one can expect 250–300 mg of purified protein from this number of frogs. If lesser quantities of the protein will suffice, we have also developed a scaled-down version of the protocol. The small-scale version is designed for 50–75 ml of packed, jellied eggs. Where applicable, parameters for the smaller version will be specified in square brackets.

Dejellying the Eggs

The first step in the purification procedure is to remove the jelly coat that surrounds the egg when it is laid. It is important to perform this procedure in a gentle fashion as the eggs become very fragile and are easily smashed when the jelly coat starts to dissolve, and as a consequence, the egg contents can be poured off with the jelly coat. To minimize breakage and loss of a large majority of the eggs, which have been so patiently collected, a detailed description of the dejellying process is presented below.

The collected eggs are combined in a large beaker so that, when dejellied, they form a thin layer on the bottom of the beaker. When the jelly coat is removed, the volume of packed eggs will be decreased by 3- to 5-fold, depending on losses due to breakage; thus, for 200–300 ml of jellied eggs, a 1-liter beaker will suffice. Any residual Barth's saline is decanted, and the eggs are washed in 500 ml of single-distilled H_2O at

[6] R. A. Laskey, A. D. Mills, and N. R. Morris, *Cell* **10**, 237 (1977).
[7] C. Dingwall, S. V. Sharnick, and R. A. Laskey, *Cell* **30**, 449 (1982).

room temperature, introduced by carefully pouring the liquid along the side of the beaker. Allow the eggs to sit in the water for 2 min, swirl the eggs gently, allow them to resettle, and carefully decant the liquid. Introduce 250 ml of 2% (w/v) cysteine, pH 7.6, at room temperature in the same manner. The cysteine solution should be prepared immediately before use. Allow the eggs to sit in this solution for a total period of 5 min, gently swirling after 3 min so that all eggs become suspended in the liquid. After the eggs have settled, carefully decant and replace with 250 ml fresh 2% cysteine, pH 7.6, at room temperature. Swirl the eggs immediately, and allow to settle for a total period of 5 min. Decant the cysteine. The jelly coat should now be dissolved, and the eggs are extremely fragile. Wash with 500 ml single-distilled H_2O at room temperature. Slowly pour the water down the side of the beaker, and do not swirl the eggs, as bumping into the sides of the beaker or other shearing forces will cause breakage. Decant the water and repeat with a second wash of 500 ml of single-distilled H_2O at room temperature. Finally, wash once in the same manner with 500 ml of Barth's modified saline at 4°. The eggs should now be completely dejellied. [For a smaller scale preparation, decrease all volumes above by one-half.]

Preparation of the Heat Supernatant

Decant the Barth's saline from the dejellied eggs, and pour them into a plastic cylinder or graduated test tube to measure their volume. Remove any excess liquid and decant the dejellied eggs into the jar of a Waring blendor [or use a very loose Dounce homogenizer]. Add 2 volumes of lysis buffer, which contains 25 mM Tris, pH 7.2 at 4°, 120 mM NaCl, 1 mM dithiothreitol (DTT), 0.2 mM phenylmethylsulfonyl fluoride (PMSF), and the phosphatase inhibitors 10 mM Na$_2$MoO$_4$, 10 mM β-glycerol phosphate, and 100 μM sodium vanadate, at 4° and homogenize the eggs at low speed. All remaining steps in the purification are performed at 4° unless otherwise indicated. Centrifuge the lysed eggs in an SS-34 rotor for 25 min at 1900 g. The mixture will stratify into three layers: a top yellow fat layer, a milky supernatant, and a large dark-green and brown pellet. Remove the milky white supernatant from under the fat layer, and clarify for 90 min at 25,000 rpm in an SW27 rotor [30 min, 35,000 rpm, SW50.1 rotor]. Three layers will again be observed: a top white lipid layer, a clear supernatant, and a golden pellet. Remove the clear supernatant from under the lipid layer and extract with an equal volume of trifluorotrichloroethane (Aldrich, Milwaukee, WI). Separate the two phases by centrifugation in an SS-34 rotor for 10 min at 3000 g. Remove the aqueous (top) phase, and repeat the trifluorotrichloroethane extraction. Finally, reclarify the aqueous phase for 90 min at 25,000 rpm in an SW27 rotor [or SW50.1

rotor, see above]. Again remove the clear supernatant, avoiding any white lipid layer at the top.

This twice-clarified supernatant is then heated to 80° for 10 min and placed on ice for 15 min. Prior to heating, add 25 μl/ml supernatant of 6 mg/ml PMSF freshly dissolved in absolute ethanol. On heating and cooling, formation of a white, flocculent precipitate in an otherwise clear supernatant should be observed. Nucleoplasmin, owing to its thermal stability, should remain soluble under these conditions, whereas a majority of the proteins in the twice-clarified supernatant will be denatured and precipitated. Indeed, this difference constitutes the primary purification step in the procedure. However, it is possible to precipitate nucleoplasmin at this stage, presumably through entrapment with other precipitated proteins. We have observed that three parameters are critical in order to avoid precipitation of nucleoplasmin during heating while effectively removing most other proteins from solution. Complete removal of lipids is essential, and the protein concentration of the solution should not be excessive (always add lysis buffer to top off the ultracentrifuge tubes). Most importantly, the pH of the lysis buffer appears to be crucial; the pH must be adjusted to 7.2 at 4°. Remove the precipitate by clarifying the solution for 20 min at 25,000 rpm in an SW27 rotor [15 min, 25,000 rpm, SW50.1 rotor]. Remove the supernatant for application to a DE-52 DEAE-cellulose column, as described below. As shown in Fig. 1, the heat supernatant should be substantially enriched in nucleoplasmin.

DEAE Column Chromatography

For all remaining steps in the purification procedure, only plasticware should be used for protein fractions containing nucleoplasmin, as the assembly factor adheres readily to glass, resulting in substantial losses of the protein. The heat supernatant is applied to a 200-ml [70-ml] bed volume DE-52 DEAE cellulose (Whatman) column (2.5 × 50 cm) [1.5 × 50 cm] previously equilibrated in 25 mM Tris, pH 7.5, 1 mM EDTA, 0.1 M NaCl. The flow rate should not exceed 35 ml/hr during sample application. The column is then washed with at least 3 column volumes of the above buffer at a flow rate of 70 ml/hr. The column is then eluted with a 900-ml [300-ml] linear gradient of increasing NaCl concentration from 0.1 M to 0.45 M in 25 mM Tris, pH 7.5, 1 mM EDTA at 70 ml/hr. Column fractions (10 ml) are analyzed by SDS–15% polyacrylamide gel electrophoresis[8] to identify those containing nucleoplasmin. As shown in the column profile and accompanying SDS–polyacrylamide gel in Fig. 2, nucleoplasmin elutes just past the midpoint of the gradient in fractions

[8] U. K. Laemmli, *Nature (London)* **227**, 680 (1970).

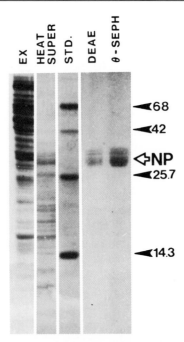

FIG. 1. Purification of *Xenopus* nucleoplasmin. SDS–15% polyacrylamide gel electrophoresis and Coomassie blue staining of proteins obtained during the purification of nucleoplasmin from unfertilized eggs of female *Xenopus laevis* injected with HCG. Lane 1, Twice-clarified egg extract; lane 2, twice-clarified egg extract after heating to 80° and centrifugation to remove precipitated proteins; lane 4, nucleoplasmin purified by DEAE-cellulose chromatography; lane 5, nucleoplasmin after subsequent purification by phenyl-Sepharose chromatography. Lane 3 contains protein molecular weight standards as described in the legend to Fig. 2.

140–146 at a conductivity of approximately 33 millimhos. The monomer subunits of nucleoplasmin exhibit a characteristic heterogeneity on the 15% acrylamide gel, migrating as a series of multiple distinct bands of 30,000–33,000 MW_{app}. If the samples are not boiled in sodium dodecyl sulfate prior to SDS–gel electrophoresis, nucleoplasmin will retain its pentameric form, migrating with a molecular weight of approximately 150,000. [Elevated ionic strengths stabilize the pentameric form; at 0.3 M NaCl and above, 4 M urea may be required during heating in the presence of SDS to dissociate the pentamer. Since nucleoplasmin should elute from the DEAE column just below 0.3 M NaCl (at ~275 mM NaCl), we have not found it necessary to add urea to the gradient fractions to detect the monomer form on SDS gels.]

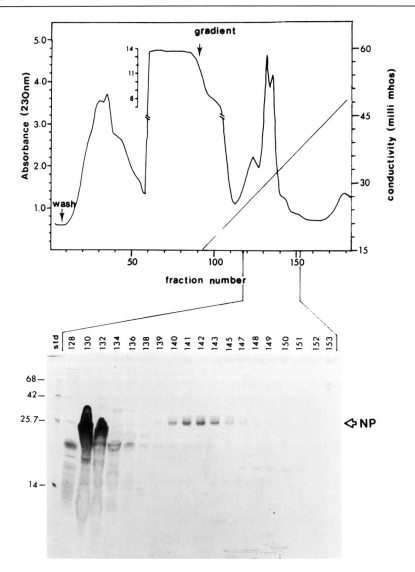

FIG. 2. DEAE-cellulose chromatography of nucleoplasmin. (Top) Absorbance profile of column fractions collected during fractionation of a heat supernatant from unfertilized *Xenopus* eggs by DE-52 DEAE-cellulose chromatography. Conductivity measurements of gradient fractions are also shown. (Bottom) Coomassie blue staining of selected column fractions analyzed on an SDS–15% polyacrylamide gel. Protein molecular weight standards include bovine serum albumin (68,000), actin (42,000), α-chymotrypsinogen (25,700), and lysozyme (14,000).

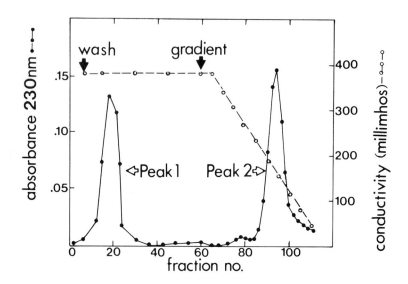

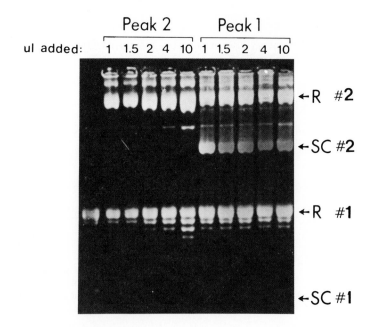

The large bed volume DEAE column, in combination with the heat step, suffices to separate nucleoplasmin from essentially all other proteins (and nucleic acids). By selectively pooling those fractions in the center of the nucleoplasmin peak, a preparation apparently homogeneous (by Coomassie blue staining) with respect to nucleoplasmin can be obtained (see Fig. 3, lane 4). However, when this material was used in nucleosome assembly assays, we found that optimal assembly required the use of excessive amounts of histone. Moreover, fully packaged material could only be obtained within a very narrow range of nucleoplasmin, histone, and DNA ratios (which varied for each nucleoplasmin preparation), necessitating tedious titrations to establish optimal conditions. We found that these difficulties could be attributed to the presence of a contaminating substance within the DEAE-purified protein that both binds positively charged histones as well as interferes with the action of topoisomerase I (used to assay the extent of nucleosome assembly, see below). Removal of this inhibitory substance (most probably a highly negatively charged carbohydrate) can be achieved by chromatography of the DEAE-purified nucleoplasmin on the hydrophobic resin, phenyl-Sepharose.

Phenyl-Sepharose Chromatography

Nucleoplasmin is sufficiently hydrophobic to bind to phenyl-Sepharose at high levels of ammonium sulfate. Thus, peak fractions containing nucleoplasmin from the DEAE column are brought to 55% saturation with solid ammonium sulfate [add 351 mg/ml solid $(NH_4)_2SO_4$ slowly with stirring plus 1 μl 1 N NaOH per g $(NH_4)_2SO_4$ added]. The material is then applied directly to a 13-ml [4-ml] bed volume phenyl-Sepharose (Pharmacia) column (1 × 20 cm) [1 × 10 cm] previously equilibrated with 20 mM Tris, pH 7.6, 2.25 M $(NH_4)_2SO_4$. Flow rate during sample appli-

FIG. 3. Separation of a topoisomerase inhibitor from nucleoplasmin by phenyl-Sepharose chromatography. Fractions from the DEAE-cellulose column containing nucleoplasmin were pooled and applied to a phenyl-Sepharose column after the addition of $(NH_4)_2SO_4$. (Top) Absorbance and conductivity profiles of the column fractions as the column was washed and eluted with a linear gradient of decreasing salt concentration. Aliquots of material that did not bind to the column (peak 1) and that eluted from the column (peak 2, nucleoplasmin) were added to a mixture of supercoiled pBR322 DNA and topoisomerase I that had been preincubated for 1 hr at 25° to allow relaxation of this DNA to occur (lane 1). Subsequently, a second, larger (~ 15 kb) supercoiled plasmid DNA was added, and the incubation was allowed to continue for 1 hr at 25°. (Bottom) One percent agarose TPE gel analysis of the DNA samples at the end of the incubation. R#1 and SC#1, relaxed and supercoiled forms of pBR322; R#2 and SC#2, relaxed and supercoiled forms of the 15-kb plasmid.

cation should not exceed one column volume per hour. The column is washed with 75 ml [20 ml] of the above buffer and then eluted with a linear gradient of decreasing $(NH_4)_2SO_4$ concentration from 2.25 to 0 M in 20 mM Tris, pH 7.6 (24.5 × 29 ml) [7 × 9 ml] at a flow rate of 15 ml/hr. Gradient fractions (1 ml) containing nucleoplasmin are identified by their absorbance at 230 nm (see Fig. 1, top).

The column profile presented in Fig. 1 reveals that a substantial amount of material absorbing at 230 nm flows through the column. When small amounts of this nonbinding material (peak 1) are added to a mixture of topoisomerase I and DNA, complete inhibition of relaxing activity results, whereas phenyl-Sepharose-purified nucleoplasmin (peak 2) has no effect on topoisomerase activity (Fig. 1, bottom). The inhibitory effect of the material in peak 1 can be neutralized by the addition of sufficient histones to the reaction (data not shown), suggesting that the substance in peak 1 also binds positively charged histones in addition to its adverse effects on topoisomerase I. The complex and unreliable behavior of the DEAE-purified protein in facilitating chromatin assembly arises from the variable amount of this topoisomerase inhibitor present in the "apparently pure" (see Fig. 1, lane 4) nucleoplasmin preparations. Since *in vitro* chromatin assembly is most conveniently monitored by the DNA supercoiling assay (see below), sufficient histone must be present in the assembly mix to neutralize the topoisomerase inhibitor, but sufficient histone must also remain to completely assemble the DNA, with sufficient (but contaminated) nucleoplasmin present to mediate the transfer.

Desalting and Storage of Purified Nucleoplasmin

Purified nucleoplasmin obtained from the phenyl-Sepharose column gradient must be desalted and concentrated prior to use in chromatin assembly assays. Dialysis of the purified protein should be avoided, as the protein adheres readily to dialysis tubing even in the presence of detergents. The most successful procedure we have found for achieving both aims in one step is to employ a Centricon 10 (Amicon) centrifugal microconcentrator. The Centricon microconcentrator is used according to the manufacturer's directions. Since the volume of pooled phenyl-Sepharose gradient fractions containing nucleoplasmin exceeds the capacity of the microconcentrator (2 ml), aliquots of the pool are successively added to the microconcentrator until the entire pooled volume has been reduced to 40 μl. The concentrated nucleoplasmin is then recovered from the Centricon by repeated washing with 10 mM Tris, pH 7.5, 1 mM EDTA, 0.1 M NaCl, 0.1% Nonidet P-40 (NP-40) to a final volume of 400 μl. The final protein sample is stored in this buffer at −70°. Recovery of protein from the low absorption YM membrane is approximately 70%.

Nucleoplasmin protein concentrations in the final sample were determined by analyzing aliquots on a 15% SDS–acrylamide gel in conjunction with a standard of calf thymus histone which had been prepared by dissolving a known amount (by weight) of pure, extensively dried histone sulfate in a known volume of H_2O. The gel was stained in fresh Coomassie Brilliant Blue R-250 obtained from Bethesda Research Laboratories. We have found Coomassie blue stain from BRL to be more sensitive and reproducible in staining intensity than that from other suppliers. After the gel was destained and dried between two layers of clear cellophane film, the gel was scanned on a DU8 spectrophotometer (Beckman) at 575 nm. Densitometer profiles were integrated to determine protein concentrations based on the absorbance of the known standard of calf thymus histone. Samples were analyzed in duplicate and within the linear response range of the stain.

Preparation of Histones

We have always found it more convenient to prepare our own histones. In this way one can define the level of acetylation and isolate core histones free from exposure to denaturing solvents. Although one can obtain satisfactory assembly with histones over a wide range of acetylation values, the most efficient assembly occurs with hyperacetylated histones (see Fig. 4). Such hyperacetylation is obtained by exposing cultured cells to sodium butyrate, which inhibits histone deacetylation.

Cell Treatment

Either monolayer or suspension cells may be used. For convenience, we have routinely employed either HeLa or HTC cells in suspension culture. Optimum acetylation of histones is obtained in vigorously replicating cells. Sodium butyrate is added from a 1 M stock solution, adjusted to pH 7.4. The cells are grown for 1.5 hr in 1.5 mM butyrate, and then the concentration of the fatty acid salt is raised to 20 mM for 12 hr. Most cells can withstand such treatment without harm and will continue to grow if the butyrate is removed. However, longer exposure can eventually kill the cells. Furthermore, some cell lines are much more sensitive than others, and it would be wise to check the viability of cells after such treatment. The cells are harvested by centrifugation or, if a monolayer culture, by scraping into phosphate-buffered saline (PBS) containing 1 mM EDTA and 20 mM sodium butyrate. Do not use trypsin. After collection, the cells are washed once in PBS containing 20 mM butyrate. They can be used directly or frozen at $-70°$.

Chromatin Isolation

Frozen cells are thawed and immediately disrupted in a Dounce homogenizer in wash medium at 4° (0.25 M sucrose, 1% Triton X-100, 5 mM sodium butyrate, 10 mM magnesium chloride, 10 mM Tris-HCl, pH 7.4, adjusted to pH 6.5 by addition of 5.2 mg/ml sodium bisulfite just before use). Additional inhibitors of proteases can be utilized if there is evidence of proteolytic damage of histones (see below). Nuclei are isolated by centrifugation at 1000 g for 10 min and washed twice in washing medium as above. The nuclei are then disrupted in 10 mM EDTA, 5 mM sodium butyrate, 10 mM Tris-HCl (pH 7.4) adjusted to pH 6.5 with bisulfite just before use, and centrifuged at 6000 g for 10 min. This pellet is washed once in the same medium. The final chromatin pellet is resuspended in water and homogenized to a viscous gellike consistency prior to dispersion by sonication. The sonication is judged complete when less than 10% of the chromatin (A_{260}) is sedimented at 10,000 g for 10 min. Failure to obtain satisfactory dispersion may reflect insufficient sonication but more usually is a consequence of inadequate extension of the chromatin in water.

Hydroxyapatite Chromatography

Core histones, free from histone H1, are prepared from dispersed chromatin by salt elution on hydroxyapatite.[9-11] An equal volume of equilibration buffer (0.3 M NaCl, 5 mM sodium butyrate, 1 mM DTT, 80 mM sodium phosphate, pH 6.8) is added to the chromatin sample, and the material is applied to a hydroxyapatite (BioGel HTP; Bio-Rad) column (1.0 ml bed volume/1.5 mg chromatin), preequilibrated with the same buffer at 4°. The amount of chromatin is determined by absorbance with a concentration of 1 mg/ml having an A_{260} value of 20. The column is washed with sufficient equilibration buffer to ensure a return to A_{260} baseline. Best results are obtained if the column is saturated with chromatin. If the hydroxyapatite is in excess, substantial losses of histone may occur, particularly if only small quantities of chromatin are used. Histone H1 is eluted with 1.5 column volumes of 0.6 M NaCl, 5 mM sodium butyrate, 1 mM DTT, 80 mM sodium phosphate, pH 6.8. Core histones are then eluted with 1.5 volumes of the same buffer, except containing 2.0 M NaCl. The elution patterns can best be determined by gel electrophoresis (see below) after precipitation of histones from salt using novo-

[9] K. S. Bloom and J. N. Anderson, *J. Biol. Chem.* **153**, 4446 (1978).
[10] R. H. Simon and G. Felsenfeld, *Nucleic Acids Res.* **6**, 689 (1979).
[11] M. Cotten, and R. Chalkley, *Nucleic Acids Res.* **13**, 401 (1985).

biocin.[12] Appropriate fractions are pooled, dialyzed against 0.15 M NaCl, 10 mM mercaptoethanol, 5 mM sodium butyrate, 10 mM Tris, pH 7.6, and stored at $-70°$.

Initial surveys of the purity of the product can be assessed by SDS electrophoresis on 15% polyacrylamide gels.[8] Histone H1 is usually contaminated with amounts of more slowly moving material, the extent of which may depend on the tissue of origin. If needed, H1 can be further purified by chromatography on Amberlite IRC 50 (Sigma) ion-exchange resin, eluting with a 0.1–1.0 M NaCl gradient in 50 mM sodium phosphate. The purity of the core histones is usually very good. If purity is less than satisfactory, one can extend the washing process in 0.6 M NaCl. Raising the ionic strength of the washing medium will result in selective losses of H2A/H2B from the final product and should be avoided. Additional bands moving more rapidly than H1, or between H2A and H4, are a sure indication of proteolysis of the histones. H1 is particularly prone to degradation during the nuclear isolation, and H3 becomes quite sensitive to endogenous proteases once it has been dissociated from the chromatin. Problems with proteolysis are frequently tissue specific, and one must respond empirically to circumstances as they arise, particularly using gel electrophoresis as a guide to whether the problem is significant. Usually, keeping the pH of nuclear manipulation at pH 6.5 is sufficient to maintain the integrity of H1, and promptness of separation coupled with low temperature will provide intact core histones. However, addition of an antiproteolysis cocktail to the various solutions has been helpful when dealing with new tissues or cell lines. The final concentrations of these inhibitors is 1 mM benzamidine, 0.1 mM PMSF 0.1 μM pepstatin, and 5 μg/ml leupeptin. The final protein concentration of the core histone fraction is determined by densitometric scanning of Coomassie blue-stained gels, in comparison with a known calf thymus histone standard, as described for nucleoplasmin above.

Assembly Assay

Nucleosome assembly is conveniently assayed by the histone/topoisomerase-catalyzed introduction of supercoils into circular DNA molecules. The circular DNA of interest is relaxed by incubating with topoisomerase I (Promega Biotec, 0.2–0.4 units/μg DNA) at room temperature for 45 min at a DNA concentration of 200 ng/μl in 50 mM Tris, pH 8.0, 1 mM EDTA, 1 mM DTT, 20% glycerol, and 50 mM NaCl. The relaxed

---

[12] M. Cotten, D. Bresnahan, S. Thompson, L. Sealy, and R. Chalkley, *Nucleic Acids Res.* **14**, 3671 (1986).

DNA (still containing the topoisomerase I) is mixed with histones and nucleoplasmin previously incubated for 45 min at room temperature in 15 mM Tris, pH 7.5, 1 mM 2-mercaptoethanol, 1 mM DTT, 1 mM sodium butyrate, 0.35 mM EDTA, 160 mM NaCl, and 0.01% NP-40. The DNA concentration in the final mix is 8 μg/ml. The mass ratios of the primary components in the assembly can be adjusted empirically to provide optimum assembly, as described in more detail below. In general the merit of the approach described here is that under optimum conditions, core histones and DNA are present at their physiological mass ratio (0.9 : 1). Assembly is continued for 60–120 min at room temperature, and then the mixture centrifuged for 5 min at 15,000 g. The supernatant is removed, and SDS is added to a final concentration of 0.2%; pelleted material is resuspended in 20 mM Tris, pH 7.5, 160 mM NaCl, 0.2% SDS. The reason for the centrifugation step at this point lies in the fact that assembly in the absence of an efficient assembly agent leads to extensive aggregation of the nucleoprotein complex. Thus, one removes inappropriately assembled material in this way. In general, assembly in the presence of nucleoplasmin minimizes or, indeed, entirely removes this problem, which is one of the attractive features of this approach.

The extent of assembly is assayed by measuring the extent of introduction of supercoils which occurs when nucleosomes are formed on closed, circular DNA molecules. For gel electrophoresis, samples are processed by adding proteinase K to a final concentration of 0.2 mg/ml followed by incubation for 1–3 hr at 37°. The samples are brought to a final concentration of 0.3 M sodium acetate and 3 volumes of ethanol added to precipitate the DNA. DNA samples are collected, dried, and dissolved in 36 mM Tris, pH 7.6, 1 mM EDTA, 30 mM NaH$_2$PO$_4$, 10% glycerol, and 0.025% bromphenol blue for analysis on a 2% horizontal agarose gel in 36 mM Tris, pH 7.6, 1 mM EDTA, 30 mM NaH$_2$PO$_4$ (TPE) at 2.6 V/cm for 24–26 hr at room temperature.[13] The buffers in the two chambers are recirculated throughout the electrophoresis to avoid ion depletion and dramatic pH changes in the separate compartments. The number of supercoils can be measured directly after visualization of the DNA by ethidium bromide staining.

Factors That Affect the Extent of Assembly

The efficiency of nucleosome assembly *in vitro* is dependent on several factors, including the following: (1) Purity of the nucleoplasmin itself (see above). (2) The nature of the histone. This is seen primarily in the extent of

[13] M. Shure and J. Vinograd, *Cell* **8**, 215 (1976).

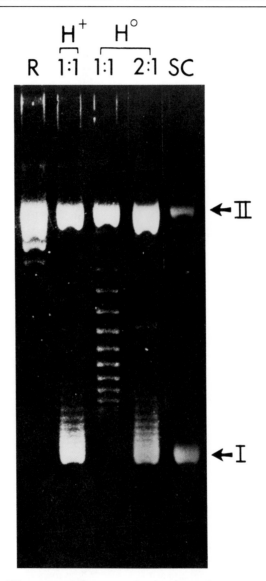

FIG. 4. Effect of histone acetylation on nucleoplasmin-mediated chromatin assembly *in vitro.* Chromatin was assembled from 400 ng relaxed pBR322 DNA (lane R), 540 ng nucleoplasmin, and 360 ng core histone (1 : 1) using our standard conditions in the presence of topoisomerase I. After 2 hr, the reaction was terminated with SDS and the DNA analyzed on a 2% TPE agarose gel. The extent of assembly using hyperacetylated (H$^+$) or control (H$^\circ$) core histones is shown. Assembly in the presence of additional control histone (720 ng core histone and 1080 ng nucleoplasmin) is also shown (2 : 1). Supercoiled (form I) pBR322 DNA is shown in lane SC.

acetylation of the histones, so that efficient assembly occurs at the more physiological core histone to DNA ratio of 0.9 : 1 only with highly acetylated histones. The effect of posttranslational modification on assembly is shown in Fig. 4. It has also been our experience that histones which are even slightly degraded show a much reduced capacity for assembly. (3) The histone/DNA input ratio. For histones with a given degree of posttranslational modification, the extent of assembly is dependent on the amount of

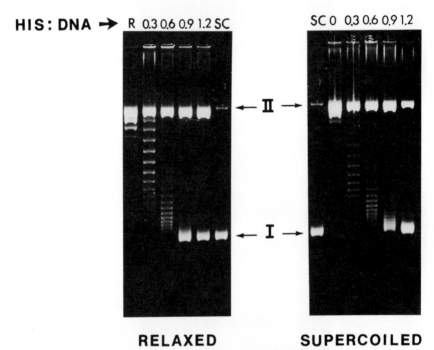

RELAXED   **SUPERCOILED**

FIG. 5. Chromatin assembly on relaxed or supercoiled DNA *in vitro*. (Left) Four hundred nanograms of pBR322 DNA, relaxed by treatment with topoisomerase I (lane R), was mixed (in the continued presence of topoisomerase I) with increasing amounts of hyperacetylated core histone previously incubated with nucleoplasmin under standard assembly conditions. The mass ratios of histone to DNA employed are shown above the lanes. The nucleoplasmin to histone mass ratio was maintained at 1.4 : 1. After 2 hr at room temperature, the reaction was terminated with SDS and the DNA analyzed on a 2% TPE agarose gel as shown. (Right) Four hundred nanograms of supercoiled pBR322 DNA (lane SC) was mixed with increasing amounts of hyperacetylated core histone previously incubated with nucleoplasmin as described above. After 2 hr at room temperature, topoisomerase I (0.4 units/μg DNA) was added, and the incubation was allowed to continue for an additional 30 min at room temperature before terminating the reactions with SDS. DNA samples were analyzed on a 2% TPE agarose gel as shown. Since the topoisomerase is added after assembly is completed, this assay monitors nucleosome formation by virtue of the resistance of supercoils in the pBR322 DNA to removal by topoisomerase I.

histone added to the reaction mix. Thus, as shown in Fig. 5, we see that at low histone to DNA ratios the amount of assembly is roughly proportional to the histone input. However, after the optimum level of assembly has been reached, additional histone has a negative affect (data not shown), presumably because of precipitation of the nucleoprotein complexes before assembly is complete. (4) The nucleoplasmin to histone ratio. In view of the role of nucleoplasmin in organizing the histone octamer it is not surprising to find that good assembly requires a stoichiometric quantity of nucleoplasmin pentamer per histone octamer added to the assembly system (see Fig. 6).

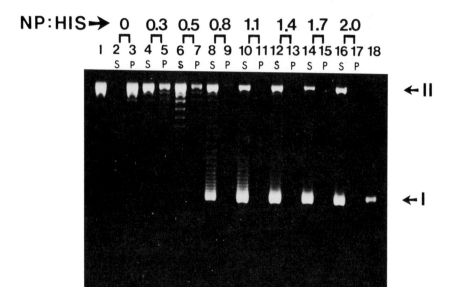

FIG. 6. *In vitro* chromatin assembly capability of egg nucleoplasmin. Four hundred nanograms of pBR322 DNA relaxed by topoisomerase I (lane 1) was incubated under standard assembly conditions with 360 ng of hyperacetylated core histones previously mixed with increasing amounts of egg nucleoplasmin (lanes 2–17). The mass ratios of nucleoplasmin to histone employed are shown above the lanes. After 2 hr at room temperature, assembly reactions were centrifuged at 12,000 *g* for 5 min, and both supernatant (S) and pelleted (P) material was analyzed on the 2% TPE agarose gel shown. Supercoiled form I pBR322 DNA is shown in lane 18. On this gel system, circular pBR322 DNA containing one to two positive or negative supercoils is not resolved from nicked circular (form II) DNA.[13] Also, parallel analysis of topoisomerase I-treated relaxed DNA (lane 1) in gels containing chloroquine over a range of concentrations indicates that at [chloroquine] = 0 (i.e., gels shown here) the center of mass of topoisomer distribution (DNA with no supercoils present under the ionic conditions of the assembly reaction) migrates at a position of + 1 supercoil.[3] Thus, the first topoisomer resolved from form II DNA in the agarose gel shown here actually contains four more negative superhelical turns than the completely relaxed reference DNA (lane 1).

The efficiency of assembly is, however, not dependent on the topological state of the circular DNA. As shown in Fig. 5, relaxed circular DNA, assembled in the presence of topoisomerase I, and supercoiled DNA, to which the enzyme was added after assembly, both show similar amounts of nucleosome formation over a range of histone concentrations.

Nucleosome Spacing on the Assembled Material

Nucleoplasmin-mediated nucleosome formation utilizing purified components *in vitro* appears to be normal by all criteria, with one exception. The spacing is found to be shorter than that seen on chromatin assembled *in vivo* or in oocyte or egg extracts (155 as opposed to ~ 180 bp). The origin of this difference is not known. We have observed the shorter spacing whether or not histone H1 was added before, during, or after assembly with the core histones, and this matter is still under active investigation.

[31] Analysis of Nucleosome Positioning by *in Vitro* Reconstitution

By BERND NEUBAUER and WOLFRAM HÖRZ

Evidence has been steadily accumulating that there is an element of nonrandomness in the locations that nucleosomes occupy on a given DNA in the nucleus. A degree of sequence specificity or at least preference in DNA–histone interactions appears to be involved. Neither the extent of sequence preference nor the exact mechanism leading to nonrandom nucleosome positioning dictated by the DNA sequence is fully understood at present (for a review, see Ref. 1). It is therefore frequently of interest to determine the nucleosome positioning potential of a certain DNA region experimentally and to compare this property to the nucleosome arrangements found *in vivo*. Such studies should shed light on the molecular basis of nucleosome positioning *in vivo* and the biological role that it plays *in vivo*. A strategy is presented in this chapter by which nucleosome positioning can be assessed for any DNA *in vitro*.

---

[1] A. A. Travers, *Trends Biochem. Sci.* **12**, 108 (1986).

Copyright © 1989 by Academic Press, Inc.
All rights of reproduction in any form reserved.

Principle of the Method

Our approach is based on the reconstitution of a single nucleosome core particle from a particular DNA. The use of single core particles eliminates histone–histone interactions between adjacent particles that might interfere with the analysis of the DNA sequence specificity of the histones. For large DNA fragments, limitation of the reconstitution to mononucleosomes can only be achieved at very low histone–DNA ratios, which makes the procedure very inefficient. Therefore, we prefer reconstitution on small DNA fragments. The optimal size of such fragments is around 250 base pairs (bp), that is, below the minimal size required for dinucleosome formation (~270 bp) and yet large enough to provide the histone octamer with many different positions to choose from. If a larger DNA region is to be analyzed, an overlapping set of 250-bp fragments can be generated, as described below.

Nucleosome reconstitution is most conveniently carried out by a histone exchange reaction (Fig. 1). The DNA in question, radioactively labeled at the 5' end, is mixed with an excess of H1-free mononucleosomes in the presence of 2 M NaCl, and the salt is subsequently removed by gradient dialysis. The reconstituted material is then fractionated by polyacrylamide gel electrophoresis under nondissociating conditions, and the mononucleosome fraction is isolated.

The analysis of the actual positions adopted by the histone octamer is based on an exonuclease III protection assay (Fig. 2). A major advantage of

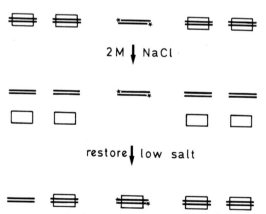

FIG. 1. Nucleosome reconstitution by a histone octamer exchange reaction. Unlabeled nucleosomes present in excess are mixed with a small amount of labeled DNA in the presence of 2 M NaCl (asterisk denotes the radioactive label). Reconstitution is accomplished by dialysis of the samples against a low-salt buffer.

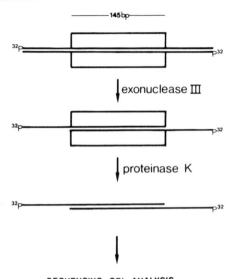

FIG. 2. Exonuclease III digestion assay for the analysis of nucleosome positioning. The left and the right boundaries of a nucleosome position can be determined independently from the size of the protected DNA fragments.

this assay is that multiple nucleosome positions simultaneously present on a DNA fragment can be distinguished.

Materials

Preparation of Histones

Nuclei are isolated from mouse liver as described by Hewish and Burgoyne,[2] with the inclusion of 1 mM phenylmethylsulfonyl fluoride (PMSF) in all buffers. For the preparation of chicken erythrocyte nuclei, blood from a decapitated chicken is collected in 0.15 M NaCl, 15 mM sodium citrate (pH 7.0). The blood cells are washed twice in the same buffer, twice in 0.15 M NaCl, sedimented at 500 g, and frozen in liquid nitrogen. Cell lysis occurs upon thawing, and nuclei are washed 3 times in buffer A (15 mM Tris-HCl, pH 7.4, 60 mM NaCl, 15 mM KCl, 0.15 mM spermine, 0.5 mM spermidine), supplemented with 0.34 M sucrose, 1 mM EDTA, 0.25 mM EGTA.

Nuclei are digested for 10 min at 37° with 150 units (U) micrococcal nuclease/ml in buffer A containing 0.2 mM EDTA, 0.2 mM EGTA,

1.4 mM CaCl$_2$ and 1 mM PMSF. Digestion is terminated by the addition of EDTA to a concentration of 4 mM, and nuclei are sedimented at 5000 g for 10 min. Chromatin is extracted in 10 mM Tris-HCl, pH 8.0, 0.2 mM EDTA, 0.1 mM PMSF, made 0.4 M in NaCl, and fractionated by sucrose gradient centrifugation. Isokinetic sucrose gradients[3] (particle density 1.51, $C_{mix} = 10\%$) are prepared in extraction buffer including 0.4 M NaCl and run at 4° for 15 hr at 40,000 rpm in a Beckman SW41 rotor. The mononucleosomal fraction is collected and stored at −20°. It is usually used as the histone source without further fractionation. When core histones free of DNA are needed, the method by Simon and Felsenfeld[4] is used, with some modifications.[5] Histone composition is routinely checked in 18% polyacrylamide gels containing sodium dodecyl sulfate (SDS).[6]

5′-End-Labeled DNA

In most cases, DNA fragments suitable for reconstitution are cloned in a suitable vector. If a region larger than 250 bp is to be analyzed, an overlapping series of fragments is required. We have prepared such a series by sonicating DNA to an average size of 250 bp, cutting a narrow size fraction from a gel, attaching DNA linkers to the ends, and subcloning the DNA fragments.[7] It is convenient to use a mixture of two different linkers in the ligation reaction and, with the appropriate vector, to select for fragments that have different linkers at their termini. This greatly facilitates the subsequent labeling of only one end, since the cloned DNA can be cut at either end of the insert. After the labeling reaction, the insert is liberated by use of the second enzyme and isolated by gel electrophoresis.

5′-End-labeling with polynucleotide kinase is performed essentially as described.[8] We usually start with about 25 μg of a pBR322 DNA clone. The DNA is linearized, phosphatase treated, phenol extracted, labeled with polynucleotide kinase, and cleaved with the second restriction nuclease. After labeling with high specific activity [^{32}P]ATP[9] and gel purification, we recover approximately 300 ng insert with a specific activity of $3-4 \times 10^7$ dpm/μg.

[2] D. R. Hewish and L. A. Burgoyne, *Biochem. Biophys. Res. Commun.* **52**, 504 (1973).
[3] K. S. McCarty, Jr., R. T. Vollmer, and K. S. McCarty, *Anal. Biochem.* **61**, 165 (1974).
[4] R. H. Simon and G. Felsenfeld, *Nucleic Acids Res.* **6**, 689 (1979).
[5] W. Linxweiler and W. Hörz, *Nucleic Acids Res.* **12**, 9395 (1984).
[6] U. K. Laemmli, *Nature (London)* **227**, 680 (1970).
[7] B. Neubauer, W. Linxweiler, and W. Hörz, *J. Mol. Biol.* **190**, 639 (1986).
[8] A. M. Maxam and W. Gilbert, this series, Vol. 65, p. 499.
[9] W. Linxweiler and W. Hörz, *Cell* **42**, 281 (1985).

Procedures

Reconstitution

1. Mix together 10 μl DNA (200 ng), 1 μl bovine serum albumin (BSA) (10 mg/ml), and 10 μl mononucleosomes (500 μg DNA/ml).

2. Transfer into dialysis tubing that is clamped shut at the bottom and open at the top. Add 600 ml high-salt solution to a mixing chamber (high-salt solution: 2 M NaCl, 10 mM Tris-HCl, pH 7.6, 1 mM EDTA, 1 mM 2-mercaptoethanol). Strips of dialysis tubing are conveniently held in narrow slits in the Lucite lid of the mixing chamber. The strips of tubing are secured in the slits with small plastic wedges, and they are weighted at the bottom to keep them immersed in the solution underneath.

3. Lower the NaCl concentration continuously, overnight, by pumping no-salt solution into the mixing chamber, while keeping the fluid level constant (no-salt solution: 10 mM Tris-HCl, pH 7.6, 1 mM EDTA, 1 mM 2-mercaptoethanol). Addition of 2.5 liter no-salt solution will result in a final NaCl concentration of less than 10 mM.

4. Recover reconstitution mixture from the dialysis tubing.

Nucleoprotein Gel Electrophoresis

Since the reconstituted material is usually heterogeneous and contaminated with residual free DNA, it is advisable to purify it by polyacrylamide gel electrophoresis under nondissociating conditions.

1. Prepare a 1-mm thick polyacrylamide slab gel (4% polyacrylamide, 0.1% bisacrylamide) between two glass plates with 40-μl sample wells. Prepare gel buffer (6.4 mM Tris, 3.2 mM sodium acetate, 0.32 mM EDTA, 20% (v/v) glycerol; adjust pH to 8.0 with acetic acid) and runing buffer (gel buffer without glycerol).

2. Add 1/5 volume glycerol to the samples and 0.1% (w/v) bromphenol blue solution to give a light blue color. Apply samples to the gel. A radioactive DNA digest can be conveniently run in parallel to serve as a relative mobility standard.

3. Run the gel at 100 V for about 4 hr at room temperature. The bromphenol blue marker will have migrated about 10 cm after that time.

4. Remove one of the glass plates and cover the gel with plastic wrap. Mark three corners with radioactive ink. Autoradiograph for 15 min in the cold. Do not clamp strongly since the gel is very soft. Place the film underneath the gel, still supported on the glass plate, and cut out the desired bands with a razor blade on a light box.

5. Recover the material by centrifuging the gel pieces through a nylon net into an Eppendorf tube containing about 50 μl of elution buffer (10 mM Tris-HCl, pH 7.6, 1 mM EDTA, 0.1 mM PMSF, 0.1 mg BSA/ml).

6. Add another 50 μl of elution buffer. Leave for 1–2 hr on ice. Add another 100 μl elution buffer. Centrifuge for 5 min in an Eppendorf centrifuge. Carefully take off supernatant. It is essential to remove all traces of polyacrylamide, since particulate matter interferes with the subsequent handling of the samples. Usually 70–90% of the radioactivity is recovered in about 100–150 μl.

Analysis of Nucleosome Positions

1. Add 1 volume exonuclease III digestion buffer (2× concentrated) containing 50 μg sonicated salmon sperm DNA/ml [exonuclease III digestion buffer (final strength): 10 mM Tris-HCl, pH 8.0, 50 mM NaCl, 3 mM MgCl$_2$, 1 mM 2-mercaptoethanol].

2. Remove control sample. Add 20 U exonuclease III/ml and incubate for 3, 6, and 12 min at 37°. The enzymatic activity of the exonuclease is taken as specified by the supplier (New England BioLabs, Boston MA).

3. To stop the reaction, pipette samples into 1/10 volume 10× concentrated proteinase K digestion buffer and incubate with 0.5 mg proteinase K/ml for 10 min at 37° [proteinase K digestion buffer (10× concentrated): 200 mM Tris-HCl, pH 8.0, 5% SDS, 50 mM EDTA].

4. Ethanol precipitate the DNA, wash with 70% ethanol, lyophilize, and dissolve samples in distilled water.

5. Electrophorese in a sequencing gel.[8] If DNA labeled at only one end is used for reconstitution, the same DNA partially degraded at purine residues provides convenient size markers.

Comments

Source of Histones

We have used both chicken erythrocyte and mouse liver nucleosomes in reconstitution experiments. Nucleosome positions obtained with the two types of histones reconstituted on mouse satellite DNA were identical. Apparently, nucleosome positioning on mouse satellite DNA does not depend on histone variants or histone modifications unique to the mouse cell. Also, in other work, no special attempts have been made to use homologous histones.

Effect of DNA Length on Nucleosome Reconstitution

In order to study the effect of DNA size on the reconstitution of mononucleosomes, we have prepared defined size fractions from total calf thymus DNA ranging from 150 to 300 bp. Total eukaryotic DNA was used in these experiments in order to average out effects of sequence specificity.

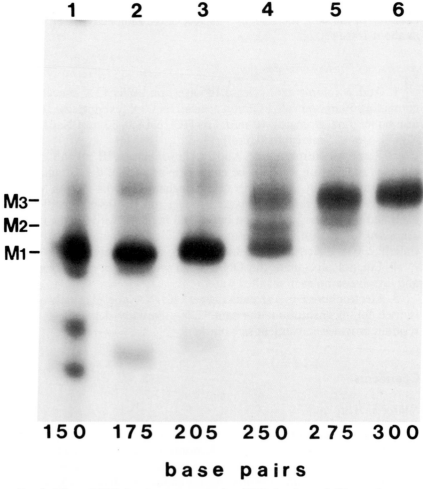

FIG. 3. Effect of DNA length on reconstitution. DNA fractions of different sizes were prepared from total calf thymus DNA, 5'-end labeled, reconstituted, and analyzed by low ionic strength polyacrylamide gel electrophoresis. Average DNA lengths are indicated for each lane. M1, M2, and M3 denote the positions of the corresponding particles (see text). (From Ref. 5.)

Nucleoprotein gel electrophoresis revealed that, with increasing DNA size, the relative amount of free DNA decreased (Fig. 3). At the same time, when DNA fragments larger than 200 bp were used, the reconstituted material was split into three bands (M1, M2, and M3), instead of the one typically found for smaller DNA (see Fig. 3). With the largest DNA tested, the most slowly migrating particle M3 was the most prominent. These findings have two immediate consequences that need to be taken into account in the reconstitution experiments. First, when relative histone affinities of different DNA fragments are to be compared, the fragments must be of the same size. Second, with larger fragments, the reconstituted material is heterogeneous.

Particles M1, M2, and M3 Obtained by Reconstitution

Nucleoprotein gel analysis of material reconstituted from total eukaryotic DNA (Fig. 3) as well as from specific DNA sequences, for example, mouse satellite DNA,[9] resolves three particles of different mobilities. As shown by digestion with exonuclease III plus nuclease S_1,[5] both M1 and M2 result from the interaction of one histone octamer with the DNA, the difference being that in M1 the octamer is located at the end of the DNA while in M2 both ends seem to be free (Fig. 4). M3 is protected against nuclease attack along its entire length, indicating interaction of the DNA with more than one histone octamer equivalent. Typical exonuclease III digestion patterns of particles M1, M2, and M3 obtained with DNA labeled at both 5′ ends are shown in Fig. 5. As discussed below, there is some overdigestion at the boundaries of the histone–DNA complexes, leading to DNA fragments shorter by 10 and 20 bp than the starting DNA.

The predominance of M1 indicates a tendency of histone octamers to align themselves with one end of the DNA. The reason for this tendency is not known. We have, however, observed that if high-affinity internal nucleosome frames are present in a DNA fragment this end effect is largely canceled,[7] and as a consequence M1 is virtually absent from the reconstitution mixtures.

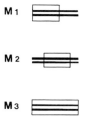

FIG. 4. Schematic diagram of particles M1, M2, and M3. See text for details.

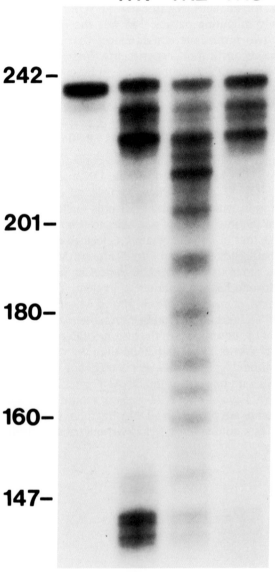

The difference in mobility between M1 and M2 is surprising. If additional histones were bound to the DNA in M2, we might have expected pause sites in the digestion with exonuclease III, like the pause at 165 bp in digestion of H1-containing mononucleosomes.[10] However, no such pause sites could be detected in the digestion of M2. It is conceivable that two DNA ends protruding from a core particle cause its retardation in gel electrophoresis. The gel system used not only separates nucleoprotein complexes on the basis of size and charge, but is also sensitive to the conformation of the complexes.

The complete protection of M3, as opposed to M1 and M2, against exonuclease III makes it virtually certain that in M3 additional histones are bound to the DNA. With the largest fragments that we have tested this might be a second octamer to yield something that resembles a compact dimer,[11] whereas with the smaller DNAs it is more likely individual histones in various combinations.

Exonuclease III Assay for Analysis of Nucleosome Positions

Nucleosome positions obtained by reconstitution can be revealed by digestion with exonuclease III. Digestion with DNase I is an alternative approach (see, e.g., Refs. 12–14). The appearance of a 10-bp ladder with end-labeled DNA in DNase I digests implies nonrandom positioning, and the actual location of the histone octamer can be deduced from the characteristic modulation of the intensity of that ladder.[15,16] However, the assignment of a nucleosome frame is not always straightforward from the signals given by DNase I, and in one case an assignment based on the use of DNase I[14] had to be revised in subsequent work employing exonuclease III.[17]

[10] A. Prunell and R. D. Kornberg, *J. Mol. Biol.* **154**, 515 (1982).
[11] K. Tatchell and K. E. Van Holde, *Proc. Natl. Acad. Sci. U.S.A.* **75**, 3583 (1978).
[12] K. Tatchell and K. E. Van Holde, *Biochemistry* **18**, 2871 (1979).
[13] R. T. Simpson and D. W. Stafford, *Proc. Natl. Acad. Sci. U.S.A.* **80**, 51 (1983).
[14] N. Ramsay, G. Felsenfeld, B. M. Rushton, and J. D. McGhee, *EMBO J.* **3**, 2605 (1984).
[15] M. Noll, *J. Mol. Biol.* **116**, 49 (1977).
[16] L. C. Lutter, *J. Mol. Biol.* **124**, 391 (1978).
[17] N. Ramsay, *J. Mol. Biol.* **189**, 179 (1986).

FIG. 5. Exonuclease III analysis of particles M1, M2, and M3 reconstituted from mouse satellite DNA. A 234-bp mouse satellite DNA fragment was reconstituted with histones. The reconstituted material was fractionated by polyacrylamide gel electrophoresis, and particles M1, M2, and M3 were isolated and treated with 25 U exonuclease III/ml for 8 min to generate limit digests. DNA was isolated and analyzed in a 6% sequencing gel. Starting DNA is shown at left. (From Ref. 9.)

A serious limitation of DNase I digestion is that more than about two alternative nucleosome positions simultaneously present on a particular DNA fragment would go undetected, since the superposition of several 10-bp ladders, coupled with the sequence preference of DNase I itself, would give results too complicated to interpret. Because of the preponderance of M1 in many reconstitutes, however, the analysis of unfractionated material with DNase I may result in the conclusion that histone cores are located in a unique position at the end of the DNA in question.

For the reasons just given, it is usually advantageous to use exonuclease III to map nucleosome positions. In doing so, it is necessary, however, to keep certain properties of this enzyme in mind. For one, exonuclease III does not attack overhanging 3' ends that are generated by a number of restriction enzymes. It is necessary to choose enzymes used for the preparation of the DNA accordingly, or to trim the ends otherwise.

A second complication arises from the fact that exonuclease III degrades double-stranded DNA until the DNA denatures, and therefore leaves behind stable, single-stranded DNA fragments in limit digests that are approximately one-half the size of the starting fragment. If the two strands of the DNA are digested at different rates by exonuclease III, larger as well as smaller fragments than expected are obtained. This is true for mouse satellite DNA, for example.[18] Care must therefore be taken to remove free DNA from reconstitution mixtures, which is accomplished by gel electrophoresis in our protocol. In addition, control digestions of free DNA should be routinely performed, and these can almost always identify suspect fragments.

Exonuclease III does have a certain sequence preference, tending to pause at G residues, while cleaving C residues quite rapidly and removing A and T residues at intermediate rates. This leads to a characteristic modulation transiently observed in the digestion of free DNA (Fig. 6). Certain fragments might again be taken to represent specific stop sites due to nucleoprotein complexes. Fortunately, however, the sequence preference of exonuclease III is slight, and the corresponding bands are quite labile and never persist in a time course experiment. This is quite contrary to stop sites caused by bound proteins. A careful comparison of the course of digestion with exonuclease between free and complexed DNA can usually relieve any ambiguities (see Fig. 6 for a typical example).

A more serious problem has to do with the stability of the barrier against exonuclease III digestion afforded by the bound protein. Usually histone octamers protect very strongly against exonuclease III, and it takes substantial increases in enzyme concentration to overcome a stop site.

[18] W. Linxweiler and W. Hörz, *Nucleic Acids Res.* 10, 4845 (1982).

minutes

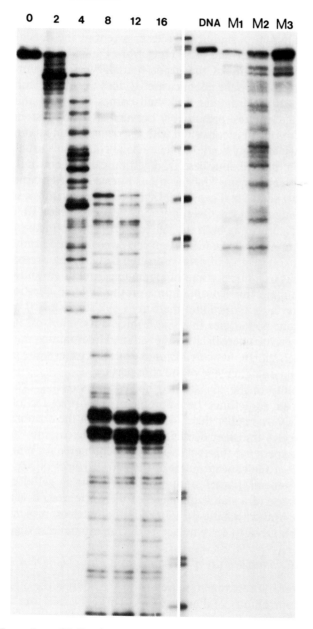

FIG. 6. Exonuclease III digestion of free versus reconstituted DNA. A 5′-end-labeled *Hae*III fragment that extends from position 597 to 830 in the pBR322 sequence (Sutcliffe, 1978) was digested with 20 U exonuclease III/ml for the times indicated and the DNA analyzed in a 6% sequencing gel (left). Limit digests of isolated particles M1, M2, and M3 with exonuclease III (20 U/ml, 15 min) are shown on the right. The autoradiogram shown is a composite of two different exposures of the same gel. (From Ref. 9.)

There is some variation, however. Terminally located histone octamers protect the DNA to a significantly lesser extent at the outside boundary. Exonuclease III can invade these cores from the end, which is in marked contrast to the stability of the inside boundary of the same particle. It should be noted that once exonuclease III does invade the core particle, it pauses again 10 bp from the end, and comes to a halt after removing another 10 nucleotides, presumably because of strong histone–DNA interactions. This result is consistent with the concept that an inner region of 100 bp is most strongly bound to the histone core.[19] It is usually advisable to do a preliminary exonuclease III digestion experiment in which a wide range of different enzyme concentrations is tested. The enzyme concentration judged to be optimal can be determined this way and is subsequently used in kinetic analyses. Overdigestion by exonuclease III can best be checked by determining both the left and the right boundary of a nucleosome position. A nucleosome position can be considered firmly established only if approximately 145 bp are protected between the two boundaries.

The results of a nucleosome mapping experiment are shown in Fig. 7. There is one major nucleosome frame on the DNA in question as well as several minor ones. Notice that there are weak stop sites 10 bp outside of the actual core boundaries in these frames. Protection of 165 bp rather than 145 bp is characteristically observed in H1-containing nucleosomes,[10] but it has been shown that core histones can also confer some protection to an extra 10 bp on either side of the core particle.

The stability of the rconstituted nucleosomes during exonuclease III protection can sometimes be a problem. The presence of significant amounts of DNA smaller than 145 nucleotides in the exonuclease digests usually indicates transient dissociation of the nucleosomes during digestion. In our experience this is occasionally encountered for otherwise stable core particles if the concentration of the core particles during digestion is too low. In general, particles are more stable during gel electrophoresis, possibly because of a stabilizing effect of the gel matrix. It is advisable in such cases of limited stability to add cold mononucleosomes instead of free salmon sperm DNA to serve as carrier during exonuclease digestion.

Measuring Relative Affinity of Histones for a Particular DNA Fragment

Our reconstitution system is well suited to address the question of the relative histone affinity of a given DNA. This is because our experiments constitute, in principle, a competition assay in which each labeled fragment is compared against the same background of average-sequence cold

[19] W. O. Weischet, K. Tatchell, K. E. Van Holde, and H. Klump, *Nucleic Acids Res.* **5**, 139 (1978).

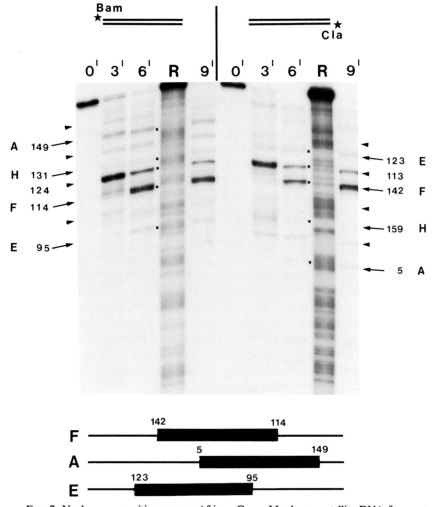

FIG. 7. Nucleosome positions on an African Green Monkey α-satellite DNA fragment reconstituted with core histones. A 271-bp fragment cloned from African Green Monkey α-satellite DNA by attaching a BamHI linker to one end and a ClaI linker to the other was labeled at either end (indicated by an asterisk) and reconstituted with H1-free mononucleosomes from mouse as the histone source. The reconstituted material was fractionated by nucleoprotein gel electrophoresis. Particle M2 was isolated and digested with exonuclease III for 0, 3, 6, and 9 min (30 U/ml) and the DNA analyzed by 6% sequencing gels. The positions of the stop sites on the α-satellite sequence as deduced from the reference (R), consisting of the same end-labeled fragment partially degraded at purine residues, are shown together with the nucleosome frames they correspond to. Arrowheads mark fragments 10 nucleotides larger than the core boundary fragments. They do not represent additional nucleosome frames but are instead due to protection of 10 extra base pairs on each side of the nucleosome core. The locations of the three strongest frames A, E, and F are shown schematically underneath. As shown by densitometry of the sequencing gels with an LKB UltroScan XL densitometer, the relative peak areas of frames A, E, and F were 7, 5, and 80%, respectively. (From Ref. 7.)

DNA present in the nucleosomes used as the histone source. If different fragments are compared, it is sufficient to monitor reconstitution by nucleoprotein gel electrophoresis and determine the relative amount of free radioactive DNA persisting after reconstitution.

At the same time, individual nucleosome frames present on any one fragment compete with each other for the histones. Their relative abundance as determined in the exonuclease III digestion experiments should reflect their relative affinities. We have demonstrated the feasibility of this approach with a DNA fragment from African Green Monkey α-satellite DNA. A 5-bp insertion introduced into the DNA at a position where three ovelapping nucleosome frames are affected was shown to change the histone affinity of these frames in a highly specific fashion.[7]

Concluding Remarks

Nucleosome reconstitution experiments have demonstrated that there is a greater specificity in histone–DNA interactions than initially suspected. Reconstitution of appropriately designed synthetic DNAs will undoubtedly contribute to the elucidation of the principles governing this specificity. The question, however, of what role positioned nucleosomes play *in vivo* is still open. Factors other than histone–DNA interactions must modulate nucleosomal arrangements in the nucleus. For example, four positioned nucleosomes present on the yeast PHO5 promoter are selectively removed on induction of the PHO5 gene.[20] Thus, one and the same DNA can exist either in a nucleosomal or in a nonnucleosomal state. The mechanism underlying this transition is not yet known. It is safe to predict, however, that reconstitution experiments with the inclusion of regulatory factors will again be instrumental in providing answers to these intriguing questions.

Acknowledgments

The work of the authors was supported by Deutsche Forschungsgemeinschaft (SFB 304) and Fonds der Chemischen Industrie.

[20] A. Almer, H. Rudolph, A. Hinnen, and W. Hörz, *EMBO J.* **5,** 2689 (1986).

Author Index

Numbers in parentheses are footnote reference numbers and indicate that an author's work is referred to although the name is not cited in the text.

Subject Index

A

Adenovirus VA RNA coding sequences, 369
Affinity chromatography
  antibodies obtained by, 226–227
  of chromatin fragments, 25
  of replicating SV40 chromosomes, 41–52
  of SV40 minichromosomes, 25
  of yeast plasmid chromatin, 7, 29, 35–40
1-Anilinonaphthalene 8-sulfonate, for rapid
    staining and quantitation of histones
    and histone peptides in polyacrylamide
    gels, 214
Antibodies. *See also* Antihistone antibodies
  as analytical reagents, 214–215
  for immunoelectron microscopy
    concentrating, 184
    immunofluorescence conditions,
      182–183
    preparation and selection of, 181–184
    prescreening, 182
    purification, 184
    slide assay, 182
    state of, 184
    storage, 184
    suspension assay, 183
  microinjection of
    into amphibian oocytes, 241–244
    into somatic cells, 245–246
Antibody–nucleosome complexes
  fixation of, 186
  preparation of, 185–186
  removal of unbound antibody from,
    186–187
  sandwich preparations, for electron
    microscopy, 187–189
Antigen–antibody reactions, inhibition
    studies, 223–225
Antigenicity studies, 263
Antihistone antibodies, 251–269
  in immunoelectron microscopy, of
    nucleosomes, 180–181

immunofluorescence prescreening, 182
indirect ELISA
  using immobilized chromatin or its
    subunits, 259–260
  using immobilized histones or peptides,
    258–259
microcomplement fixation assay,
    260–261
monoclonal, 218, 262–263
preparation of, 217
solid-phase radioimmunoassay, 260–
    261B

B

Benzamidine, in solution for isolation of
    nuclei and chromatin, 4
Bio-4-dUTP, in isolation of replicating
    SV40 chromosomes, 52
Bio-19-SS-dUTP, 42
  in affinity isolation of replicating SV40
    chromosomes, 41–52
  incorporation of, into replicating SV40
    chromosomes, 45–49
  synthesis of, 43–45
Biotin–cellulose, synthesis of, 49–50
Bisimidoesters, cross-linking of histones
    with, 550–551
Bis(N-hydroxysuccinimide esters), cross-
    linking of histones with, 550–551

C

Calcium ion, in solution for isolation of
    nuclei and chromatin, 4
Carbonate dehydratase, M_r, estimates of,
    from corrected electrophoretic
    mobilities, 472
Carcinogen, distribution of, in cellular or
    viral chromatin and DNA, mapping,
    249–250